Basins of the Rio Grande Rift: Structure, Stratigraphy, and Tectonic Setting

Edited by

G. Randy Keller
Department of Geological Sciences
University of Texas at El Paso
El Paso, Texas 79968

and

Steven M. Cather
New Mexico Bureau of Mines
and Mineral Resources
Campus Station
Socorro, New Mexico 87801

SPECIAL PAPER

291

1994

Published by The Geological Society of America, Inc.
3300 Penrose Place, P.O. Box 9140, Boulder, Colorado 80301

Printed in U.S.A.

GSA Books Science Editor Richard A. Hoppin

Library of Congress Cataloging-in-Publication Data

Basins of the Rio Grande rift : structure, stratigraphy, and tectonic setting / edited by G. Randy Keller and Steven M. Cather.
p. cm. — (Special paper / Geological Society of America ; 291)
Includes bibliographical references and index.
ISBN 0-8137-2291-8
1. Basins (Geology)—Southwest, New. 2. Geology, Stratigraphic--Cenozoic. 3. Geology, Structural—Southwest, New. I. Keller, G. Randy (George Randy), 1946– . II. Cather, Steven M. III. Series. Special papers (Geological Society of America) ; 291.
QE627.5.S84B37 1994
551.8'0978—dc20 94-21616
CIP

Cover: Doug West; *Arroyo Salado* © 3/1993.

The cover image depicts the view westward up the Rio Salado a short distance upstream of its confluence with the Rio Grande in the Socorro Basin. Bluffs in the middle ground are basin-fill deposits consisting of the fluviolacustrine Santa Fe Group and younger piedmont, eolian, and terrace deposits. Ladron Peak, a footwall uplift on the west flank of the rift, is on the skyline.

Doug West, an artist specializing in Southwestern landscapes, resides in Socorro, New Mexico.

10 9 8 7 6 5 4 3 2 1

Contents

Geological Society of America
Special Paper 291
1994

Introduction

G. Randy Keller
Department of Geological Sciences, The University of Texas at El Paso, El Paso, Texas 79968
Steven M. Cather
New Mexico Bureau of Mines and Mineral Resources, Campus Station, Socorro, New Mexico 87801

After the definitive paper by Chapin (1971), the Rio Grande rift has gradually become recognized as a major continental rift zone. It now joins the Kenya rift, the Baikal rift, and the Rhine graben as classic late Cenozoic examples of such features. These well-studied rifts provide a great deal of insight on how extension effects the evolution of the continental lithosphere, but it is necessary to bear in mind that they are all parts of much larger zones of extension (the Western Cordillera and the East African, Baikal, and Central European rift zones, respectively). The Rio Grande rift shares many geological and geophysical characteristics with other rift zones (e.g., Keller et al., 1991) but stands out because of the extensive development of well-exposed sedimentary basins. The relatively recent establishment and incision of through-going drainage has exposed a large amount of the basin fill, thus providing a good geologic record to study. In addition, petroleum exploration in the area has provided the recent release of a significant amount of seismic reflection data. Many of these seismic profiles are contained in this volume. Also the scientific community has been successful in establishing a large data base of high-quality radiometric dates in the Rio Grande rift area. These data are also a major resource used repeatedly in this volume.

This volume focuses on recent research results and the basins that characterize the Rio Grande rift as it extends from central Colorado to west Texas and the state of Chihuahua, Mexico (Fig. 1). For example, Chapin and Cather discuss the Upper Arkansas graben in central Colorado just north of the headwaters of the Rio Grande, and Dickerson and Muehlberger discuss the Big Bend region of west Texas, which is more than 1,100 km downstream. Every major basin in the rift (Fig. 1) is addressed to some degree. Crustal structure, the role of older structures, and relations with nearby features are also discussed. The fact that these papers, most of which were initially given at a symposium during the 1991 joint meeting of the Rocky Mountain and South-Central sections of the Geological Society of America, speak to such a broad area and range of topics attests to the scientific interest in this complex rift zone.

The northern and central portions of the rift have been the most intensively studied; a fact that is reflected in this volume. Chapin and Cather provide an overview of this portion of the rift and discuss geologic evidence that indicates a significant increase in extension occurs between the San Luis Basin in the north (8 to 12%) and the Socorro area on the south (50%). Geophysical data are in general agreement with this observation (Cordell, 1982). Chapin and Cather suggest opening of the rift by clockwise rotation of the Colorado Plateau about a Euler pole in northeastern Utah with the most rapid regional extension in the middle to late Miocene. They also propose that the rift may have developed in a terrain that was already elevated as a result of Laramide crustal thickening.

The San Luis Basin (Fig. 1) is the most northern feature specifically addressed in this volume. Brister and Gries as well as Kluth and Schaftenaar present analyses of industry reflection data and deep drilling results in this large topographic basin. Its western part is underlain by a late Laramide basin, and to the east the rift-related sedimentary fill overlies a foundered Laramide uplift and reaches a thickness of 5 to 6 km. The strata in this area are tilted down to the east.

The next basin to the south, the Española Basin (Fig. 1), historically has been one of the best-studied basins of the Rio Grande rift. Salyards, Ni, and Aldrich present new data that indicate significant counterclockwise rotation of local structural blocks has occurred within the Española Basin during the Neogene. These new results follow closely on the heels of recent structural papers by Biehler et al. (1991) and Cather (1992) and illustrate that even well-studied areas can continue to produce important interpretations.

The Albuquerque Basin (Fig. 1) is located south of the Española Basin and is the focus of five papers in this volume. Overviews of the structure and stratigraphy of this basin deduced from drilling and seismic reflection data are presented

Keller, G. R., and Cather, S. M., 1994, Introduction, *in* Keller, G. R., and Cather, S. M., eds., Basins of the Rio Grande Rift: Structure, Stratigraphy, and Tectonic Setting: Boulder, Colorado, Geological Society of America Special Paper 291.

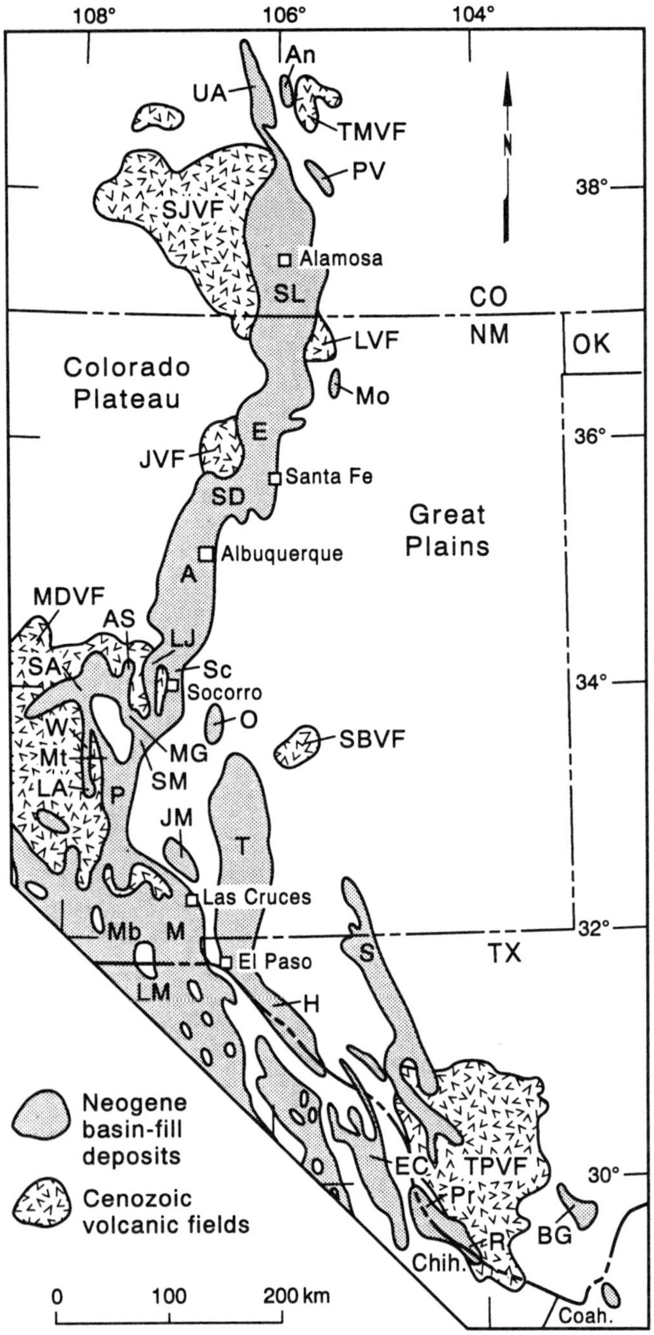

Figure 1. Map showing basins of the Rio Grande rift and Cenozoic volcanic fields. *Basins* (north to south); UA, Upper Arkansas; An, Antero; PV, Pleasant Valley; SL, San Luis; Mo, Moreno; E, Española; SD, Santo Domingo; A, Albuquerque; Sc, Socorro; LJ, La Jencia (together, the Socorro and La Jencia basins and the intervening Lemitar Mountains comprise the early rift Popotosa basin); AS, Abbe Springs (also termed northern Milligan Gulch Basin); SA, San Agustin; O, Oscura; MG, Milligan Gulch; SM, San Marcial; Mt, Monticello; W, Winston; LA, Las Animas; P, Palomas; T, Tularosa; JM, Jornada del Muerto; Mb, Mimbres; M, Mesilla; LM, Los Muertos; H, Hueco; S, Salt; EC, El Cuervo; Pr, Presidio; BG, Black Gap; R, Redford. *Volcanic fields:* TMVF, Thirtynine Mile; SJVF, San Juan; LVF, Latir; JVF, Jemez; MDVF, Mogollon-Datil; SBVF, Sierra Blanca; TPVF, Trans-Pecos.

in papers by Lozinsky and Russell and Snelson. Twelve deep exploration wells, several of which are more than 4 km deep, have been drilled in this basin, and about 200 km of seismic reflection data are presented here. Two major subbasins are present and are separated by a northeast-trending accommodation zone. The northern basin contains more than 6.5 km of Eocene and younger sediments and volcanics. May and Russell have studied the synrift basin fill and found it to be more than 4 km thick. May, Kelley, and Russell document the isostatic uplift of footwall blocks due to tectonic denudation along major normal faults in the Albuquerque Basin.

The development of extensional structures in the southwest Albuquerque Basin and adjoining Ladron Peak area is addressed by Lewis and Baldridge, who propose that some features previously attributed to Laramide shortening may have instead developed during rifting. To the south near Socorro, Cather, Chamberlin, Chapin, and McIntosh analyze the tilt history of the Lemitar Mountains during Neogene extension and relate it to depositional episodes and the development of unconformities.

Older features in the flanking areas of the central portion of the rift were the focus of three papers. Barrow and Keller present an analysis of the Estancia Basin (Fig. 1), which is located just east of the Albuquerque Basin. Their seismic reflection data show that this basin is primarily due to divergent, left-lateral strike-slip movements during the Pennsylvanian. There are only minor rift-related faults present, and there are major Proterozoic features evident in the seismic data. The Joyita Hills are an exposed fault block south of the Albuquerque Basin. Beck and Chapin show that this exposure provides excellent documentation of multiple episodes of reactivation of Proterozoic and Phanerozoic structures, some of which controlled the development of rifting. The Pennsylvanian movements indicated are in agreement with those proposed for the Estancia Basin area.

The Mogollon-Datil volcanic field is a feature that is generally considered to be prerift in age although radiometric dating there indicates this voluminous volcanism overlaps in time with the inception of extension. This situation suggests that the perception of the Rio Grande being a volcanically "dry" rift is dependent on one's view of when rifting began. Schneider and Keller conducted a study of the deep structure of the Mogollon-Datil volcanic field area. They found the crust to be transitional in nature and to contain a large batholith. This result suggests a gradual transition in crustal character exists between the central portion of the rift and the Mogollon Plateau. Harrison conducted an analysis of the little-studied Winston graben, which is a rift feature within the eastern Mogollon-Datil volcanic field. The location of this graben is within the broad transition between the rift and the volcanic plateau as indicated by the geophysical data.

Adams and Keller conducted an integrated geophysical analysis of the southern portion of the rift in New Mexico, which included new gravity and seismic reflection data. Their results show that the broadening of the surface expression of the rift south of Socorro, New Mexico, is accompanied by a broadening of the deep crustal structure anomaly. Their gravity results also show that immediately south of Socorro, a decrease in the size and number of basins suggests a local decrease in the amount of extension before the rift again widens dramatically in southernmost New Mexico. Mack, James, and Salyads analyzed intrabasinal faulting in this area and documented evidence for fault control of sedimentation during the late Pliocene and early Pleistocene.

At El Paso, Texas, the surface expression of the rift takes a conspicuous bend to a northwest-southeast trend (Fig. 1). The Hueco bolson is the major rift feature at this latitude and was analyzed by Collins and Raney. They documented the presence of two subbasins that contain as much as 3 km of fill and two major fault scarps with late Pleistocene-Holocene movement. In Texas, the rift structures continue to the southeast into the Big Bend area where Dickerson and Muehlberger studied the basins. Their results show that major rift structures extend southward into Mexico toward an undefined terminus in Chihuahua and Coahuila.

The research represented by the papers in this volume is the product of diverse methods applied over a large geographic area. Nonetheless, several commonalities have emerged among the interpretations of these varied topics of investigation. These commonalities may represent "regional truths" for the Rio Grande rift and are summarized below.

Initial sedimentation in most basins of the rift began in late Oligocene–early Miocene, but rapid regional extension did not begin until middle Miocene and continued until late Miocene. Strain rates and basin subsidence have been relatively modest since the late Miocene. In this view, extension has been weak to moderate in most areas of the Rio Grande rift during the past 30 m.y., with the exception of this episode of rapid regional extension (at about 16 to 10 Ma). This contrasts with the views of some previous workers who have interpreted a two-fold history of rapid extension for the rift, separated by a tectonic lull in the middle and late Miocene.

Most of the structural depressions of the rift are half grabens; full grabens are relatively rare. Polarity of the half grabens changes across accommodation zones that are typically at high angles to the trend of the rift. Development of half grabens and accommodation zones are both commonly controlled by preexisting structures, and many half grabens of the rift have been tectonically inverted from Laramide upthrust blocks. Geological and geophysical data indicate that extension increases southward along the rift to near Socorro. Extension appears to abruptly decrease directly south of Socorro, and then begins again to increase systematically to the south. Miocene extension in the northern and central portions of the rift was left oblique.

REFERENCES CITED

Biehler, S., and 12 others, 1991, A geophysical model of the Española Basin, Rio Grande rift, New Mexico: Geophysics, v. 56, p. 340–353.

Cather, S. M., 192, Suggested revisions to the Tertiary tectonic history of north-central New Mexico: New Mexico Geological Society, 43rd Field Conference Guidebook, p. 109–122.

Chapin, C. E., 1971, The Rio Grande rift, Part 1: Modifications and additions: New Mexico Geological Society, 22nd Field Conference Guidebook, p. 191–202.

Cordell, L., 1982, Extension in the Rio Grande rift: Journal of Geophysical Research, v. 87, p. 8561–8569.

Keller, G. R., and 7 others, 1991, A comparative study of the Rio Grande and Kenya rifts: Tectonophysics, v. 197, p. 355–371.

Manuscript Accepted by the Society October 13, 1993

Printed in U.S.A.

Geological Society of America
Special Paper 291
1994

Tectonic setting of the axial basins of the northern and central Rio Grande rift

Charles E. Chapin and Steven M. Cather
New Mexico Bureau of Mines and Mineral Resources, Campus Station, Socorro, New Mexico 87801

ABSTRACT

The axial basins of the northern and central Rio Grande rift evolved since late Oligocene as a chain of half grabens between the Colorado Plateau on the west and the interior of the craton on the east. The basins range from 80 to 240 km in length and from 5 to 95 km in width, with an average width of approximately 50 km. Basin-fill sedimentary deposits range up to 5 to 6 km in thickness. The oldest dated volcanic rocks interbedded with syn-rift sediments are about 26 Ma. Extension in the rift was left oblique and increases southward from 8 to 12% in the San Luis Basin, to about 17% in the northern Albuquerque Basin, to at least 28% in the southern Albuquerque Basin, to an average of 50% in the Socorro area. The most rapid phase of regional extension was middle to late Miocene. Sedimentary deposits of this age are volumetrically dominant within the rift basins and on the rift flanks (Ogallala and Bidahochi Formations).

Basin asymmetry shifts back and forth from east-tilted to west-tilted across complex northeast-trending accommodation zones that developed along preexisting transverse structural lineaments. These accommodation zones lie on small circles about the Miocene Euler pole of rotation for the Colorado Plateau relative to the stable craton. Rotation about this Euler pole in northeastern Utah caused synchronous episodes of Miocene extension and sedimentation in the Rio Grande rift and in the central Wyoming–northern Colorado area, largely by the collapse of Laramide uplifts. Late Cenozoic rotation of the Colorado Plateau was 1.0 to 1.5° clockwise.

Structural subdivision of basin floors within the rift by longitudinal master faults increased with increasing extension until, at some point in excess of 28% extension, one or more longitudinal fault blocks was uplifted isostatically through the basin surface to form an intrabasin tilted-block mountain range that subdivided the basin. Faulting and subsidence tended to migrate towards the basin axis as newer master faults cut off or merged downdip with, earlier faults. This deactivated the outer longitudinal blocks and left them stranded as shallow suballuvial benches covered by relatively thin sections of the older syn-rift sediments.

Recent paleobotanical studies indicate that elevations prior to rifting were similar to present elevations. This suggests that crustal thickening during Laramide compression resulted in regional uplift during the early Cenozoic. Exhumation of the escarpment along the Rocky Mountains–High Plains boundary by erosion of approximately 400 to 600 m of Cenozoic and Cretaceous strata may be a consequence late Cenozoic climate change.

Chapin, C. E., and Cather, S. M., 1994, Tectonic setting of the axial basins of the northern and central Rio Grande rift, *in* Keller, G. R., and Cather, S. M., eds., Basins of the Rio Grande Rift: Structure, Stratigraphy, and Tectonic Setting: Boulder, Colorado, Geological Society of America Special Paper 291.

INTRODUCTION

The purpose of this paper is to describe the structural and stratigraphic architecture of the axial basins of the central and northern Rio Grande rift in the context of the regional tectonic framework. There has been much recent progress in our understanding of the axial basins of the rift. Industry seismic lines and drill-hole data are now available in the San Luis, Española, and Albuquerque Basins and several studies based on these lines are reported in this volume. Detailed doctoral studies of drill cuttings and cores in the Albuquerque Basin by Lozinsky (1988) and the San Luis Basin by Brister (1990) have greatly increased our knowledge of the syn-rift sedimentary fill and have improved interpretation of the seismic lines. An integrated geophysical study of the east half of the Española Basin by the Summer of Applied Geophysical Experience (SAGE) field course sponsored by Los Alamos National Laboratory has provided detailed subsurface data in that basin for the first time (Biehler et al., 1991).

Important progress has also been made in mapping and outcrop studies of synrift sedimentary deposits, radiometric dating of volcanic rocks and ash beds, fission-track dating of uplifts flanking the basins, studies of accommodation zones and the influence of Precambrian basement structures, reevaluation of paleoelevation estimates based on fossil flora, and several geophysical studies that have contributed both to knowledge of individual basins and to the regional structure of the Rio Grande rift.

We begin with a brief outline of the regional setting and geographical extent of the rift followed by a detailed discussion of when rifting began. The subsurface geometries of several basins are then summarized, followed by a discussion of the role of accommodation zones, a synthesis of the nature of synrift sedimentary deposits, and a summary of the current controversy over late Cenozoic versus Laramide epeirogenic uplift.

REGIONAL SETTING

For much of its length, the Rio Grande rift is a north-trending break in the continental lithosphere that separates the Colorado Plateau on the west from the interior of the craton on the east. From near Socorro, New Mexico, northward to the vicinity of Leadville, Colorado, a distance of approximately 550 km, the Rio Grande rift consists mainly of four axial basins aligned in a right-stepping echelon pattern between uplifted and outward-tilted shoulders. From south to north, these basins are the Albuquerque, Española, San Luis, and Upper Arkansas Basins (Fig. 1). Many basins of the northern and central Rio Grande rift have resulted from the collapse of Laramide uplifts (Tweto, 1979; Sales, 1983; Cather, 1983, 1989, 1992). South of the Albuquerque Basin, the rift bifurcates and widens into a series of parallel basins and intrarift tilted-block uplifts (Chapin, 1971, 1979) that resembles the Basin and Range province.

Considerable controversy has arisen over the years as to where the Rio Grande rift terminates to the south. Opinions differ mainly on whether the rift south of the Albuquerque Basin is merely part of the Basin and Range province or whether it is a structural entity in itself. Certainly the geologic history of the two provinces is closely linked, but as pointed out by Seager and Morgan (1979) the modern Rio Grande rift can be distinguished from the adjacent southern Basin and Range province by its higher heat flow, more abundant Pliocene and Quaternary faulting, exceptionally deep basins, and late Quaternary volcanism. During the Miocene when major extension was occurring throughout much of the western United States and northern Mexico, however, the boundary between the rift and the southern Basin and Range province may have been more subtle and difficult to define.

The northward termination of the rift is also problematical. Although the longitudinally linked basins of the Rio Grande rift terminate northward near Leadville in central Colorado, rift-age normal faults and isolated extension basins continue northward through north-central and northwestern Colorado, central Wyoming, and northeastern Utah. As will be discussed below, extensional basin development in these areas was synchronous with the main phase of subsidence in the basins of the Rio Grande rift, which requires that existing hypotheses for the regional kinematic development of the Southern Rocky Mountains and Rio Grande rift be revised.

Late Cenozoic normal faults in the Rio Grande rift and southern and central Rocky Mountains are oriented approximately tangential to the margin of the Colorado Plateau. The average trend of normal faults in New Mexico is north-northeast, but becomes dominantly north-northwest in southern and central Colorado, northwest in northern Colorado, and east-west in central Wyoming and northeast Utah. The timing of faulting adjacent to the eastern and northeastern margins of the Colorado Plateau may be assessed by radiometric and paleontologic data from sedimentary basins within this zone of extension. These age constraints are briefly summarized in the following paragraphs. Implicit in our analysis is the assumption that sedimentation accompanied basin subsidence rather than predating it as has been interpreted by some workers in northern Colorado. The association between basin subsidence and sedimentation has long been recognized in the Rio Grande rift, and has been documented in several of the better-studied extensional basins to the north by recent workers (e.g., Split Rock Basin, Love, 1970; Browns Park Basin, Hansen, 1986).

In the basins of the Rio Grande rift, deposits of late Oligocene–Miocene age form the bulk of the basin-fill; Plio-Pleistocene deposits are volumetrically subordinate. Maximum thickness of late Oligocene–Miocene deposits typically exceeds those of the Plio-Pleistocene by at least a factor of seven, and in places by as much as a factor of thirty (cf. Lozinsky, 1988; Brister, 1990; Hawley and Lozinsky, 1993; Cather et al., this volume). Within the upper Oligocene–Miocene deposits,

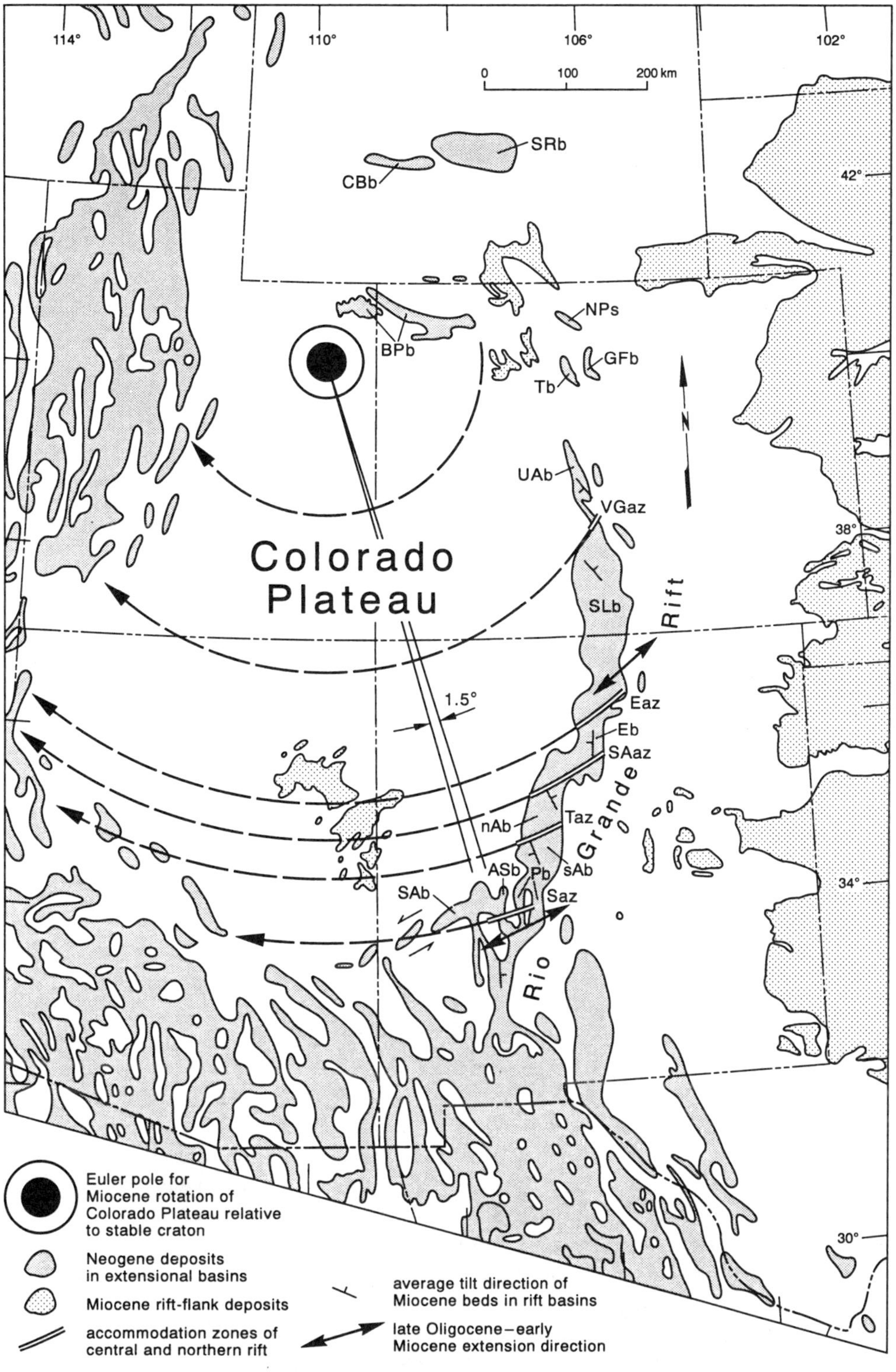

Figure 1. Map showing distribution of Neogene basin-fill and rift-flank deposits and selected structural features. Abbreviations for basins (from north to south) are CBb, Circle Bar Basin; SRb, Split Rock Basin; BPb, Browns Park Basin; NPs, North Park syncline; Tb, Troublesome Basin; GFb, Granby-Fraser basin; UAb, Upper Arkansas Basin; SLb, San Luis Basin; Eb, Española Basin; nAb, northern Albuquerque Basin; sAb, southern Albuquerque Basin; Pb, Popotosa basin; ASb, Abbe Springs Basin. Abbreviations for accommodation zones are VGaz, Villa Grove; Eaz, Embudo; SAaz, Santa Ana; Taz, Tijeras; Saz, Socorro. See text for discussion.

strata of middle and late Miocene age are volumetrically dominant, due to enhanced extension and basin-subsidence rates during this time (see discussion in Cather et al., this volume).

Of the late Cenozoic extensional basins north of central Colorado, the Split Rock Basin in Wyoming displays the greatest thickness of Neogene deposits, about 1 km (Love, 1970). The Split Rock Basin was formed by collapse of the Laramide Granite Mountains along east-west–trending normal faults. The basin contains fluviolacustrine and eolian deposits (Split Rock and Moonstone Formations) that range in age from early to late Miocene based on paleontological data (Love, 1970). Most of these deposits, however, are of middle to late Miocene age (K. M. Flanagan, personal communication, 1993). Although the Split Rock Basin has undergone fluvial degradation and erosion since the latest Miocene, Quaternary normal-fault scarps indicate that crustal extension is continuing today (Case, in preparation).

To the west of the Split Rock Basin, the Circle Bar Basin formed by collapse of the southern part of the Laramide Wind River uplift by normal displacement along the Continental fault (Steidtmann and Middleton, 1991). The western part of the Wind River uplift, however, remains uncollapsed. The Circle Bar beds compose part of the basin fill and contain a tuffaceous unit that has yielded a zircon fission-track age of 13.5 ± 2.4 Ma (Steidtmann and Middleton, 1986, 1991).

The Browns Park Formation near its type locality is about 550 m thick and was deposited in an extensional basin (Browns Park Basin in Fig. 1) formed by the collapse of the eastern part of the Laramide Uinta uplift (Hansen, 1984, 1986). Deposition began in the late Oligocene and continued until the late Miocene, based on paleontologic and radiometric constraints (Izett, 1975; Hansen, 1984, 1986; Luft, 1985). Further normal faulting has occurred since the end of deposition of the Browns Park Formation (Hansen, 1986, p. 51).

As much as 550 m of the North Park Formation crops out along the North Park syncline (Hail, 1965), a large drag(?) fold in the hanging wall of the Spring Creek normal fault (Izett, 1975; Tweto, 1979). The North Park Formation has been constrained paleontologically to be of late Miocene age (Izett, 1975). Additional exposures of Miocene deposits (Izett, 1975) crop out in the area intervening between the North Park syncline and the Browns Park Basin, along the northern and western flanks of the Park Range. These deposits have been variously attributed to the Browns Park or North Park Formations (Izett, 1975). Many of these deposits do not appear to be confined to structural basins, thus they are depicted as rift-flank deposits in Figure 1. This is probably an oversimplification.

The Troublesome Formation crops out in two structural basins along the east and west margins of Middle Park in northern Colorado (Troublesome and Granby-Fraser basins, Fig 1), and is as much as 450 m thick. Radiometric and fossil data indicate an age range of early to late Miocene for the unit (Izett, 1975). The Troublesome and Granby-Fraser basins owe their origin, at least in part, to structural inversion along Laramide reverse faults (Izett, 1975).

Previous workers have postulated that the pole of rotation for the opening of the Rio Grande rift is located in central Colorado (Hamilton, 1981, 1988), northern Colorado (Cordell, 1982), or northeastern Colorado (Spence and Gross, 1990). A geometric requirement common to all of these Euler pole locations is that extension in the Rio Grande rift to the east of the Colorado Plateau be accompanied by contemporaneous crustal shortening to the north of the plateau. Evidence for such late Cenozoic crustal shortening, however, is largely lacking east of the central Uinta Mountains. As noted above, abundant evidence for Miocene extension is present along the eastern margin of the Colorado Plateau and in a zone to the north of the plateau westward to about 110°W longitude (Fig. 1). West of about 110°W, Laramide uplifts to the north of the plateau (western Uinta uplift, western Wind River uplift, Rock Springs uplift) have survived largely unscathed by the effects of late Cenozoic extensional collapse; east of 110°W Laramide uplifts have commonly collapsed along favorably oriented older structural zones that are approximately tangential to the Colorado Plateau margin. Late Cenozoic structural relief along collapsed Laramide uplifts increases southward along the eastern margin of the plateau from a minima at about 40°N latitude. Structural relief along late Cenozoic normal faults generally does not exceed that of precursor Laramide reverse faults north of central Colorado. It is probably not coincidental that the area where late Cenozoic normal displacement begins to fully counteract Laramide reverse displacement in central and southern Colorado also marks the northern terminus of the longitudinally linked axial basins of the Rio Grande rift.

Based on the evidence cited above, we infer that the Euler pole of Miocene rotation of the Colorado Plateau relative to the stable craton is located in northeastern Utah (approximately 110°W, 40°N; Fig. 1). This interpretation differs considerably from those of previous workers and has numerous implications, some of which are intuitively appealing and others which may prove controversial.

The Eulerian geometry depicted in Figure 1 implies that Miocene extension along the eastern side of the Colorado Plateau was left oblique, particularly in Colorado. Evidence for such left-oblique extension has been noted by Kelley (1982), who described en echelon structures and left-lateral offsets and drag features of late Cenozoic age in the central Rio Grande rift. The orientation of upper Oligocene to upper Miocene dikes in New Mexico (Aldrich et al., 1986) indicates locally variable extension directions ranging from southwest-northeast to northwest-southeast within the rift, presumably due to control of dike intrusion by older structures. Dikes that formed over the tops of mid-Tertiary plutons, however, such as those near Socorro and Questa (Fig. 1), reflect the regional northeast-southwest extension direction rather than the orientation of preexisting structures (Aldrich et al., 1986, p. 6207). The extension directions at Socorro and Questa are subparallel to the accommodation zones of the rift (see below), which are transverse structural zones that divide domains of opposing stratal tilt within the rift. The location and orientation of these accommodation zones are generally well constrained and appear to be approximately parallel to small circles of rotation about the proposed Euler pole and, hence, the tectonic transport direction (Fig. 1). These relations imply that Miocene extension was not orthogonal to the trend of the rift and that left-oblique slip must have occurred on at least some of the northerly trending normal faults of the rift. This would counteract a small portion of the right-lateral slip of the preceding Laramide deformation in the Southern Rocky Mountains (Chapin and Cather, 1981). Several aspects of the left-oblique extension in the Rio Grande rift are similar to the oblique-slip deformation in parts of the East African rift (Scott et al., 1992).

Palinspastic restoration using seismic-reflection profiles allows estimation of extension within several basins of the central and northern Rio Grande rift. These include the San Luis Basin (8 to 12%; Kluth and Schaftenaar, this volume), the northern Albuquerque Basin (at least 17%; Russell and Snelson, 1990), and the southern Albuquerque Basin (at least 28%; Russell and Snelson, 1990). Using these estimates of extension in relation to the proposed Euler pole, we estimate 1.0 to 1.5° of clockwise rotation of the Colorado Plateau has occurred during the development of the Rio Grande rift. Assuming 1.5° of clockwise plateau rotation, north-south extension along the northeast margin of plateau varies from approximately zero measured along the meridian of 110°W longitude near the central Uinta Mountains to about 6 km measured along 107°W longitude through the eastern part of the Split Rock Basin.

The Euler pole of rotation depicted in Figure 1 was operative from late Oligocene to at least the late Miocene, as shown by the above-described timing of development of extensional basins within and to the north of the Rio Grande rift. The kinematics of the Colorado Plateau rotation following the main phase of middle to late Miocene rapid extension, however, are more obscure. Paleostress indicators for the interval 0 to 5 Ma show that extension was west-northwest to east-southeast in the Rio Grande rift in New Mexico (Aldrich and Laughlin, 1984; Aldrich et al., 1986). Extension in such an orientation is compatible with an Euler pole location in Colorado, as has been proposed by earlier workers (Hamilton, 1981, 1988; Cordell, 1982; Spence and Gross, 1990). As noted previously, however, such a kinematic model would require contemporaneous crustal shortening along the entire northern margin of the Colorado Plateau. Although modern crustal shortening is occurring in the northwest part of the Colorado Plateau (Wong and Humphrey, 1989), normal faulting similar in distribution to that during Miocene extension has continued in the eastern Uinta Mountains (Hansen, 1986, and personal communication, 1993) and in central Wyoming (Case, in preparation). It is perplexing to us that the Miocene kinematic history of the rift and its related structures is more clearly apparent than is that of Plio-Pleistocene. Because post-Miocene extensional strain in the Rio Grande rift–Southern Rocky Mountains area has been relatively modest, much detailed work may be necessary to discern the regional aspects of Plio-Pleistocene deformation.

Thus, in simplest terms, the Rio Grande rift can be thought of as the chain of longitudinally linked extensional basins and related fault zones that forms the western boundary of the North American craton from central Colorado through New Mexico and continuing southeastward along the Texas-Mexico border at least as far as the Big Bend country. Although late Cenozoic faulting and development of isolated basins of Miocene age in northern Colorado, northeastern Utah, and central Wyoming are kinematically related to development of the Rio Grande rift, we do not advocate their inclusion in the rift (sensu stricto) because of the lack of interconnectivity between basins and the broad locus over which extensional deformation occurred in these areas.

ONSET OF RIFTING

Burke (1980, p. 42) defined rifts as "elongate depressions overlying places where the entire thickness of the lithosphere has ruptured in extension." Mohr (1982, p. 293) utilized morphology and surface structure alone to define rift valley (or just "rift") as: "a parallel-sided, down-faulted valley, tens of kilometers in width and at least a few hundred kilometers in length." Mohr also stated that: "Graben and rift valleys are commonly asymmetric about their longitudinal axes, and where a faulted margin faces a warped margin, a so-called half graben is developed." He defined a rift system as "a set of rift valleys united through closely parallel, en-echelon or anastomosing links." Certainly the Rio Grande rift has fit these definitions for the past 15 m.y. The question is, Was the Rio Grande rift an intracontinental rift during the early stages of its development?

Lipman and Mehnert (1975) placed the beginning of rifting at about 26 Ma (radiometric dates cited herein have been adjusted for current IUGS [International Union of Geological Sciences] constants) on the basis of an angular unconformity along the west side of the San Luis Basin. The unconformity was cut on middle Tertiary regional ash-flow sheets (31 to 27 Ma) and overlain by basalt flows as old as 25 to 27 Ma (Fig. 2) that interfinger with alluvial-fan deposits of the Los Pinos Formation (herein considered part of the Santa Fe Group). This age is in excellent agreement with the 25.7-Ma age of the Amalia Tuff erupted from the Questa caldera and interbedded in the Los Pinos Formation (Fig. 2) along the southwest side of the San Luis Basin (Lipman et al., 1986; Manley, 1984). In the Upper Arkansas Basin, Tweto (1979) placed the beginning of rifting at younger than the Gribbles Park Tuff, which was deposited in east-trending paleovalleys that crossed the site of the Upper

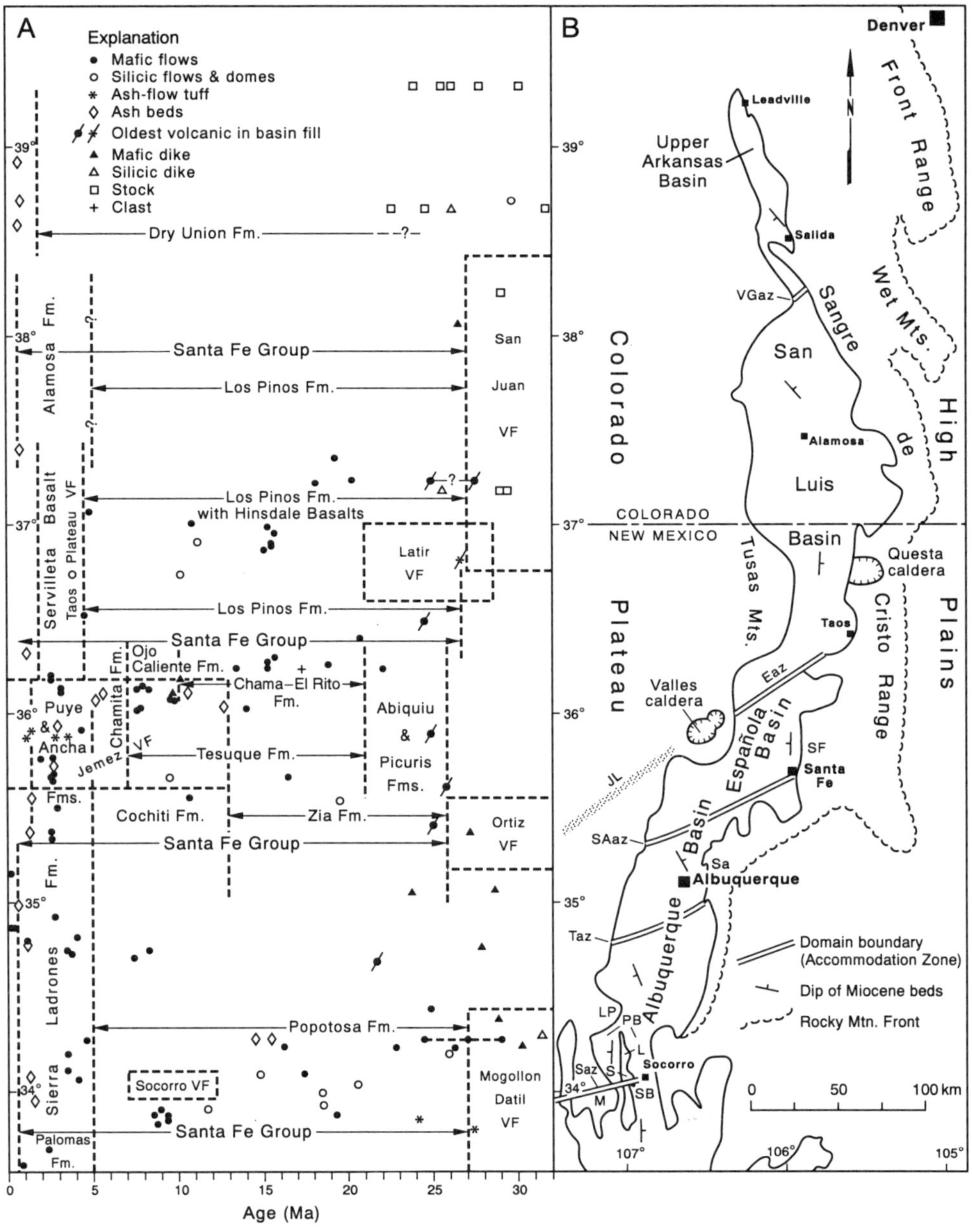

Figure 2. A, Latitude versus age plot of radiometric ages of igneous rocks and stratigraphic nomenclature for axial basins of the northern and central Rio Grande rift. Major volcanic fields are shown as time-space boxes without plotting individual ages except for rocks that constrain boundaries of syn-rift sedimentary units. References and details of the radiometric ages are available in a databank at the New Mexico Bureau of Mines and Mineral Resources. Older K-Ar ages have been adjusted for the current IUGS constants using the table of Dalrymple (1979). Mafic flows include some andesites. Modified from Chapin (1988, 1989). See Chapin (1988) for discussion of stratigraphic nomenclature. B, Sketch map showing axial basins of the northern and central Rio Grande rift (same scale as A). All basins are half grabens with asymmetry changing across accommodation zones that separate domains of west-tilted versus east-tilted strata. Abbreviations for accommodation zones are VGaz, Villa Grove; Eaz, Embudo; SAaz, Santa Ana; Taz, Tijeras; Saz, Socorro. Abbreviations for other features are SF, Santa Fe Range; Sa, Sandia Mountains; LP, Ladron Peak; PB, Popotosa basin; L, Lemitar Mountains; S, Socorro Peak; M, Magdalena Mountains; SB, Socorro Basin; JL, Jemez lineament. The Morenci lineament extends southwestward from Socorro. Base modified from Tweto (1978) and Woodward and others (1978).

Arkansas Basin (Epis and Chapin, 1975). Recent $^{40}Ar/^{39}Ar$ dating by McIntosh et al. (1992b) has established the radiometric age of the Gribbles Park Tuff at 32.76 ± 0.14 Ma rather than the 29-Ma K-Ar age cited by Tweto. Lack of dated volcanic rocks within the sedimentary fill of the Upper Arkansas graben prevents establishment of a good bracket on the beginning of rifting at this latitude. However, intrusion of several high-silica plutons along the margins of the Upper Arkansas graben between about 33 to 24 Ma may reflect the beginning of rifting (southern Sawatch Range, 29 to 30 Ma; Shannon, 1988, 1989; Nathrop topaz rhyolite, 29 Ma; Van Alstine, 1969; Climax area, 33 to 24 Ma; Bookstrom et al., 1988).

In the Española Basin, mafic flows dated at 24.6 to 25 Ma (Fig. 2) are interbedded with the Picuris Formation and the Jarita Member of the Los Pinos Formation (Bachman and Mehnert, 1978; Baldridge et al., 1980). For the southern Albuquerque Basin, the best evidence for the age of the oldest synrift fill comes from a 22-Ma basalt flow interbedded in strongly tilted lower basin fill along the southeast flank of the basin (Bachman and Mehnert, 1978). In the Popotosa basin near Socorro (Fig. 1), basal deposits of the Santa Fe Group are bracketed between the 16.2-Ma andesite of Silver Creek and the 27.0-Ma La Jara Peak Basaltic Andesite (Machette, 1978; Cather and others, this volume). In the Abbe Springs Basin, 30 km southwest of the Albuquerque Basin, 26-Ma dacite flows at Arroyo Montosa are interbedded with lower Santa Fe fanglomerates (Osburn and Chapin, 1983).

Middle Tertiary ash flows crossed the rift prior to the development of sufficient topographic relief to prevent their crossing. Precise $^{40}Ar/^{39}Ar$ dating (McIntosh and others, 1992a) indicates that four major ash-flow sheets were emplaced across the Popotosa basin in the short interval 28.8 to 27.4 Ma. Large volumes of basaltic andesite lavas are interbedded with these ash-flow sheets to form a distinctly bimodal suite of rocks. The truly catastrophic eruption rates during this brief 1.4-m.y. interval, after a 3.3-m.y. lull in volcanism (McIntosh et al., 1992a), suggests a relationship to the onset of rifting. Angular unconformities between successive ash-flow sheets, separated by wedge-shaped accumulations of mafic lavas (Chamberlin, 1983), indicate that rapid extension and domino-style faulting began during the early part of the 28.8 to 27.4-Ma interval. Rapid distension of the crust and infilling of strike valleys with mafic lavas prevented development of significant topographic relief and permitted successive ash flows to cross the rift. This precise dating of the onset of rifting near Socorro agrees well with less-precise dating in the San Luis and Upper Arkansas Basins, as discussed previously. Also, the 25.7-Ma Questa caldera on the southeast side of the San Luis Basin was intensely deformed concurrent with, and after, caldera formation by closely spaced listric normal faults in a style and timing very similar to that south of the Albuquerque Basin (Lipman, 1983; Meyer and Foland, 1991). The subdued, early Miocene topography in the Questa area allowed the Amalia Tuff to be deposited as far west as the Tusas Mountains on the west side of the present San Luis Basin and as far east as the High Plains (Lipman, 1983). Major topographic relief probably did not develop along the rift until the early middle Miocene (15 to 16 Ma), when rapid extension caused formation of numerous closed basins (e.g., Cather et al., this volume).

The beginning of rifting coincided with an interval of intense pyroclastic volcanism from caldera centers, both within and outside the early rift basins. If these volcanic rocks are proven to be rift related, then the Rio Grande rift is not so devoid of volcanic rocks as some workers have presumed (e.g., Baldridge et al., 1984). Elston and Bornhorst (1979) conclude that the Rio Grande rift did not begin until after a mid-Miocene lull in volcanic activity that ended about 15 Ma. They assign the copious late Oligocene–early Miocene pyroclastic activity and basaltic andesite volcanism to a period of back-arc extension. However, regardless of the cause of extension, sedimentary basins were accumulating bolson-type sediments along the entire length of the central and northern Rio Grande rift by late Oligocene–early Miocene as demonstrated by radiometric dating of interbedded volcanic rocks (Fig. 2). Similar timing has been demonstrated for the beginning of sedimentation in many fault-bounded basins along the projected trend of the Rio Grande rift in northern Colorado (Izett, 1975; Tweto, 1979) as well as in the southern Rio Grande rift (Chapin and Seager, 1975; Seager et al., 1984). Throughout its history, the Rio Grande rift opened as a north-trending zone of extension marked by a linear chain of sedimentary basins between the Colorado Plateau on the west and the High Plains on the east (Fig. 1). Thus, using any reasonable definition of a continental rift, it seems difficult to deny that the Rio Grande rift was a structural and sedimentological entity from late Oligocene to present. Although causes, rates, and styles of extension have varied during the long history of the Rio Grande rift, a perusal of literature on continental rifts around the world reveals that such variations are common (Mohr, 1982; Williams, 1982).

BASIN GEOMETRY

As defined by the present distribution of basin-fill sediments, the axial basins in the northern and central Rio Grande rift range from 80 to 240 km in length and from 5 to 95 km in width, with an average width of approximately 50 km. Basin-fill deposits range up to 5 to 6 km in thickness. The basins are asymmetric half grabens, hinged down on one side with major fault boundaries on the opposing side. The sense of asymmetry shifts back and forth from basin to basin and in places within a basin (Figs. 1, 2). The transverse boundaries between east-tilted versus west-tilted basins, or basin segments, are enigmatic structural zones referred to as transfer faults or accommodation zones (see below). Structural relief, as determined by differences in elevation of Precambrian rocks cropping out in mountain ranges bordering the rift and the Precambrian-Phanerozoic interface beneath rift basins, ranges up to 12 km, and averages about 5 km. For example, the top of the Precambrian basement lies at depths as great as 8,700 m below sea level in the Albuquerque Basin (Black, 1982; Lozinsky, 1988 and this volume; Russell and Snelson, 1990) yet Precambrian rocks crop out as high as 3,300 m above sea level in the adjacent Sandia Mountains. Except for the Upper Arkansas Basin, structural relief is generally greater along the east side of the Rio Grande rift (Chapin, 1971), although much of this relief in some areas is Laramide.

Within the above framework, however, there is an interesting progression of structural styles from the San Luis Basin southward to the Popotosa basin at Socorro due to the effects of increasing crustal extension. The San Luis Basin at the latitude of Alamosa is a simple east-tilted half graben with a gentle homocline on the hinged western margin and a high-angle master fault against the Sangre de Cristo Range on the east (Brister, 1990; Brister and Gries, this volume; Kluth and Schaftenaar, this volume; Tandon, 1992). Sedimentary deposits of the late Oligocene to Quaternary Santa Fe Group dip 0 to 12° eastward and are not cut by major antithetic or synthetic faults within the basin. Kluth and Schaftenaar (this volume) estimate 8 to 12% extension across the rift from structure sections based on seismic-reflection lines. From reprocessing the same seismic lines to search for basement structure, Tandon (1992) estimates that the basin-bounding Sangre de Cristo fault continues at dips of 40 to 60° to depths of at least 26 to 28 km and does not sole out at midcrustal depths as suggested by Kluth and Schaftenaar (this volume). The San Luis Basin overlies more complex Laramide structure and its simple eastward tilt is slightly modified by a longitudinal central basin high, the Alamosa "horst." This central basin high is partly inherited from underlying Laramide structure and partly due to arching along the west edge of the more strongly tilted eastern subbasin, the Baca basin (Brister, 1990). Even with this modest complication, the San Luis Basin is the simplest major half graben in the Rio Grande rift for which we have subsurface control.

The next basin to the south, the Española Basin, contains a relatively simple half graben, tilted down to the west with a hinged margin on the eastern, Sangre de Cristo side and bounded on the west by the Pajarito fault zone. Syn-rift sediments of the Santa Fe Group dip 6 to 8° westward and thicken from an eroded, homoclinal edge along the Santa Fe Range portion of the Sangre de Cristo Mountains to as much as 3 km against the Pajarito fault zone (Cather, 1992). The Pajarito fault displaces 4- to 5-Ma dacites of the Tschicoma Formation less than 300 m and the 1.1-Ma Bandelier Tuff less than 150 m (Biehler et al., 1991). The Pajarito fault merges to the north with the northeast-trending Embudo fault zone, which acts as an accommodation zone between the east-tilted San Luis Basin to the north and the west-tilted Española Basin to the south. A narrow intrabasin graben, the Velarde graben, subsided in early Pliocene time along the Pajarito-Embudo fault zone; Manley (1979) estimates as much as 360 m of stratigraphic offset across the graben faults. Thus, the broad, early Española Basin has become narrower and more complex with time.

The Española half graben is unusual in that it is structurally and topographically higher on the hinged side of the basin instead of the master fault side, presumably due to Laramide crustal thickening beneath the Santa Fe Range, which forms the basin hinge. Anomalously old fission-track apatite ages (50 to 70 Ma; Kelley and Duncan, 1986; Kelley et al., 1992) in the Santa Fe Range indicate that the early Española Basin may have extended as far east as the modern summit of the range, as initially suggested by Denny (1940). Uplift and exhumation of the Precambrian core of the range from beneath a relatively thin covering of synrift sediments would explain the preservation of these Laramide cooling ages as well as the unusually fine-grained nature of the Miocene Tesuque Formation (Santa Fe Group) along the present mountain front.

The Albuquerque Basin is the most explored and most complex of the axial basins in the central and northern Rio Grande rift. It consists of a northern half graben tilted down to the east with the Sandia Mountains looming above its deep side and a southern half graben tilted down to the west and bordered by the Lucero uplift and Ladron Peak. The two half grabens are separated by the Tijeras accommodation zone that crosses the basin as a southwestward projection of the Tijeras fault zone, a major strike-slip fault zone bordering the Sandia Mountains on the south. The half grabens of the Albuquerque Basin differ from basins to the north in having wide, structurally high, suballuvial benches between the flanking uplifts and major intrabasin faults across which syn-rift sediments thicken abruptly (Russell and Snelson, 1990; May and Russell, this volume). The master faults are interpreted by the above authors to be listric with dips decreasing from 60 to 65° at shallow depths to as low as 15 to 17° at depths of about 10 km. Stratigraphic throws of 3,050 to 6,100 m have been estimated for the largest of these faults from seismic reflection lines interpreted by Russell and Snelson (1990).

Basinward migration of faulting with time has cut off and stranded long, narrow blocks between earlier formed faults. This appears to be the origin of the suballuvial benches. Benches on the hinged sides of the half grabens are relatively wide and shallow and contain relatively thin sections of the lower Santa Fe sediments. Master fault benches tend to be narrow and structurally lower as a result of rapid subsidence in late Miocene time (Russell and Snelson, 1990; May and Russell, this volume). Syn-rift sedimentary deposits thicken by as much as 3 km across master faults, largely by abrupt thickening of the middle unit of the Santa Fe Group (Hawley and Haase, 1992). A basalt flow penetrated by the Shell Isleta #2 well and dated by K-Ar at 16.1 Ma (May and Russell, this volume) is overlain by three times as much Santa Fe as occurs between it and the base of the syn-rift section in the Albuquerque Basin. A similar relationship has been documented in the Popotosa basin to the south where the Santa Fe Group is much thicker above the 16.2-Ma basaltic andesite of Silver Creek than beneath it (Cather and Kugler, in preparation). Most of these deposits in both basins are of middle and late Miocene age; Plio-Pleistocene deposits are typically less than a few hundred meters thick.

Russell and Snelson (1990) estimate a minimum of 17% extension for the northern Albuquerque Basin and at least 28% for the southern Albuquerque Basin. The northern half graben is divided longitudinally into three relatively broad blocks by three master faults and the western hinge (Russell and Snelson, 1990). The southern half graben is divided longitudinally into six major blocks, some of which are quite narrow, by seven master faults (Russell and Snelson, 1990). In comparison, Kluth and Schaftenaar (this volume) estimate only 8 to 12% extension across the San Luis Basin, which is a simple half graben bounded by one high-angle master fault. Thus, there appears to be a relationship between the amount of extension and the amount of internal subdivision of basins by major faults. The early rift Popotosa basin, which joins the Albuquerque Basin at its south end, has undergone an average of 50% extension (Chamberlin and Osburn, 1984; Chapin, 1988) and as outlined in the following paragraphs is another step up in the spectrum of complexity.

The Popotosa basin developed in the transition zone where the Rio Grande rift broadens into a series of parallel basins and intrarift tilted-block uplifts. The rift also bifurcates here with the San Agustin arm trending southwestward to near the Arizona border (Chapin, 1971). The early rift Popotosa basin was at least 35 km wide east-west and approximately 45 km long between the Albuquerque Basin on the north and the Socorro accommodation zone on the south. The Popotosa basin was split longitudinally into the parallel La Jencia and Socorro Basins by uplift of the Lemitar Mountain block (Fig. 3) through the surface of the basin in late Miocene time (Denny, 1940; Bruning, 1973; Chapin and Seager, 1975; Chamberlin, 1983; Cather et al., this volume). Excellent exposures of Precambrian to upper Oligocene strata that floored the Popotosa basin, as well as syn-rift clastic sedimentary rocks of the Popotosa Formation, are present along the uptilted edge of the La Jencia half graben in the Lemitar Range. This provides one of the best opportunities anywhere in the rift for examination of the deeper portions of a basin.

The Popotosa basin experienced two major episodes of rapid extension, the first in late Oligocene (largely 28.6 to 27.4 Ma), the second in middle to late Miocene (16 to 10 Ma; Cather et al., this volume). Both episodes were associated with major volcanism. The first episode occurred during the peak of ignimbrite volcanism in the Mogollon-Datil volcanic field and was accompanied by major basaltic andesite volcanism within and adjacent to the area of the basin. Heat flow was undoubtedly very high and the upper crust extended rapidly above a shallow brittle-ductile transition. A domain of west-tilted, narrow fault blocks began to form between the Socorro accommodation zone and the south end of the Albuquerque

Figure 3. View of Lemitar intrabasin, tilted-block mountain range from top of Socorro Peak. View is to northwest with Strawberry Peak (a remnant of a late Miocene silicic dome) at left center, Colorado Plateau on center skyline, Bear Mountains on left skyline, and Ladron Peak at upper right. La Jencia Basin in left middle distance is a half graben formed by down-to-the-west tilting of the 15 to 25-km-wide Lemitar block. Dark ridge at left center and western-most hogbacks in Lemitar Range are late Oligocene–early Miocene basal syn-rift sedimentary rocks of Popotosa Formation. Light-colored strata at lower right are late Miocene playa deposits in upper Popotosa Formation. Polvadera Mountain at right-middle distance consists of west-dipping late Eocene and Oligocene volcanic rocks overlying Pennsylvanian limestone and shales dipping about 45° westward (light band) above broad exposures of Precambrian granitic and gabbroic rocks. The entire stratigraphic section is repeated across the Silver Creek fault that runs longitudinally the length of the Lemitar Range (see Fig. 1 of Cather et al., this volume). Stratal tilts within the Lemitar block range from approximately 48 to 70° for late Oligocene ash-flow tuffs to 28 to 45° for basal syn-rift strata of the Popotosa Formation.

Basin after eruption of the 28.6-Ma Vicks Peak Tuff and prior to eruption of the 28.0-Ma Lemitar Tuff (Chamberlin, 1983; Chapin, 1989). A similar domain of east-tilted fault blocks formed concurrently south of the Socorro accommodation zone. Progressive domino-style rotation of these fault blocks is indicated by systematic decreases in present-day dips of progressively younger strata, ranging from about 48 to 70° for the 28.6-Ma Vicks Peak Tuff to approximately 28 to 45° for basal strata (upper Oligocene–lower Miocene) of the syn-rift Popotosa Formation (Chamberlin, 1983; Chapin, 1989; Cather et al., this volume). This episode of close-spaced faulting and strong rotation of strata resulted in approximately 25% extension in only about 1.2 m.y. as the Popotosa basin began to form. As discussed by Cather et al. (this volume), domino-style extension with strong rotation of strata is present in only a few localities in the Rio Grande rift, primarily in local areas of very high heat flow associated with major volcanism.

The second episode of major extension in the Popotosa basin occurred between about 16 and 10 Ma, more or less concurrently with extrusion of large volumes of silicic lavas in the Magdalena Peak (18.4 to 13.5 Ma) and Socorro Peak (12.1 to 7.2 Ma) volcanic centers and eruption of moderate volumes of basaltic andesite lavas (16 Ma) in the Silver Creek area (Fig. 2). Cather et al. (this volume) estimate that this second episode of faulting resulted in about 30% extension, roughly equal to the first episode but significantly slower. However, this second episode had a dramatic effect on the Popotosa basin by uplift of the Lemitar Mountain block through the surface of the basin and by rapid elevation of Ladron Peak (Fig. 3). Fission-track apatite ages reported by Kelley et al. (1992), and May et al. (this volume) indicate rapid cooling between 14.1 ± 2.0 and 9.0 ± 1.5 Ma for Ladron Peak and between 12.5 ± 2.0 and 6.1 ± 1.8 Ma for the Lemitar Mountains. In contrast, the Joyita Hills block on the northeast flank of the basin cooled largely between 27 and 22 Ma while the Magdalena Mountains on the southwest flank cooled mainly between about 27 and 16 Ma (Kelley et al., 1992).

The floors of the axial basins became increasingly subdivided into parallel, longitudinal blocks as extension increased from 8 to 12% in the San Luis Basin to at least 28% in the southern Albuquerque Basin, yet at today's ground surface the basins appear simple and broad. Apparently, at some point beyond 28% extension, perhaps aided by high heat flow, tectonic denudation by movement on moderate to low-angle master faults within the Popotosa basin caused sufficient isostatic uplift to bring the upper edge of the 15- to 25-km-wide La Jencia half graben (Lemitar Range) to the surface and subdivide the basin. It is interesting to look at the map and cross sections of the southern Albuquerque Basin interpreted by Russell and Snelson (1990, this volume) from Shell's seismic-reflection profiles and visualize how a 15-km-wide block, such as between the Jeter and Rio Puerco faults, could be tilted up to break the surface of the basin and form an intrabasin tilt-block range. In such a scenario, intrabasin faults would evolve into mountain-front faults as deformation proceeded, as has been described in the Lemitar Mountains by Cather et al. (this volume). Are the bounding faults of the Lemitar Range (the Silver Creek and Socorro Canyon–Loma Peleda faults) exposed examples of what the Santa Fe–Coyote and Rio Puerco faults are like in the subsurface of the Albuquerque Basin? And did the allochthonous upper plates of low-angle detachment faults mapped in the Joyita Hills by Beck and Chapin (this volume) slide eastward in response to uplift of the Lemitar and Ladron blocks? The Joyita Hills block is a southern extension of the Hubbell bench, a "hinge bench" left stranded along the east side of the southern Albuquerque Basin as extension moved basinward. Beck and Chapin (this volume) have mapped allochthonous plates containing Paleozoic through upper Oligocene strata that have moved an unknown distance eastward and southeastward across the Joyita Hills block. Several other examples of younger-on-older, low-angle detachment faults have been observed in recent years along the east side of the Popotosa basin. Age constraints are poor but some faults cut upper Oligocene volcanic strata (Smith, 1983; Smith et al., 1983). Answers to the above questions must await further work, but much recent progress has been made in understanding the axial basins of the Rio Grande rift.

ACCOMMODATION ZONES

The basins of the Rio Grande rift are half grabens whose asymmetry reverses from basin to basin across complex transverse structural zones called transfer faults (Gibbs, 1984) or accommodation zones (Bosworth, 1985). Detailed studies of accommodation zones have been published by Bosworth et al. (1986), Rosendahl (1987), Chapin (1989), and Faulds et al. (1990). In the central and northern Rio Grande rift, the best documented of these structures are the Socorro accommodation zone (Chapin et al., 1978; Chapin, 1989) south of the Popotosa basin, the Embudo fault zone (Muehlberger, 1979) between the San Luis and Española Basins, and the Tijeras "transfer zone" (Russell and Snelson, 1990) that separates the northern and southern half grabens of the Albuquerque Basin. The Santa Ana accommodation zone (Cather, 1992) and the Villa Grove accommodation zone are poorly exposed and less constrained, as are several such zones in the southern rift.

The transverse structures formed where the north-trending rift broke across preexisting northeast-trending lineaments of probable Precambrian ancestry (Figs. 1, 2). As noted above, the accommodation zones of the northern and central Rio Grande rift developed along older structural trends that were approximately parallel to small circles of rotation between the Colorado Plateau and the stable craton during Miocene extension. Thus, the length of the half grabens within the rift were determined primarily by the spacing of major, favorably oriented lineaments. Structurally complex constrictions formed at the lineament crossings; basin-fill sedimentary deposits are commonly thin on these constrictions and may be strongly deformed. The transverse structures tend to leak magmas and to be conduits for ascending hydrothermal fluids; they also act as low-permeability sills that constrict and alter ground-water flow patterns (Bryan, 1938; Anderholm, 1983; Coons and Kelly, 1984). As shown on Figure 2A, syn-rift magmatism has been most continuous where the rift crosses the Morenci lineament in the Socorro area and the Jemez lineament in northern New Mexico. (See Chapin and others [1978] for a map showing major lineaments in Colorado and New Mexico and a discussion of the Morenci lineament. Aldrich and Laughlin [1984] describe the Jemez lineament and discuss its Cenozoic history.)

Accommodation zones have been recognized during recent years as being fundamental features of all extended terranes (Bosworth, 1986; Lister et al., 1986a, b; Rosendahl, 1987), yet they are poorly understood, especially in terms of the kinematics of their development. A considerable diversity exists in the observable characteristics of accommodation zones, both within the Rio Grande rift and elsewhere. Most controversial is the amount of strike-slip movement. The Socorro accommodation zone, which was originally interpreted as a "transverse shear zone" (Chapin et al., 1978), has upon detailed mapping been found to have no observable strike-slip displacement (Chapin, 1989). In fact, it is invisible on geologic maps except in terms of a 50-km-long by 2-km-wide boundary between a northern domain of west-tilted strata and a southern domain of east-tilted strata. No major faults parallel the accommodation zone and it crosses the boundaries of three Oligocene calderas without displacing them. Detailed mapping of an accommodation zone in the Black Mountains of Arizona and Eldorado Mountains of Nevada by Faulds et al. (1990) has likewise failed to document significant strike-slip movement parallel to the accommodation zone. How these accommodation zones "accommodate" tilting of strata as much as 60 to 70° in opposite directions remains poorly understood.

In contrast to the Socorro accommodation zone, the Embudo accommodation zone between the San Luis and Española Basins is marked by a conspicuous, through-going fault, the Embudo fault. Displacement across the Embudo fault changes from normal southwest of Pilar, where it bounds the Velarde graben, to reverse northeast of Pilar, where the Picuris Mountains are being actively thrust over the south end of the San Luis Basin (Muehlberger, 1979). Structural relief across the latter segment was estimated by Muehlberger to be at least 3,000 m. Muehlberger described the Embudo fault as an active intracontinental transform that must have "a significant but as yet undemonstrated strike-slip component."

The Tijeras accommodation zone is known mainly from seismic reflection profiling and drill-hole data (Russell and Snelson, 1990). May and Russell (this volume) reported that the synrift sedimentary section thickens abruptly to the north of the Tijeras zone indicating "a major dip slip as well as probable strike-slip component of motion between the two sub-basins." The Shell Isleta #2 well drilled about 6 km north of the Tijeras accommodation zone bottomed at 6,482 m in Eocene strata after penetrating 288 m of Eocene sediments beneath 1,787 m of Oligocene volcaniclastic strata, for a combined pre-rift sedimentary section in excess of 2,075 m (Lozinsky, 1988, and this volume). Cather (1992) postulated that normal faulting along the Tijeras accommodation zone formed the southern margin of the Eocene Galisteo basin. Eocene strata beneath the northern Albuquerque Basin are as much as 454 m thick but appear to be thin or absent to the south of the Tijeras zone (Lozinsky, 1988, and this volume; Cather, 1992). The Oligocene volcaniclastic strata also become thinner to the south of the accommodation zone (Lozinsky, 1988, and this volume). Thus, the Tijeras accommodation zone appears to have been structurally active through much of Cenozoic time (Cather, 1992).

SYN-RIFT SEDIMENTARY DEPOSITS

Syn-rift sedimentary and volcanic deposits of the Rio Grande rift are collectively known as the Santa Fe Group (Fig. 2). For a review of the long and often controversial usage of the term Santa Fe see Hawley (1978), Tedford (1981, 1982) and Chapin (1988). Present practice is to use Santa Fe as a group term that includes all the syn-rift basin fill, both volcanic and sedimentary, ranging in age from late Oligocene to Quaternary, but excluding deposits that postdate entrenchment of the Rio Grande in early to middle Pleistocene time. Such broad usage of the term Santa Fe is obviously advantageous in the San Luis and Albuquerque Basins where most of the basin fill is not exposed and in the Upper Arkansas Basin where the older basin fill is not exposed and the younger fill has not been subdivided nor studied in detail. In the Española Basin, where numerous studies have been published and the basin fill has been subdivided and dated in much detail (Galusha and Blick, 1971; Manley, 1979; Ingersoll et al., 1990), the various formations and their members make up the Santa Fe Group. Eventually, as more detailed stratigraphic studies are published and map coverage becomes more complete, the Santa Fe Group will probably be elevated to supergroup status and some of the thicker and more complex formations will become groups.

Sedimentary fill of the early closed basins of the Rio Grande rift consists mainly of broad, coalesced alluvial-fan deposits that grade basinward into fine-grained alluvial-plain and playa deposits. Following a widespread episode of rapid deformation and unconformity development in the late Miocene (e.g., Cather et al., this volume), through drainage was established in early Pliocene (Bachman and Mehnert, 1978) and the bordering piedmont deposits interfingered with axial-river sands and gravels. In some areas, broad pediments have been cut on tilted fault-block uplifts; the sedimentary veneers on these pediments are also included in the Santa Fe Group when of appropriate age. Eolian dune sands are also present in some basins.

Strata of the Santa Fe Group are typically poorly indurated and pinkish buff to gray in color. However, older basin fills in areas that have experienced diagenetic reactions induced by alkaline, saline basin waters are very well indurated and reddish brown to brick red in color (Chapin and Lindley, 1986). The compositions of clasts vary markedly from one alluvial fan or piedmont slope to another depending upon the source regions. For example, in the Española Basin, the Tesuque Formation in the eastern half of the basin consists mostly of pinkish buff, arkosic sandstones and siltstones derived from basement rocks exposed in the Sangre de Cristo Range (Galusha and Blick, 1971; Manley, 1979), whereas the laterally equivalent Los Pinos, Chama–El Rito, and Ojo Caliente Formations in the northern and western portions of the basin contain abundant volcanic detritus derived from the San Juan and Latir volcanic fields (Manley, 1981; May, 1984; Ekas et al., 1984, Ingersoll et al., 1990). Similar relationships are present in other basins.

Following the cessation of rapid extension in the late Miocene (e.g., Cather et al., this volume), decreased rates of basin subsidence and increased runoff due to enhanced orographic effects allowed integration of drainages to form the ancestral Rio Grande. In the Socorro area, basalt flows underlying and interbedded with ancestral Rio Grande sands indicate that the central segment of the Rio Grande developed before 4.1 Ma and possibly before 4.5 Ma. An early Pliocene age of inception of through drainage has been documented in most rift basins in New Mexico. However, the northern segment of the Rio Grande, north of Questa, and the southern segment, southeast of El Paso, were not integrated with the central segment until middle Pleistocene (Wells et al., 1987; Gile et al., 1981). Well-sorted, axial-river deposits of sand and gravel accumulated concurrently with relatively poorly sorted piedmont-slope deposits shed from the basin margins and graded to the through-going river. These contrasting facies intertongued in a complex manner due to changes in climate, periods of aggradation versus incision of the river, avulsion of the axial-river channel, basin subsidence, and progradation of opposing piedmont slopes. Axial river deposits and their intertonguing piedmont facies are known as the Camp Rice and Palomas Formations in southern New Mexico (Hawley, 1975; Lozinsky and Hawley, 1986), the Sierra Ladrones Formation in the Socorro area and most of the Albuquerque Basin (Machette, 1978; Lozinsky and Tedford, 1991), and the Puye Formation in the western Española Basin (Griggs, 1964; Smith et al., 1970). Terminology for deposits of similar age in the northern Albuquerque Basin varies from upper Buff Member of the Santa Fe Formation (Bryan and McCann, 1937; Lambert, 1968); to Ceja Member of the Santa Fe Formation (Kelley, 1977); to various informal terms (Tedford, 1982), or simply upper Santa Fe Group.

The Ancha Formation (Spiegel and Baldwin, 1963) of the Española Basin and the Servilleta Formation (Lambert, 1966) of the southern San Luis Basin have a similar early Pliocene to middle Pleistocene age and depositional setting. Lipman and Mehnert (1979) redefined the Servilleta to include only the distinctive olivine tholeiite basalts of the Taos Plateau and relegated the interbedded sedimentary rocks to correlative upper Santa Fe units, such as the Ancha Formation. The interbedded lacustrine and fluvial deposits of the Alamosa Formation that underlie much of the northern San Luis Basin have been dated at 0.7 to 2.5 Ma in outcrop and drill core (Rogers, 1984; Rogers et al., 1992) and probably extend back to early Pliocene in age. In the Mapco-Amoco well 29 km northeast of Alamosa (Burroughs, 1981; Brister, 1990), a 625-m section of Alamosa Formation was penetrated and a Mio-Pliocene pollen age was obtained from underlying sediments of the Santa Fe Group undifferentiated (Gries, 1985). The Alamosa Formation should also be included in the Santa Fe Group using the broad definition of Spiegel and Baldwin (1963). The correlation charts of Hawley (1978) provide a concise summary of stratigraphic nomenclature and an excellent regional perspective on correlations and ages of units.

Post–Santa Fe deposits, laid down after the initial incision of the Rio Grande in the early to middle Pleistocene, consist of inset alluvium on a stepped sequence of valley-border surfaces, terrace deposits, eolian sands, paleosols, basalt flows, glacial moraines and outwash gravels, and arroyo and river valley alluvium. Alternate incision and aggradation by the Rio Grande during glacial and interglacial stages formed a complex inner-valley stratigraphy. Summaries of Quaternary stratigraphy are available by Hawley et al. (1976) and Gile et al. (1981) for New Mexico and by Scott (1975) for central Colorado.

Stratigraphic units of the upper Santa Fe Group (Camp Rice, Palomas, Sierra Ladrones, Puye, Ancha, Alamosa, etc.) are commonly bounded above and below by unconformities. The upper unconformities separate these units from a complex of late Pleistocene and Holocene deposits that postdate incision of the basin fill by the Rio Grande. The lower unconformities separate these units from older portions of the basin fill that are usually more strongly tilted and better indurated. These lower unconformities developed during and after the second major episode of extension that occurred between approximately 16 and 10 Ma (Cather et al., this volume). The oldest dated axial-river deposits of the ancestral Rio Grande that overlie the lower unconformity are 4.1 to 4.6 Ma (Bachman and Mehnert, 1978). Thus, the unconformities at the base of upper Santa Fe units typically separate axial river and intertonguing flood-plain and piedmont deposits from older closed-basin deposits. Where basins are still closed, as in the northern San Luis Basin and parts of several basins in the southern Rio Grande rift, sedimentation has been relatively continuous throughout the Neogene and the upper limit of the Santa Fe Group is difficult to establish.

No seismic profiles or deep drill holes exist in the Upper

Arkansas Basin so its sedimentary fill and subsurface configuration are the least known of the axial basins. The syn-rift sedimentary deposits were named the Dry Union Formation by Tweto (1961, p. B133) for exposures south of Leadville. The best exposures, however, are in the Salida area at the south end of the basin. Here, southwest-tilted strata of the Dry Union Formation (Santa Fe Group) are exposed adjacent to the major west-northwest–trending cross fault that constricts the basin (Fig. 2). Van Alstine (1970, 1974) and Taylor (1975) report the occurrence in the Dry Union strata of vertebrate fossils interpreted as correlative with a 12-Ma Valentine fauna and a 7-Ma Ash Hollow fauna from the Ogallala of Nebraska. G. E. Lewis (in Van Alstine, 1970) states that some of the fossils are similar to those in the upper Miocene part of the Tesuque Formation near Santa Fe.

A narrow trough, 5 to 6 km wide, projects southward approximately 11 km from the west end of the Salida cross fault and contains about 1,800 m of west-tilted Dry Union Formation (Knepper, 1976; Van Alstine, 1968). The accommodation zone separating the east-tilted San Luis half graben from the west-tilted Upper Arkansas half graben is located farther south between the small communities of Alder and Villa Grove. We refer to it in this paper as the Villa Grove accommodation zone; its location and orientation are poorly constrained in this alluvium-covered valley. Estimates of the maximum thickness of Dry Union Formation in the Upper Arkansas Basin range from 1,500 m to 300 m (Knepper, 1976; Scott, 1975; Scott et al., 1975; Van Alstine, 1970, 1974). From gravity data, Gridley and others (1991) estimate thicknesses of 1,500 m near Salida to about 900 m in the Leadville area. The western half of the basin is heavily mantled by Pleistocene deposits that include glacial moraines and outwash gravels. Lenses of Bishop ash (0.7 Ma) and Pearlette Type-O or Lava Creek B ash (0.6 Ma) have been found in lower Pleistocene alluvium (Scott et al., 1975). The Pleistocene deposits rest on extensive pediments carved across gently west-tilted beds of the Dry Union Formation.

TIMING OF UPLIFT

Most models of continental rifts assume broad uplift of flanking areas due to rise of the asthenosphere and accompanying thermal expansion as the lithosphere is extended (Burke, 1980; Cordell, 1978; Eaton, 1986, 1987; Morgan and Ramberg, 1987). This regional uplift is accentuated by narrow mountain ranges with high topographic and structural relief that border the rift basins and are caused by isostatic uplift of footwall blocks of master faults. (Buck, 1988; Wernicke and Axen, 1988; May et al., this volume). But what if the land surface was already at high elevations, as in the case of a high plateau? Would local relief, due largely to collapse of the rift basins and isostatic uplift of footwall blocks, be simply superimposed on a preexisting regional uplift? This question has been raised about the Rio Grande rift because of new interpretations of paleoelevations based upon fossil flora.

A key to this argument is the elevation of the Florissant Lake Beds in central Colorado at the time of deposition. The Florissant beds were deposited about 34 to 35 Ma in a shallow valley on the late Eocene erosion surface (Epis and Chapin, 1975). The late Eocene surface is a broad surface of low relief that beveled Laramide uplifts and extended across the sedimentary fills of Laramide basins. In central Colorado, it has been partially exhumed from beneath Oligocene volcanic rocks. From a detailed study of fossil plant remains in the Florissant beds, MacGinitie (1953) concluded that the elevation at the time of deposition was not more than 300 to 900 m. He based this estimate on paleofloristics, which is essentially a comparison of where the closest living relatives of these plants are found today. The Florissant beds are now at an elevation of about 2,500 m. This apparent uplift of 1,600 to 2,200 m since 34 to 35 m.y. ago, coupled with geomorphic evidence of major canyon cutting in late Miocene–Pliocene time (Scott, 1975), led Epis and Chapin (1975) to conclude that a major epeirogenic uplift had occurred in the late Cenozoic. This conclusion seemed to fit well with abundant evidence for major, late Cenozoic displacements across normal faults bounding many of the ranges along the Rio Grande rift and with expectations of broad, regional uplift caused by rifting.

The interpretation of major late Cenozoic epeiric uplift was reinforced by a study of several fossil flora along the Rio Grande rift by Axelrod and Bailey (1976). Using the paleofloristic method that MacGinitie had used, they concluded that a fossil flora in the Galisteo Formation (about 40 Ma) near Bernalillo grew at an elevation of approximately 550 m and a fossil flora in the Tesuque Formation (about 14 Ma) near Santa Fe grew at an elevation of approximately 700 m. The Bernalillo flora is now at an elevation of 1,800 m and the Tesuque Formation at 1,830 m. Because both flora had been uplifted about the same amount, they concluded that approximately 1,200 m of epeiric uplift had occurred since 14 Ma. This seemed to agree well with the late Miocene–Pliocene culmination of rifting postulated by Chapin and Seager (1975) and with the more than 1,200 m of uplift of the northern Sangre de Cristo Range since 7 Ma reported by Taylor (1975).

Then, in a pair of elegantly reasoned papers, Eaton (1986, 1987) analyzed the structure and topography of the Southern Rocky Mountains and concluded that the Rio Grande rift is the crestal graben of a mammoth, north-trending, bilaterally symmetrical, continental ridge formed in late Cenozoic by profound epeirogenic uplift driven by lithospheric thinning. He named it the Alvarado Ridge after one of its earliest explorers. In his analysis, Eaton used the elevation estimates of MacGinitie (1953), the conclusions of Epis and Chapin (1975) concerning uplift of the late Eocene surface, and the evidence for timing of uplift developed by many of the authors described in the preceding paragraphs. In addition, Eaton pointed out that the distribution of the middle Miocene to early Pliocene Ogallala Formation coincided with the north-south extent of the east flank of the Alvarado Ridge and concluded that this sedimentary apron should reflect the time of uplift.

Ogallala deposition began as early as 17 Ma in northern Colorado, but the oldest known Ogallala in New Mexico and west Texas is about 12 Ma (Izett, 1975; Scott and Taylor, 1986; Hawley, 1984). Deposition apparently ended by 4 to 5 Ma and was followed by development of an extensive caliche cap rock and incision of drainages in Pliocene-Recent times (Hawley, 1984; Gustavson and Winkler, 1988; Reeves, 1972). Eaton and many other workers have interpreted the Ogallala as being formed by coalescing alluvial fans of debris eroded from the rising Southern Rocky Mountains. However, recent sedimentological studies by Gustavson and Winkler (1988) show that much of this alluvial sediment has been reworked into extensive eolian deposits.

Some geophysicists have been troubled by the hypothesis of late Cenozoic uplift. Morgan (written communication, 1984, 1987) argues that the thick crust of the Great Plains (50 km) and the Colorado Plateau (42 to 45 km) could not have been that thick when the region was at or below sea level during deposition of Upper Cretaceous marine formations. And further, that it makes more sense to thicken the crust during Laramide compression than during late Cenozoic rifting, which should result in lithospheric thinning (Morgan, 1987). Molnar and England (1990, p. 29) have recently discussed the problem that ". . . high altitudes of most mountain ranges have commonly been ascribed to late Cenozoic uplift, without reference to when crustal thickening and other tectonic processes occurred." They point out that late Cenozoic global climate change towards lower temperatures could have produced many of the features used to infer late Cenozoic uplift.

New studies of the Florissant fossil flora by Meyer (1986, 1992), Wolfe (1992), Gregory (1992), and Gregory and Chase (1992) have focussed attention again on the question of when did the Southern Rocky Mountains attain their high elevation. These studies utilize methods developed by Wolfe (1978, 1990, 1992, 1994) using physical characteristics of leaves to estimate mean annual temperature. Meyer (1986, 1992) classified MacGinitie's collection of leaves from the Florissant beds in terms of vegetation type and estimated an elevation of 2,450 m at deposition, compared to 2,500 m today. He used only one quantitative physiognomic factor, percent smooth-margined species, which Wolfe (1978) found to have a linear relationship with mean annual temperature in mesic climates where availability of water is not a limiting factor. Wolfe (1990) recorded 31 physiognomic characteristics of leaves from 73 modern forests and observed that 70% or more of the variation corresponded to changes in temperature and/or precipitation. He developed a multivariate analysis that calculates mean annual temperature to within 0.6°C and also estimates mean annual precipitation. By comparison to paleoclimate estimates of the coast, Wolfe (1992) calculated a paleoelevation for Florissant of 2.7 to 2.9 km. Gregory (1992) made a new collection at Florissant of leaves from 29 species and developed a multiple regression model using five leaf physiognomic characteristics scored by Wolfe on 86 modern vegetation sites. Gregory's estimate of mean annual temperature, when combined with coeval mean annual temperature at sea level and the terrestrial lapse rate, indicates a 2.3 to 3.3 km elevation for Florissant at 34 to 35 Ma.

These new paleobotanical data have forced a comprehensive review of geomorphological evidence for the timing of uplift and questioning of previous assumptions. One result has been the realization that the frontal ranges of the Southern Rocky Mountains (Front Range, Wet Mountains, Sangre de Cristo Range) have a two-sided character. The west flanks facing the basins of the northern Rio Grande rift have young topography, prominent Quaternary fault scarps, and abundant alluvial fans (Fig. 4). The east flanks that face the High Plains lack large Pliocene-Quaternary alluvial fans and have few Quaternary fault scarps (Fig. 5). Furthermore, the east-facing topographic escarpment has resulted mainly from exhumation and differential erosion of Laramide anticlinal structures from beneath Paleogene sedimentary deposits that buried the east flank during and following carving of the late Eocene surface. A similar conclusion was reached many years ago by several Wyoming geologists (Moore, 1959; Love, 1960; Denson and Bergendahl, 1961; Blackstone, 1975) from relationships on the sedimentary "gangplank" leading from the High Plains near Cheyenne, Wyoming, to near the crest of the Laramie Range. The Dakota hogback (Fig. 5) along the toe of the east-facing escarpment of the Southern Rocky Mountains, so conspicuous from northern Colorado to south of Las Vegas, New Mexico, should be largely buried by alluvial fans if significant late Cenozoic block faulting had caused the escarpment.

Three interfluves between major rivers exiting the east flank of the Southern Rocky Mountains contain relationships that substantiate a history of exhumation and differential erosion rather than faulting for the topographic step bordering the High Plains. The "gangplank" between the South Platte and North Platte Rivers has been mentioned previously. Here, the White River Formation (of upper Eocene age)[1] rests unconformably on Cretaceous and older rocks and rises gently to near the crest of the Laramie Range where it is found resting on Precambrian rocks at scattered localities within the range at elevations up to 2,400 m (Blackstone, 1975; Evanoff, 1990; Knight, 1953). The next interfluve to the south, between the South Platte and Arkansas Rivers, contains upper Eocene strata that can be projected up dip from buttes in the Castle Rock area (Fig. 6) to near the crest of the Front Range (Jacob

[1]The Eocene-Oligocene boundary used in this paper is about 33.4 Ma as determined by the work of McIntosh et al., 1992b.

Figure 4. Aerial view of Sierra Blanca massif (part of Sangre de Cristo Range) on east side of San Luis Basin. View is to southeast; Spanish Peaks at upper left. Note triangular facets along base of mountain and prominent alluvial fans. Fission-track apatite ages (Kelley et al., 1992) indicate that Precambrian rocks comprising the Sierra Blanca massif cooled rapidly through the 100°C isotherm between about 20 and 12 Ma. Holocene fault scarps cutting alluvial fans attest to the youthful nature of this topography.

Figure 5. Aerial view of east flank of Front Range at Morrison, Colorado. View is to northwest. Dakota Sandstone (Cretaceous) forms prominent ridge at right. Flatirons of Fountain Formation (Pennsylvanian) at Red Rocks Park visible at lower right. Note lack of large alluvial fans that permits exposures of uptilted sedimentary formations flanking the Front Range. Remnants of late Eocene erosion surface are responsible for gentle topography within the range. Contrast this exhumed Laramide structure with the youthful, fault-generated topography of Figure 3.

and Albertus, 1985; Leonard and Langford, 1994). Here, ash flows that deposited the 36.6-Ma Wall Mountain Tuff were able to sweep eastward from a source in the Sawatch Range, cross the Front Range, and fill shallow paleovalleys cut into the Eocene Dawson Arkose of the Denver Basin (Epis and Chapin, 1975).

Figure 6. View looking southwest at east flank of Front Range (on skyline) from a butte capped by the 36.6-Ma Wall Mountain Tuff west of Castle Rock, Colorado. Pikes Peak, barely visible at left center, towers above late Eocene erosion surface that beveled the Front Range (also called Rampart Range in this area). Boulders in foreground are fragments of Wall Mountain Tuff from quarry. Buttes in middle distance are also capped by Wall Mountain Tuff. Note how Wall Mountain Tuff can be projected to near crest of Front Range. As estimated by Jacob and Albertus (1985, Fig. 26) and Leonard and Langford (1994), post-Laramide displacement on the range-bounding faults is minor, probably 20 to 75 m, possibly less. The Wall Mountain Tuff rests on Dawson Arkose of Eocene and Paleocene age here in the Denver Basin. Florissant is beyond the skyline northwest of Pikes Peak. There the Wall Mountain Tuff rests on Precambrian rocks of the core of the Front Range. The late Eocene erosion surface sloped gently eastward from the Front Range onto the High Plains across a broad piedmont constructed of arkosic debris shed from the range. The prominent topographic escarpment that now marks the High Plains–Front Range boundary was created mainly by differential erosion in late Cenozoic time.

The third major interfluve occurs between the Arkansas and Canadian Rivers and is marked by the prominent southeast-trending mesa that is surmounted by Raton Pass. The basalt capping the mesa occurs at elevations as high as 2,920 m and occupies a paleovalley on an erosion surface that projects westward across the Park Plateau to an equivalent surface at an elevation of about 3,000 m cut into the Sangre de Cristo Mountains (Levings, 1951; O'Neill, 1988). The highest basalt at Raton Pass has a K-Ar age of 7.4 Ma (Stormer, 1972). The oldest basalts in the Ocate volcanic field, which straddles the High Plains–Sangre de Cristo boundary southwest of Raton, range in age from 8.3 to 5.7 Ma and also rest on remnants of a widespread surface, or surfaces, as high as 3,140 m (O'Neill and Mehnert, 1988). Progressively younger basalts, ranging in age from 4.7 to 0.8 Ma, occur at systematically lower elevations (as low as 1,830 m). Thus, an enormous volume of sedimentary rocks has been removed along the east side of the Sangre de Cristo Range in the last 5 m.y. It is probably not coincidental that the Ogallala Formation ceased aggrading at about this same time. The surface on which the 7- to 8-m.y.-old basalts rest is carved across diverse rock types and crosses the High Plains–Southern Rocky Mountains boundary with little displacement (O'Neill, 1988). Removal of a comparable amount of material along the east side of the Front Range in

Colorado would easily produce the topographic escarpment that marks the Rocky Mountain front.

The late Eocene paleoelevation determinations of Meyer (1986, 1992), Wolfe (1992), and Gregory (1992), if correct, rule out a late Cenozoic origin for the present high regional elevation in the Southern Rocky Mountains–Rio Grande rift area. These paleoelevation determinations are tied to the High Plains (east flank of Eaton's, 1986, 1987, Alvarado Ridge) by the three interfluves which indicate that little relative displacement between the High Plains and the frontal ranges of the Rocky Mountains has occurred since late Eocene. If the present 2,500-m elevation of Florissant was largely attained during Laramide crustal shortening, the basins of the Rio Grande rift may be basically collapse structures with little accompanying thermotectonic uplift of regional scale. Additional paleobotanical studies and, if possible, an independent determination based on isotopic data, are needed to test the paleoelevation hypothesis.

CONCLUSIONS

The axial basins of the central and northern Rio Grande rift and the related Miocene basins of northern Colorado, northeastern Utah, and central Wyoming are, for the most part, structurally inverted from Laramide uplifts. The principal exceptions are the North Park syncline that developed within the Laramide North Park Basin (Izett, 1975); the shallow, western part of San Luis Basin that overlies the Eocene Monte Vista basin (Brister, 1990; Brister and Gries, this volume); and the northern Albuquerque Basin that overlies part of a large, northeast-trending half graben of Eocene age (the Galisteo basin; Cather, 1992). The high regional elevation of the Rio Grande rift, Southern Rocky Mountains, and Great Plains appears to predate the development of the rift, and may be related to Laramide crustal thickening (Gregory, 1992; Gregory and Chase, 1992).

In the most general sense, the tectonic and stratigraphic development of the rift system and its flanking areas may be divided into three phases (Fig. 7). Sedimentation began in most basins of the rift and in related extensional basins to the north in late Oligocene–early Miocene. Strain rates were modest during this early phase of rifting, except where volcanism and high heat flow caused local focusing of extensional strain (e.g., Meyer and Foland, 1991; Cather et al., this volume). Sediment transport by fluvial systems in areas flanking the rift (Great Plains, Colorado Plateau) was able to keep pace with sediment supply from the slowly rising footwall uplifts, resulting in a mildly erosive regime on the rift flanks.

Beginning in middle Miocene, rates of extensional strain began to increase in the rift. Increased subsidence in the half grabens of the rift caused topographic closure and development of playa systems in many basins. As strain rates further increased in the late Miocene, sedimentation offlapped toward narrow zones of maximum subsidence near the faulted margins of half grabens, causing widespread erosion of older sediments on hanging-wall dipslopes (Cather et al., this volume). Sedimentation rates appear to have been greatest during this phase of rift development (e.g., Lozinsky, this volume).

Deposits on the flanking areas of the Rio Grande rift system are dominantly of middle Miocene to early Pliocene age. Deposition of the Ogallala Formation began about 17 Ma in the north and about 12 Ma in the south and ceased aggrading in both areas about 4 to 5 Ma. To the west on the Colorado Plateau, the Bidahochi Formation (Figs. 1, 7) began to accumulate about 12 to 9 Ma and ceased about 4 Ma (Love, 1989). Both the Ogallala and Bidahochi Formations were broad, sheetlike units that were not confined to discrete structural basins. The age range of these units suggests that they were genetically related to the middle to late Miocene episode of rapid extension recorded in the basins within and to the north of the Rio Grande rift. We infer that increased sediment supply from rapidly rising footwall uplifts overwhelmed the transportational capacity of fluvial systems draining the rift-flank areas, resulting in aggradation (Fig. 7). Our analysis differs from that of Eaton (1986, 1987) in that we envision rift-flank aggradation was driven by the rise of narrow (10 to 25 km) footwall uplifts, and not by regional epeirogenic uplift (i.e., Eaton's Alvarado Ridge).

Rapid extension within and to the north of the rift began to wane in the late Miocene. Decreased subsidence rates relative to sediment supply rates within basins of the Rio Grande rift caused onlap and aggradation of Pliocene deposits, producing major unconformities between the upper and lower Santa Fe Group and enabling the eventual integration of drainages to form the ancestral Rio Grande (Cather et al., this volume). Following a prolonged period of net aggradation, fluvial incision of rift-flank deposits began in the Pliocene. Incision may have been the result of erosional degradation of the source-area footwall uplifts and/or the effects of late Cenozoic climate change (e.g., Molnar and England, 1990). Incision of the ancestral Rio Grande and its tributaries began in early to middle Pleistocene and has continued episodically to the present, producing a complex inner-valley stratigraphy inset against the older, aggradational deposits of the Santa Fe Group.

ACKNOWLEDGMENTS

The paper was improved by reviews from E. Evanoff, K. M. Gregory, J. W. Hawley, G. H. Mack, and W. R. Seager. Warren Hamilton also reviewed the paper but did not agree with many of our conclusions. We gratefully acknowledge the many geologists and students whose work in the Southern Rocky Mountains/Rio Grande rift area has made this paper possible. The recent paleobotanical studies by K. M. Gregory and H. W. Meyer and fission-track apatite dating by S. A. Kelley have been especially helpful in reexamining the uplift history of the Southern Rocky Mountains.

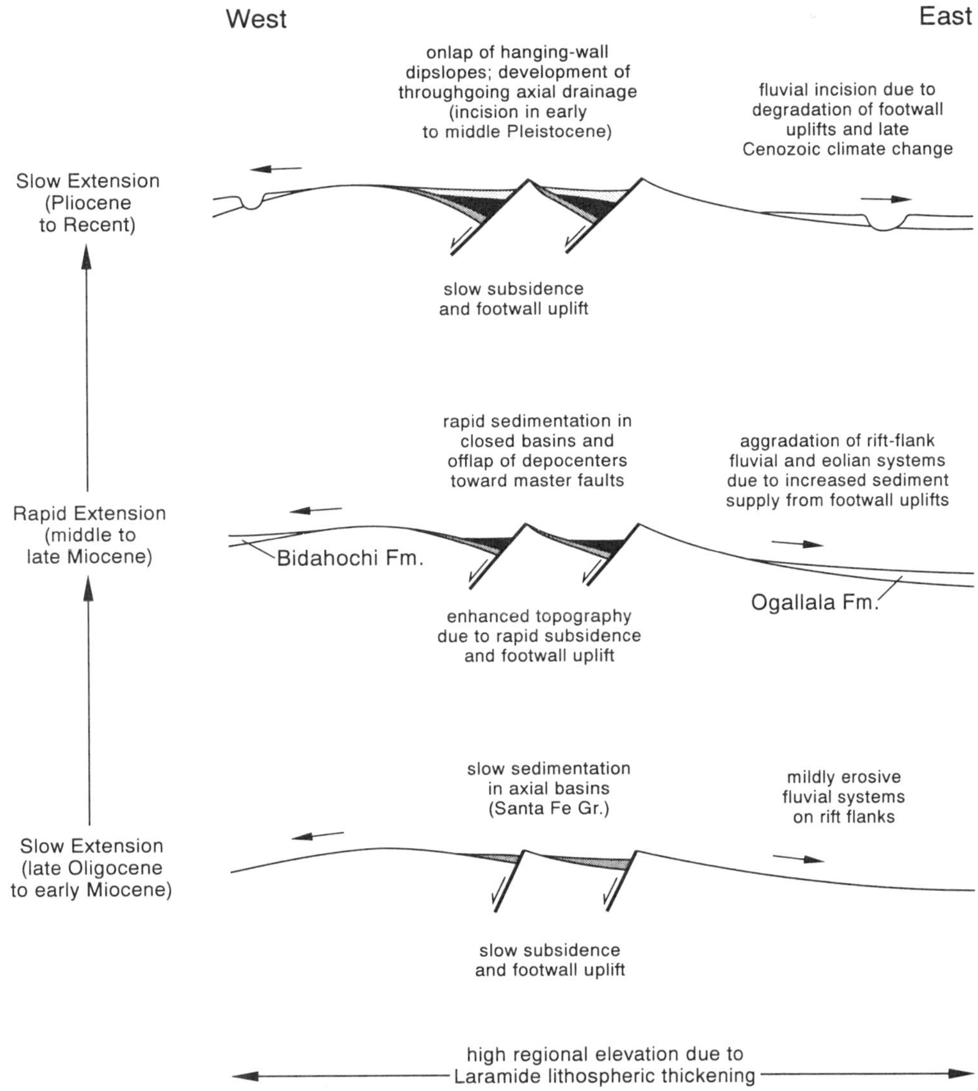

Figure 7. Schematic east-west cross sections through central New Mexico showing generalized structural and stratigraphic development of rift and rift-flank areas.

REFERENCES CITED

Aldrich, M. J., Jr., and Laughlin, A. W., 1984, A model for the tectonic development of the southeastern Colorado Plateau boundary: Journal of Geophysical Research, v. 89, p. 10207–10218.

Aldrich, M. J., Jr., Chapin, C. E., and Laughlin, A. W., 1986, Stress history and tectonic development of the Rio Grande rift, New Mexico: Journal of Geophysical Research, v. 91, p. 6199–6211.

Anderholm, S. K., 1983, Hydrogeology of the Socorro and La Jencia basins, Socorro County, New Mexico: New Mexico Geological Society, Guidebook 34, p. 303–310.

Axelrod, D. I., and Bailey, H. P., 1976, Tertiary vegetation, climate, and altitude of the Rio Grande depression, New Mexico–Colorado: Paleobiology, v. 2, p. 235–254.

Bachman, G. O., and Mehnert, H. H., 1978, New K-Ar dates and the late Pliocene to Holocene geomorphic history of the central Rio Grande region, New Mexico: Geological Society of America Bulletin, v. 89, p. 283–292.

Baldridge, W. S., Damon, P. E., Shafiqullah, M., and Bridwell, R. J., 1980, Evolution of the central Rio Grande rift, New Mexico: New potassium-argon ages: Earth and Planetary Science Letters, v. 51, p. 309–321.

Baldridge, W. S., Olsen, K. H.,and Callender, J. F., 1984, Rio Grande rift: Problems and perspectives: New Mexico Geological Society, Guidebook 35, p. 1–12.

Biehler, S., and 12 others, 1991, A geophysical model of the Española Basin, Rio Grande rift, New Mexico: Geophysics, v. 56, p. 340–353.

Black, B. A., 1982, Oil and gas exploration in the Albuquerque Basin: New Mexico Geological Society, Guidebook 33, p. 313–324.

Blackstone, D. L., Jr., 1975, Late Cretaceous and Cenozoic history of Laramie Basin region, southeast Wyoming, *in* Curtis, B. F., ed., Cenozoic history of the Southern Rocky Mountains: Geological Society of America Memoir 144, 249–279.

Bookstrom, A. A., Carten, R. B., Shannon, J. R., and Smith, R. P., 1988, Origins of bimodal leucogranite-lamprophyre suites, Climax and Red

Mountain porphyry molybdenum systems, Colorado: Petrologic and strontium isotopic evidence, *in* Drexler, J. W., and Larson, E. E., ed., Cenozoic volcanism in the Southern Rocky Mountains revisited: Golden, Colorado School of Mines Quarterly, v. 83, no. 2, pt. 3, p. 1–22.

Bosworth, W., 1985, Geometry of propagating continental rifts: Nature, v. 316, p. 625–627.

Bosworth, W., 1986, Comment *on* 'Detachment faulting and the evolution of passive continental margins': Geology, v. 14, p. 890–891.

Bosworth, W., Lambiase, J., and Keisler, R., 1986, A new look at Gregory's rift: The structural style of continental rifting: EOS Transactions of the American Geophysical Union, v. 67, no. 29, p. 577, 582–583.

Brister, B. S., 1990, Tertiary sedimentation and tectonics: San Juan sag–San Luis Basin region, Colorado and New Mexico [Ph.D. thesis]: Socorro, New Mexico Institute of Mining and Technology, 267 p.

Bruning, J. E., 1973, Origin of the Popotosa Formation, north-central Socorro County, New Mexico [Ph.D. thesis]: Socorro, New Mexico Institute of Mining and Technology, 142 p.

Bryan, K., 1938, Geology and ground-water conditions of the Rio Grande depression in Colorado and New Mexico, *in* Rio Grande joint investigations in the upper Rio Grande Basin in Colorado, New Mexico, and Texas: Washington, D.C., U.S. Natural Resources Planning Board, v. 1, pt. 2, p. 197–225.

Bryan, K., and McCann, F. T., 1937, The Ceja del Rio Puerco—A border feature of the Basin and Range province in New Mexico, Part I, Stratigraphy and structure: Journal of Geology, v. 45, p. 801–828.

Buck, W. R., 1988, Flexural rotation of normal faults: Tectonics, v. 7, p. 959–973.

Burke, K., 1980, Intracontinental rifts and aulacogens, *in* Continental tectonics: Washington, D.C., National Academy of Sciences, p. 42–49.

Burroughs, R. L., 1981, A summary of the geology of the San Luis Basin, Colorado–New Mexico with emphasis on the geothermal potential for the Monte Vista graben: Colorado Geological Survey Special Publication 17, 30 p.

Cather, S. M., 1983, Laramide Sierra uplift: Evidence for major pre-rift uplift in central and southern New Mexico: New Mexico Geological Society, Guidebook 34, p. 99–101.

Cather, S. M., 1989, Post-Laramide tectonic and volcanic transition in west-central New Mexico: New Mexico Geological Society, Guidebook 40, p. 91–97.

Cather, S. M., 1992, Suggested revisions to the Tertiary tectonic history of north-central New Mexico: New Mexico Geological Society, Guidebook 43, p. 109–122.

Chamberlin, R. M., 1983, Cenozoic domino-style crustal extension in the Lemitar Mountains, New Mexico: A summary: New Mexico Geological Society, Guidebook 34, p. 111–118.

Chamberlin, R. M., and Osburn, G. R., 1984, Character and evolution of extensional domains in the Socorro area of the Rio Grande rift, central New Mexico: Geological Society of America Abstracts with Programs, v. 16, p. 467.

Chapin, C. E., 1971, The Rio Grande rift, Part I: Modifications and additions: New Mexico Geological Society, Guidebook 22, p. 191–201.

Chapin, C. E., 1979, Evolution of the Rio Grande rift: A summary, *in* Riecker, R. E., ed., Rio Grande rift: Tectonics and magmatism: Washington, D.C., American Geophysical Union, p. 1–5.

Chapin, C. E., 1988, Axial basins of the northern and central Rio Grande rift, *in* Sloss, L. L., ed., Sedimentary cover—North American craton: U.S.: Geological Society of America, The Geology of North America, v. D-2, p. 165–170.

Chapin, C. E., 1989, Volcanism along the Socorro accommodation zone, Rio Grande rift, New Mexico, *in* Chapin, C. E., and Zidek, J., eds., Field excursions to volcanic terranes in the western United States, Volume 1: Southern Rocky Mountains: New Mexico Bureau of Mines and Mineral Resources Memoir 46, p. 46–57.

Chapin, C. E., and Cather, S. M., 1981, Eocene tectonics and sedimentation in the Colorado Plateau–Rocky Mountain area: Arizona Geological Society Digest, v. 14, p. 173–198.

Chapin, C. E., and Lindley, J. I., 1986, Potassium metasomatism of igneous and sedimentary rocks in detachment terranes and other sedimentary basins: Economic implications, *in* Beatty, B., and Wilkinson, P.A.K., eds., Frontiers in geology and ore deposits of Arizona and the southwest: Arizona Geological Society Digest, v. 16, p. 118–126.

Chapin, C. E., and Seager, W. R., 1975, Evolution of the Rio Grande rift in the Socorro and Las Cruces areas: New Mexico Geological Society, Guidebook 26, p. 297–321.

Chapin, C. E., Chamberlin, R. M., Osburn, G. R, White, D. W., and Sanford, A. R., 1978, Exploration framework of the Socorro geothermal area, New Mexico, *in* Chapin, C. E., and Elston, W. E., eds., Field guide to selected cauldrons and mining districts of the Datil-Mogollon volcanic field, New Mexico: New Mexico Geological Society Special Publication 7, p. 115–129, map in pocket.

Coons, L. M., and Kelley, T. E., 1984, Regional hydrogeology and the effect of structural control on the flow of ground water in the Rio Grande trough, northern New Mexico: New Mexico Geological Society, Guidebook 35, p. 241–244.

Cordell, L., 1978, Regional geophysical setting of the Rio Grande rift: Geological Society of America Bulletin, v. 89, p. 1073–1090.

Cordell, L., 1982, Extension in the Rio Grande rift: Journal of Geophysical Research, v. 87, p. 8561–8569.

Dalrymple, G. R., 1979, Critical tables for conversion of K-Ar ages from old to new constants: Geology, v. 7, p. 558–560.

Denny, C. S., 1940: Santa Fe Formation in the Española Valley, New Mexico: Geological Society of America Bulletin, v. 51, p. 677–694.

Denson, N. M., and Bergendahl, M. H., 1961, Middle and upper Tertiary rocks of southeastern Wyoming and adjoining areas: U.S. Geological Survey Professional Paper 424-C, p. C168–C172.

Eaton, G. P., 1986, A tectonic redefinition of the Southern Rocky Mountains: Tectonophysics, v. 132, p. 163–193.

Eaton, G. P., 1987, Topography and origin of the Southern Rocky Mountains and Alvarado Ridge, *in* Coward, M. P., Dewey, S. F., and Hancock, P. L., eds., Continental extensional tectonics: Geological Society Special Publication 28, p. 355–369.

Ekas, L. M., Ingersoll, R. V., Baldridge, W. S., and Shafiqullah, M., 1984, The Chama–El Rito Member of the Tesuque Formation, Española Basin, New Mexico: New Mexico Geological Society, Guidebook 35, p. 137–143.

Elston, W. E., and Bornhorst, T. J., 1979, The Rio Grande rift in context of regional post-40 m.y. volcanic and tectonic events, *in* Riecker, R. E., ed., Rio Grande rift: Tectonics and magmatism: Washington, D.C., American Geophysical Union, p. 416–438.

Epis, R. C., and Chapin, C. E., 1975, Geomorphic and tectonic implications of the post-Laramide, late Eocene erosion surface in the Southern Rocky Mountains, *in* Curtis, B. F., ed., Cenozoic history of the Southern Rocky Mountains: Geological Society of America Memoir 144, p. 45–74.

Evanoff, E., 1990, Early Oligocene paleovalleys in southern and central Wyoming: Evidence of high local relief on the late Eocene unconformity: Geology, v. 18, p. 443–446.

Faulds, J. E., Geissman, J. W., and Mawer, C. K., 1990, Structural development of a major extensional accommodation zone in the Basin and Range province, northwestern Arizona and southern Nevada: Implications for kinematic models for continental extension, *in* Wernicke, B., ed., Basin and Range extensional tectonics near the latitude of Las Vegas, Nevada: Geological Society of America Memoir 176, p. 37–76.

Galusha, T., and Blick, C., 1971, Stratigraphy of the Santa Fe Group, New Mexico: American Museum of Natural History Bulletin, v. 144, no. 1, 127 p.

Gibbs, A. D., 1984, Structural evolution of extensional basin margins: Geological Society of London, v. 141, p. 609–620.

Gile, L. H., Hawley, J. W., and Grossman, R. B., 1981, Soils and geomorphology in the Basin and Range area of southern New Mexico—Guidebook to the Desert Project: New Mexico Bureau of Mines and Mineral Resources Memoir 39, 222 p.

Gregory, K. M., 1992, Late Eocene paleoaltitude, paleoclimate, and paleogeography of the Front Range region, Colorado [Ph.D. thesis]: Tucson, University of Arizona, 246 p.

Gregory, K. M., and Chase, C. G., 1992, Tectonic significance of paleobotanically estimated climate and altitude of the late Eocene erosion surface, Colorado: Geology, v. 20, p. 581–585.

Gridley, J. M., Adams, D. C., and Keller, G. R., 1991, Structural interpretation of the upper Arkansas graben area, Colorado, from gravity data: Geological Society of America Abstracts with Programs, v. 23, no. 4, p. 27.

Gries, R. R., 1985, Seismic lines in the San Luis Valley, south-central Colorado, *in* Gries, R. R., and Dyer, R. C., eds., Seismic exploration of the Rocky Mountain region: Denver, Rocky Mountain Association of Geologists, p. 267–274.

Griggs, R. L., 1964, Geology and groundwater resources of the Los Alamos area, New Mexico: U.S. Geological Survey Water-Supply Paper 1753, 107 p.

Gustavson, T. C., and Winkler, D. A., 1988, Depositional facies of the Miocene-Pliocene Ogallala Formation, northwestern Texas and eastern New Mexico: Geology, v. 16, p. 203–206.

Hail, W. J., Jr., 1965, Geology of northwestern North Park, Colorado: U.S. Geological Survey Bulletin 1188, 133 p.

Hamilton, W. B., 1981, Plate-tectonic mechanism of Laramide deformation, *in* Boyd, D. W., and Lillegraven, J. A., eds., Rocky Mountain foreland basement tectonics: Laramie, University of Wyoming Contributions to Geology, v. 19, p. 87–92.

Hamilton, W. B., 1988, Laramide crustal shortening: Geological Society of America Memoir 171, p. 27–39.

Hansen, W. R., 1984, Post-Laramide tectonic history of the eastern Uinta Mountains, Utah, Colorado, and Wyoming: The Mountain Geologist, v. 21, p. 5–29.

Hansen, W. R., 1986, Neogene tectonics and geomorphology of the eastern Uinta Mountains in Utah, Colorado, and Wyoming: U.S. Geological Survey Professional Paper 1356, 78 p.

Hawley, J. W., 1975, Quaternary history of Doña Ana County region, south-central New Mexico: New Mexico Geological Society, Guidebook 26, p. 139–150.

Hawley, J. W., Compiler, 1978, Guidebook to Rio Grande rift in New Mexico and Colorado: New Mexico Bureau of Mines and Mineral Resources Circular 63, 241 p.

Hawley, J. W., 1984, The Ogallala Formation in eastern New Mexico, *in* Whetstone, G. A., ed., Proceedings of the Ogallala Aquifer Symposium II Lubbock, Texas, June, 1984: Texas Tech University Water Resources Research Center, p. 157–176.

Hawley, J. W., and Haase, C. S., 1992, Hydrogeologic framework of the northern Albuquerque Basin: New Mexico Bureau of Mines and Mineral Resources Open-File Report 387, 165 p.

Hawley, J. W., and Lozinsky, R. P., 1993, Hydrogeologic framework of the Mesilla Basin in New Mexico and western Texas: New Mexico Bureau of Mines and Mineral Resources Open-File Report 323, 73 p.

Hawley, J. W., Bachman, G. O., and Manley, K., 1976, Quaternary stratigraphy in the Basin and Range and Great Plains provinces, New Mexico and western Texas, *in* Maheny, W. C., ed., Quaternary stratigraphy of North America: Stroudsberg, Dowden, Hutchinson, and Ross, Inc., p. 235–274.

Ingersoll, R. V., Cavazza, W., Baldridge, W. S., and Shafiqullah, M., 1990, Cenozoic sedimentation and paleotectonics of north-central New Mexico: Implications for initiation and evolution of the Rio Grande rift: Geological Society of America Bulletin, v. 102, p. 1280–1296.

Izett, G. A., 1975, Late Cenozoic sedimentation and deformation in northern Colorado and adjoining areas, *in* Curtis, B. F., ed., Cenozoic history of the Southern Rocky Mountains: Geological Society of America Memoir 144, p. 179–209.

Jacob, A. F., and Albertus, R. G., 1985, Thrusting, petroleum seeps, and seismic exploration: Front Range south of Denver, Colorado, *in* Macke, D. L., and Maughan, E. K., eds., Rocky Mountain Section field trip guide 1985: Society of Economic Paleontologists and Mineralogists, p. 77–96.

Kelley, S. A., and Duncan, I. J., 1986, Late Cretaceous to middle Tertiary tectonic history of the Rio Grande rift: Journal of Geophysical Research, v. 91, p. 6246–6262.

Kelley, S. A., Chapin, C. E., and Corrigan, J., 1992, Late Mesozoic to Cenozoic cooling histories of the flanks of the northern and central Rio Grande rift, Colorado and New Mexico: New Mexico Bureau of Mines and Mineral Resources Bulletin 145, 39 p.

Kelley, V. C., 1977, Geology of Albuquerque Basin, New Mexico: New Mexico Bureau of Mines and Mineral Resources Memoir 33, 59 p.

Kelley, V. C., 1982, The right-relayed Rio Grande rift, Taos to Hatch, New Mexico: New Mexico Geological Society, Guidebook 33, p. 147–151.

Knepper, D. H., Jr., 1976, Late Cenozoic structure of the Rio Grande rift zone, central Colorado, *in* Epis, R. C., and Weimer, R. J., eds., Studies in Colorado field geology: Golden, Professional Contributions of the Colorado School of Mines, no. 8, p. 421–430.

Knight, S. H., 1953, Summary of the Cenozoic history of the Medicine Bow Mountains, Wyoming: Wyoming Geological Association, Guidebook 8, p. 65–76.

Lambert, P. W., 1966, Notes on the late Cenozoic geology of the Taos-Questa area, New Mexico: New Mexico Geological Society, Guidebook 17, p. 43–50.

Lambert, P. W., 1968, Quaternary stratigraphy of the Albuquerque area, New Mexico [Ph.D. thesis]: Albuquerque, University of New Mexico, 257 p.

Leonard, E. M., and Langford, R. P., 1994, Post-Laramide deformation along the eastern margin of the Colorado Front Range—A case against significant faulting: The Mountain Geologist (in press).

Levings, W. S., 1951, Late Cenozoic erosional history of the Raton Mesa region: Golden, Colorado School of Mines Quarterly, v. 46, no. 3, 111 p., 16 plates.

Lipman, P. W., 1983, The Miocene Questa caldera, northern New Mexico: Relation to batholith emplacement and associated molybdenum mineralization, *in* The genesis of Rocky Mountain ore deposits: Changes with time and tectonics: Denver Region Exploration Geologists Society, p. 133–149.

Lipman, P. W., and Mehnert, H. H., 1975, Late Cenozoic basaltic volcanism and development of the Rio Grande depression in the Southern Rocky Mountains, *in* Curtis, B. F., ed., Cenozoic history of the Southern Rocky Mountains: Geological Society of America Memoir 144, p. 119–154.

Lipman, P. W., and Mehnert, H. H., 1979, The Taos Plateau volcanic field, northern Rio Grande rift, New Mexico, *in* Riecker, R. E., ed., Rio Grande rift: Tectonics and magmatism: Washington, D.C., American Geophysical Union, p. 289–311.

Lipman, P. W., Mehnert, H. H., and Naeser, C. W., 1986, Evolution of the Latir volcanic field, northern New Mexico, and its relation to the Rio Grande rift, as indicated by potassium-argon and fission-track dating: Journal of Geophysical Research, v. 91, p. 6329–6345.

Lister, G. S., Etheridge, M. A., and Symonds, P. A., 1986a, Detachment faulting and the evolution of passive continental margins: Geology, v. 14, p. 246–250.

Lister, G. S., Etheridge, M. A., and Symonds, P. A., 1986b, Reply *to* Comment *on* 'Detachment faulting and the evolution of passive continental margins': Geology, v. 14, p. 891–892.

Love, D. W., 1989, Bidahochi Formation: An interpretive summary: New Mexico Geological Society, Guidebook 40, p. 273–280.

Love, J. D., 1960, Cenozoic sedimentation and crustal movement in Wyoming: American Journal of Science, v. 258A, p. 204–214.

Love, J. D., 1970, Cenozoic geology of the Granite Mountains area, central

Wyoming: U.S. Geological Survey Professional Paper 495-C, 154 p.

Lozinsky, R. P., 1988, Stratigraphy, sedimentology, and sand petrology of the Santa Fe Group and pre–Santa Fe Tertiary deposits in the Albuquerque Basin, central New Mexico [Ph.D. thesis]: Socorro, New Mexico Institute of Mining and Technology, 298 p.

Lozinsky, R. P., and Hawley, J. W., 1986, The Palomas Formation of south-central New Mexico—A formal definition: New Mexico Geology, v. 8, p. 73–78.

Lozinsky, R. P., and Tedford, R. H., 1991, Geology and paleontology of the Santa Fe Group, southwestern Albuquerque Basin, Valencia County, New Mexico: New Mexico Bureau of Mines and Mineral Resources Bulletin 132, 35 p.

Luft, S. J., 1985, Airfall tuff in the Browns Park Formation, northwestern Colorado and northeastern Utah: The Mountain Geologist, v. 22, p. 110–127.

McGinitie, H. D., 1953, Fossil plants of the Florissant Beds, Colorado: Carnegie Institution of Washington Publication 599, 198 p.

Machette, M. N., 1978, Geologic map of the San Acacia Quadrangle, Socorro County, New Mexico: U.S. Geological Survey Geologic Quadrangle Map GQ 1415, scale 1:24,000.

Manley, K., 1979, Stratigraphy and structure of the Española Basin, Rio Grande rift, New Mexico, in Riecker, R. E., ed., Rio Grande rift: Tectonics and magmatism: Washington, D.C., American Geophysical Union, p. 71–86.

Manley, K., 1981, Redefinition and description of the Los Piños Formation of north-central New Mexico: Geological Society of America Bulletin, part I, v. 92, p. 984–989.

Manley, K., 1984, Brief summary of the Tertiary geologic history of the Rio Grande rift in northern New Mexico: New Mexico Geological Society, Guidebook 35, p. 63–66.

May, S. J., 1984, Miocene stratigraphic relations and problems between the Abiquiu, Los Piños, and Tesuque Formations near Ojo Caliente, northern Española Basin: New Mexico Geological Society, Guidebook 35, p. 129–135.

McIntosh, W. C., Chapin, C. E., Ratté, J. C., and Sutter, J. F., 1992a, Time-stratigraphic framework for the Eocene-Oligocene Mogollon-Datil volcanic field, southwest New Mexico: Geological Society of America Bulletin,v. 104, p. 851–871.

McIntosh, W. C., Geissman, J. W., Chapin, C. E., Kunk, M. J., and Henry, C. D., 1992b, Calibration of the latest Eocene-Oligocene geomagnetic polarity time scale using $^{40}Ar/^{39}Ar$ dated ignimbrites: Geology, v. 20, p. 459–463.

Meyer, H. W., 1986, An evaluation of the methods for estimating paleoaltitudes using Tertiary floras from the Rio Grande rift vicinity, New Mexico and Colorado [Ph.D. thesis]: Berkeley, University of California, 217 p.

Meyer, H. W., 1992, Lapse rates and other variables applied to estimating paleoaltitudes from fossil flora: Palaeogeography, Palaeoclimatology, Palaeoecology, v. 99, p. 71–99.

Meyer, J., and Foland, K. A., 1991, Magmatic-tectonic interaction during early Rio Grande rift extension at Questa, New Mexico: Geological Society of America Bulletin, v. 103, p. 993–1006.

Mohr, P., 1982, Musings on continental rifts, *in* Palmason, G., ed., Continental and oceanic rifts: American Geophysical Union and Geological Society of America Geodynamics Series, v. 8, p. 293–309.

Molnar, P., and England, P., 1990, Late Cenozoic uplift of mountain ranges and global climate change: Chicken or egg?: Nature, v. 346, p. 29–34.

Moore, F., 1959, The geomorphic evolution of the east flank of the Laramie Range, Colorado and Wyoming [Ph.D. thesis]: Laramie, University of Wyoming, 123 p.

Morgan, P., 1984, The thermal structure and thermal evolution of the continental lithosphere: Physics and Chemistry of the Earth, v. 15, p. 107–193.

Morgan, P., 1987, Regional vertical movements and development of the Rio Grande rift: Geological Society of America Abstracts with Programs, v. 19, p. 778.

Morgan, P., and Ramberg, I. B., 1987, Physical changes in the lithosphere associated with thermal relaxation after rifting: Tectonophysics, v. 143, p. 1–11.

Muehlberger, W. R., 1979, The Embudo fault between Pilar and Arroyo Hondo, New Mexico: An active intracontinental transform fault: New Mexico Geological Society, Guidebook 30, p. 77–82.

O'Neill, M. J., 1988, Late Cenozoic physiographic evolution of the Ocate volcanic field: U.S. Geological Survey Professional Paper 1478-B, 15 p.

O'Neill, J. M., and Mehnert, H. H., 1988, Petrology and physiographic evolution of the Ocate volcanic field, north-central New Mexico: U.S. Geological Survey Professional Paper 1478-A, 30 p.

Osburn, G. R., and Chapin, C. E., 1983, Nomenclature for Cenozoic rocks of northeast Mogollon-Datil volcanic field, New Mexico: New Mexico Bureau of Mines and Mineral Resources Stratigraphic Chart 1, 7 p., 1 sheet.

Reeves, C. C., Jr., 1972, Tertiary-Quaternary stratigraphy and geomorphology of west Texas and southeastern New Mexico: New Mexico Geological Society, Guidebook 23, p. 108–117.

Rogers, K. L., 1984, A paleontological analysis of the Alamosa Formation (south-central Colorado: Pleistocene: Irvingtonian): New Mexico Geological Society, Guidebook 35, p. 151–155.

Rogers, K. L., and 11 others, 1992, Pliocene and Pleistocene geologic and climatic evolution of the San Luis Valley of south-central Colorado: Palaeogeography, Palaeoclimatology, and Palaeoecology, v. 94, p. 55–86.

Rosendahl, B. R., 1987, Architecture of continental rifts with special reference to East Africa: Annual Reviews of Earth and Planetary Sciences, v. 15, p. 445–503.

Russell, L. R., and Snelson, S., 1990, Structural style and tectonic evolution of the Albuquerque Basin segment of the Rio Grande rift, *in* Pinet, B., and Bois, C., eds., The potential of deep seismic profiling for hydrocarbon exploration: Paris, Editions Technip, p. 175–207.

Sales, J. K., 1983, Collapse of Rocky Mountain basement uplifts, *in* Lowell, J. D., ed., Rocky Mountain foreland basins and uplifts: Denver, Rocky Mountain Association of Geologists, p. 79–97.

Scott, D. L., Etheridge, M. A., and Rosendahl, B. R., 1992, Oblique-slip deformation in extensional terrains: A case study of the Lakes Tanganyika and Malawi rift zones: Tectonics, v. 11, p. 998–1009.

Scott, G. R., 1975, Cenozoic surfaces and deposits in the Southern Rocky Mountains, *in* Curtis, B. F., ed., Cenozoic history of the Southern Rocky Mountains: Geological Society of America Memoir 144, p. 227–248.

Scott, G. R., and Taylor, R. B., 1986, Map showing late Eocene erosion surface, Oligocene-Miocene paleovalleys and Tertiary deposits in the Pueblo, Denver, and Greeley $1° \times 2°$ Quadrangles, Colorado: U.S. Geological Survey Miscellaneous Investigations Series Map I-1626, scale 1:250,000.

Scott, G. R., Van Alstine, R. E., and Sharp, W. N., 1975, Geologic map of the Poncha Springs Quadrangle, Chaffee County, Colorado: U.S. Geological Survey Miscellaneous Field Studies Map MF-658, scale 1:62,500.

Seager, W. R., and Morgan, P., 1979, Rio Grande rift in southern New Mexico, west Texas, and northern Chihuahua, *in* Riecker, R. E., ed., Rio Grande rift: Tectonics and magmatism: Washington, D.C., American Geophysical Union, p. 87–106.

Seager, W. R., Shafiqullah, M., Hawley, J. W., and Marvin, R. F., 1984, New K-Ar dates from basalts and the evolution of the southern Rio Grande rift: Geological Society of America Bulletin, v. 95, p. 87–99.

Shannon, J. R., 1988, Geology of the Mount Aetna cauldron complex, Sawatch Range, Colorado [Ph.D. thesis]: Golden, Colorado School of Mines, 434 p.

Shannon, J. R., 1989,The Mount Aetna cauldron: Structures in deeply eroded ring zones, *in* Chapin, C. E., and Zidek, J., eds., Field excursions to volcanic terranes in the western United States, Volume I: Southern Rocky Mountain region: New Mexico Bureau of Mines and Mineral Resources Memoir 46, p. 286–293.

Smith, C. T., 1983, Structural problems along the east side of the Socorro constriction, Rio Grande rift: New Mexico Geological Society, Guidebook 34, p. 103–109.

Smith, C. T., and 8 others, 1983, First day road log from Socorro to Mesa del Yeso, Joyita Hills, Johnson Hill, Cerros de Amado, Lomas de las Cañas, Jornado del Muerto, Carthage, and return to Socorro: New Mexico Geological Society, Guidebook 34, p. 1–28.

Smith, R. L., Bailey, R. A., and Ross, C. S., 1970, Geologic map of the Jemez Mountains, New Mexico: U.S. Geological Survey Miscellaneous Geologic Investigations Map I-571, scale 1:125,000.

Spence, W., and Gross, R. S., 1990, A tomographic glimpse of the upper mantle source of magmas of the Jemez lineament, New Mexico: Journal of Geophysical Research, v. 95, p. 10829–10849.

Spiegel, Z., and Baldwin, B., 1963, Geology and water resources of the Santa Fe area, New Mexico: U.S. Geological Survey Water-Supply Paper 1525, 258 p.

Steidtmann, J. R., and Middleton, L. T., 1986, Eocene-Pliocene stratigraphy along the southern margin of the Wind River Range, Wyoming: Revisions and implications from field and fission-track studies: The Mountain Geologist, v. 23, p. 19–25.

Steidtmann, J. R., and Middleton, L. T., 1991, Fault chronology and uplift history of the southern Wind River Range, Wyoming: Implications for Laramide and post-Laramide deformation in the Rocky Mountain foreland: Geological Society of America Bulletin, v. 103, p. 472–485.

Stormer, J. C., Jr., 1972, Ages and nature of volcanic activity on the southern High Plains, New Mexico and Colorado: Geological Society of America Bulletin, v. 83, p. 2443–2448.

Tandon, K., 1992, Deep structure beneath the San Luis Basin in Colorado from reprocessing of an industry reflection survey [Ph.D. thesis]: Ithaca, New York, Cornell University, 285 p.

Taylor, R. B., 1975, Neogene tectonism in south-central Colorado, *in* Curtis, B. F., ed., Cenozoic history of the Southern Rocky Mountains: Geological Society of America Memoir 144, p. 211–226.

Tedford, R. H., 1981, Mammalian biochronology of the late Cenozoic basins of New Mexico: Geological Society of America Bulletin, Part I, v. 92, p. 1008–1022.

Tedford, R. H., 1982, Neogene stratigraphy of the northwestern Albuquerque Basin: New Mexico Geological Society, Guidebook 33, p. 273–278.

Tweto, O., 1961, Late Cenozoic events of the Leadville district and upper Arkansas Valley, Colorado: U.S. Geological Survey Professional Paper 424-B, p. B133–B135.

Tweto, O., 1978, Tectonic map of the Rio Grande rift system in Colorado: New Mexico Bureau of Mines and Mineral Resources Circular 163, Plate 1, scale 1:1,000,000.

Tweto, O., 1979, The Rio Grande rift system in Colorado, *in* Riecker, R. E., ed., Rio Grande rift: Tectonics and magmatism: Washington, D.C., American Geophysical Union, p. 33–56.

Van Alstine, R. E., 1968, Tertiary trough between the Arkansas and San Luis valleys, Colorado: U.S. Geological Survey Professional Paper 600-C, p. C158–C160.

Van Alstine, R. E., 1969, Geology and mineral deposits of the Poncha Springs NE Quadrangle, Chaffee County, Colorado: U.S. Geological Survey Professional Paper 626, 52 p.

Van Alstine, R. E., 1970, Allochthonous Paleozoic blocks in the Tertiary San Luis–Upper Arkansas graben, Colorado: U.S. Geological Survey Professional Paper 700-B, p. B43–B51.

Van Alstine, R. E., 1974, Geology and mineral deposits of the Poncha Springs SE Quadrangle, Chaffee County, Colorado: U.S. Geological Survey Professional Paper 829, 19 p., scale 1:24,000.

Wells, S. G., Kelson, K. I., and Menges, C. M., 1987, Quaternary evolution of fluvial systems in the northern Rio Grande rift, New Mexico and Colorado: Implications for entrenchment and integration of drainage systems: Friends of the Pleistocene—Rocky Mountain Cell Field Trip, Guidebook, Albuquerque, p. 55–69.

Wernicke, B., and Axen, G. J., 1988, On the role of isostasy in the evolution of normal fault systems: Geology, v. 16, p. 848–851.

Williams, L.A.J., 1982, Physical aspects of magmatism in continental rifts, in Palmason, G., ed., Continental and oceanic rifts: American Geophysical Union and Geological Society of America Geodynamics Series, v. 8, p. 193–222.

Wolfe, J. A., 1978, A paleobotanical interpretation of Tertiary climates in the Northern Hemisphere: American Scientist, v. 66, p. 694–703.

Wolfe, J. A., 1990, Palaeobotanical evidence for a marked temperature increase following the Cretaceous/Tertiary boundary: Nature, v. 343, p. 153–156.

Wolfe, J. A., 1992, An analysis of present-day terrestrial lapse rates in the western conterminous United States and their significance to paleoaltitudinal estimates: U.S. Geological Survey Bulletin 1964, 35 p.

Wolfe, J. A., 1994, A method of obtaining climatic parameters from leaf assemblages: U.S. Geological Survey Bulletin 2040 (in press).

Wong, I. G., and Humphrey, J. R., 1989, Contemporary seismicity, faulting, and the state of stress in the Colorado Plateau: Geological Society of America Bulletin, v. 101, p. 1127–1146.

Woodward, L. A., and 6 others, 1978, Tectonic map of Rio Grande rift region in New Mexico, Chihuahua, and Texas: New Mexico Bureau of Mines and Mineral Resources Circular 163, Plate 2, scale 1:1,000,000.

Manuscript Accepted by the Society October 13, 1993

Printed in U.S.A.

Geological Society of America
Special Paper 291
1994

Depth and geometry of the northern Rio Grande rift in the San Luis Basin, south-central Colorado

Charles F Kluth,* and Carl H. Schaftenaar*
Chevron U.S.A., Inc., Denver, Colorado 80201

ABSTRACT

The deepest part of the northern Rio Grande rift is just northwest of the Great Sand Dunes, on the eastern side of the San Luis Valley, in south-central Colorado. Approximately 150 km (95 m) of high-quality, 20-fold Common Depth Point seismic data indicate that the basin is filled with approximately 6.4 km (21,000 ft) of Tertiary sediments, mostly Oligocene or younger in age. These data, combined with published data, provide a basis for interpreting the structural geometry of the rift.

The internal structural geometry of this part of the rift is surprisingly uncomplicated. Sympathetic and antithetic faulting is not widespread, or major, within the deep part of the basin. The internal geometry of the sedimentary packages, however, indicates a complicated movement history on the faults that are present within the rift.

The bounding fault zone between the rift and the Sangre de Cristo Mountains has approximately 9.2 km (30,000 ft) of vertical separation and may be very complicated. Although we were not able to resolve the bounding fault zone with our seismic data, the data constrains the minimum angle of the fault zone to be approximately 45°. Our modeling suggests that a 60° angle is the most likely orientation of the fault zone. The influence of earlier, low-angle extension recognized nearby and elsewhere in the Rio Grande rift is not directly evident in our data.

Assuming 60° antithetic shear in the hanging wall, which is supported by the seismic data, a depth of detachment (flattening) is estimated to be approximately 16 km, in the brittle-ductile transition zone estimated from heat-flow data.

Cross sections drawn using the constraints of the new seismic data, require only 8 to 12% extension across the rift at this latitude.

INTRODUCTION

The Rio Grande rift is an intracontinental rift system that trends generally north-south from central Colorado through central New Mexico and into northern Mexico (Fig. 1). The northern part of the rift is characterized by a northerly trend that, in many places, parallels earlier structures (Chapin, 1979; Tweto, 1979; Lindsey and others, 1983). In the vicinity of Great Sand Dunes National Monument and Crestone, Colorado (Fig. 2), gravity, seismic, and well data indicate that the rift is a large, complex, east-tilted, half graben with a central horst (Behrendt and Bajwa, 1974; Cordell, 1978; Burroughs, 1981; Gries, 1985). The deepest part of the half graben along the western flank of the Sangre de Cristo Mountains has been estimated to contain 6 to 9 km of Tertiary fill (Gaca and Karig, 1965; Davis and Keller, 1978).

*Present addresses: Kluth, Chevron Overseas Petroleum, Inc., P.O. Box 5046, San Ramon, California 94583; Schaftenaar, 1068 Stonemeadow Drive, Parker, Colorado 80134.

Kluth, C. F., and Schaftenaar, C. H., 1994, Depth and geometry of the northern Rio Grande rift in the San Luis Basin, south-central Colorado, *in* Keller, G. R., and Cather, S. M., eds., Basins of the Rio Grande Rift: Structure, Stratigraphy, and Tectonic Setting: Boulder, Colorado, Geological Society of America Special Paper 291.

Geologic relationships and radiometric age data indicate that uplift of the region and the initial stages of rifting in southern Colorado began approximately 30 to 27 m.y. ago (Tweto, 1979; Lindsey et al., 1983; Chapin, 1988). Jones (1991) reports early low angle faults that date approximately 26 to 24 Ma in southern Colorado. In northern New Mexico, Hagstrum and Lipman (1986) interpret rifting to postdate the 26-Ma Amalia Tuff. It appears that rifting proceeded in two stages, an earlier phase that included lower angle normal faults and a later stage that included higher angle normal faults (Morgan and Golombek, 1984). Our data pertain to the geometry of the San Luis Basin as the result of the second phase of extension. There is abundant evidence that faulting associated with the rift has continued into Holocene time (Tweto, 1979; McCalpin, 1982).

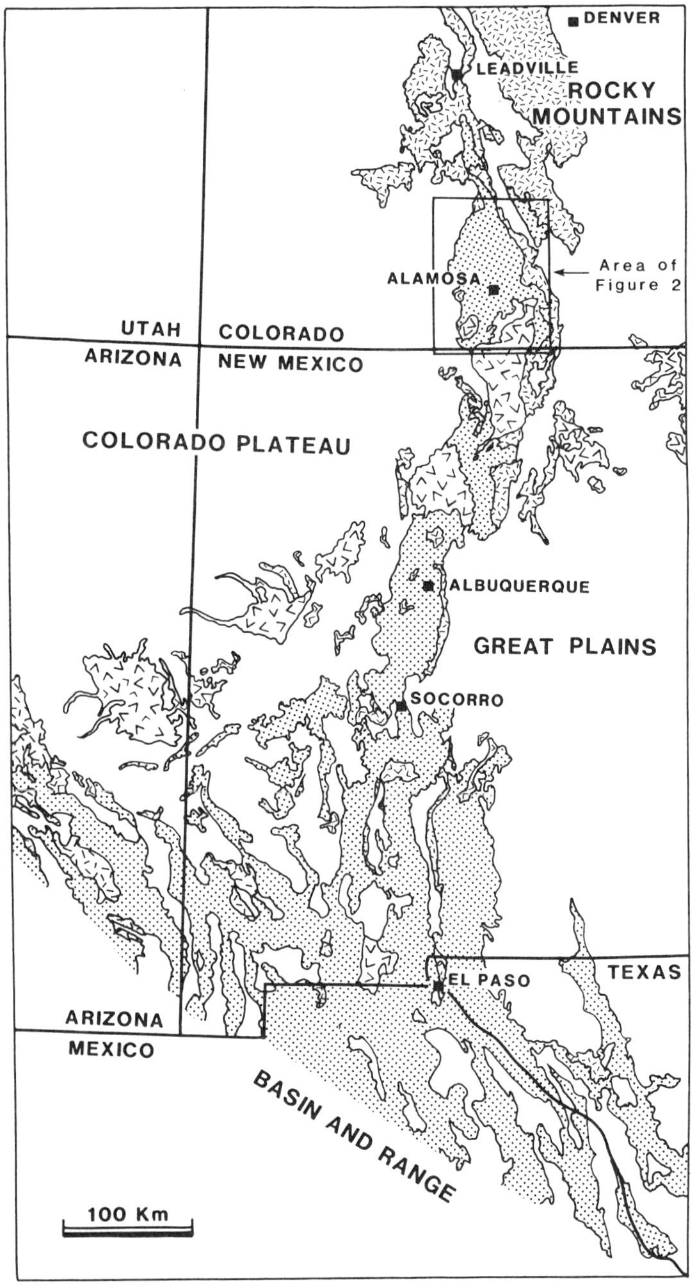

Figure 1. Location map of the Rio Grande rift. Stippled pattern represents Tertiary and Quaternary extension-related basins. "V" pattern represents late Tertiary and Quaternary volcanic rocks. Short-dashed pattern represents Precambrian rocks in the cores of Late Cretaceous to Tertiary basement-cored uplifts. Unornamented areas represent areas not directly related to Rio Grande rift extensional structures. (After Baldridge and others, 1984.)

The rocks that fill the San Luis Valley part of the rift are stratigraphically complex and poorly exposed. They are fine to coarse, orange-red-brown sandstones interbedded with shales, conglomerates, and volcanics (Brister and Gries, this volume). The section has been subdivided according to age and the presence of volcanic and volcaniclastic materials (Powell, 1958; Huntley, 1979a, b). The units interfinger in an apparently complex fashion and details of the nomenclature are reviewed by Huntley (1979a, b) and Brister and Gries (this volume). None of the units crops out from below the Tertiary and Quaternary alluvium in our study area.

This paper presents an interpretation of the depth and structural geometry of the northern Rio Grande rift in southern Colorado as constrained by our new seismic data. Our data were acquired just north of Great Sand Dunes National Monument. It includes the deepest part of the rift (Baca graben) in the San Luis Basin. We then used this geometry, along with published data, to construct balanced cross sections across the basin.

DATA

The seismic reflection data presented here are 20-fold, Common Depth Point data acquired with a conventional vibrator crew. A seismic-velocity profile was constructed from sonic-velocity logs from wells in the area (Table 1). Since no single well penetrates the entire section, a composite velocity model was constructed using logs from several wells. The coherence of the velocity profiles in parts of the section penetrated by more than one well and the relatively straightforward velocity/depth relationship provide a measure of confidence that the model is applicable. This profile was used to migrate the seismic data and to convert the data to depth for use in constructing the structural cross sections. The velocity profile shows gradually increasing velocity with depth. Seismic data were recorded for 6 sec (all times are two-way travel time), but there are no coherent data below approximately 4 sec, so the lines shown here are displayed only to 4 sec.

Line 1

Line 1 is approximately 24 km (14 mi) long and was shot at right angles to the structural grain of the half graben (Fig. 2). It was shot across the deepest part of the basin and ties the strike line (Line 3) as discussed below.

The reflections on the southwestern (left) end of Line 1

(Fig. 3) form a coherent package that dips gently eastward. This end of the line is just east of the central basin horst (Alamosa horst) suggested by the gravity data (Behrendt and Bajwa, 1974) and confirmed by well control. This coherent package of reflections is underlain, at approximately 2.3 sec, by a much less coherent band of reflections. The package of coherent reflections dips gently eastward and the package thickens uniformly, but with minor breaks, to the vicinity of point 1A (see Fig. 3). In the vicinity of point 1A, the dip on the reflections flattens and reverses to a westerly direction. The zone of less-coherent reflections down to approximately 3.6 sec, while not strong, apparently continues dipping easterly to the vicinity of point 1B. Between these poor-coherence reflections and coherent data at approximately 2.6 sec, there is a zone of poor-quality data with flat dips. The flat-dipping data, if real, diverge from the overlying westerly dipping package. This divergence may represent an equivalent of the angular unconformity interpreted on data to the south by Burroughs (1981), Gries (1985), and Brister and Gries (this volume).

The gentle southwesterly dip in the package above 2.5 sec continues from 1B to the vicinity of point 1C where dip increases and data quality decreases in some zones. From point 1C northeastward to the end of the line, the reflections dip more steeply southwestward, the time interval between them decreases, and the data quality deteriorates toward the mountain front.

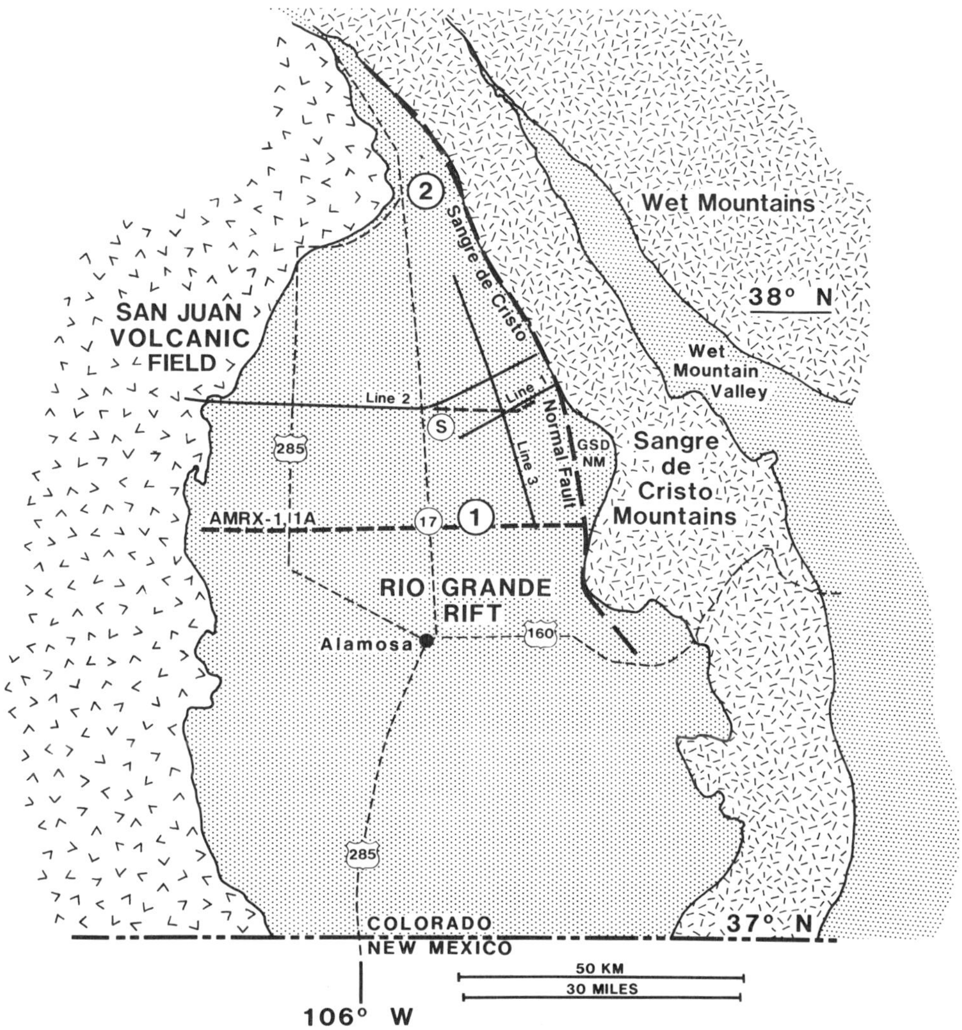

Figure 2. Location map of the northern Rio Grande rift in south-central Colorado. Stipple pattern represents Tertiary sediments. The geographic name for this part of the rift is the San Luis Valley. The "V" pattern represents the volcanic rocks of the San Juan volcanic field. The short-dashed pattern represents Precambrian to Tertiary rocks associated with the Late Cretaceous to Tertiary basement cored uplifts. Federal and state highways are shown by dashed lines. "GSDNM" shows the location of Great Sand Dunes National Monument. Heavy dashed line labeled "1" is the location of the AMRX1, 1A seismic data of Gries (1985). Heavy-dashed line labeled "S" is location of seismic data of Stoughton (1977). Label "2" is the location of seismic data of Davis and Stoughton (1979).

TABLE 1. VELOCITY MODEL USED IN CONVERTING SEISMIC DATA FROM TIME TO DEPTH FOR USE IN CONSTRUCTING THE GEOLOGIC CROSS SECTIONS*

Depth		Interval Velocity	
(m)	(ft)	(m/sec)	(ft/sec)
0	0		
460	1,510	1,841	6,040
1,125	3,689	2,657	8,716
1,933	6,341	3,233	10,608
2,095	9,530	3.888	12,756
4,004	13,136	4,396	14,424
5,154	16,909	4,600	15,092
6,357	20,856	4,812	15,788

*This model is based on sonic velocity log data from several nearby wells: Amoco-Mapco 1-32 State (T40N,R12E,32); Couger 1 Cros (T39N,R11E,3); Reserve 1-33 NBH Land (T40N,R11E,33); Amerada 1-16 F State (T39N,R10E,16).

Line 2

Line 2 (Fig. 4; see Fig. 2 for location) was shot approximately perpendicular to the structural grain of the half graben and continues to the west across the entire width of the San Luis Basin. The western end of the line was shot on exposed volcanic rocks on the San Juan volcanic field.

We did not obtain coherent data at the west (left) end of Line 2, probably due to the presence of volcanic rocks with high seismic velocities on the surface. Where the line enters the area with alluvial material on the surface at point 2A (Fig. 4) the data improve substantially. Our seismic lines and cross sections therefore are shown in Figure 4 only where we had interpretable data. At point 2A, the reflections above 2.9 sec form an eastward-thinning wedge that onlaps the central basin horst-block of Precambrian basement at approximately point 2D.

At point 2B, reflections near the top of the section form the lower part of an eastward thickening wedge-shaped package that continues across the central-basin horst. In the vicinity of point 2B the reflections in the upper package are concordant with the underlying package, but small angular discordances such as those in the vicinity of point 2C indicate that the packages should be considered separately. In the vicinity of point 2D the line crosses the central-basin horst. The reflections are disrupted by apparent normal-sense displacements in the area of 2E to 2F. At point 2F the reflection near the top of the basement reaches its deepest level, approximately 3.3 to 3.4 sec.

The reflections reverse dip and the time interval between them begins to decrease at point 2G. At approximately that point, data quality begins to decrease. The steeply west dipping reflection a 1.7 sec below point 2H may be a normal fault plane within the Sangre de Cristo normal fault zone.

Line 3

Line 3 (Fig. 5) trends north-northwest and was shot to tie Line 1, Line 2, the Amoco-Mapco 1-32 State well, and AMRX-1 seismic data shown by Gries (1985) and Brister and Gries (this volume) along the strike of the basin. It illustrates the geometry of the deepest part of the San Luis Basin along strike and ties together the other data to permit a three-dimensional view of the basin geometry.

Line 3 ties AMRX-1 seismic line (Gries, 1985; Brister and Gries, this volume) at point 3A and includes a concordant north (right)-dipping package of reflections above 1.7 sec. Below 1.7 sec the reflections dip more gently northward than the upper package. This angular discordance is interpreted to be the angular unconformity in the section described by Burroughs (1981), Gries (1985), and Brister and Gries (this volume). The data tie to the Amoco-Mapco 1-32 State well (Huntley, 1979a,b; Burroughs, 1981; Gries, 1985) at point 3B. Data from this well provide limited control on the age of the rocks represented by the reflections on our seismic data. Further details of the ages of rocks within the basin are provided by Brister and Gries (this volume).

Reflections of the upper package cut progressively downward to the north and truncate reflections in the lower package. The lower package of reflections, those below the unconformity, appear to be completely cut out in the vicinity of point 3C. From that point northward, we are not able to identify the unconformity in the seismic data.

Line 3 ties Line 1 at point 3D, near the deepest part of the basin. Between point 3C and point 3D the package of reflections expands northward. Just to the north of point 3D, the dip of the reflections flattens and reverses and the basin becomes shallower in that direction. The package of reflections thins with only minor interruption, northward to the vicinity of point 3E, where the dip direction reverses into a smaller subbasin shown by Behrendt and Bajwa (1974) and Davis and Stoughton (1979).

INTERPRETATION

Our interpretation of the geometry of the northern part of the San Luis Basin is shown in the structure sections constructed after depth conversion of the seismic two-way time data (Figs. 3c, 4c, 5c). In all three seismic sections (Figs. 3b, 4b, 5b), stratigraphic geometry within the syn-rift section is illustrated by four reflections, identified as 1 through 4. These reflections are also shown in the structure sections (Figs. 3c, 4c, 5c). Using the new stratigraphic information provided by Brister (1990) and Brister and Gries (this volume) we assign reflection 1 (Figs. 3, 4, 5) to the lower part of the Alamosa Formation. Reflections 2, 3, and 4 are within the lower Santa Fe Group (see Brister and Gries, this volume, for details of the stratigraphy).

Line 1

Using well control in the area, we interpret the coherent reflections on the southwestern (left) end of Line 1 at approximately 2.3 sec to represent Tertiary volcanic and clastic rocks and the poor-coherence reflections to represent the upper part of the Precambrian basement. The basement on the southwestern end of Line 1 is approximately 3,200 m below the surface of the valley (Fig. 3c). The basement dips gently northeastward toward the deepest part of the basin. The overlying marker horizons also dip northeastward at progressively lower angles higher in the section, indicating continued growth of the basin and continued sedimentation through time. Minor faulting disrupts the markers on the east-dipping flank, but the basin has surprisingly little internal deformation (large-scale antithetic faulting, etc.).

The thickness of the rift-fill sequence between the lowest marker horizon and the basement increases in the deepest part of the basin. This indicates a history of rapid movement on the Sangre de Cristo fault zone during the early history of rift development. Influx of clastic material from nearby high areas filled the rift depression. The westerly dip on the reflections near the Sangre de Cristo fault zone suggests that the Sangre de Cristo Mountains provided abundant clastic material to the

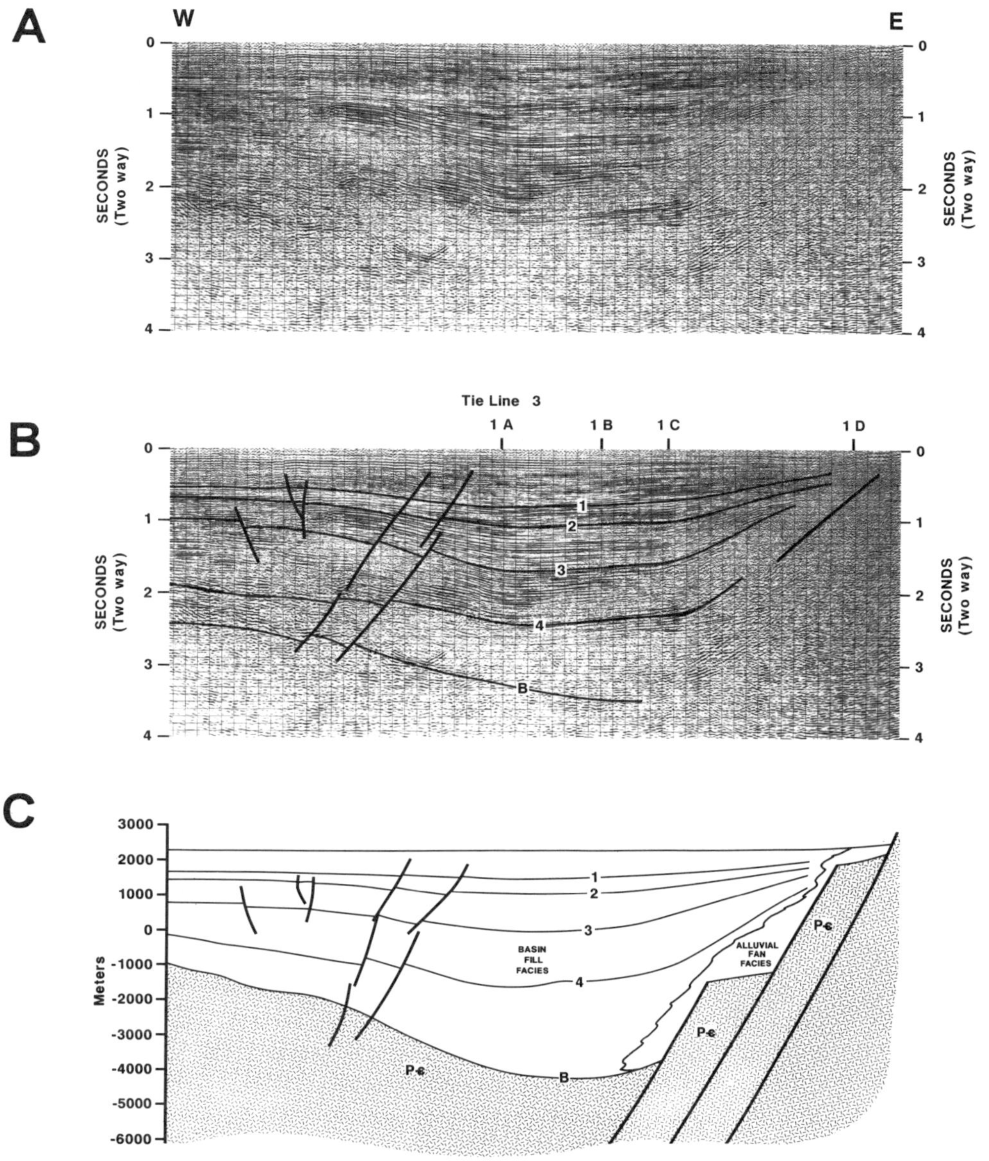

Figure 3. Line 1 (see Fig. 2 for location). A, uninterpreted, migrated seismic data; B, interpretation of seismic data; C, geologic cross section constructed using the seismic and gravity data. Lines labeled 1A through 1D are reference marks described in the text.

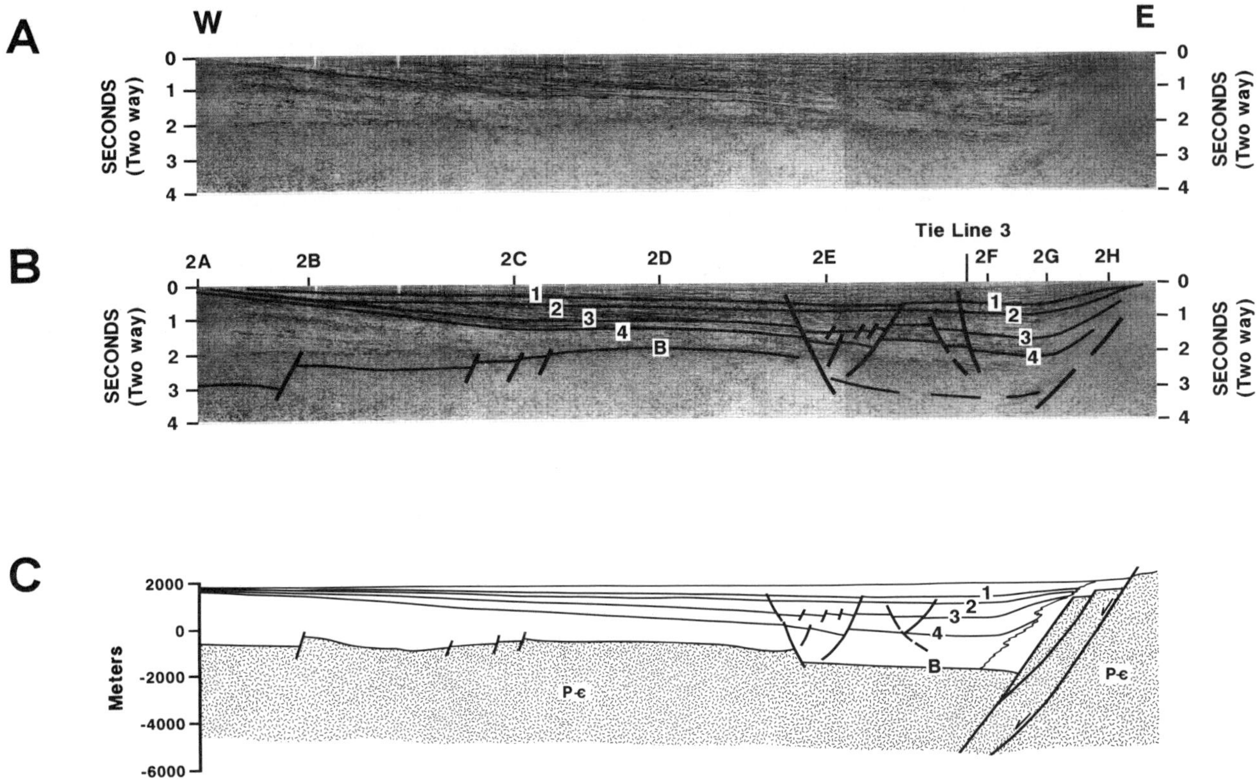

Figure 4. Line 2 (see Fig. 2 for location). A, uninterpreted, migrated seismic data; B, interpretation of seismic data; C, geologic cross section constructed using the seismic and gravity data. Lines labeled 2A through 2H are reference marks described in the text.

basin throughout its history. We interpret the data to indicate that the volume of clastic material shed from the adjacent mountains and highlands, to the east and west of the basin during the development of the rift completely filled any topographic low that might have formed adjacent to the bounding normal fault zone. This scenario is different than in areas such as Badwater, the topographically lowest point in Death Valley, California, which is adjacent to the bounding normal fault. The implication of this relationship is that the topography in the basin, near the bounding fault, has always sloped to the west, away from the mountain front, as it does today. This may have limited the development of ponds or lakes along the mountain front and thus decreased the likelihood of nonmarine source rocks for hydrocarbon exploration. The shaly material in nearby wells has very low (much less than 1%) total organic carbon content.

A possible additional factor in producing west-sloping topography during the development of the rift is uplift of the footwall. The crest of the narrow Sangre de Cristo Range to the east includes several peaks above 4,265 m (14,000 ft). This suggests possible uplift of the range as a rift shoulder. The change of the reflections on the seismic data from easterly to westerly dip near the deepest part of the basin (see point 1B in Fig. 3 and point 2G in Fig. 4) may reflect the effect of the footwall or regional uplift. It may also reflect slight antilistric shape of the fault zone or differential compaction of the sediments. None of these factors is constrained at present. Further interpretation must await additional data.

At its deepest point on the structure section drawn from the seismic and gravity data on Line 1, the basement is at approximately 4,250 m below sea level and approximately 7 km below the present land surface. The marker horizons dip more steeply to the west, and the interval between them decreases as they approach the eastern flank of the basin. This geometry reflects thinning in the alluvial-fan section, possibly over intermediate fault blocks of basement rocks in the Sangre de Cristo fault zone.

The strong reflection at approximately 0.6 seconds under point 1D on the seismic data may be a fault plane within the normal fault zone. It does not correspond to a fault on the surface, but its geometry suggests that it may be part of the fault zone buried by alluvium. The fault blocks shown within the Sangre de Cristo zone on the cross section (Fig. 3c) are schematic since our data generally did not image reflections from the zone.

While we were not able to positively identify the Sangre

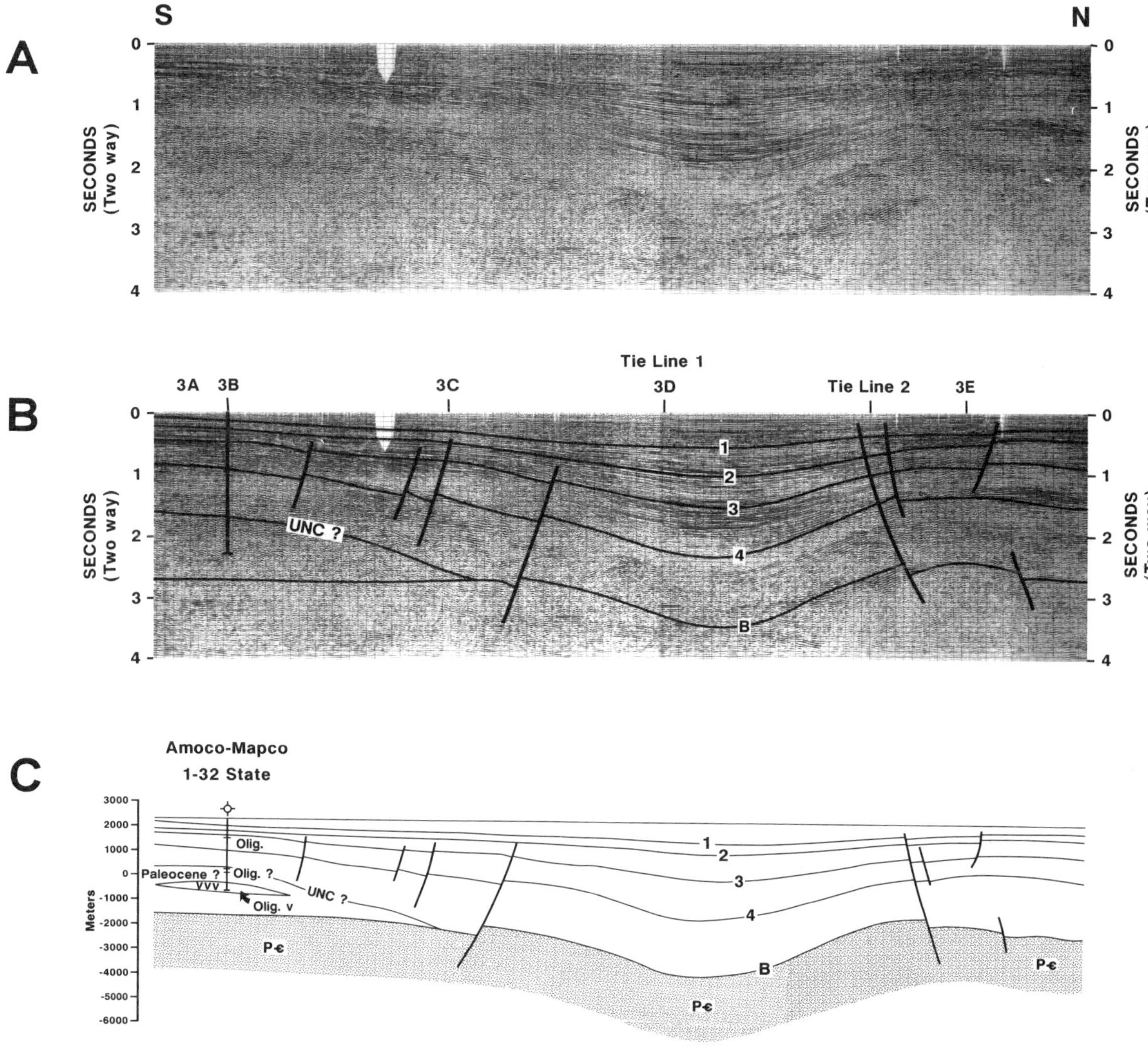

Figure 5. Line 3 (see Fig. 2 for location). A, uninterpreted, migrated seismic data; B, interpretation of seismic data; C, geologic cross section constructed using the seismic and gravity data. Lines labeled 3A through 3E are reference marks described in the text.

de Cristo fault zone in our data, we are able to constrain its position and angle. The minimum angle of the fault zone is constrained to be approximately 45° by a line connecting the most easterly seismic event that we interpret as basement on our data and that of Gries (1985) and the surface of the outcrop of the fault. We have shown a 60° dip on the Sangre de Cristo fault zone as permitted by the seismic and by preliminary two-dimensional models of the gravity data. We also suggest that a portion of the fault zone is beneath the alluvial fans of the Sangre de Cristo Mountains. This geometry is different than that in the Albuquerque Basin, part of the Rio Grande rift in central New Mexico. In that area, Russell and Snelson (1990, this volume) interpreted basin-bounding faults dipping approximately 30° to 40°.

Jones (1991) reported low-angle (15 to 40°) normal faults and associated mineralization in the Sangre de Cristo Range in southern Colorado, south of our study area. We see no evidence for the earlier phase (Morgan and Golombek, 1984) of low-angle faulting in our data. Presumably these faults would be located in the hanging wall of the later, high-angle faults and below the basement-rift fill contact, where we recorded no usable data. The low-angle normal faults may have reactivated earlier thrust faults as suggested to the south by Russell and Snelson (1990, this volume). The more steeply dipping reflections below the unconformity to the south (Burroughs, 1981; Gries, 1985; Brister and Gries, this volume) may represent more steeply dipping beds on the hanging wall of low-angle normal faults. The unconformity above these rotated beds would mark abandonment of the low-angle faults and the onset of movement and relatively less tilting on high-angle normal faults. The amount of extension due to low-angle faulting in the area of our study is unknown.

Line 2

The basement dips very gently eastward at the western (left) end of Line 2 (Fig. 4a, b, c), where the line emerges from the area where the surface is covered by volcanic rocks. Basement in the vicinity of point 2A (Fig. 4c) is approximately 2050 m below the surface of the valley.

The rocks represented by the eastward-thinning wedge of reflections at point 2A (beneath point 2A, the top of this wedge is approximately 0.1 sec, the base is approximately 2.0 sec) are approximately 1,500 m thick and thin onto the western side of the central basin horst in the vicinity of point 2C. Brister and Gries (this volume) interpret pre-rift volcanic rocks to be present on the eastern side of the rift. This indicates that some of the rocks represented by this package of reflections were deposited to the east of the horst. This wedge may represent the eastern margin of a Cretaceous or early Tertiary basin that has later been rotated so that the margin is now nearly flat lying or has gentle easterly dip. Alternatively this wedge may represent the eastward thinning distal margin of the San Juan volcanic field that filled in the western part of the San Luis Basin. Based on the work of Brister and Gries (this volume) we assign the rocks of this eastward-thinning sequence to the Conejos and Blanco Basin Formations.

Above the eastward-thinning wedge, at point 2A, is the thin edge of an eastward-thickening wedge of rocks that is probably of middle Oligocene and younger age (see Brister and Gries, this volume). This package of rocks thickens continuously to the east, across the central basin horst and into the deepest part of the basin (point 2G). These rocks represent the filling of the large half graben that is the manifestation of the Rio Grande rift at this latitude. These rocks belong to the Santa Fe Group of Brister and Gries (this volume).

The basement is interpreted to be broken by faults that are upthrown to the east on the west side of the central basin horst. On the horst, the basement is approximately 2,450 m below the surface of the valley. This feature is deeper on our data than it appears farther to the south, on line AMRX-1 (Gries, 1985), despite similar gravity signatures in both areas (Behrendt and Bajwa, 1974). Different lithologies and densities in basement and/or basin-fill rocks are possible reasons for the difference in the gravity between the two areas, but this problem has not yet been investigated.

On the east side of the horst, the basement dips gently eastward and is cut by faults with east sides downthrown to produce the deepest part of the basin. With the exception of the slight westerly dip of the west side of the central basin horst, the basement surface is interpreted to dip gently east. For purposes of calculating depth to detachment (discussed below) we infer that the central horst block, the lower, eastward thinning wedge of the sedimentary rocks, and the gentle easterly dip of the basement are related to earlier movements. We have used the edge of the eastward-thickening package of sediments as the "hinge" edge of the half graben.

The sedimentary rocks east of the central basin horst are cut by relatively minor faults, but no large-scale structures are present in that area. The rocks reverse dip and their thickness begins to decrease in the vicinity of point 2G. As in Line 1, we interpret a coarse alluvial fan facies to cover intermediate level fault blocks along the east side of the basin. We attribute the deterioration of seismic data quality on the eastern (right) ends of Lines 1 and 2 to dispersion of energy by the coarse material in the proximal alluvial fan. The zone of rapid deterioration of the data may indicate either the facies change from the coarse, proximal fan to the finer-grained distal fan or that there is complex structure within the fault zone.

Our interpretation is that the basement is represented by the zone of reflections at approximately 3.3 sec at point 2F. This area is the deepest part of the basin where it is crossed by Line 2. We acknowledge the possibility that the basement may be somewhat deeper, and that the more steeply east dipping reflections may represent rocks below an angular unconformity as shown to the south by Gries (1985) and Brister and Gries (this volume).

Line 3

The basement at the southern (left) end of Line 3 (Fig. 5b, 5c) is represented by reflections 2.7 sec beneath point 3A and is approximately 4,000 m below the surface of the valley. The marker units from the seismic section are part of a northward-dipping and northward-thickening package of sedimentary rocks. Palynomorph data indicates that these rocks are mostly of Neogene age (Huntley, 1979a; Burroughs, 1981; Gries 1985, S. N. Nelson, written communication, 1988).

The significant deep well in the area, the Amoco-Mapo 1-32 State, is located at point 3B. Details of the well, including interpretations of age of deposition of sedimentary rocks, have been reported by Huntley (1979a), Burroughs (1981), and Gries (1985). Briefly, the well was drilled to 2,860 m (9380 ft). It drilled a section of Tertiary rocks to total depth, ending in an Oligocene igneous sill (Huntley, 1979a; Burroughs, 1981; Gries, 1985; Brister and Gries, this volume). No palynomorphs were recovered in the vicinity of the angular unconformity, reported by Burroughs (1981) and Gries (1985). Oligocene forms are present, however, a few hundred meters (several hundred feet) above the discordance. Palynomorphs whose age is latest Paleocene or younger (*Juglaus* sp., S. N. Nelson, written communication, 1988) were recovered below the unconformity. One specimen of one form (*Momipites wyomingensis*) was recovered from samples still deeper in the well. This palynomorph indicates a Paleocene to Eocene age, but the assemblage is too sparse to assess the possibility that this form represents a reworking of the sedimentary section or that it represents the true age of deposition (S. N. Nelson, personal communication, 1988). Brister and Gries (this volume) interpret Oligocene volcanic rocks at the base of the stratigraphic section at this point, suggesting that the older palynomorphs are indeed reworked. The angular unconformity apparently cuts down section to the north, so that by point 3C, the unequivocally younger rocks rest directly on the Precambrian basement.

The basin fill along this cross section gradually thickens into the deepest part of the basin. In the vicinity of point 3D the basement reflections dip northward at approximately 3.2 to 3.4 sec. We interpret the basement to be approximately 6,300 m below the surface of the valley floor (approximately 3,900 m below sea level). North of point 3D, the dip on the rocks flattens and reverses and the section thins. At the north end of the cross section, the basement is approximately 4,700 m below the surface.

Relatively small faults cut the basement surface and much of the basin-fill section. They are interpreted on the basis of reflection character to be normal faults with their downthrown side away from the deepest part of the basin. This somewhat surprising geometry may reflect extension in the strike direction of the hanging wall of the Sangre de Cristo fault, perhaps related to irregularities in the geometry or splays of the fault zone.

These data and published data (Gries, 1985; Stoughton, 1977; Davis and Stoughton, 1979) allow the interpretation of the three-dimensional geometry of the Rio Grande rift in south-central Colorado (Fig. 6). The deepest part of the rift is on its east side adjacent to the Sangre de Cristo fault zone. The basin becomes shallower westward onto a central-basin horst, the Alamosa horst. The basin then deepens westward from the horst into the western subbasin (Monte Vista graben; Brister and Gries, this volume). The deep (eastern) part of the basin

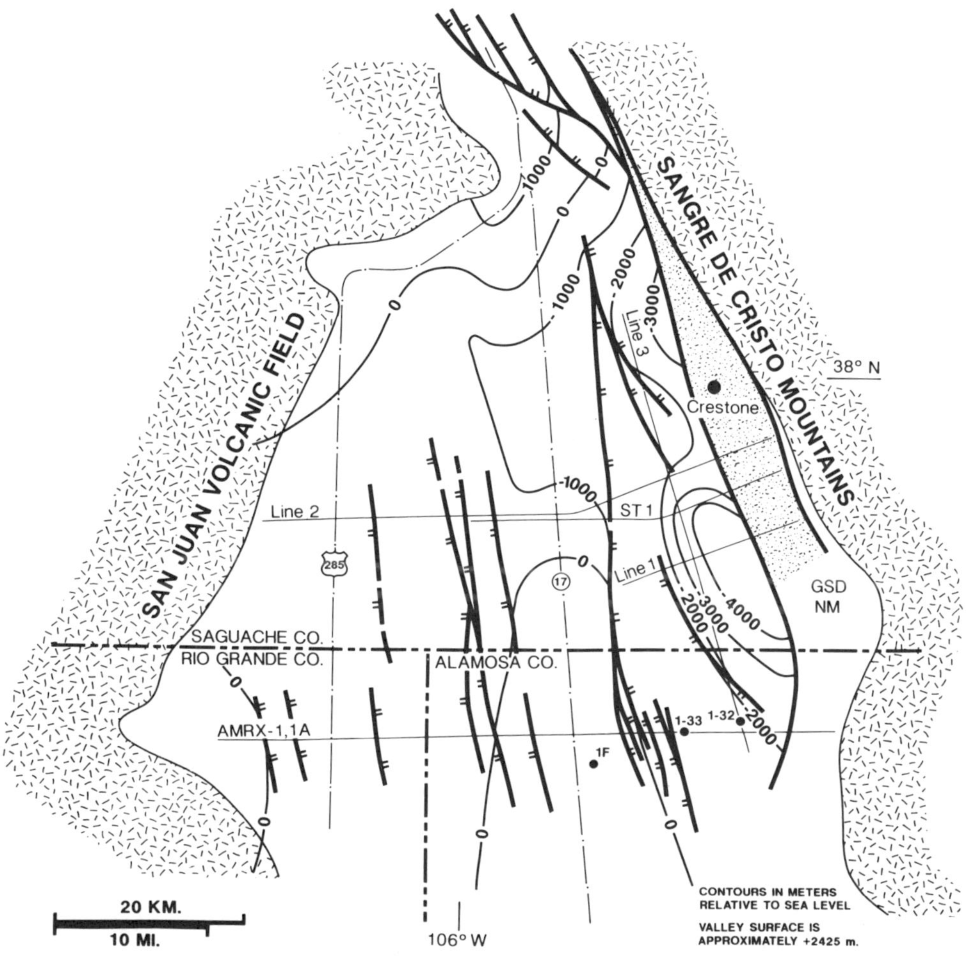

Figure 6. Structure contour map of the top of the basement surface in the northern Rio Grande rift, south-central Colorado. This map incorporates the interpretations of the seismic data and of published data (Gries, 1985, Line AMRX-1, 1A; Stoughton, 1977, Line ST-1; Davis and Stoughton, 1979). Contours are in meters relative to sea level. The land surface in this part of the rift ranges from 2,300 to 2,450 m above sea level. The stippled area on the west side of the Sangre de Cristo Mountains represents the Sangre de Cristo fault zone.

becomes shallower at the northern end of the area that we studied and then deepens into another subbasin at the northern, narrow end studied by Davis and Stoughton (1979).

DEPTH TO DETACHMENT

The rotation of the hanging-wall block suggests that the Sangre de Cristo fault may have a listric shape. The depth to the detachment or flat part of the fault can be calculated using methods given by P. Verrall (personal communication, 1981; see Dula, 1991), Gibbs (1983, 1984), and White and others (1986). The methods vary chiefly in the assumption of the angle of shearing in the hanging wall (see White et al., 1986; Dula, 1991). Using the graphical method of P. Verrall (personal communication, 1981), and Gibbs (1983, 1984), and assuming net vertical shear in the hanging wall, the depth of detachment for the Sangre de Cristo fault is approximately 35 to 38 km. Using a graphical method (P. Verrall, personal communication, 1987) based on the equations of White and others (1986) and a 60° antithetic shear angle (based on the angle of antithetic faults on the seismic data) the depth to detachment is approximately 16 km. This depth would correspond approximately to the transition zone from brittle deformation (shallower) to more ductile behavior (deeper), based on heat-flow data (Barrett and others, 1976; Repplier and Fargo, 1981; Morgan and Golombek, 1984). At present, if the fault is curved, we interpret the shallower depth of detachment to be more realistic geologically, because of the angle of antithetic faulting interpreted on Lines 1 and 2.

A critical factor in the determination of depth to detachment in listric faults is the determination of a regional elevation to use as the upper surface of the synorogenic wedge. This is very difficult to do with confidence in our area because of the possibility of uplift and deformation of the footwall. Therefore, we view the interpretation of these depth-to-detachment calculations cautiously.

The wide hanging-wall block with a relatively small amount of tilting could also result from movement on a planar fault and be indistinguishable on our data from the geometry resulting from a curved fault that flattened at great depth. The deep geometry of the Sangre de Cristo fault zone remains unresolved.

We suggest, though, that the upper part of the Sangre de Cristo fault has a dip of approximately 60° in the San Luis Basin. The structure sections based on these data also support the interpretation that development of the basin requires a relatively small amount of extension, approximately 10 to 12%. The steep angle for the upper part of the fault and the suggestion of midcrustal detachment depths, produce a deep basin with a relatively small amount of extension.

SUMMARY

Our data indicate that the deepest part of the San Luis Basin is filled with approximately 6.4 km of sedimentary and possibly some volcanic rocks. Age data from the Amoco-Mapco well, which is tied by our seismic survey, indicate that the thickest part, and possibly all, of the basin-fill section is late Oligocene or younger in age. The basement surface beneath the basin-fill displays relatively minor deformation except by the Sangre de Cristo fault zone that bounds the east side of the rift. The Sangre de Cristo fault zone is interpreted to be a complex fault zone with large-displacement faults buried beneath the alluvial fans shed from the Sangre de Cristo Mountains. The zone has a minimum dip of approximately 45° and is interpreted to be 60°. Vertical displacement on the Sangre de Cristo fault, from the top of basement outcrops in the adjacent mountains to the top of basement in the deepest part of the basin, is a minimum of approximately 8.2 km (27,000 ft). Depth-to-detachment calculations, based on the structure sections that utilize the depth-converted seismic data suggest that the bounding fault zone remains steep to great depths, flattening at the presumed depth of the brittle-ductile transition zone (approximately 16 km). The structure sections require approximately 10 to 12% extension across the width of the rift at this latitude.

ACKNOWLEDGMENTS

We thank Bruce Smith and Raina Powell for their work on the early phases of this project, and Sheldon Nelson for palynological interpretations. We gratefully acknowledge Chevron U.S.A., Inc. for allowing us to publish these data and interpretations. We thank Bruce Bryant, Lin Cordell, Scott Baldridge, Brian Brister, Chuck Chapin, Steve Cather, and an anonymous reviewer for comments that improved the quality of this paper.

REFERENCES CITED

Baldridge, W. S., Olsen, K. H., and Callender, J. E., 1984, Rio Grande rift: Problems and perspectives, *in* Baldridge, W. S., Dickerson, P. W., Riecker, R. E., and Riolek, J., eds., Rio Grande rift: New Mexico Geological Society, Northern New Mexico, Guidebook, p. 1–12.

Barrett, J. K., Pearl, R. H., and Pennington, A. J., 1976, Map showing thermal springs, wells and heat-flow contours in Colorado: Colorado Geological Survey Information Series 4, scale 1:1,000,000.

Behrendt, J. C., and Bajwa, L. Y., 1974, Bouguer gravity and generalized elevation maps of Colorado: U.S. Geological Survey Map GP-896, scale 1:1,000,000.

Brister, B. S., 1990, Tertiary sedimentation and tectonics: San Juan sag—San Luis Basin region, Colorado and New Mexico [Ph.D. thesis]: Socorro, New Mexico Institute of Mining and Technology, 267 p.

Burroughs, R. L., 1981, A summary of the geology of the San Luis Basin, Colorado–New Mexico, with emphasis on the geothermal potential of the Monte Vista graben: Colorado Geological Survey Special Publication 17, 30 p.

Chapin, C. E., 1979, Evolution of the Rio Grande rift—A summary, *in* Riecker, R. E., ed., Rio Grande rift: Tectonics and magmatism: American Geophysical Union, p. 1–5.

Chapin, C. E., 1988, Axial basins of the northern and central Rio Grande rift, *in* Sloss, L. L., ed., Sedimentary cover—North American craton: U.S.:

Geological Society of America, The Geology of North America, v. D-2, p. 165–170.

Cordell, L., 1978, Regional geophysical setting of the Rio Grande rift: Geological Society of America Bulletin, v. 89, p. 1073–1090.

Davis, G. H., and Keller, G. R., 1978, Subsurface structure of San Luis Valley, *in* Hawley, J. W., compiler, Guidebook to Rio Grande rift in New Mexico and Colorado: New Mexico Bureau of Mines and Mineral Resources, Circular 163, p. 28.

Davis, T. L., and Stoughton, D., 1979, Interpretation of seismic reflection date from the northern San Luis Valley, south-central Colorado, *in* Reicker, R. E., ed., Rio Grande rift: Tectonics and magmatism: American Geophysical Union, p. 185–194.

Dula, W. F., 1991, Geometric models of listric normal faults and rollover folds: American Association of Petroleum Geologists Bulletin, v. 75, p. 1609–1625.

Gaca, J. R., and Karig, D. E., 1965, Gravity survey in the San Luis Valley area, Colorado: U.S. Geological Survey Open-File Report 66-46, 21 p.

Gibbs, A. D., 1983, Balanced cross-section construction from seismic sections in areas of extensional tectonics: Journal of Structural Geology, v. 5, p. 153–160.

Gibbs, A. D., 1984, Structural evolution of extensional basin margins: Journal of the Geological Society of London, v. 141, p. 609–620.

Gries, R. R., 1985, Seismic lines in the San Luis Valley, south-central Colorado, *in* Rice, R. R., and Dyer, R. C., eds., Seismic exploration in the Rocky Mountain region: Rocky Mountain Association of Geologists and Denver Geophysical Society, p. 267–274.

Hagstrum, J. T., and Lipman, P. W., 1986, Paleomagnetism of the structurally deformed Latir volcanic field, northern New Mexico: Relations to formation of the Questa Caldera and development of the Rio Grande rift: Journal of Geophysical Research, v. 91, p. 7383–7402.

Huntley, D., 1979a, Cenozoic faulting and sedimentation in northern San Luis Valley, Colorado: Summary: Geological Society of America Bulletin, Part I, v. 90, p. 8–12.

Huntley, D., 1979b, Cenozoic faulting and sedimentation in northern San Luis Valley, Colorado: Geological Society of America, Part II, v. 90, p. 135–153.

Jones, D. M., 1991, Low-angle normal faulting and gold mineralization in the Sangre de Cristo Range of southern Colorado: Geological Society of America Abstracts with Programs, v. 25, p. A246.

Lindsey, D. A., Johnson, B. R., and Andriessen, P.A.M., 1983, Laramide and Neogene structure of the northern Sangre de Cristo Range, south-central Colorado, *in* Lowell, J. D., ed., Rocky Mountain foreland basins and uplifts: Rocky Mountain Association of Geologists, p. 219–228.

McCalpin, J. P., 1982, Quaternary geology and neotectonics of the west flank of the northern Sangre de Cristo Mountains, south-central Colorado: Colorado School of Mines Quarterly, v. 77, no. 3, 97 p.

Morgan, P., and Golombek, M. P., 1984, Factors controlling the phases and styles of extension in the northern Rio Grande rift, *in* Baldridge, W. S., Dickerson, P. W., Riecker, R. E., and Zidek, J., eds., Rio Grande rift: New Mexico Geological Society, Northern New Mexico, Guidebook, p. 13–19.

Powell, J. W., 1958, Ground-water resources of the San Luis Valley, Colorado: U.S. Geological Survey Water-Supply Paper 1379, p. 274.

Repplier, F. N., and Fargo, R. L., 1981, Geothermal gradient map of Colorado: Colorado Geological Survey Map Series 20, scale 1:1,000,000.

Russell, L. R., and Snelson, S., 1990, Structural style and tectonic evolution of the Albuquerque Basin segment of the Rio Grande rift, *in* Pinet, B., and Bois, C., eds., The potential of deep seismic profiling for hydrocarbon exploration: Paris, Editions Technip, p. 175–207.

Stoughton, D., 1977, Interpretation of seismic reflection data from the San Luis Valley, south-central Colorado [M.S. thesis]: Golden, Colorado School of Mines, p. 100.

Tweto, O., 1979, The Rio Grande rift system in Colorado, *in* Riecker, R. E., ed., Rio Grande rift: Tectonics and magmatism: Washington, D.C., American Geophysical Union, p. 33–56.

White, N. J., Jackson, J. A., and McKenzie, D. P., 1986, The relationship between the geometry of normal faults and that of the sedimentary layers in the hanging wall: Journal of Structural Geology, v. 8, p. 897–909.

Manuscript Accepted by the Society October 13, 1993

Printed in U.S.A.

Geological Society of America
Special Paper 291
1994

Tertiary stratigraphy and tectonic development of the Alamosa basin (northern San Luis Basin), Rio Grande rift, south-central Colorado

Brian S. Brister*
New Mexico Institute of Mining and Technology, Socorro, New Mexico 87801
Robbie R. Gries*
Independent Geologist, Denver, Colorado 80203

ABSTRACT

Analysis of borehole and reflection seismic data from the Alamosa basin (northern San Luis Basin, Rio Grande rift) reveals tectonic development in response to three Tertiary events, each with an associated package of rocks distinguished by mineralogy and petrology. Eocene redbeds of the Blanco Basin Formation (0 to 696 m thick) are micaceous, sandy mudstone and coarse arkosic sandstone units containing lithic pebbles derived from granitic basement rock. They were deposited in a late Laramide basin formed during wrench-fault-related segmentation of the early Laramide San Luis–Brazos uplift. The western half of the younger, rift-related Alamosa basin is superposed over this late Laramide basin. Initiation of Oligocene volcanism is marked by andesitic lava flows and volcaniclastic rocks of the Conejos Formation (0 to 2,300 m thick), also limited in extent to the western half of the Alamosa basin. Ash-flow tuffs (380 to 580 m thick) correlative to 29 to 27 Ma tuffs of the San Juan volcanic field cap the Conejos Formation in the western half of the basin and rest directly on denuded Precambrian basement in the eastern half of the basin. These tuffs exist in deep wells across the Alamosa basin and together represent a basinwide time marker. Extension related to the Rio Grande rift resulted in eastward-tilting of the entire basin area following emplacement of the ash-flow tuffs. Filling the resulting half graben is the upper Oligocene–middle Pleistocene Santa Fe Group (as much as 5.6 km thick) composed of variegated mudstones and coarse lithic sandstones and conglomerates. Lithic fragments in the Santa Fe Group represent two sources: variable-composition volcanic rocks from the San Juan volcanic field to the west (majority) and plutonic-metamorphic-sedimentary, basement-derived rocks from the Sangre de Cristo Range to the east (minority). An angular unconformity within the Santa Fe Group documents strong early tilting due to movement on the Sangre de Cristo fault zone during an early phase of rifting (late Oligocene–early Miocene). The rift-related geometry of the crust beneath the Alamosa basin is that of two east-tilted crustal blocks creating two second-order half grabens within the basin.

*Present addresses: Brister, Burnett Oil Company, Inc., 801 Cherry Street, Suite 1500, Fort Worth, Texas 76102; Gries, Pease Oil and Gas Company, 633 17th Street, Suite 1500, Denver, Colorado 80202.

Brister, B. S., and Gries, R. R., 1994, Tertiary stratigraphy and tectonic development of the Alamosa basin (northern San Luis Basin), Rio Grande rift, south-central Colorado, *in* Keller, G. R., and Cather, S. M., eds., Basins of the Rio Grande Rift: Structure, Stratigraphy, and Tectonic Setting: Boulder, Colorado, Geological Society of America Special Paper 291.

INTRODUCTION

The Alamosa basin is a subbasin of the San Luis Basin of south-central Colorado and north-central New Mexico. The San Luis Basin is one of a series of similar features in the Rio Grande rift (Fig. 1), a north-trending intracontinental rift that extends from north of Leadville, Colorado, to El Paso, Texas, and beyond (Chapin, 1971). The San Luis Basin is more than 200 km long from north to south. Its northern physiographic limit is at Poncha Pass, Colorado (Upson, 1939), and its southern limit is near the Embudo constriction in New Mexico (Kelley, 1956). The San Luis Basin is bordered by the San Juan and Tusas Mountains on the west and the Sangre de Cristo Range on the east. At the latitude of Alamosa, Colorado, it is about 70 km across. This paper examines a 5,520 km^2 area in the San Luis Basin, north of the San Luis Hills, termed the "Alamosa basin" by Burroughs (1981). Its stratigraphy, structure, and tectonic history are interpreted from subsurface data, including borehole samples and geophysical surveys.

The Alamosa basin has had a complex history. The region has been recurrently uplifted and down dropped during several tectonic events (Sales, 1983). As a Rio Grande rift basin, the Alamosa basin today takes the form of an asymmetric, first-order half graben with stratigraphic units tilted eastward from their pre-rift orientations (Fig. 2). This half graben is bounded along the Sangre de Cristo Range by the Sangre de Cristo fault zone (Personius and Machette, 1984, p. 87). The basin is divided into a western ("Monte Vista") graben and eastern ("Baca") graben by a basement high termed the "Alamosa horst" (Burroughs, 1981). Both grabens are second-order half grabens, but the Baca graben was the site of the greatest degree of tilting and thickest sedimentary infilling during rifting due to its proximity to the Sangre de Cristo fault zone.

The purpose of this paper is to present an interpretation of

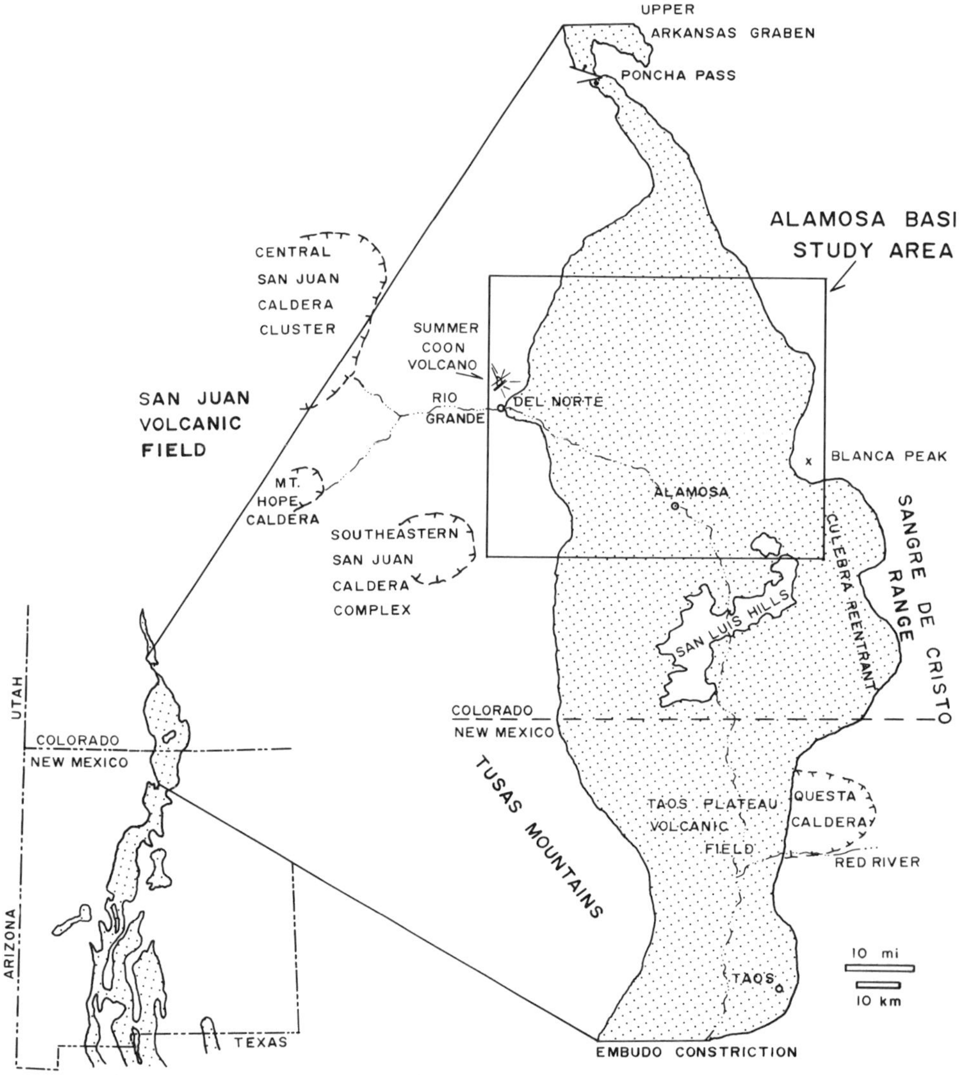

Figure 1. Location map showing Alamosa basin study area, San Luis Basin, Rio Grande rift, and geographic features discussed in text. Stippled pattern denotes rift-basin fill.

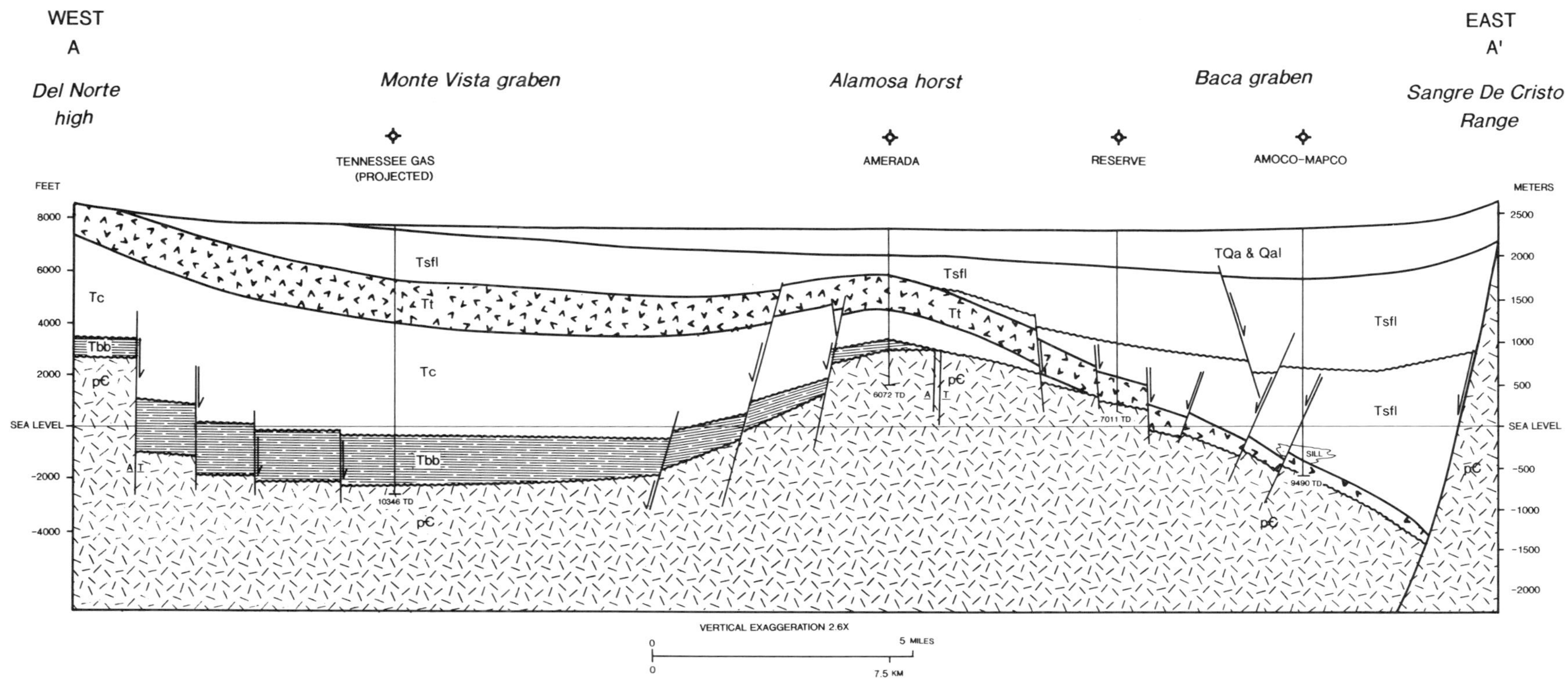

Figure 2. Interpretive cross section A–A′ across the San Luis Basin; location of section indicated in Figure 3. Symbols: TQa and Qal, Alamosa Formation (Plio-Pleistocene) and Quaternary alluvium; Tsfl, lower Santa Fe Group (Mio-Pliocene); Tt, ash-flow tuffs of San Juan volcanic field (Oligocene); Tc, Conejos Formation and equivalents (Oligocene); Tbb, Blanco Basin Formation (Eocene); pЄ, granite-gneiss basement (Precambrian); TD, total depth. Figure modified from Gries and Brister (1989).

the tectonic development of the Alamosa basin based on new insight into the stratigraphy of the basin fill. The physical characteristics of the basin stratigraphy, determined from petrologic analysis of subsurface samples, define lithostratigraphic units that can be correlated using reflection seismic lines. Combining new subsurface stratigraphic and structural information has led to a new interpretation of the timing of episodes of basin development.

PREVIOUS WORK

The classic references on the location and physiographic setting of the basin are Siebenthal (1910b) and Upson (1939). Overviews that discuss the relation of the San Luis Basin to the Rio Grande rift are Chapin (1971, 1979, 1988), Cordell (1978), Hawley (1978), Keller et al. (1984),and Tweto (1979). Ingersoll et al. (1990) present a detailed study of the sedimentation and paleotectonics of the southern San Luis Basin based on outcrop-derived data and regional correlation. Notable geophysical investigations of the northern San Luis Basin are Cordell (1978), Gries (1985a), Keller et al. (1984), and Stoughton (1977).

Huntley (1979) was the first to publish a correlation of stratigraphic units between boreholes suggesting that the basin fill was entirely of Tertiary age. Burroughs (1981) noted that some of the more wide ranging volcanic units of the region could be correlated in borehole geophysical logs.

Reflection seismic lines published by Gries (1985a) were of high quality, but a lack of detailed petrologic data limited interpretation of the various features visible on the lines. Attempts by Gries to resolve this problem utilizing radiometric and palynological dating techniques were unsuccessful. This paper applies the results of a petrologic study of borehole samples to reinterpret seismic data in the basin. Preliminary results were discussed in Gries and Brister (1989).

STRATIGRAPHY

The sequence of lithostratigraphic units in the Alamosa basin varies greatly depending upon location. Regional tectonic events had different effects on the western and eastern halves of the basin, therefore the two halves have significantly different stratigraphic sections. The stratigraphy is illustrated by a west-to-east cross section (Fig. 2, location shown in Fig. 3).

Precambrian basement (p€)

The stratigraphically lowest unit is Precambrian basement. In general, the basement rocks are granitic in composition but have a gneissic texture and have high electrical resistivity (100 to 2,000 ohmms) and densities typical of granitic rocks (greater than 2.6 gm/cc). Cuttings are commonly stained orange by hematite, especially along grain boundaries. This may be a result of weathering when the basement was exposed subaerially during several Phanerozoic orogenic events as will be discussed below.

Basement rocks contain quartz, orthoclase, microcline, perthite, plagioclase, muscovite, biotite, and amphibole. Alignment of micas is common in most samples. Most quartz grains are strained, monocrystalline types that reflect a probable metamorphic origin. Some quartz grains are coarsely polycrystalline with nonsutured boundaries. Although these grains would usually be expected to come from metaquartzites, they may also be found in finer-grained granites and schists (Folk, 1974).

Many metamorphic rock types have been described from exposures in the nearby Sangre de Cristo Range (Johnson, 1969), including granites, gneisses of various compositions, amphibolites, and metasediments. Such variation probably also exists beneath the Alamosa basin, but has not been found by sparse drilling. The top of the Precambrian basement is an unconformable surface of erosion, upon which Tertiary units were deposited.

Blanco Basin Formation (Tbb)

The Blanco Basin Formation (Eocene) consists of nonvolcanic, alluvial redbeds unconformably overlying the Precambrian basement in wells in the western half of the Alamosa basin (Monte Vista graben). Total thickness varies across the basin from 0 to 696 m. The Blanco Basin Formation is composed of sandy, micaceous mudstone and coarse arkosic sandstone and conglomerate.

The mineral composition of the coarser sedimentary rocks is identical to that of the Precambrian basement and indicates that basement rocks were their primary source. Sandstones in some wells also contain a small percentage of sedimentary rock fragments of Paleozoic and/or Mesozoic provenance. Rarely, fine- to medium-grained, frosted, rounded quartz grains exist in the samples, perhaps eroded from Jurassic eolian sandstone formations such as the Junction Creek Formation. In Figure 4, the composition of the Blanco Basin Formation is compared with younger Tertiary sandstones in the basin.

Those units sampled that were interpreted as "sands" from geophysical logs are arkosic, coarse, pebbly sandstones and conglomerate beds containing granitic pebbles. The log characteristics of these bodies are variable, but individual sand bodies tend to be "cylinder shapes" (terminology of Rider, 1986) with lesser bell and funnel shapes. Contacts against surrounding mudstone beds range from abrupt to gradational.

Blanco Basin mudstones are generally reddish brown (Hue 10R, saturation 4 to 6, value 3 to 4) but may vary to red, green, gray, maroon, and purple. A single cutting chip may display mottling of two or more of these colors suggesting that coloration may be dependent upon diagenesis. The mudstones have not yielded fossils or pollen, and thus the exact age of the formation is unknown.

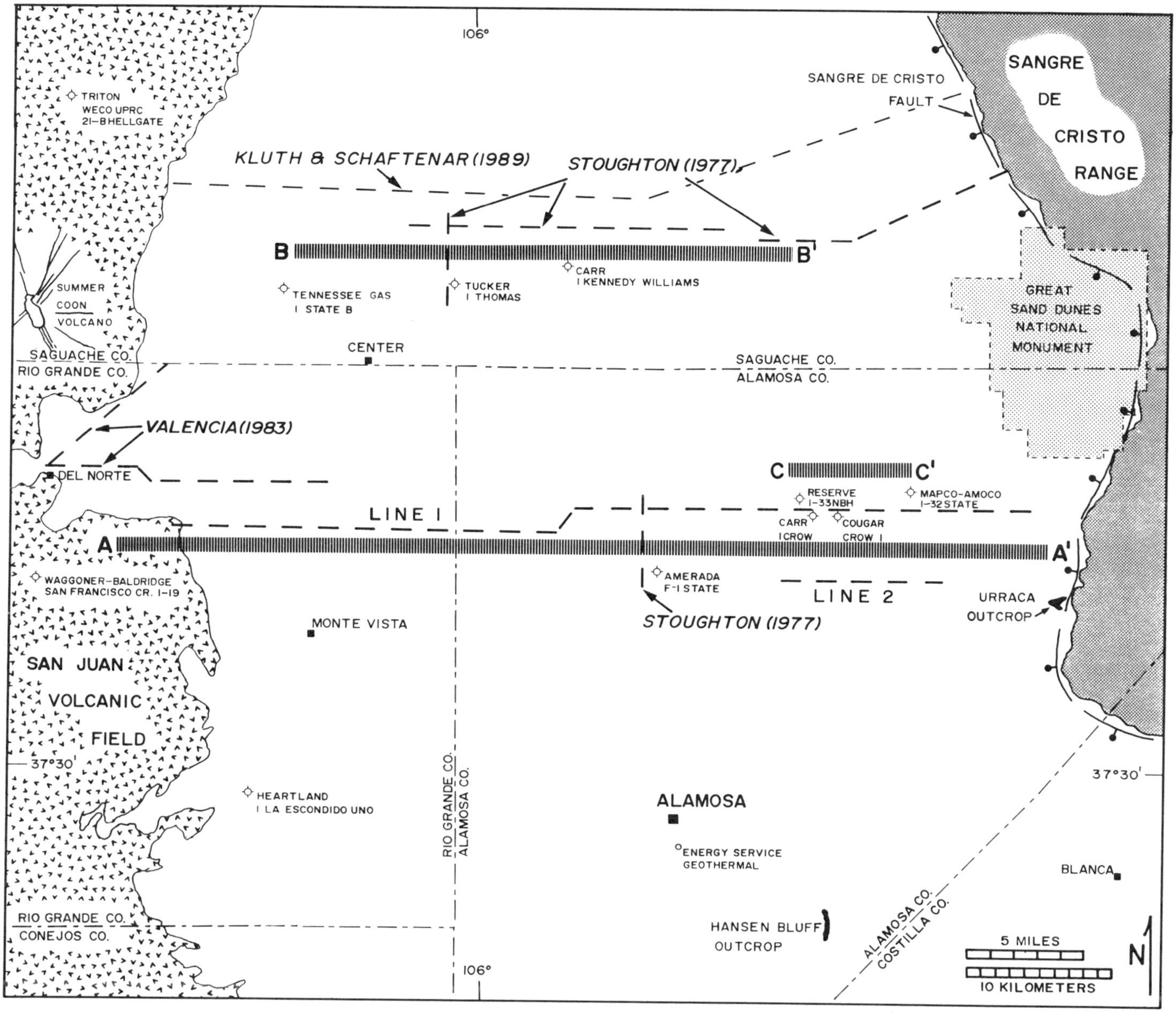

Figure 3. Map of the Alamosa basin study area showing oil and gas drilling, seismic lines illustrated or discussed in this paper, and cross sections A–A′, B–B′, and C–C′.

The formation here termed Blanco Basin Formation in the western part of the Alamosa basin has been considered to be Eocene in age by other workers based on its color, degree of consolidation and cementation, and depositional facies characteristics. Tweto (1979) correlated it with the "Eocene Echo Park Alluvium" (Echo Park Formation of Epis and Chapin, 1974). Covarrubias (1988) termed the formation "Eocene red beds." Baltz (1965) believed the formation to be similar to the Blanco Basin Formation of Larsen and Cross (1956) although he did not specifically assign that name.

Huntley (1979) and Burroughs (1981) applied the name Vallejo Formation due to the apparent similarity of these beds to the "fluvial red-beds" described by Upson (1941) in the Culebra reentrant southeast of the study area. However, we believe that the age of the Vallejo Formation at its type locality is Miocene or younger based on regional correlation to units with similar petrologic characteristics. Regardless of the name chosen, both Huntley and Burroughs believed their "Vallejo" beds in the subsurface to be pre-Oligocene.

The best evidence for the identity of the redbeds in the Monte Vista graben comes from recent drilling in the San Juan sag (Gries, 1985b) west of the Alamosa basin. The San Juan sag is separated from the Alamosa basin by the Laramide wrench-fault-related Del Norte high (Gries, 1989; Brister,

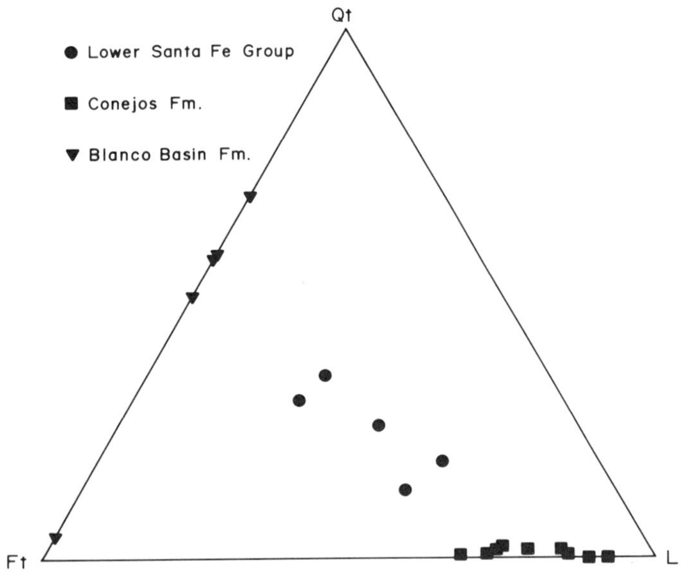

Figure 4. Ternary diagram of sandstone compositions from subsurface units in the Alamosa basin. Qt, total quartz, including polycrystalline grains + quartzite + chert; Ft, total feldspar; L, sedimentary + volcanic + metamorphic lithic grains. Each symbol represents at least 500 points counted in a single sample. See Brister (1990) for data tables and sample locations.

1990). The redbeds of the Monte Vista graben can be correlated in the subsurface across the Del Norte high, into the San Juan sag, and eventually to outcrops some 45 km to the west along the southwestern flank of the San Juan Mountains where they have been assigned the name Blanco Basin Formation (Cross and Larsen, 1935). Their age in the San Juan sag is generally believed to be Eocene because, like the Blanco Basin Formation to the southwest, they are bracketed by unconformable contacts with the Paleocene Animas Formation below and the Oligocene Conejos Formation above. The petrologic characteristics of the redbeds encountered in the Monte Vista graben are similar to those of the Blanco Basin Formation in general and are consistent with the petrologic variability between, and within, Blanco Basin outcrops (Brister, 1992).

The Blanco Basin Formation has been drilled in six of the wells shown in Figure 3. The thickest sections penetrated are 696 m in the Triton/Weco/UPRC 21-B Hellgate well, and 643 m in the Tennessee Gas Transmission 1-State B well. A possible explanation for the great thickness in these wells compared to typical thickness of the formation of about 175 m is that they exist adjacent to a major Laramide basin-bounding fault zone separating the Laramide precursor basin of the Monte Vista graben from the Del Norte high.

Wells drilled west of this fault zone on the Del Norte high contain thinner Blanco Basin Formation. They are the Waggoner-Baldridge San Francisco Creek 1-19 well with 159 m, and the Heartland #1 La Escondido Uno well with only 49 m. The Amerada F-1 State well in the vicinity of the Alamosa horst drilled 115 m, which is the easternmost-known Blanco Basin equivalent rocks. The Tucker #1 Thomas well bottomed in the Blanco Basin Formation, drilling only 34 m as estimated from a description by Powell (1958).

Conejos Formation (Tc)

The Conejos Formation is a series of intermediate-composition volcaniclastic rocks and lava flows (35 to 30 Ma) that were derived from volcanoes active in the San Juan volcanic field from 35 to 30 Ma (Lipman et al., 1970). The Conejos volcanic rocks are high-K, subalkalic, and commonly range from andesite to quartz latite (silicic dacite) in composition (Lipman, 1989). Lithologic units in the formation vary depending upon distance from vents and/or periodic extrusive activity. These include volcanic breccias of both "hot" and "cold" origin and emplacement (these range from mono- to heterolithologic and include, but are not limited to, debris-avalanche and debris-flow deposits), lava flows, autobrecciated flows, stream-laid conglomerate and sandstone, and organic-rich lacustrine claystone. Ash-flow tuffs are rare (Lipman, 1975), but some of the volcaniclastic deposits are tuffaceous, containing relict glass shards and rare reworked welded tuff fragments. The above description applies well to the package of rocks marked Conejos Formation on Figure 2. As seen in Figure 4, Conejos sandstones from wells in the Alamosa basin are lithic rich; this lithic component is generally 100% intermediate-composition volcanic rock fragments.

Only one Conejos vent has been documented in the study area: that of the Summer Coon volcano (Lipman, 1968) located on the western edge of the basin north of the town of Del Norte. Other sources for Conejos deposits in the Alamosa basin are located to the southwest, where vents have been documented in the Platoro, Colorado, area (Lipman, 1975), and possibly the south, from vents in the San Luis Hills (Burroughs, 1971, 1972, 1981; Thompson and Machette, 1989). The vents responsible for the Bonanza Tuff and related andesitic breccias (Steven and Lipman, 1976) comprise a possible northern source for Conejos-equivalent volcaniclastic detritus.

The Conejos units in the Alamosa basin were probably deposited on the distal fringes of vent complexes. In general, those wells along the western side of the Alamosa basin have a high percentage of flows and coarse volcaniclastic rocks, but this percentage decreases eastward in favor of finer-grained deposits. The Conejos Formation also decreases in thickness eastward to a zero edge over the central part of the basin. The thickest section drilled in the study area is 2,300 m in the Triton/WECO/UPRC 21-B Hellgate well, which is situated on the northern flank of the Summer Coon volcano. Most other wells in the western half of the Alamosa basin have penetrated 1.3 to 1.5 km of the Conejos Formation. The easternmost section of the Conejos Formation is in the Amerada F-1 State well, which had only 400 m.

Ash-flow tuffs (Tt)

A package of interbedded ash-flow tuffs and tuffaceous clastic rocks marked Tt on Figure 2 overlies the Conejos Formation. This package is important because it: (1) exists throughout the basin; (2) separates Rio Grande rift-related deposits above it from pre-rift formations below it; (3) has been radiometrically dated (in the San Juan Mountains) and has been demonstrated to represent a short interval of time; and (4) shows distinctive (although nonunique) identifiable characteristics on borehole geophysical logs and reflection seismic lines. These characteristics used in conjunction with petrologic examination have allowed the utilization of the package as a time marker in the stratigraphic sequence (Brister, 1990).

The ash-flow tuffs in this sequence are part of a series of volcanic rocks, 26 to 30 m.y. old (Steven and Lipman, 1976), which originated in the eastern San Juan volcanic field. Only the most voluminous flows made their way to the Alamosa basin area. The key to their identity are outcrops in the eastern foothills of the San Juan Mountains where the tuffs dip at angles of less than 10° into the basin. The 29.5 to 28.4 Ma Treasure Mountain Tuff (Lipman and Steven, 1970; Lipman, 1989), erupted from the southeastern San Juan caldera complex, and the 28.4-Ma Masonic Park Tuff (Lipman et al., 1970; Steven et al., 1974; Lipman, 1989), erupted from the Mount Hope caldera, contributed the majority of the material to the tuff package in the southern part of the Alamosa basin. Likewise, the 27.75-Ma Fish Canyon and 27.35-Ma Carpenter Ridge tuffs (Olson et al., 1968; Lipman, 1989) from the central caldera cluster dominate the tuff sequence to the north.

Figure 5 is a cross section constructed from borehole geophysical logs drawn such that the datum is the top of the tuff package. In each well shown, there are no lava flows or welded tuffs above this horizon. The figure illustrates which units have been positively identified as ash-flow tuffs using criteria outlined above. Tentative correlations have been drawn between the logs based primarily on log response rather than petrographic criteria. Resistivity curve profiles for the tuff units may mimic welding profiles. It is assumed that any given ash-flow tuff has similar log response between two wells over a relatively short distance, thus pattern recognition of log curves was helpful in correlation.

It can be observed that the westernmost (which are also the northernmost) wells generally have thin Masonic Park Tuff and Treasure Mountain Tuff compared to those wells farther south and east. These wells are near the distal edge of the

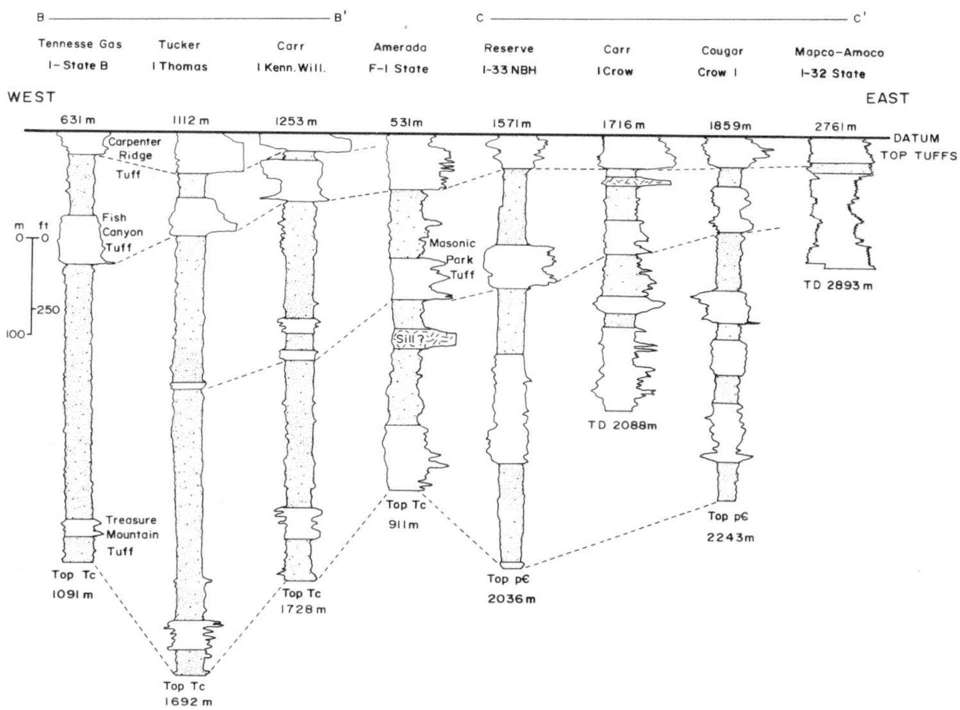

Figure 5. Well-log comparison of Oligocene ash-flow-tuff package in Alamosa basin illustrating tentative correlations; drillholes indicated are on Figure 3. Log curves are spontaneous potential (left) and resistivity (right); exception is Mapco-Amoco 1-32 State well (gamma ray on left; density on right). Datum is top of the ash-flow tuffs/base of Santa Fe Group (depth of datum listed above heavy line). Unit underlying tuffs and depth of contact indicated at bottom. A similar package of tuffs to that in the Amerada F-1 State well is present in the Energy Service #1 Alamosa Geothermal well but electric logs are not available. Stippled areas are unwelded tuffs and tuffaceous sediments; unstippled areas are welded units.

Masonic Park and Treasure Mountain tuffs that were erupted from the Mount Hope and southeastern caldera complex (see Fig. 1). The nearby Summer Coon volcano of Conejos age was probably topographically high enough to stand in the way of the ash flows, creating a shadow. For the same reasons, wells to the southeast generally lack a recognizable section of Carpenter Ridge Tuff.

Prior to this study, it was believed that the upper Oligocene San Juan tuffs pinch out in the Monte Vista graben (Tweto, 1979) or over the Alamosa horst (Burroughs, 1981). The explanation for such pinching out was that the Alamosa horst prevented tuffs from reaching the eastern half of the basin. Figure 5 shows that if a topographically high area existed, it provided no impediment to ash flows moving across it. In fact, it is likely that the ash flows may have reached eastward beyond the present-day Sangre de Cristo Range (Burroughs, 1981; Scott, 1975). Also, there is no apparent thinning of ash-flow tuffs in wells (Amerada F-1 State, and Energy Service Geothermal) over the horst area. There is, however, overall thinning of the intertuff clastic deposits eastward in the wells, suggesting increasing distance from the major source of these sediments and/or a positive paleotopographic gradient to the east. The effect of this sedimentation was to blanket and subdue existing topography. The top of the tuffs in the subsurface is an east-dipping surface today, due to postemplacement tilting.

Santa Fe Group (Tsfl and Ta)

The Santa Fe Group includes all pre–middle Pleistocene sediments above the upper Oligocene tuffs. It can be roughly divided into two units, upper and lower Santa Fe Group, based upon lithologic criteria. The lower Santa Fe Group is ubiquitous in the subsurface of the basin and represents large-scale sedimentary response to half-graben development from late Oligocene to Pliocene. The upper Santa Fe Group, named the Alamosa Formation, represents lacustrine and fluvial sedimentation ranging in age from Pliocene to middle Pleistocene. Cross section B′–C′ (Fig. 6), summarized from well data, illustrates the stratigraphy of the Santa Fe Group in the Alamosa basin.

Lower Santa Fe Group. The lower Santa Fe Group in the Alamosa basin has been called the Santa Fe Formation (Siebenthal, 1910b; Powell, 1958). However, Spiegel and Baldwin (1963) began the modern usage of "Santa Fe" as a group term "that includes all the synrift basin fill, both volcanic and sedimentary, ranging in age from late Oligocene to Quaternary, but excluding deposits that postdate entrenchment of the Rio Grande in middle Pleistocene time" (Chapin, 1988, p. 169). Burroughs (1981) has suggested that the lower Santa Fe Group may be divided into formations based on composition and provenance. No attempt has been made in this study to subdivide the lower Santa Fe Group because of lack of sufficient subsurface data. The age of lower Santa Fe Group sediments in the Alamosa basin is poorly constrained, but ranges from about 26 Ma to about 4.5 Ma. The older age is approximately the end of ash-flow volcanism in the eastern San Juan Mountains (Steven et al., 1967; Lipman, 1989). The younger age marks the beginning of tholeiitic basalt volcanism and construction of the Taos Plateau volcanic field that blocked surface drainage in the southern San Luis Basin (Lipman and Mehnert, 1979) and was responsible for the widespread fluvio-lacustrine system of the Alamosa Formation.

Grain size and composition of the lower Santa Fe Group vary depending primarily upon proximity to sources and character of dispersal systems. Well samples from the study area are of claystone, sandstone, and conglomerate. The claystone and sandy mudstone in the lower Santa Fe Group are compact but soft, nonfissile, and micaceous. They deepen in color with increasing depth in the basin, but are everywhere variegated. Typical colors include tan, pink, buff, orange, brick red, light olive, and light gray to black. Most of these colors are indicative of an oxidizing environment of deposition. A coal seam was penetrated in fine-grained units in the Mapco-Amoco #1-32 State well at 1,774 m depth. There are no known bedded evaporites in the sequence, but scattered selenite crystals can be found in the cuttings.

The lowermost Santa Fe beds have been reported to be early Tertiary in age on the basis of pollen analyses (Huntley, 1976; Gries, 1985a). However, the pollen assemblages are sparse, not indicative of any specific Tertiary age, and contain a few specimens of probable Eocene age that may have been reworked. Lithologic and stratigraphic criteria discussed above provide strong evidence that these sediments postdate upper Oligocene ash-flow tuff volcanism. A rhyolite sill(?) in the Mapco-Amoco 1-32 State well at the base of the Santa Fe Group but above the Oligocene ash-flow tuff package has yielded a whole-rock K-Ar age of 22.2 Ma (R. Gries, unpublished data).

Some generalized observations of Santa Fe Group sandstones from the basin indicate that Precambrian detritus tends to increase in abundance eastward in the basin towards the Sangre de Cristo Range. All sandstones examined are lithic-rich with a majority of these fragments being of volcanic origin. Sandstone samples from the deepest part of the Baca graben are tuffaceous, but contain significant amounts of Precambrian material. On a ternary diagram (Fig. 4), these sandstones span the range in composition between samples of the Blanco Basin Formation and the Conejos Formation.

The lower Santa Fe Group sediments indicate provenance from two primary sources. The San Juan volcanic field to the west and northwest provided an influx of intermediate-composition volcanic debris including ash-flow-tuff clasts. Such deposits of late Oligocene–Pliocene age derived from the San Juan Mountains are usually referred to as the Los Pinos Formation (Butler, 1946, 1971; Manley, 1981), which Chapin (1988) has included in the Santa Fe Group. Interbedded volcanic flows are rare in the Santa Fe Group of the Alamosa basin; however, Hinsdale basalts ranging in age from 25.7 to

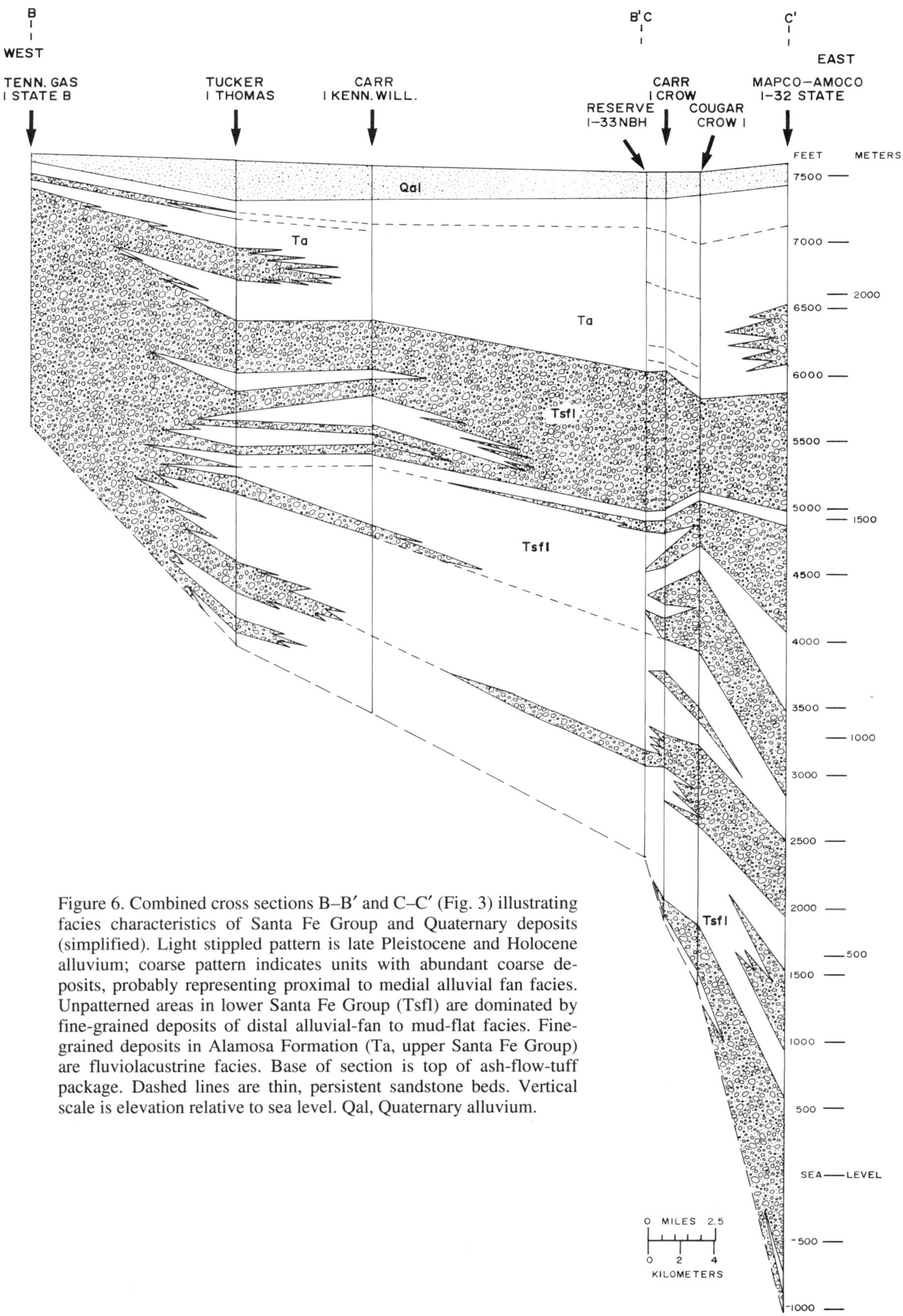

Figure 6. Combined cross sections B–B′ and C–C′ (Fig. 3) illustrating facies characteristics of Santa Fe Group and Quaternary deposits (simplified). Light stippled pattern is late Pleistocene and Holocene alluvium; coarse pattern indicates units with abundant coarse deposits, probably representing proximal to medial alluvial fan facies. Unpatterned areas in lower Santa Fe Group (Tsfl) are dominated by fine-grained deposits of distal alluvial-fan to mud-flat facies. Fine-grained deposits in Alamosa Formation (Ta, upper Santa Fe Group) are fluviolacustrine facies. Base of section is top of ash-flow-tuff package. Dashed lines are thin, persistent sandstone beds. Vertical scale is elevation relative to sea level. Qal, Quaternary alluvium.

26.4 Ma are interbedded with these sediments in the San Luis Hills along the southern border of the basin (Thompson and Machette, 1989). In the San Juan volcanic field the Hinsdale basalts range in age from 26 to 5 Ma (Lipman, 1975, 1969; Steven et al., 1974).

The second source of sediment was the rising Sangre de Cristo Range to the east, which was a source of detritus from Precambrian granitic and metamorphic rocks and Paleozoic limestones and clastic rocks. The Sangre de Cristo Range was at least partially covered by volcanic rocks in the late Oligocene and thus was also a source for some volcanic material.

Another possible source of sediment that has been suggested is stream flow from the Upper Arkansas graben into the Alamosa basin (Hanna and Harmon, 1989). A possible connection between the San Luis Basin and Upper Arkansas graben has been postulated due to the presence of the Mio-Pliocene Dry Union Formation (Tweto, 1961) in a narrow graben north of Poncha Pass, Colorado (Van Alstine, 1968, 1970; Knepper and Marrs, 1971, Taylor, 1975). However, neither surface mapping nor gravity surveys indicate significant Dry Union/Santa Fe deposits at Poncha Pass, but instead, demonstrate that a topographic and/or structural barrier existed in the area during the Miocene and Pliocene (Knepper, 1976).

Rare outcrops of the lower Santa Fe Group exist along the basin margins. In one location along the Sangre de Cristo fault zone, a fault slice of lower Santa Fe Group is exposed at the surface (marked “Urraca outcrop” on Fig. 3; see Gries and Vandersluis, 1989, p. 35). This deposit contains interbedded fluvial conglomerates and pebbly sandstones with beds less than a meter thick. Pebble imbrications suggest a south-southwest stream-flow direction that paralleled the Sangre de Cristo fault zone at this location. Current indicators show that the braided stream was draining parallel to the axis of the paleobasin. The rocks are well indurated and reddish brown, probably due to postdepositional cementation and oxidation. Such reddening of Santa Fe Group sediments has been reported elsewhere in the Rio Grande rift by Chapin and Lindley (1986) and has already been noted above for nearby deep-basin well samples. Composition of the pebbles in the outcrop reflect two provenances. Pebbles from the Sangre de Cristo Range include Precambrian granitic and metamorphic rocks and minor amounts of Paleozoic arkose, graywacke, siltstone, shale, limestone, and chert. Pebbles derived from the San Juan volcanic field include volcanic rocks of intermediate composition and minor flow-banded rhyolite and ash-flow tuff.

Alamosa Formation. The Alamosa Formation (Siebenthal, 1910a) is a fluviolacustrine formation deposited conformably upon the uppermost beds of the lower Santa Fe Group in the Alamosa basin. The depositional environment of the Alamosa Formation was dominated by reducing conditions (Huntley, 1979) as demonstrated by the predominance of gray, black, and green claystones. Well samples contain fossil debris including ostracods, bones, peat and wood fragments, and mollusc shell fragments. Organic material taints water from some wells and is a source of methane, the discovery of which has helped stimulate intermittent oil and gas wildcat drilling in the basin (Gries, 1985a). Toward the top of the section are persistent, poorly cemented sandstone horizons that can be correlated over broad areas of the basin. These mark the beginning of a return to drier and/or higher energy conditions in the basin. These beds have been extensively drilled in the basin because they are fresh-water bearing and artesian. They are the primary source of ground water for agriculture in the San Luis Valley.

Rogers (1984) sampled some 20 m of surface outcrop of the Alamosa Formation at Hansen's Bluff near the town of Alamosa (Fig. 3) and found the beds to be Pleistocene (0.6 to 0.9 Ma). The outcrop contains volcanic ash, various fish, bird, and mammalian bones, and fresh-water mollusc shells. The great thickness of underlying Alamosa deposits (the formation ranges up to 550 m thick) suggests that the formation is as old as Pliocene. The Hansen's Bluff deposits are overlain by Quaternary alluvium and probably predate the capture of drainage in the Alamosa basin by the Rio Grande in middle to late Pleistocene.

INTERPRETATION OF SEISMIC DATA

Seismic characteristics

Figure 7 is a synthetic seismogram from the Tennessee Gas 1-B State well, which contains all the stratigraphic units in the basin. The strongest reflections at about 0.6 seconds (two-way travel time) mark the late Oligocene welded ash-flow tuffs and the base of the Santa Fe Group. This strong reflection package was interpreted in every seismic line examined in the Alamosa basin. Because the top of the tuffs is the transition between pre-rift and syn-rift deposits, any faults that offset these reflections are a result of the rifting events. Faults that offset units below these reflections, but not the reflections themselves, are pre-rift in origin. Thus, this reflection package is important for deciphering the tectonic history of the basin.

Another strong reflection on Figure 7 exists at about 1.25 seconds. This is a local reflection associated with a 75-m-thick lava flow within the Conejos Formation. Such reflections in the Conejos are generally not far ranging. A weak double reflection exists at about 1.95 seconds and marks the top of the Precambrian basement.

An interpreted seismic line across the Alamosa basin (Fig. 8) illustrates the seismic characteristics of the lithostratigraphic units present. The top of the late Oligocene tuffs is the strong reflection that is not visibly faulted in the western half of the basin, but is faulted and tilted down to considerable depth in the Baca graben to the east. Well control helps to distinguish this reflection from internal reflections of the Santa Fe Group in the deeper parts of the Baca graben.

Below the strong tuff reflections, the Conejos and Blanco Basin Formations contain reflections that vary from strong and

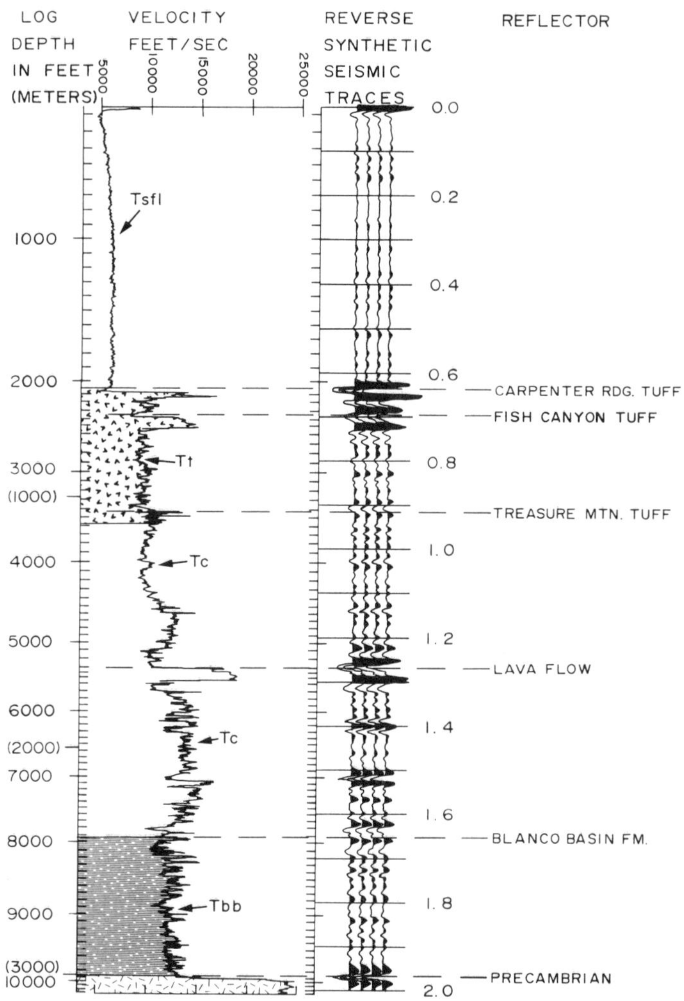

Figure 7. Synthetic seismogram from the Tennessee Gas Transmission 1 State B well in the Monte Vista graben. For other synthetic seismograms in the area, see Gries (1985a).

continuous, to weak and discontinuous. In some cases a weak reflection exists at the contact between the two formations; however, well control is generally necessary to correctly identify the contact. The Conejos–Blanco Basin package of reflections thins eastward to zero over the central high of the Alamosa basin.

Above the strong tuff reflections is a complex series of reflections corresponding to the Santa Fe Group. The Alamosa Formation is distinguished by low-amplitude, weak reflections typical of unconsolidated fine-grained lithologies. The transition to the lower Santa Fe Group is marked by a change to higher amplitude, more continuous reflections characteristic of interbedded coarse and fine clastic lithologies. The lower Santa Fe Group is characterized by two seismic packages, illustrated on Figure 9. The upper package is relatively flat lying and unfaulted, and traceable to some extent across the basin. The lowermost package dips moderately to the east, is

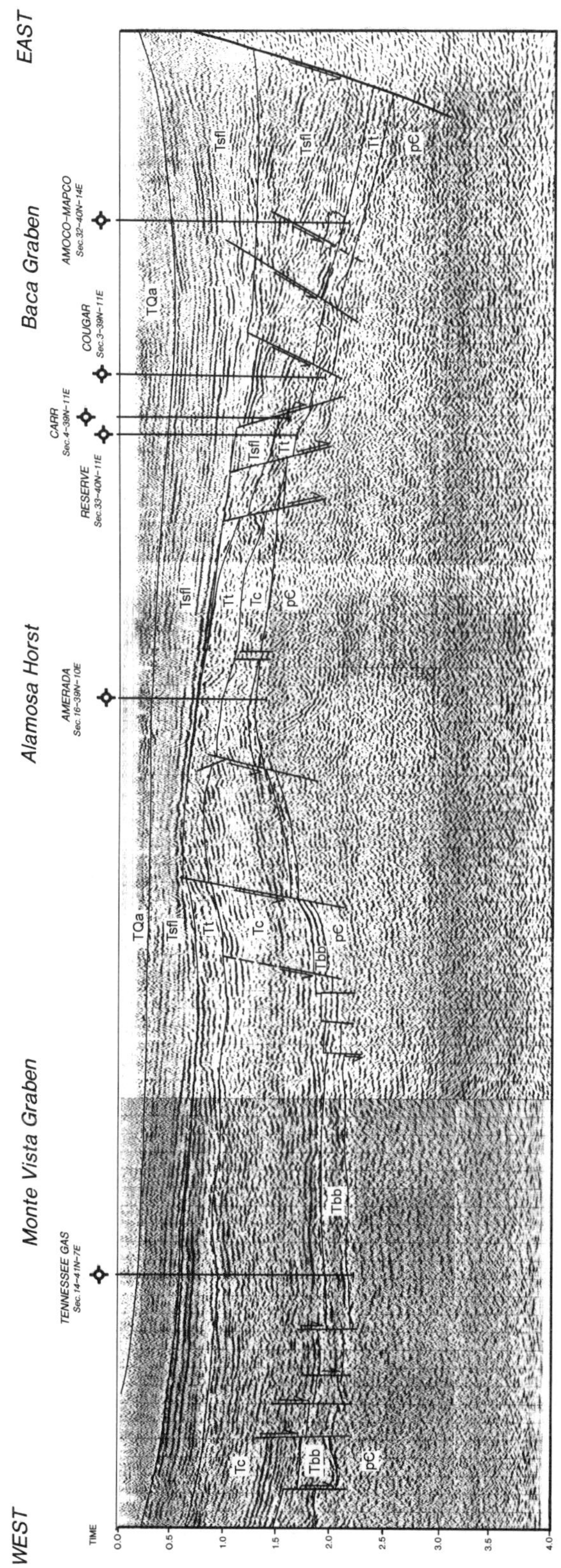

Figure 8. Seismic line 1, crosses the Alamosa basin from west to east (modified from Gries, 1985a; Gries and Brister, 1989). See Figure 3 for location and Figure 2 for abbreviations. Vertical scale in two-way travel time; see Figure 2 for approximate depth conversion.

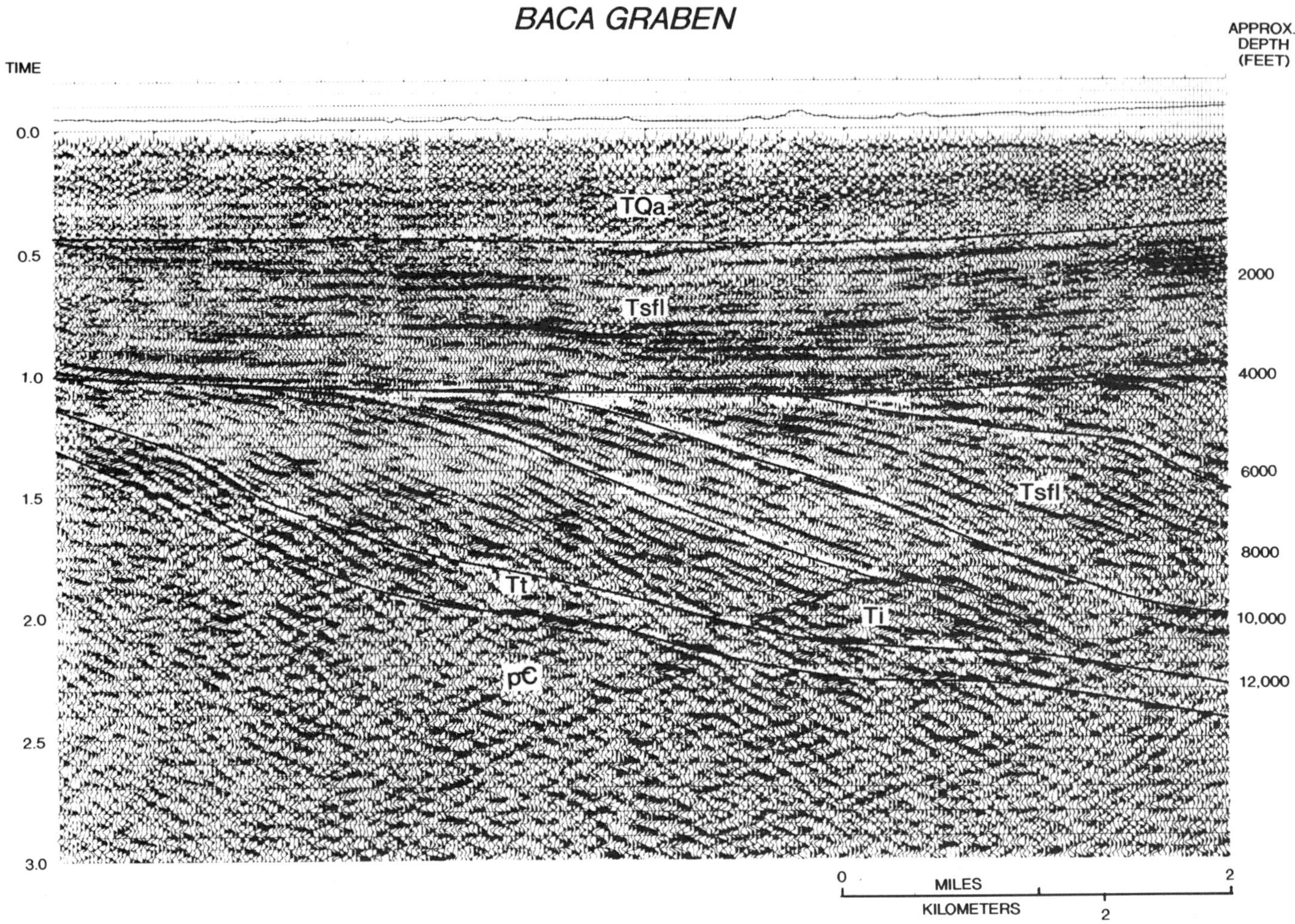

Figure 9. Seismic line 2, showing angular unconformity within the lower Santa Fe Group (after Gries, 1985a; Gries and Brister, 1989). See Figure 3 for location and Figure 2 for abbreviations. Ti, rhyolite intrusion.

highly faulted, and appears limited in extent to the Baca graben. The intervening angular unconformity within the lower Santa Fe Group can be seen on a number of seismic lines from the southern part of the Baca graben, and has some obvious implications for its early tectonic development. Although an increase in bed dip in the Baca graben is suggested in Figure 6 starting at about 1,400 m elevation, there is not an abrupt change in sample composition to mark this boundary. The mudstones and claystones common to the lower Santa Fe Group do not seem to radically change color across this boundary, although they do become reddened towards the base of the lower package.

Isochron maps

The presence of a recognizable, ubiquitous horizon that essentially separates pre-rift and syn-rift rock units in the Alamosa basin has great implications for reconstructing its tectonic history. The top and base of these intervals are readily identified on the 350+ km of seismic lines in the basin. The top of the rift-related interval is simply a horizontal datum at 0 seconds. The base of the rift interval and the top of the pre-rift interval is the uppermost strong reflection of the tuff package. The base of the pre-rift Tertiary interval is the Precambrian double reflection.

Isochron maps made for these intervals from published and unpublished seismic lines are Figures 10 and 11. Published seismic lines used in constructing the isochron maps are indicated on Figure 3. The maps are not intended to show detailed depth and structure of the basin; rather, they illustrate general trends of "thickening" or increasing of the two-way time intervals in the grabens and "thinning" or decreasing of the intervals across the central basement high. No depth conversion is offered because of the scarcity of well control. Faults or fault zones that offset reflections within the intervals are indicated by dashed lines. These maps compare favorably with gravity studies conducted in the area (e.g., Keller et al., 1984).

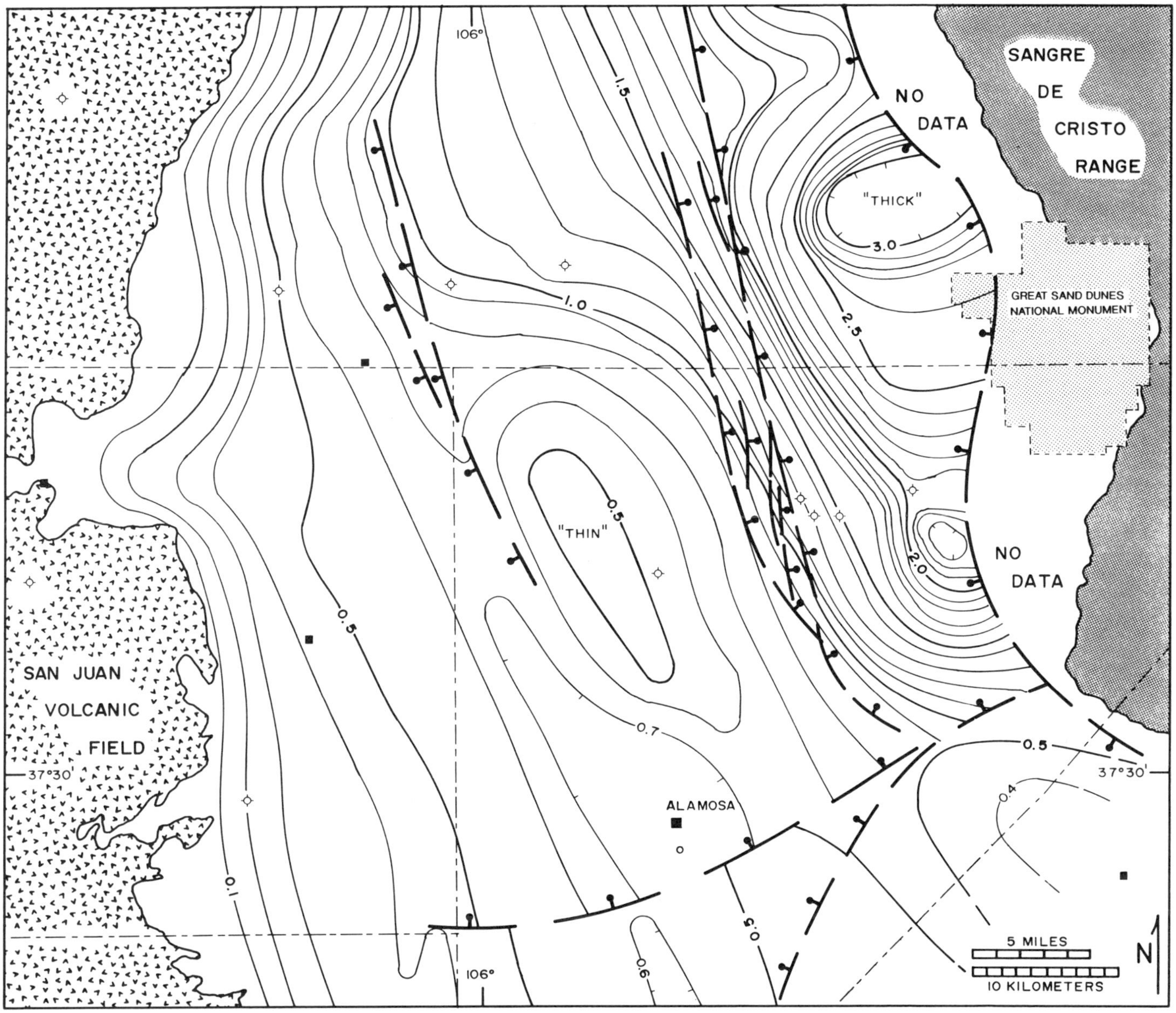

Figure 10. Isochron map: contoured two-way travel time interval between surface datum and top of Oligocene ash-flow tuff reflections. Fault zones depicted have demonstrated syn-rift displacement. Contour interval 0.1 seconds.

Rift-basin geometry

The rift-related architecture of the Alamosa basin is essentially that of a first-order, east-tilted half graben (Figs. 8, 9, 10). All pre-rift lithostratigraphic units have been tilted eastward; however, tilting within the Baca graben is more pronounced than in the Monte Vista graben. The western hinge of the first-order rift graben exists within the San Juan volcanic field, west of the depositional basin (Lipman, 1975; Phillips, 1985). The western edge of the depositional basin is where volcanic units of the San Juan volcanic field dip into the basin beneath the Santa Fe Group. Faulting along this margin is minor but has a distinctive style (Lipman, 1969). Faults generally have only a few meters or tens of meters of normal displacement and fault planes dip steeply to the west. However, due to the eastward dip of the beds, the net displacement of the volcanic rocks over the region is downward to the east.

The eastern boundary of the basin is the Sangre de Cristo fault zone. In plan view, it is concave towards the basin adjacent to the Great Sand Dunes National Monument. Farther south, the fault zone is convex toward the basin, forming the distinct promontory of the Blanca Peak massif. Based strictly on well control, the thickness of graben fill adjacent to the fault zone is at least 3 km. A seismic line by Stoughton (1977)

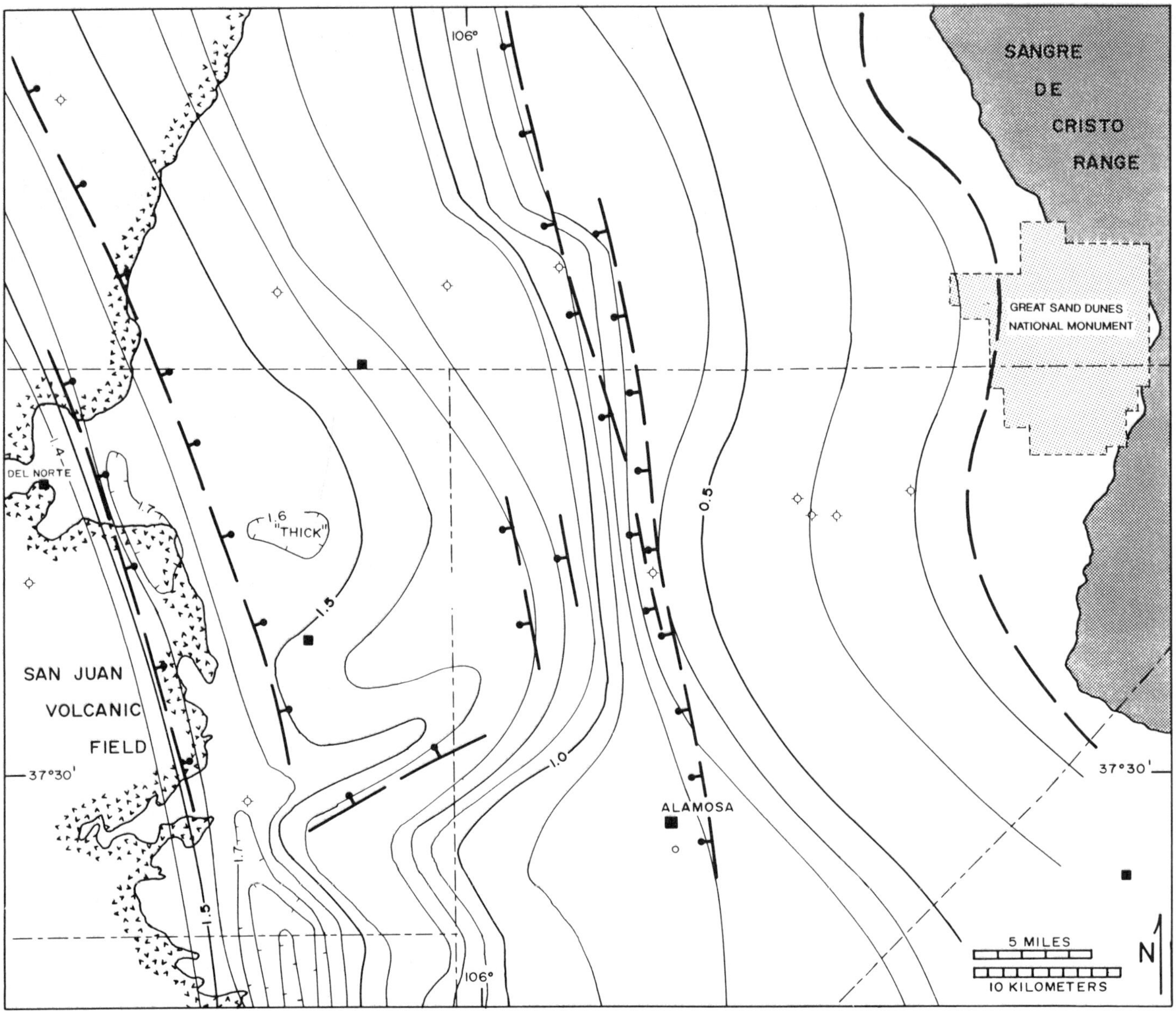

Figure 11. Isochron map: contoured two-way travel time interval between top of Oligocene ash-flow-tuff reflections and top of Precambrian reflections. Fault zones depicted have demonstrated pre-rift displacement. Contour interval 0.1 seconds.

indicates that the greatest thickness of fill exists in the Baca graben adjacent to the Sangre de Cristo fault zone near the Great Sand Dunes National Monument. A reasonable estimation of the depth of the basin at this locality is about 5.6 km. When the difference in elevation between the valley floor and the top of the Sangre de Cristo Range immediately adjacent to the dunes area (about 2 km) is added, the total vertical displacement of basement at the Sangre de Cristo fault zone is about 7.6 km. Basement relief has been estimated from gravity studies to be 7 km (Davis and Keller, 1978), approximately the estimate from seismic data in this paper. The fault plane dips steeply (>45°) to the west and remains steeply dipping to the depths of resolution on seismic lines.

The southern boundary of the Alamosa basin is a probable accommodation zone between the Alamosa basin and the San Luis Hills, an intrabasin horst block that brings Oligocene volcanic rocks to the surface. The Baca graben is truncated at this zone, accounting for the great increase in isochron gradient at that location in Figure 10. South of this zone, the eastern graben is deflected some 20 km eastward to form the Culebra reentrant (Fig. 1).

The Alamosa basin is divided into the two second-order

half grabens by the central basement high, the Alamosa horst, which is not symmetrical. This north-trending feature is marked by high gravity and conspicuous thinning of syn-rift sediments. The eastern side of the "horst" is marked by a combination of strong eastward tilt of the basement and normal faults. The Alamosa horst is essentially the uptilted western edge of the east-tilted basement-cored crustal block flooring the Baca graben. The western margin of the "horst" is a hinge zone marked by monoclinal folding and less pronounced normal faulting of the ash-flow-tuff package (Figs. 2 and 8).

The high-gravity gradient on the western side of the Alamosa horst (e.g., Keller et al., 1984) is not the result of major rift-related block faulting as suggested by Tweto (1979) because faults along its western edge (Fig. 10) show relatively minor displacement of syn-rift units. Instead, the gravity gradient is attributable to pre-rift structural relief that developed along the eastern margin of the Paleogene precursor basin of the Monte Vista graben.

The syn-rift seismic interval over the horst area is "thinnest" west of the Amerada F-1 State well, but "thickens" northward along the trend of the horst. The basement high is not prominent on seismic lines in the northern part of the basin in the vicinity of the Carr #1 Kennedy Williams well (Stoughton, 1977). The positive gravity anomaly on Bouguer and residual gravity anomaly maps of Keller et al. (1984), thought to be a northern extension of the Alamosa horst, probably represents dense basement rock in this area. The only manifestation of the Alamosa horst in this location as seen in Figure 10 is a flattening of the east-sloping surface on top of the ash-flow-tuff package. Along the northern border of the study area, the Monte Vista graben and Alamosa horst essentially merge into an east-sloping bench, whereas the Baca graben continues to be a deep, narrow trough extending into the northernmost part of the San Luis Basin.

Pre-rift basin

Figure 11, the contoured two-way time interval between the top of the upper Oligocene tuffs and the top of the Precambrian, supports the observation that the area of the present Monte Vista graben was a depositional basin during the Paleogene. This basin was bounded on the east by a north-trending fault zone, paralleling a basement shoulder. This fault zone does not coincide exactly with the modern western margin of the Alamosa horst, but as mentioned above, is probably in part responsible for the steep gravity gradient between the present Monte Vista graben and the Alamosa horst.

The faults bounding the basin on the west, at about the longitude of Del Norte, are north-northwest trending. These faults created the structural relief that distinguishes the Del Norte high. Both the Blanco Basin and Conejos Formations appear to thicken eastward across this fault zone, indicating that it was active during their deposition. This fault zone was also active during rifting, but with an opposite sense of displacement. The Paleogene basin appears to end southward at about the latitude of Alamosa. Its northern extent is probably in the vicinity of the town of Saguache, Colorado.

DISCUSSION: TECTONIC DEVELOPMENT

Pre-Tertiary tectonic setting

No post-Precambrian, pre-Eocene, strata exist in the boreholes studied. During the early and middle Paleozoic, the San Luis Basin region was part of a broad highland that lay to the south of the east-west–trending central Colorado sag of Eardley (1951, Plates 3, 4). Unfortunately, little evidence of the uplift or of deposition along its flanks has survived later tectonic events (Ross and Tweto, 1980). Parts of this highland were reactivated in Pennsylvanian to Permian time as the Uncompahgre–San Luis highland (Tweto, 1980), apparently bounded on the east by reactivated Precambrian structures (Sutherland, 1972; De Voto, 1980). This uplift was responsible for the coarse synorogenic deposits of the Sangre de Cristo Formation exposed in the Sangre de Cristo Range east of the Alamosa basin (De Voto, 1980). Denudation of the uplift continued until the Jurassic, when Middle to Late Jurassic nonmarine deposits lapped onto the old highland, eventually inundating it (Berman et al., 1980). Cretaceous nonmarine and marine strata were deposited over the area (Haun and Weimer, 1960) until onset of Laramide tectonism in latest Cretaceous time.

Laramide history

The Laramide orogenic event, extending from late Campanian into Eocene time, was a period of uplift and erosion in a broad region including the San Luis Basin–Sangre de Cristo Range area and the Brazos uplift (Tweto, 1975), collectively referred to here as the San Luis–Brazos uplift. Two pulses of uplift are recorded in the synorogenic sedimentary deposits of the San Juan sag and San Juan Basin. The first pulse, marked by the Animas Formation (Upper Cretaceous–Paleocene), initiated uplift of the region and stripping of post-Precambrian strata from the San Luis–Brazos uplift. The Animas Formation contains volcanic detritus recording nearby Laramide volcanic activity, but it becomes increasingly arkosic upwards in its section, indicating that the uplift was being unroofed to expose its Precambrian core (Brister and Chapin, 1994). The known fault style of the eastern part of the uplift was that of west-dipping reverse faults that flatten at depth (Lindsey et al., 1983) and low-angle thrust faults (Schavran, 1984). The Del Norte high is probably a remnant of the western flank of the uplift.

An angular unconformity between the Animas and the Blanco Basin Formations indicates that a second pulse of tectonic activity began in late Paleocene, extending into the Eocene (Cather and Chapin, 1990). North-northeast translation of the Colorado Plateau during late Laramide time resulted in wrenching in a north-south zone along the axis of the Southern

Rocky Mountains in New Mexico and Colorado (Chapin and Cather, 1981; Chapin, 1983) and north-south compression within uplifts in Wyoming (Gries, 1983, 1990). This episode involved reactivation of the western margin of the San Luis–Brazos uplift due to development of a wrench-fault system, creation of a basin over the western half of the former uplift, and rejuvenation of sedimentation westward into the San Juan sag–San Juan Basin areas (Brister and Chapin, 1994).

The Blanco Basin Formation depocenter, which was probably wrench-fault related (Brister and Chapin, 1994), formed within the western part of San Luis–Brazos uplift. This basin is classified as an Echo Park–type basin (terminology of Chapin and Cather, 1981) or as an axial basin (Dickinson et al., 1988). The Eocene basin was approximately 20 to 25 km wide and at least 60 km long in a north-northwest trend with its thickest preserved deposits existing along the fault system on the west side. It may have had an open drainage connection to the San Juan sag to the west.

The effect of 30+ m.y. of Laramide uplift and erosion of the San Luis–Brazos uplift was to subdue the mountain chain and fill the adjacent basins, resulting in the development of a wide-ranging, low-relief, geomorphic surface in the region (Steven, 1975), referred to as the late Eocene erosion surface of Epis and Chapin (1975). This surface is now buried deeply beneath Oligocene volcanic strata; it was developed on the Blanco Basin Formation in the western half of the basin and the eroded Precambrian basement in the eastern half. An unconformity marks the late Eocene erosion surface between the top of the Blanco Basin Formation and volcaniclastic rocks of the basal Conejos Formation. There is only minor angularity visible between the two formations on seismic lines. Most drill cuttings from the lowermost Conejos Formation contain minor amounts of basement detritus, either removed from remnant high areas or reworked from Blanco Basin sediments. Oligocene stream channels eroded into the Eocene surface in Colorado have been described by Epis et al. (1980).

Oligocene volcanism

The Oligocene was a time of widespread andesitic volcanism in a north-trending band along the axis of the Southern Rocky Mountains. This period of volcanism has been attributed to rising magmas in a continental-arc tectonic setting during a change from a flat-dipping to a steep-dipping subduction zone along the western North American margin (Lipman, 1983a). During this event, several kilometers of volcanic rocks were deposited in the San Juan volcanic field. As illustrated in Figures 2 and 8, the faults bounding the Paleogene basin in the western half of the Alamosa basin were reactivated during Conejos deposition.

By 29 Ma, a period of extension had begun in southern Colorado, marked by north-northeast–trending dikes in the San Luis Hills mapped by Bartlett (1984) and dated by Burroughs (1972). Volcanism accompanying this early period of extension was primarily associated with caldera formation in the San Juan volcanic field and emplacement of regional ash-flow sheets, such as those deposited across the Alamosa basin. This early extension preceded development of rift-related half grabens.

Neogene rifting

Indicators of the beginning of the Rio Grande rift event have been discussed in some detail by Chapin (1971, 1979, 1988). Initiation of rifting in the Alamosa basin is fairly well constrained by the stratigraphy of its western margin. Structures directly associated with rifting were not active until after emplacement of the uppermost Oligocene ash-flow tuff (Carpenter Ridge Tuff, 27.35 Ma, Lipman, 1989). Several lines of evidence support this conclusion. First, there is no evidence that the pre-27-Ma ash-flow sheets ponded in areas soon to become grabens. Secondly, sediments deposited between the ash-flow sheets thin eastward, indicating that the Baca graben had not yet begun to form. Where these sediments are thickest is in an area of structural sagging along the western edge of the present Monte Vista graben. There is no evidence that sagging in that vicinity continued after initiation of rifting. Also, where faults cut the Oligocene tuffs, they cut across the entire package. As seen on reflection seismic lines, the lower tuffs in the package are not more highly faulted than upper tuffs. This indicates that the entire tuff package was emplaced prior to significant local rift faulting.

The rift event began at about 27 Ma. This conclusion is supported by several observations. At about this time, the composition of regional volcanism changed from dacitic to basaltic. The Hinsdale basalts (as old as 26.8 Ma) lie atop the ash-flow tuffs of the San Juan volcanic field in angular unconformity (Lipman and Mehnert, 1975; Lipman, 1975, 1976). Following emplacement of the basalts, the eastern edge of the San Juan volcanic field was uplifted, tilted eastward, and deeply eroded. Other dated evidence for initiation of rifting is the 26.5-Ma Amalia Tuff (Lipman et al., 1986; Lipman, 1988) interbedded with the Los Pinos Formation in the Tusas Mountains along the west side of the southern San Luis Basin (Lipman, 1983b; Manley, 1981).

During early rifting the eastern half of the Alamosa basin was progressively tilted eastward along the Sangre de Cristo Range. The Sangre de Cristo fault zone was probably controlled by the thrust root zones of the Laramide San Luis uplift (Sales, 1983). Seismic lines and cross sections illustrate episodic faulting on the basin-bounding faults during the rift event. This is demonstrated in angular relationships within the lower Santa Fe Group and the cyclic pattern of alternating coarse and fine deposition. Streams flowing from the San Juan volcanic field carrying a coarse volcanic load emptied into the Alamosa basin, developing alluvial fans at their mouths, perhaps much like the modern alluvial fan constructed where the Rio Grande enters the San Luis Basin at the town of Del

Norte. A broad piedmont extended as a veneer eastwards across the western half of the basin towards the eastern (Baca) narrow graben.

The Sangre de Cristo Range also supplied material to the basin. Shorter, steeper alluvial fans were constructed along the eastern margin of the Baca graben. Rift half grabens such as the Baca graben tend to develop axial drainage systems that closely parallel the faulted margin of the graben due to increased subsidence along the margin (Leeder and Gawthorpe, 1987; Mack and Seager, 1990). The sediment load from alluvial fans is redistributed within such axial stream systems. This probably accounts for the mix of Precambrian and volcanic detritus in the lower Santa Fe Group of the Baca graben. The fault-slice of lower Santa Fe Group cropping out on the western edge of Blanca Peak (Urraca outcrop, Fig. 3) contains current indicators suggesting stream deposition parallel to the basin margin. Recent fission-track dating of apatite in Precambrian rocks making up Blanca Peak indicate rapid uplift of the Blanca massif between 28 and 18 Ma (Kelley, 1990). This probably coincided with rapid subsidence and eastward tilting of the Baca half graben.

The angular unconformity visible in Figure 9 is evidence for a change in fault activity and degree of tilting of the Baca graben. Sediments above this unconformity are only gently tilted, and not as intensely faulted as the lowermost package of Santa Fe rocks. The basin-spanning coarse units above the lower Santa Fe Group (Fig. 6) represents response of drainage systems to this period of truncation of previously deposited sediments. By about 4.5 Ma, the through drainage from the northern to southern parts of the San Luis Basin was impeded by the intervening Servilleta flood basalts of the Taos Plateau. Internal drainage created conditions for fluviolacustrine sedimentation common to the Alamosa Formation (Rogers et al., 1992).

Elsewhere in the rift at this time, basins were being integrated into the ancestral Rio Grande. By middle Pleistocene time, integration of the upper Rio Grande drainage system with the lower Rio Grande system (Gulf of Mexico terminus) in the Trans-Pecos Texas region below El Paso, caused lowering of base levels, dissection of basins, and stranding of geomorphic surfaces (Gile et al., 1981). The Rio Grande integrated basins as far north as Red River of New Mexico (Wells et al., 1987). The Servilleta basalts north of the Red River and the San Luis Hills horst served as a hydrologic divide until middle to late Pleistocene when the Rio Grande was able to erode a gorge through these barriers. This event occurred after 0.69 Ma, the youngest documented age of the Alamosa Formation (Rogers et al., 1985), but before 0.3 Ma (Wells et al., 1987).

ACKNOWLEDGMENTS

This paper resulted from a dissertation by the senior author at the New Mexico Institute of Mining and Technology, which was directed by Charles Chapin and funded by the New Mexico Bureau of Mines and Mineral Resources. The authors wish to thank Charles Chapin and Chuck Kluth for their critical reviews. We also thank Steve Cather, Matthew Golombek, John Hawley, Raymond Ingersoll, and Peter Lipman for suggestions that improved the manuscript. Amarex Inc. and Reserve Oil provided seismic data and Randy Ray contributed the synthetic seismogram.

REFERENCES CITED

Baltz, E. H., 1965, Stratigraphy and history of Raton Basin and notes on San Luis Basin, Colorado–New Mexico: Bulletin of the American Association of Petroleum Geologists, v. 49, p. 2041–2075.

Bartlett, R. D., 1984, Geology of an Oligocene-age acid hot-spring, San Luis Hills, Conejos and Costilla Counties, Colorado [M.S. thesis]: Fort Collins, Colorado State University, 111 p.

Berman, A. E., Poleschook, D., and Dimelow, T. E., 1980, Jurassic and Cretaceous systems of Colorado, *in* Kent, H. C., and Porter, K. W., eds., Colorado geology: Denver, Colorado, Rocky Mountain Association of Geologists, p. 111–128.

Brister, B. S., 1990, Tertiary sedimentation and tectonics: San Juan sag–San Luis Basin region, Colorado and New Mexico [Ph.D. thesis]: Socorro, New Mexico Institute of Mining and Technology, 267 p.

Brister, B. S., 1992, The Blanco Basin Formation (Eocene), San Juan Mountains region, Colorado and New Mexico, *in* Lucas, S. J., and 3 others, eds., San Juan Basin IV: New Mexico Geological Society 43rd Annual Field Conference, Guidebook, p. 321–331.

Brister, B. S., and Chapin, C. E., 1994, Sedimentation and tectonics of the Laramide San Juan sag, southwestern Colorado: The Mountain Geologist,v. 31, p. 2–18.

Burroughs, R. L., 1971, Geology of the San Luis hills, south-central Colorado, *in* James, H. L., ed., Guidebook of the San Luis Basin, Colorado: New Mexico Geological Society 22nd Field Conference, Guidebook, p. 277–287.

Burroughs, R. L., 1972, Geology of the San Luis hills, south-central Colorado [Ph.D. thesis]: Albuquerque, University of New Mexico, 140 p.

Burroughs, R. L., 1981, A summary of the geology of the San Luis Basin, Colorado and New Mexico, with emphasis on the geothermal potential for the Monte Vista graben: Colorado Geological Survey Special Publication 17, 30 p.

Butler, A. P., Jr., 1946, Tertiary and Quaternary geology of the Tusas–Tres Piedras area, New Mexico [Ph.D. thesis]: Cambridge, Massachusetts, Harvard University, 188 p.

Butler, A. P., Jr., 1971, Tertiary volcanic stratigraphy of th eastern Tusas Mountains, southwest of the San Luis Valley, Colorado–New Mexico, *in* James, H. L., ed., Guidebook of the San Luis Basin: New Mexico Geological Society, 22nd Annual Field Conference, Guidebook, p. 289–300.

Cather, S. M., and Chapin, C. E., 1990, Paleogeographic and paleotectonic setting of Laramide sedimentary basins in the central Rocky Mountain region: Alternative interpretation: Geological Society of America Bulletin,v. 102, p. 256–258.

Chapin, C. E., 1971, The Rio Grande rift; Part I, Modifications and additions, *in* James, H. L., ed., Guidebook of the San Luis Basin, Colorado: New Mexico Geological Society, 22nd Annual Field Conference, Guidebook, p. 191–201.

Chapin, C. E., 1979, Evolution of the Rio Grande rift, A summary, *in* Riecker, R. E., ed., Rio Grande rift: Tectonics and magmatism: Eos Transactions of the American Geophysical Union, p. 1–5.

Chapin, C. E., 1983, An overview of Laramide wrench faulting in the Southern Rocky Mountains with emphasis on petroleum exploration, *in* Lowell, J. D., ed., Rocky Mountain foreland basins and uplifts: Rocky Mountain Association of Geologists, p. 169–179.

Chapin, C. E., 1988, Axial basins of the northern and central Rio Grande rifts, *in* Sloss, L. L., ed., Sedimentary cover, North American craton: U.S.: Geological Society of America, p. 165–170.

Chapin, C. E., and Cather, S. M., 1981, Eocene tectonics and sedimentation in the Colorado Plateau–Rocky Mountain region, *in* Dickinson, W. R., and Payne, W. D., eds., Relations of tectonics to ore deposits in the Southern Cordillera: Arizona Geological Society Digest, v. 14, p. 173–189.

Chapin, C. E., and Lindley, J. I., 1986, Potassium metasomatism of igneous and sedimentary rocks in detachment terranes and other sedimentary basins: Economic implications: Arizona Geological Society Digest, v. XVI, p. 118–126.

Cordell, L., 1978, Regional geophysical setting of the Rio Grande rift: Geological Society of America Bulletin, v. 89, p. 1073–1090.

Covarrubias, O., 1988, An integrated geological and geophysical study of the structural boundary between the San Luis Basin and the San Juan sag, south-central Colorado [M.S. thesis]: Golden, Colorado School of Mines, 72 p.

Cross, W., and Larsen, E. S., 1935, A brief review of the geology of the San Juan region of southwestern Colorado: U.S. Geological Survey Bulletin 843, 138 p.

Davis, G. H., and Keller, G. R., 1978, Subsurface structure of San Luis Valley, *in* Hawley, J. W., ed., Guidebook to Rio Grande rift in New Mexico and Colorado: New Mexico Bureau of Mines and Mineral Resources Circular 163, p. 28.

De Voto, R. H., 1980, Pennsylvanian stratigraphy and history of Colorado, *in* Kent, H. C., and Porter, K. W., eds., Colorado geology: Rocky Mountain Association of Geologists, p. 71–101.

Dickinson, W. R., and 6 others, 1988, Paleogeographic and paleotectonic setting of Laramide sedimentary basins in the central Rocky Mountain region: Geological Society of America Bulletin, v. 100, p. 1023–1039.

Eardley, H. J., 1951, Structural geology of North America: New York, Harper and Row, 743 p.

Epis, R. C., and Chapin, C. E., 1974, Stratigraphic nomenclature of the Thirtynine Mile volcanic field, central Colorado: U.S. Geological Survey Bulletin 1395-C, 23 p.

Epis, R. C., and Chapin, C. E., 1975, Geomorphic and tectonic implications of the post-Laramide, late Eocene erosion surface in the Southern Rocky Mountains, *in* Curtis, B. F., ed., Cenozoic history of the Southern Rocky Mountains: Geological Society of America Memoir 144, p. 45–74.

Epis, R. C., Scott, G. R., Taylor, R. B., and Chapin, C. E., 1980, Summary of Cenozoic geomorphic, volcanic and tectonic features of central Colorado and adjoining areas, *in* Kent, H. C., and Porter, K. W., eds., Colorado geology: Rocky Mountain Association of Geologists, p. 135–156.

Folk, R. L., 1974, Petrology of sedimentary rocks: Austin, Texas, Hemphill Publishing Co., 184 p.

Gile, L. H., Hawley, J. W., and Grossman, R. B., 1981, Soils and geomorphology in the Basin and Range area of Southern New Mexico—Guidebook to the Desert Project: New Mexico Bureau of Mines and Mineral Resources Memoir 39, 222 p.

Gries, R., 1983, North-south compression of Rocky Mountain foreland structures, *in* Lowell, J. D., ed., Rocky Mountain foreland basins and uplifts: Denver, Colorado, Rocky Mountain Association of Geologists, p. 9–32.

Gries, R. R., 1985a, Seismic lines in the San Luis Valley, south-central Colorado, *in* Gries, R. R., and Dyer, R. C., eds., Seismic exploration of the Rocky Mountain region: Rocky Mountain Association of Geologists and Denver Geophysical Society, p. 267–274.

Gries, R. R., 1985b, San Juan sag: Cretaceous rocks in a volcanic-covered basin, south-central Colorado: The Mountain Geologists, v. 22, p. 167–179.

Gries, R. R., 1989, San Juan sag: Oil and gas exploration in a newly discovered basin beneath the San Juan volcanic field, *in* Lorenz, J. C., and Lucas, S. G., eds., Energy frontiers in the Rockies: Albuquerque Geological Society, p. 69–78.

Gries, R. R., 1990, Rocky Mountain foreland structures: Changes in compression direction through time, *in* Letouzey, J., ed., Petroleum and tectonics in mobile belts: Paris, Editions Technip, p. 129–148.

Gries, R. R., and Brister, B. S., 1989, New interpretation of seismic lines in the San Luis Valley, south-central Colorado, *in* Harmon, E. J., ed., Water in the valley: Colorado Ground-Water Association, 8th Annual Field Trip, Guidebook, p. 241–254.

Gries, R. R., and Vandersluis, G. D., 1989, Laramide and Cenozoic geology: Rocky Mountain Association of Geologists Field Trip Road Log, 90 p.

Hanna, T. M., and Harmon, E. J., 1989, An overview of the historical, stratigraphic, and structural setting of the aquifer system of the San Luis Valley, *in* Harmon, E. J., ed., Water in the valley: Colorado Groundwater Association, 8th Annual Field Trip, Guidebook, p. 1–34.

Haun, J. D., and Weimer, R. J., 1960, Cretaceous stratigraphy of Colorado, *in* Weimer, R. J., and Haun, J. D., eds., Guide to geology of Colorado: Geological Society of America, Rocky Mountain Association of Geologists, Colorado Scientific Society, p. 58–65.

Hawley, J. W., 1978, Guidebook to Rio Grande rift in New Mexico and Colorado: New Mexico Bureau of Mines and Mineral Resources Circular 163, 241 p.

Huntley, D., 1976, Ground water recharge to the aquifers of northern San Luis Valley, Colorado: A remote sensing investigation [Ph.D. thesis]: Golden, Colorado School of Mines, 240 p.

Huntley, D., 1979, Cenozoic faulting and sedimentation in northern San Luis Valley, Colorado: Summary: Geological Society of America Bulletin, v. 90, p. 8–12.

Ingersoll, R. V., Cavazza, W., Baldridge, W. S., Shafiqullah, M., 1990, Cenozoic sedimentation and peleotectonics of north-central New Mexico: Implications for initiation and evolution of the Rio Grande rift: Geological Society of America Bulletin, v. 102, p. 1280–1296.

Johnson, R. B., 1969, Geologic map of the Trinidad Quadrangle south-central Colorado: U.S. Geological Survey Miscellaneous Investigations Map I-558, scale 1:250,000.

Keller, G. R., Cordell, L., Davis, G. H., Peeples, W. J., and White, G., 1984, A geophysical study of the San Luis Basin, *in* Baldridge, W. S., and 3 others, eds., Rio Grande rift: Northern New Mexico: New Mexico Geological Society, 35th Annual Field Conference, Guidebook, p. 51–57.

Kelley, S. A., 1990, Late Mesozoic to Cenozoic cooling history of the Sangre de Cristo Mountains: New Mexico Geological Society, 41st Annual Field Conference, Guidebook, p. 123–132.

Kelley, V. C., 1956, The Rio Grande depression from Taos to Santa Fe, *in* Rosenzweig, A., ed., Southeastern Sangre de Cristo Mountains: New Mexico Geological Society, 7th Annual Field Conference, Guidebook, p. 109–114.

Kluth, C. F., and Schaftenaar, C. H., 1989, Seismic record section and interpretation of Line 2, *in* Callender, J. F., Hawley, J. W., and Keller, G. R., eds., Rio Grande rift: 28th International Geological Congress Field Trip, Guidebook T318, p. 25.

Knepper, D. H., 1976, Late Cenozoic structure of the Rio Grande rift zone, central Colorado, *in* Epis, R. C., and Weimer, R. J., eds., Studies in Colorado field geology: Golden, Colorado School of Mines Professional contributions, no. 8, p. 421–430.

Knepper, D. H., and Marrs, R. W., 1971, Geological development of the Bonanza–San Luis Valley–Sangre de Cristo Range area, south-central Colorado, *in* James, H. L., ed., Guidebook of the San Luis Basin, Colorado: New Mexico Geological Society, 22nd Annual Field Trip, Guidebook, p. 249–264.

Larsen, E. S., and Cross, W., 1956, Geology and petrology of the San Juan region southwestern Colorado: U.S. Geological Survey Professional Paper 258, 303 p.

Leeder, M. R., and Gawthorpe, R. L., 1987, Sedimentary models for extensional tilt-block/half graben basins: Geological Society of London Special Publication 28, p. 139–152.

Lindsey, D. A., Johnson, B. R., and Andriessen, P.A.M., 1983, Laramide and Neogene structure of the northern Sangre de Cristo Range, south-central Colorado, *in* Lowell, J. D., ed., Rocky Mountain foreland basins and uplifts: Denver, Colorado, Rocky Mountain Association of Geologists, p. 219–228.

Lipman, P. W., 1968, Geology of Summer Coon volcanic center, eastern San Juan Mountains, Colorado, *in* Epis, R. C., ed., Cenozoic volcanism in the Southern Rocky Mountains: Colorado School of Mines Quarterly, v. 63, p. 211–236.

Lipman, P. W., 1969, Alkalic and tholeiitic basaltic volcanism related to the Rio Grande depression, southern Colorado and northern New Mexico: Geological Society of America, v. 80, p. 1343–1354.

Lipman, P. W., 1975, Evolution of the Platoro caldera complex and related volcanic rocks, southeastern San Juan Mountains, Colorado: U.S. Geological Survey Professional Paper 852, 128 p.

Lipman, P. W., 1976, Geologic map of the Del Norte area, eastern San Juan Mountains, Colorado: U.S. Geological Survey Miscellaneous Investigations Series Map I-952, scale 1:48,000.

Lipman, P. W., 1983a, Tectonic setting of the mid to late Tertiary in the Rocky Mountain region—A review, *in* The genesis of Rocky Mountain ore deposits: Changes with time and tectonics: Denver Region Exploration Geologists Society, p. 125–131.

Lipman, P. W., 1983b, The Miocene Questa caldera, northern New Mexico: Relation to batholith emplacement and associated molybdenum mineralization, *in* The genesis of Rocky Mountain ore deposits: Changes with time and tectonics: Denver Region Exploration Geologists Society, p. 133–149.

Lipman, P. W., 1988, Evolution of silicic magma in the upper crust: The mid-Tertiary Latir volcanic field and its cogenetic granitic batholith, northern New Mexico, U.S.A.: Transactions of the Royal Society of Edinburgh: Earth Sciences, v. 79, p. 265–288.

Lipman, P. W., ed., 1989, Oligocene-Miocene San Juan volcanic field, Colorado, *in* Chapin, C. E., and Zidek, J., eds., Field excursions to volcanic terranes in the western United States, Volume I: Southern Rocky Mountains: New Mexico Bureau of Mines and Mineral Resources Memoir 46, p. 303–380.

Lipman, P. W., and Mehnert, H. H., 1975, Late Cenozoic basaltic volcanism and development of the Rio Grande depression in the Southern Rocky Mountains, *in* Curtis, B. F., ed., Cenozoic history of the Southern Rocky Mountains: Geological Society of America Memoir 144, p. 119–154.

Lipman, P. W., and Mehnert, H. H., 1979, The Taos Plateau volcanic field, northern Rio Grande rift, New Mexico, in Riecker, R. E., ed., Rio Grande rift, Tectonics and magmatism: EOS Transactions of the American Geophysical Union, p. 289–310.

Lipman, P. W., and Steven, T. A., 1970, Reconnaissance geology and economic significance of the Platoro caldera, southeastern San Juan Mountains, Colorado, *in* Geological Survey research 1970: U.S. Geological Survey Professional Paper 700-C, p. 19–C29.

Lipman, P. W., Steven, T. A., and Mehnert, H. H., 1970, Volcanic history of the San Juan Mountains, Colorado, as indicated by potassium-argon dating: Geological Society of America Bulletin, v. 81, p. 2329–2352.

Lipman, P. W., Mehnert, H. H., and Naeser, C. W., 1986, Evolution of the Latir volcanic field, northern New Mexico and its relation to the Rio Grande rift, as indicated by potassium-argon and fission track dating: Journal of Geophysical Research, v. 91, p. 6329–6345.

Mack, G. H., and Seager, W. R., 1990, Tectonic control on facies distribution of the Camp Rice and Palomas Formations (Pliocene-Pleistocene in the southern Rio Grande rift): Geological Society of America Bulletin, v. 102, p. 45–53.

Manley, K., 1981, Redefinition and description of the Los Pinos Formation of north-central New Mexico: Geological Society of America Bulletin, Part 1, v. 92, p. 984–989.

Olson, J. C., Hedlund, D. C., and Hansen, W. R., 1968, Tertiary volcanic stratigraphy in the Powderhorn–Black Canyon region, Gunnison and Montrose Counties, Colorado: U.S. Geological Survey Bulletin 1251-C, 29 p.

Personius, S. F., and Machette, M. N., 1984, Quaternary and Pliocene faulting in the Taos Plateau region, northern New Mexico, *in* Baldridge, W. S., and 3 others, eds., Rio Grande rift: Northern New Mexico: New Mexico Geological Society, 35th Annual Field Conference, Guidebook, p. 83–90.

Phillips, D. H., 1985, A geophysical investigation of the southeastern San Juan Mountains, Colorado [M.S. thesis]: Golden, Colorado School of Mines, 69 p.

Powell, W. J., 1958, Ground-water resources of the San Luis Valley, Colorado: U.S. Geological Survey Water-Supply Paper 1379, 284 p.

Rider, M. H., 1986, The geological interpretation of well logs: New York, Halstead Press, 175 p.

Rogers, K. L., 1984, A paleontological analysis of the Alamosa Formation (south-central Colorado: Pleistocene: Irvingtonian), *in* Baldridge, W. S., Dickerson, P. W., Riecker, R. E., and Zidek, J., eds., Rio Grande rift: Northern New Mexico: New Mexico Geological Society, 35th Annual Field Conference, Guidebook, p. 151–155.

Rogers, K. L., and 7 others, 1985, Middle Pleistocene (late Irvingtonian-Nebraskan) climatic changes in south-central Colorado: National Geographic Review, v. 1, p. 535–563.

Rogers, K. L., and 11 others, 1992, Pliocene and Pleistocene geologic and climatic evolution in the San Luis Valley of south-central Colorado: Palaeogeography, Palaeoclimatology, Palaeoecology, v. 94, p. 55–96.

Ross, R. J., and Tweto, O., 1980, Lower Paleozoic sediments and tectonics in Colorado, *in* Kent, H. C., and Porter, K. W., eds., Colorado geology: Denver, Colorado, Rocky Mountain Association of Geologists, p. 47–56.

Sales, J. K., 1983, Collapse of Rocky Mountain basement uplifts, *in* Lowell, J. D., ed., Rocky Mountain foreland basins and uplifts: Denver, Colorado, Rocky Mountain Association of Geologists, p. 79–97.

Schavran, G., 1984, Structural features of a Laramide fold and thrust belt, east flank of the Sangre de Cristo Range, Colorado [M.S. thesis]: Golden, Colorado School of Mines, 67 p.

Scott, G. R., 1975, Cenozoic surfaces and deposits in the Southern Rocky Mountains, *in* Curtis, B. F., ed., Cenozoic history of the Southern Rocky Mountains: Geological Society of America Memoir 144, p. 227–248.

Siebenthal, C. E., 1910a, The San Luis Valley, Colorado: Science, v. 31, p. 744–746.

Siebenthal, C. E., 1910b, Geology and water resources of the San Luis Valley, Colorado: U.S. Geological Survey Water-Supply Paper 240, 128 p.

Spiegel, Z., and Baldwin, B., 1963, Geology and water resources of the Santa Fe area, New Mexico: U.S. Geological Survey Water-Supply Paper 1525, 258 p.

Steven, T. A., 1975, Middle Tertiary volcanic field in the Southern Rocky Mountains, *in* Curtis, B. F., ed., Cenozoic history of the Southern Rocky Mountains: Geological Society of America Memoir 144, p. 75–94.

Steven, T. A., and Lipman, P. W., 1976, Calderas of the San Juan volcanic field, southwestern Colorado: U.S. Geological Survey Professional Paper 958, 35 p.

Steven, T. A., Mehnert, H. H., and Obradovich, J. D., 1967, Age of volcanic activity in the San Juan Mountains, Colorado: U.S. Geological Survey Professional Paper 575-D, p. D47–D55.

Steven, T. A., Lipman, P. W., Hail, W. J., Barker, F., and Luedke, R. G., 1974, Geologic map of the Durango Quadrangle, southwestern Colorado: U.S. Geological Survey Miscellaneous Geologic Investigations Map I-764, scale 1:250,000.

Stoughton, D., 1977, Interpretation of seismic reflection data from the San Luis valley, south-central Colorado [M.S. thesis]: Golden, Colorado School of Mines, 100 p.

Sutherland, P. K., 1972, Pennsylvanian stratigraphy, southern Sangre de Cristo Mountains, New Mexico, *in* Mallory, W. W., and 25 others, eds., Geologic atlas of Rocky Mountain region: Denver, Colorado, Rocky Mountain Association of Geologists, p. 139–142.

Taylor, R. B., 175, Neogene tectonism in south-central Colorado, *in* Curtis, B. F., ed., Cenozoic history of the Southern Rocky Mountains: Geological Society of America Memoir 144, p. 211–226.

Thompson, R. A., and Machette, M. N., 1989, Geologic map of the San Luis Hills area, Conejos and Costilla Counties, Colorado: U.S. Geological Survey Miscellaneous Investigations Series Map I-1906, scale 1:50,000.

Tweto, O., 1961, Late Cenozoic events of the Leadville district and Upper Arkansas Valley, Colorado: U.S. Geological Survey Professional Paper 424-B, p. B133–135.

Tweto, O., 1975, Laramide (Late Cretaceous–early Tertiary) orogeny in the Southern Rocky Mountains, *in* Curtis, B. F., ed., Cenozoic history of the Southern Rocky Mountains: Geological Society of America Memoir 144, p. 1–44.

Tweto, O., 1979, The Rio Grande rift system in Colorado, *in* Riecker, R. E., ed., Rio Grande rift: Tectonics and magmatism: EOS Transactions of the American Geophysical Union, p. 33–56.

Tweto, O., 1980, Tectonic history of Colorado, *in* Kent, E. C., and Porter, K. W., eds., Colorado geology: Denver, Colorado, Rocky Mountain Association of Geologists, p. 5–9.

Upson, J. E., 1939, Physiographic subdivisions of the San Luis Valley, southern Colorado: Journal of Geology, v. 47, p. 721–736.

Upson, J. E., 1941, The Vallejo Formation: New early Tertiary red-beds in southern Colorado: American Journal of Science, v. 239, p. 577–589.

Valencia, R. S., 1983, An integrated geophysical study of the west-central part of San Luis Valley, Colorado [M.S. thesis]: Golden, Colorado School of Mines, 92 p.

Van Alstine, R. E., 1968, Tertiary trough between the Arkansas and San Luis Valleys, Colorado: U.S. Geological Survey Professional Paper 600-C, p. C158–C160.

Van Alstine, R. E., 1970, Allochthonous Paleozoic blocks in the Tertiary San Luis–Upper Arkansas graben, Colorado: U.S. Geological Survey Professional Paper 700-B, p. B43–51.

Wells, S. G., Kelson, K. I., and Menges, C. M., 1987, Quaternary evolution of fluvial systems in the northern Rio Grande Rift, New Mexico and Colorado: Implications for entrenchment and integration of drainage systems, *in* Menges, C. M., ed., Quaternary tectonics, landform evolution, soil chronologies and glacial deposits—northern Rio Grande rift of New Mexico: Friends of the Pleistocene–Rocky Mountain Cell Field Trip, Guidebook, p. 55–69.

Manuscript Accepted by the Society October 13, 1993

Printed in U.S.A.

Geological Society of America
Special Paper 291
1994

Variation in paleomagnetic rotations and kinematics of the north-central Rio Grande rift, New Mexico

Stephen L. Salyards,* and James F. Ni
Geophysics Program, Department of Physics, New Mexico State University, Las Cruces, New Mexico 88003
M. James Aldrich, Jr.
Earth and Environmental Sciences Division, EES-1, Los Alamos National Laboratory, Los Alamos, New Mexico 87545

ABSTRACT

Previous paleomagnetic work in the north-central Rio Grande rift (RGR) of northern New Mexico has demonstrated the presence of counterclockwise paleomagnetic rotations, possibly due to left-lateral shearing across the rift. To improve the spatial coverage and resolution of the intrarift rotation, we analyzed the late Tertiary sediments within the exposed Española Basin for paleomagnetic rotations. From five sites broadly distributed across the rift, the amount of rotation varies from little on the eastern edge to over 20° counterclockwise in the central and western rift. If interpreted as tectonic net rotation axes, the amount of rotation varies from 10 to 24°. Gently northeast plunging net rotation axis orientations in the eastern part of the rift indicate dominantly extension-related deformation. Steeper axis inclinations near the center and western part of the rift and Embudo accommodation zone indicate the presence of increased horizontal shearing in these regions. The paleomagnetic data show that the north-central Rio Grande rift is not rotating as a unit but rather consists of a number of smaller independently, counterclockwise-rotating blocks.

INTRODUCTION

The Rio Grande rift (RGR) is an intracontinental rift that extends from central Colorado into northern Mexico and consists of series of en echelon, right-stepping basins. Kelley (1977, 1979, p. 57) concluded that the rift is undergoing oblique extension citing as evidence: (1) right en echeloning of the bordering uplifts, ramps, and relay faults; (2) left drag on several bounding faults; (3) left-oblique faults striking approximately N60°E (such as the Tijeras-Cañoncito and Embudo fault zones); and (4) left offset of two Jurassic wedges across the Albuquerque Basin. Based upon left-lateral offset along and compressional deformation across the northeast-trending Embudo fault zone (Fig. 1), Muehlberger (1979) inferred that a large diamond-shaped crustal block (the Española block) bounded by the Embudo fault zone, Pajarito (rift western boundary fault)–La Bajada–San Francisco–Sandia fault systems, Tijeras-Cañoncito fault zone, and Pecos-Picuris fault (rift eastern boundary fault) is rotating counterclockwise as a rigid mass.

Brown and Golombek (1985, 1986) studied the paleomagnetism of volcanic rocks in and adjacent to the exposed portion of the Española Basin within the rift and concluded that significant counterclockwise paleomagnetic rotation has occurred in the Tertiary volcanics, which lie on the western edge and southern end of the Española block. Their data show the greatest amount of counterclockwise rotation, about 19 ± 11°, in the Oligocene volcanics of the Ortiz Mountains at the south end of the Española block (Fig. 1). However, because the data do not allow rotations on the various subblocks in the rift to be determined, Muehlberger's kinematic model cannot be fully assessed.

In this study we present new paleomagnetic measurements across the north-central RGR (exposed Española Basin) and combine our data with the results of previous paleomag-

*Present address: Department of Earth and Space Sciences, University of California at Los Angeles, Los Angeles, California 90024.

Salyards, S. L., Ni, J. F., and Aldrich, M. J., Jr., 1994, Variation in paleomagnetic rotations and kinematics of the north-central Rio Grande rift, New Mexico, *in* Keller, G. R., and Cather, S. M., eds., Basins of the Rio Grande Rift: Structure, Stratigraphy, and Tectonic Setting: Boulder, Colorado, Geological Society of America Special Paper 291.

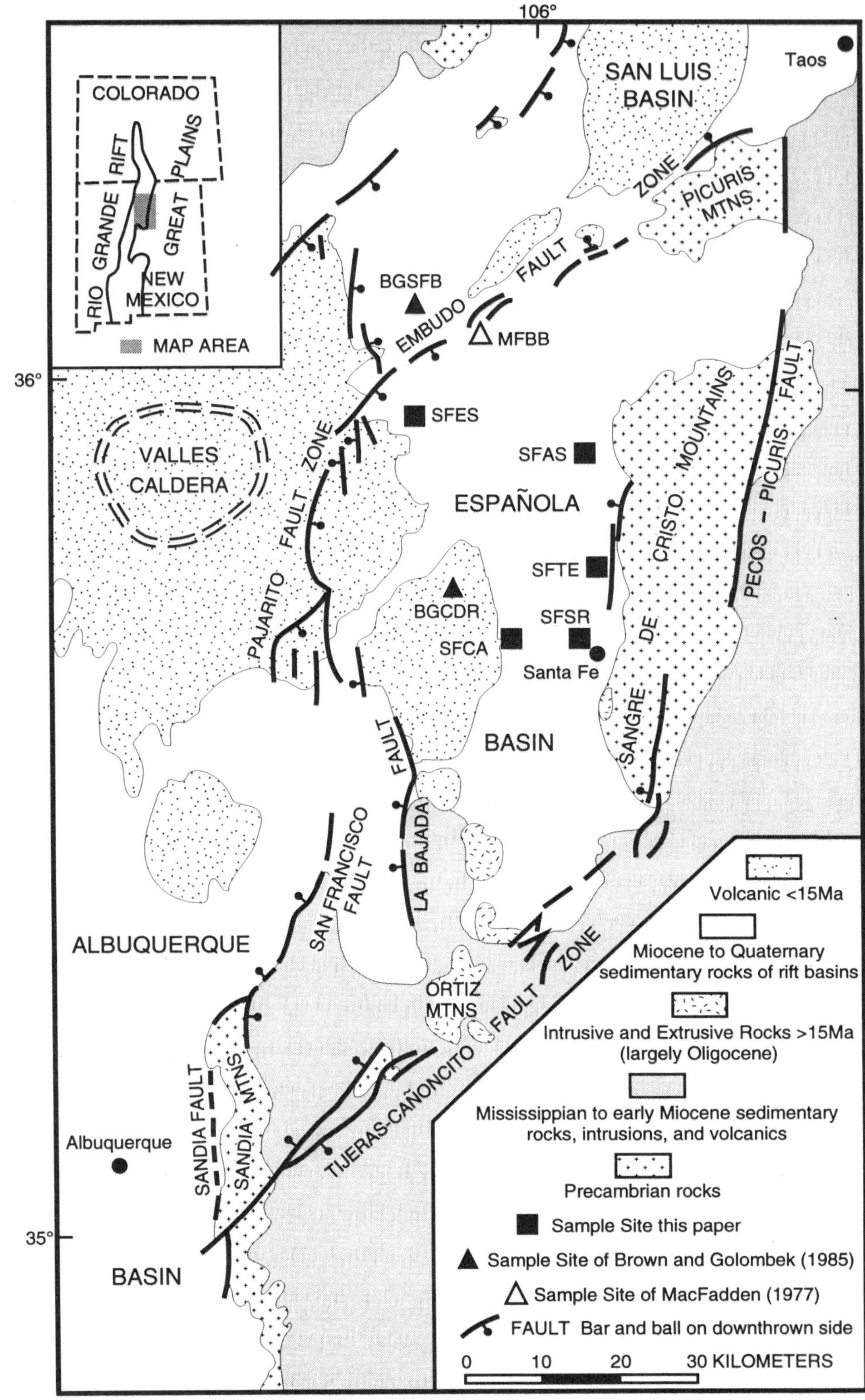

Figure 1. A generalized geologic map showing the basic geologic units and geologic structures in the Española Basin. Labeled squares show the locations of sampling discussed in this study. (Modified from Manley, 1979.)

netic investigations to show that the rigid block rotation hypothesis is not applicable to deformation in the north-central RGR. Instead, our paleomagnetic data show that rotations about vertical axes vary significantly within the Española block, from essentially zero on the east side to large rotations in the western part of the rift. Additionally, our data indicate that for three of the four sites in the center of the active rift, rotation has been about nonvertical axes (oblique to the surface of the earth).

GEOLOGIC SETTING

The Española Basin, one of the right-relay basins of the central Rio Grande rift (Kelley, 1979), is bordered on the east, principally in depositional contact (Manley, 1979), by the Sangre de Cristo Mountains (Fig. 1). The western margin is less clear because it is covered by volcanic rocks of the Jemez Mountains (Manley, 1979). The portion of the Española Basin east of the Pajarito fault zone lies within the currently active rift. The area of the Española Basin between the Pajarito fault zone and a series of north-south–trending normal faults near the center of the rift, has been named the "Velarde graben" by Manley (1979). However, recent geophysical investigations indicate that the only real graben structure trends parallel to the Embudo fault zone and does not extend south along the Pajarito fault zone (W. S. Baldridge, personal communication, 1990; Biehler et al., 1991, Figs. 1 and 11). Golombek et al. (1983), using gravity data and geologic data including offsets on the 1.1- to 1.5-Ma Bandelier Tuff (Spell et al., 1989), show that this area has been the most active region of the basin in the last 5 m.y., extending at a rate of 0.1 to 0.2 mm/yr.

The Santa Fe Group comprises the dominant basin fill and includes the Tesuque and Chamita Formations (Manley, 1979). The Tesuque Formation is further subdivided into several members of local extent (see Galusha and Blick, 1971). Radiometric dates indicate that it ranges in age between about 18 Ma (Ekas et al., 1984; May, 1984) and 10 Ma (Manley, 1979). Barghoorn (1981) presented three possible matches of magnetic reversal stratigraphy in the Tesuque Formation to the magnetostratigraphic time scale because of the lack of agreement between the radiometric dates and paleontological age. The overlying Chamita Formation ranges in age from about 12 Ma (Aldrich and Dethier, 1990) to 4.5 Ma (MacFadden, 1977).

For this study we collected paleomagnetic samples from five sites distributed across the RGR (Fig. 1). Of these five sites, one (SFTE) is located near the Sangre de Cristo Mountains that border the rift on the east. Two (SFSR, SFAS) are located a few kilometers farther into the rift. A fourth site (SFCA) is centrally located in the rift, while the fifth site (SFES) is near the active western boundary fault.

Site descriptions

The age of each member of the Tesuque Formation was determined from the type section magnetostratigraphic work of Barghoorn (1981). The sampling sites in the Tesuque Formation are in the Nambé (SFTE, SFSR) and Skull Ridge (SFAS) Members. For this stratigraphic interval at the base of the Tesuque Formation, Barghoorn's preferred match, his hypothesis B, and his second choice, hypothesis C (Barghoorn, 1981, Fig. 10), fall in the same age range, between 17 and 14.6 Ma.

Samples from sites SFES and SFCA are from the top of the Chamita Formation. Based on MacFadden's (1977) study, which indicates that deposition of the Chamita Formation ceased at about 5 Ma, these samples are probably early Pliocene.

Site SFTE. This site is a roadcut located along New Mexico Route 592 about 6.2 km east of the intersection of New Mexico Route 59. This roadcut is near the base of the Tesuque Formation in the Nambé Member of Galusha and Blick (1971). Its proximity to the mountain front is apparent in the proximal nature of the sediments. Sediment size varies from small boulders to silt clay layers with an abundance of cross-bedding and channel lenses incised into the underlying beds. Clast compositions are mostly granitic, reflecting the composition of the adjacent source in the Sangre de Cristo Mountains.

The 96 samples collected from the outcrop were taken from the finer-grained layers because experience has shown that finer-grained sediments are both easier to sample and more likely to be magnetically stable. The average vertical distance between individual samples is approximately half a meter. However, from the sieving of the samples, described later, we found them to be more poorly sorted than those at the other sites.

The bedding at this site dips moderately to the west with an average bedding orientation of N25°E, 15°W, although, as discussed below, an initial bedding dip of 8°W is likely at this location. The section sampled is cut by three faults with maximum vertical offsets of 4 m. Sediments can be reliably correlated across all of the faults. Comparison of mean magnetic directions from each fault-bounded block show no significant differences allowing us to treat the site as a single group.

Site SFAS. Samples from this site were collected over a 2-km-long stretch of the south bank of Arroyo Seco between Pojoaque and Chimayo beginning at about 2 km from New Mexico Route 4. The outcrops are probably the Skull Ridge Member of the Tesuque Formation of Galusha and Blick (1971) as mapped by Kelley (1978). Although only slightly farther from the mountain front than SFTE, the sediments are, on the average, the finest grained of any sampled in this study. These medium to fine sands show some cross-bedding, but channel lenses and erosional unconformities are relatively rare. Some sheets of coarse sand to pebbles are present in places and several prominent volcanic ashes are present in the local section. Sediment composition is compatible with a source from the granite of the Sangre de Cristo Mountains.

This site had the largest percentage of samples (54 of the 68) with a good-quality magnetic signature. The samples are representative of the entire section with the exception of the few coarse clastic sheets. The sample grain-sizes are the smallest for the sites sampled, and their variation in size is significantly less than at any of the other four sites (Fig. 2).

The average bedding orientation is N30°E, 9°W. The site is cut by at least four normal faults, but the stratigraphic offsets are not obvious due to the lack of significant variability in the lithology. Magnetic directions of the fault-bounded blocks are similar, allowing all of the samples to be included as a single site.

Site SFSR. These samples were collected from exposures in the banks of the Santa Fe River beginning 0.5 km west of U.S. Highway 285 within the city of Santa Fe. Kelley (1978) assigns this part of the section to the Nambé Member of the Tesuque Formation of Galusha and Blick (1971). Sediment size varies from clay to small cobble, but individual layers are usually well sorted. These sediments were all poorly cemented and only soft-sediment samples were acquired from this site. Cross-bedding, channel lenses, and erosional contacts reflect the fluvial nature of the sediments and their relative proximity to the range front.

The bedding dips westward with an average attitude of N23°W, 12°W. No faults were encountered through the section, although a fault is present 0.25 km east of the site.

Site SFES. This section in the Chamita Formation, is located 4.5 to 5.5 km west of U.S. Highway 84 on U.S. Forest Service Road 144 just northwest of Española where the road climbs westward onto a high erosional surface. The top of the sampled section is unconformably overlain by the Puye Formation. The sediments are dominantly fine sands, silts, and clays (Fig. 2) with few sedimentological indicators of fluvial action. The bedding dips moderately to the west with a general attitude of N05°E, 10°W. Although no faults are seen crossing the section, the site lies 5 km south of the Embudo fault and close to north-south trending, steeply dipping faults that have complex history of both dip-slip and strike-slip displacements (Aldrich and Dethier, 1990).

Although the 86 samples that were collected at this site are fine grained, they are less stable than samples from other sites of similar grain size. Their grain size is representative of the section.

Site SFCA. Located 11 km west of Santa Fe, this site is on the eastern escarpment of the Cerros del Rio basalt field. The site can be reached from the road to Buckman where Alamo Creek joins Cañada Ancha. The section, which dips gently to the west at about 7°, is predominantly well sorted, coarse sand that is cross-bedded and probably has a fluvial origin. The top of the section is mostly mud and silt with matrix supported granule- to pebble-size clasts. An ash-fall tuff, overlain by Cerros del Rio Basalts, occurs 0.2 m above the stratigraphically highest samples. Twenty-three samples were collected throughout the section but the majority of the ten stable samples came from the finer-grained sediments of the upper 5 m of the section. No faults are present in the immediate vicinity of the site.

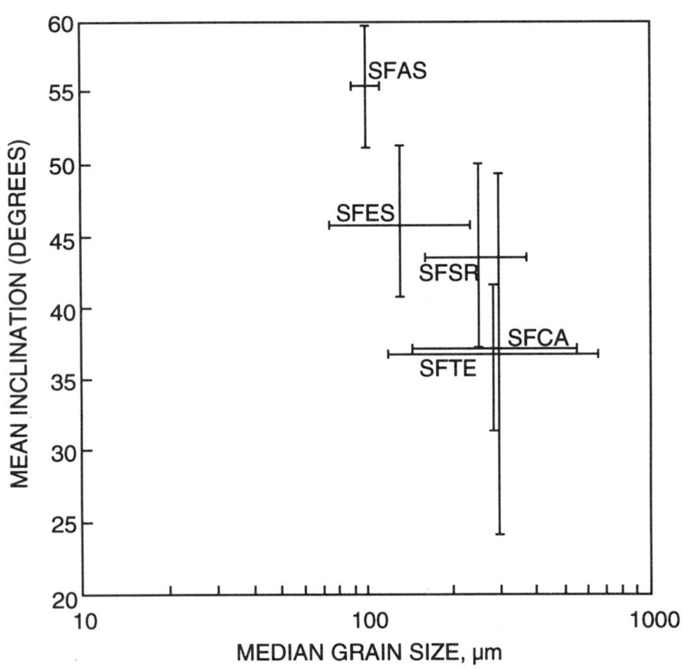

Figure 2. Diagram showing the effect of grain size on mean magnetic inclination. Vertical error bars are obtained from α_{95} from the sites, and horizontal error bars are log-normal one standard deviation of the median grain sizes of the samples.

SAMPLING METHODS

At each site samples were collected over enough stratigraphic section to be able to average secular variation (Table 1). Samples were collected using three different sampling methods: soft-sediment sampling, traditional paleomagnetic drilling, and block sampling. Most samples were collected using the soft-sediment technique, which works well for poorly consolidated and unconsolidated sediments. A nonmagnetic stainless steel tube is pushed or lightly pounded into the outcrop. The tube's orientation is measured, and the tube is removed from the outcrop. A teflon plunger is used to push the sediment sample from the metal sample tube into a quartz-glass vial that holds the sample through the measurement sequence. While this method worked well for the younger sediments at SFES and SFCA and the softer layers at the other sites, most of the samples from SFTE and SFAS were collected using traditional paleomagnetic core drilling. After drilling, the core was marked, oriented, labelled, and removed. Before measurement, the core was cut to the proper length using a copper blade trim saw. The few particularly friable samples were collected as oriented block samples and drilled and trimmed in the laboratory.

The sampling objective was to collect at least 60 samples over 50 m of section. The stratigraphic separation between samples averages 1.2 m, and redundant samples were collected at roughly half of the horizons. From Barghoorn's (1981) correlations, the average sedimentation rate is around

TABLE 1. IN SITU AND CORRECTED SITE MEAN PALEOMAGNETIC DIRECTIONS*

Site	Thickness of Section Sampled (m)	Stratigraphic Unit	n/N	In Situ (Geographic) D (°)	In Situ (Geographic) I (°)	Tilt Corrected (Stratigraphic) D (°)	Tilt Corrected (Stratigraphic) I (°)	α_{95}	κ
SFTE	63	Tesuque Formation, Nambé Member	61/93	2.4	41.5	-4.5	36.7	5.2	13
SFAS	58	Tesuque Formation, Skull Ridge Member	54/68	1.1	59.6	-10.6	55.4	4.4	20
SFSR	64	Tesuque Formation, Nambé Member	31/52	0.7	46.8	-13.9	43.5	6.5	17
SFES	55	Chamita Formation	46/86	1.2	44.3	-12.3	45.7	5.1	18
SFCA	59	Chamita Formation	10/23	155.4	-41.3	150.2	-37.1	12.7	14

*n = number of samples used in calculation; N = total number of samples processed; D = declination; I = inclination of magnetic direction; α_{95} = 95 percent confidence region about the mean direction; κ = Fisher distribution precision parameter.

25 cm/ka. At this rate, a 50-m section would represent about 200 ka, clearly enough time to adequately average geomagnetic secular variation, which appears to have a dominant periodicity of about 1 to 3 ka (Sternberg, 1982). At site SFCA limited exposure and large grain size restricted the number of layers that were suitable for sampling, but enough section was included to average multiple secular variation cycles.

For each pair of duplicate samples one was demagnetized using alternating-field demagnetization and the other by thermal demagnetization. For most horizons the directions were in good agreement; the only consistent exceptions being locations that showed an isothermal remanent magnetization (IRM) from a lightning strike or a chemical overprint. In nearly every case when an IRM was present, the sample was not completely overprinted and a primary remanence could be identified.

Samples were demagnetized by either stepwise alternating-field demagnetization to at least 110 mT or thermal demagnetization to at least 600°C. Some of the thermally demagnetized samples were first treated with alternating-fields up to 10 mT to remove the weak viscous overprint or to identify an IRM. Least-squares lines were fit to the demagnetization paths using the method of Kirschvink (1980). For some samples what is probably a slight hematite chemical overprint kept the demagnetization paths from intersecting the origin. In these cases the origin was not included in the least-squares fit. In all cases the objective was to demagnetize the sample to less than 5% of its natural remanent magnetization (NRM). Magnetic instability in some of the samples produced chaotic demagnetization paths and magnetic intensity increases making it impossible to obtain the 5% objective. These samples, or that part of the demagnetization path, were not used in the data analysis.

ROCK MAGNETIC CHARACTERIZATION

Magnetic stability of the samples varied widely, but did show a correlation of decreased stability with increasing grain size. The sites with the coarser grained sediments, SFTE, SFCA, and SFSR, had fewer stable samples than sites SFAS and SFES where much of the section contains fine sands and silts. Figure 3 shows demagnetization paths from four samples representing each site, both AF and thermal demagnetization, and both normal and reversed polarity.

Based upon our rock magnetic studies and the samples' behavior under the two different types of demagnetization the carrier of the primary magnetization appears to be magnetite with some hematite. A drastic difference between samples demagnetized using AF and thermal techniques show that a significant chemical overprint is present at the stratigraphically lower sites, particularly SFAS. AF demagnetization results in a decreasing magnetic moment and a generally linear and single component path (e.g., Fig. 2, sample SFSR 242.2). Many thermally demagnetized samples have paths like that of sample SFAS 14.07A in Figure 2. The thermal steps from 100 to 275°C quickly remove a significant magnetic component, often with an increase in the intensity of the sample. A consistent, linear magnetic component is usually removed between 275°C and 500 or 525°C with the demagnetization path moving toward the origin. At higher temperatures the path may continue moving toward the origin (e.g., SFES 1.3) or may loose its linear path (e.g., SFAS 14.07A). The absence of a similar component in the AF demagnetization shows that the low-temperature component is magnetically stiff, and we interpreted this as a chemical overprint of secondary hydrous iron oxides that are destroyed at low temperatures. We interpreted the component between 275 and 500°C as the primary detrital magnetization carried by magnetite and the high-temperature component to be a hematite component that is probably both detrital and chemical in nature. Lack of petrographic work makes this last conclusion more speculative. Because AF demagnetization appears not to remove the chemical overprint in most cases these samples were not used in the calculation of the mean directions for the sites.

The magnetic field directions of all samples for each site are plotted in Figure 4. The Fisher cone at 95% confidence about the mean is calculated by inverting the reversed polarity sample directions (normal polarity samples for SFCA) so that

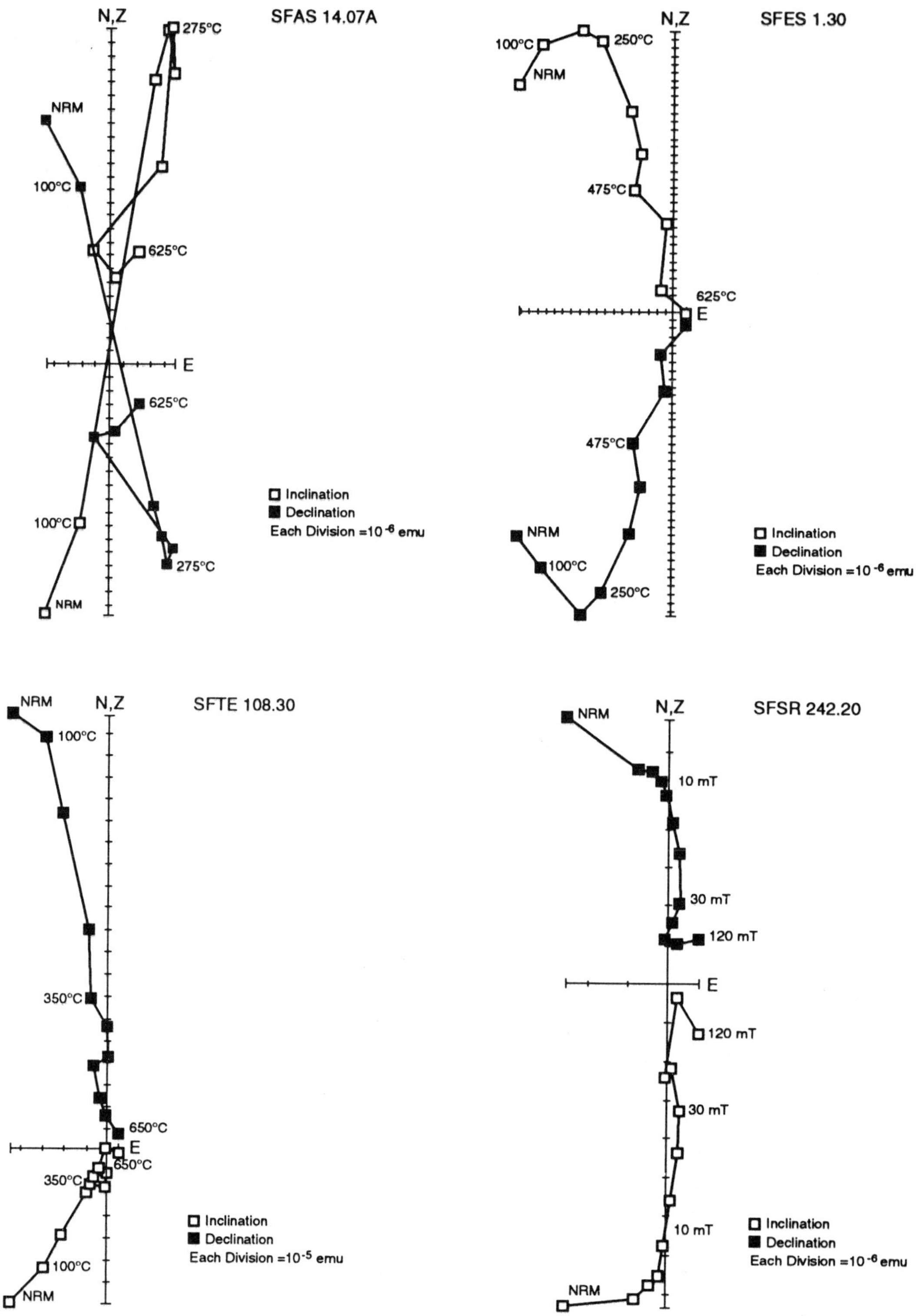

Figure 3. Four typical, orthogonal demagnetization diagrams showing typical samples from the Tesuque Formation. Both alternating field and thermal demagnetization are included. Demagnetization paths begin with the natural remanent magnetization (NRM). Vertical axis is the shared vertical (Z) and north (N) axis.

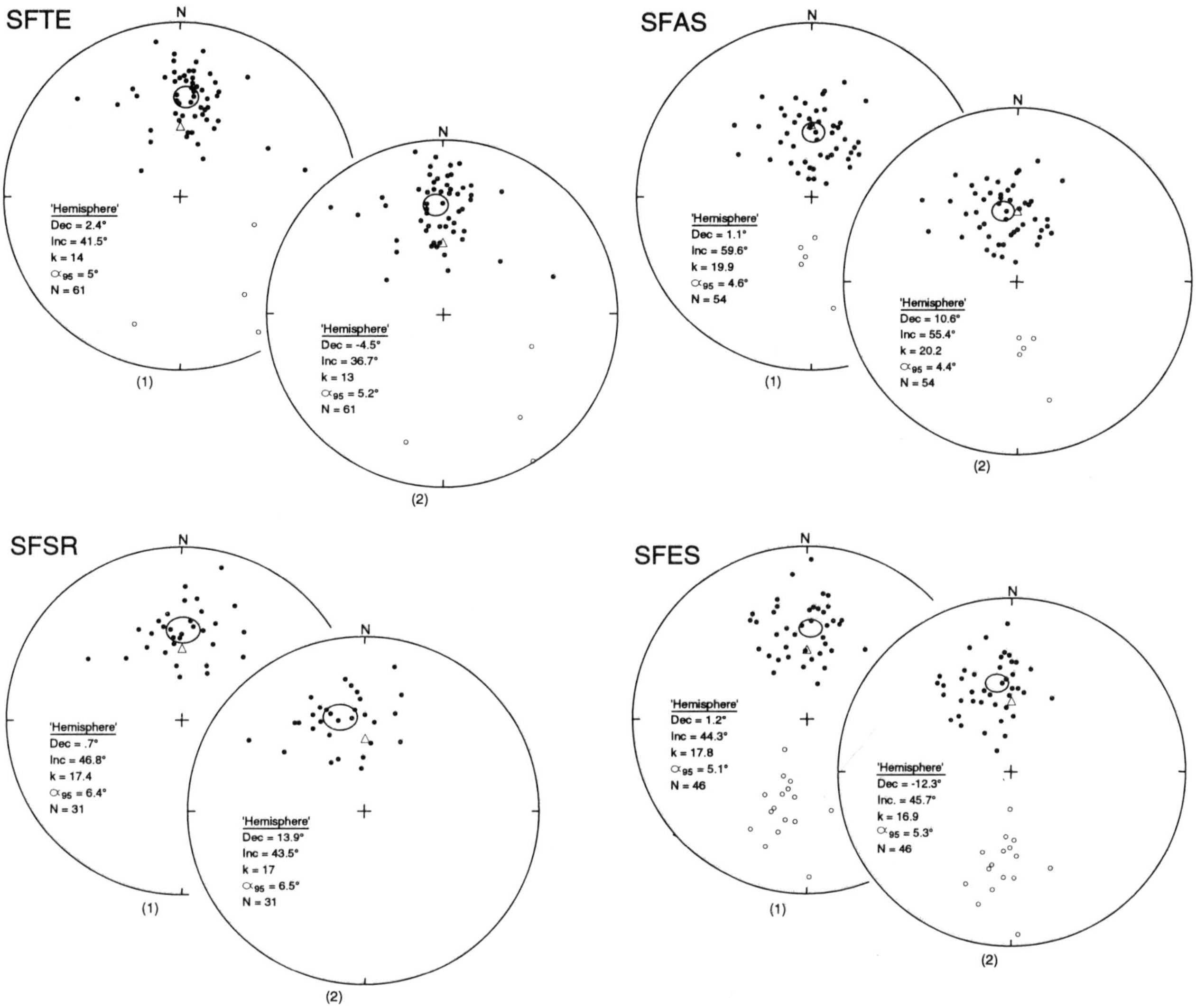

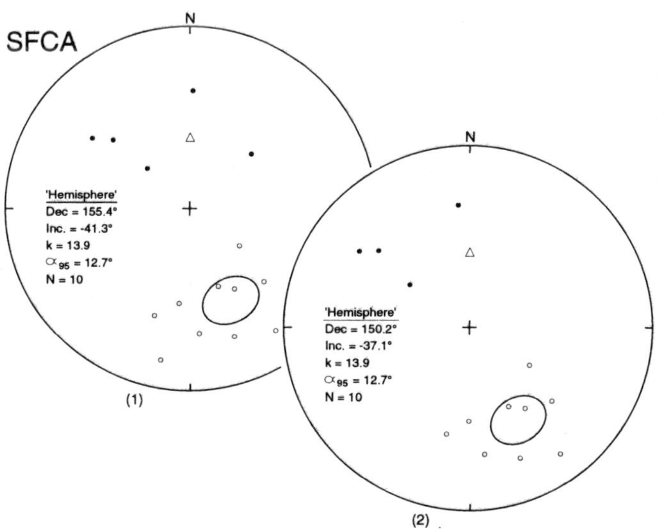

Figure 4. Equal-area plots showing sample direction and Fisher 95% confidence limits on the mean (largest open circle) for sites sampled. Plots labeled 1 show in situ data, and plots labeled 2 show tilt-correction data. Solid circles are lower-hemisphere data and open circles are upper-hemisphere data. Triangles designate modern dipole field direction.

all polarities are in the same hemisphere. The mean directions and the Fisher statistics for each site are listed in Table 1. For horizons with redundant samples the average direction of the samples from that horizon was used to calculate the mean direction at that site. The number "n" in Table 1 gives the number of horizons averaged to calculate the mean direction.

The nonparallel polarity mean directions are one aspect of concern in the results. Although not exactly antiparallel, reversed and normal mean directions at sites SFTE, SFAS, and SFCA do pass the reversal test at the 95% confidence level. At site SFSR this cannot be tested because of the absence of stable reversed samples. The normal and reversed samples at site SFES do not pass the reversal test at the 95% confidence level. Although attributable to an unrecognized overprint, or an unremoved or overlapping chemical overprint, a tectonic explanation also exists, which we discuss below.

The inclinations of the mean paleomagnetic directions at four of the sites show shallowing ranging from about 10° at SFES to 20° at SFTE with respect to the magnetic field direction, corresponding to the paleopole for that age. Studies by King and Rees (1966), and Verosub (1977) demonstrate that several depositional and post-depositional processes, which can influence the paleomagnetic orientation, have most of their affect on the inclination and little affect on the declination. Because vertical axis tectonic rotations are calculated from the declinations alone, the inclination shallowing is not a significant problem. But it is a problem for net rotation axes and, in general, for interpretation.

Calculation of the vertical-axis rotation must take into account the fact that both the magnetic pole and the mean magnetic direction at all five sites lie west of north. Pole positions of Irving and Irving (1982), and the interpolated positions of Harrison and Lindh (1982) yield magnetic declinations for northern New Mexico that are within 1° of each other for the last 10 m.y. Before then, the magnetic declinations diverge so at 20 Ma the Harrison and Lindh (1982) declination is 3° west of the Irving and Irving (1982) field direction. Needing to make a choice, we used the pole of Harrison and Lindh (1982) favoring their greater number of site means for each mean pole and their weighting system to properly average secular variation. For sites SFTE, SFAS, and SFSR, those sites whose sediments are older than 10 Ma, the rotation from the pole of Irving and Irving (1982) can be found by adding about 2.25° to our rotations.

Because the ages of the stratigraphic section at each site fall between published paleomagnetic poles, we interpolated between poles to obtain the pole position and the magnetic field direction. The interpolated paleopole latitude positions are: (SFTE, 86.5°N; SFAS, 86.9°N; SFSR, 86.7°N; SFCA, 88.8°N; and SFES, 88.8°N. The paleopole longitude for all sites is 151.1°E since we interpolate directly from the 20-Ma paleopole and the present geographic pole. The published 95% confidence limit of 3.6° was used even though the uncertainty decreases as the interpolated position approaches the modern pole.

ANALYSIS OF TECTONIC ROTATIONS

Vertical axis rotations

As an initial test of the Muehlberger model, we calculated the vertical axis rotations. Calculations were done using the difference between the tilt-corrected site mean direction and the pole corresponding to that age. The errors on the rotations were found using the method of Demarest (1983). The paleomagnetic rotations vary from no resolvable rotation (–0.6 ± 5.1°) on the eastern edge of the basin at site SFTE to 28.5 ± 10.6° at SFCA (Table 2). The results do not support the Muehlberger (1979) model, which demands that all parts of the Española block rotate uniformly.

Net axis rotations

Further analysis of the data was done with the "net rotation axis" method of MacDonald (1980) in which the total rotation at a site is represented as rotation about a nonvertical axis. This method presents two simple interpretations: (1) the present orientation of the rocks reflects rotation about a single, nonvertical, nonhorizontal axis; (2) the net rotation axis represents separate near-vertical and near-horizontal rotations and provides a measure of their relative importance. A steep axis would indicate more horizontal shearing and vertical axis rotation, a shallow axis would indicate more tilting and horizontal axis rotation. The net rotation axis may not be interpretable in either of these two simple cases and may instead represent a complex series of uninterpretable rotations.

In a model of simultaneous rotations about a horizontal axis (tilting) and a vertical axis (e.g., McKenzie and Jackson, 1986), the net rotation axis is oblique. As the deformation continues, the

TABLE 2. VERTICAL AND NET ROTATION AXIS ROTATIONS*

Site	Vertical Rotation Axis (°)	Net Rotation Axes: Rotation Axis (°)	Net Rotation Axes: Rotation (°)
SFTE	-0.6 ± 5.1	7/025	-11
SFAS	-5.0 ± 6.0	31/032	-10
SFSR	-8.3 ± 7.2	34/013	-24
SFES	-11.1 ± 5.1	Not solvable	
SFCA	-28.5 ± 10.6	65/001	-18
MFBB	-13.2 ± 3.5	62/041	-15
BGCDR	-17.1 ± 20.5	76/015	-17
BGSFB6	-29.2 ± 11.1	55/254	-35
BGSFB54	19.5 ± 9.1	20/002	54
BGSFB55	-45.4 ± 5.2	65/281	-49
BGSFB56	-90.0 ± 11.1	72/320	-93

*Orientations give plunge/trend. Negative rotations are counterclockwise viewed down the plunge. BGCDR and BGSFB6-56 were calculated from data in Brown and Golombek (1985). MFBB calculated from data in MacFadden (1977).

net rotation axis rotates smoothly with the horizontal axis making a net rotation axis a kind of average rotation axis that is meaningful only for small amounts of vertical axis rotation.

For each site the net rotation axis was calculated using a modified version of the method of MacDonald (1980). Briefly, our method uses the preservation of angular relationships within a rigidly rotating, nondeforming block. Any two vectors within the block will maintain the angle between the vectors regardless of the change in the orientation of the vectors. Now, although there are an infinite number of possible multiple rotation combinations that can produce an observed change in vector orientation, the consequences of all rotations can be summarized using the net-tectonic-rotation axis. In our solution method, the net-tectonic-rotation axis is included as one of the vectors, but with the property that its position remains constant. The rotation axis is calculated by solving three simultaneous equations. First, the scalar product of the initial position of a vector and the rotation axis must equal the scalar product of the final position of that vector and the rotation axis. This is equivalent to holding the angle between a vector and the rotation axis constant. Since there are two vectors with initial and final positions, this gives two equations. The third equation is the relationship that the vector for the rotation axis must be of unit length.

In our situation the two vectors are the magnetic vector and the pole to bedding. From measurement in the field, we know the final direction of the pole to bedding and from the mean magnetic direction of the paleomagnetic samples (not corrected to the untilted direction) we know the final direction of the magnetic vector. An estimate of the initial orientation of each vector (the original bedding attitude and original paleomagnetic inclination and declination) was required. Because the inclination of the beginning direction of the magnetic vector requires knowing the tilt correction, when the bedding is corrected about a nonhorizontal axis to an original correction position that is not horizontal the problem becomes slightly more difficult. One more quantity, the original position of the magnetic vector, becomes an unknown. As we discuss below, in our estimate of the original position of this vector direction we use the declination corresponding to the paleopole position and so that, at a minimum, the ratio of the two vector components in the horizontal plane is known. This is enough that we can now calculate the total vector using the directions of the pole to bedding. Since the angle between the bedding pole and the magnetic direction must remain constant, equating the scalar products and using the relationship for a unit vector provide our means of calculating the original direction of the magnetic vector.

For the bedding orientation one estimate of the initial position is to invoke original horizontality and assume that the initial position of the pole to the bedding plane was near vertical. However, the strata from which the samples were taken are parts of alluvial fans and, therefore, the assumption of original horizontality of bedding may be unrealistic. To assess this assumption we determined the deviation in the original bedding orientation where sufficient data could be obtained.

Cross-beds were used as paleoslope indicators to determine the original strike and dip of bedding. Studies by Potter and Olson (1954), Potter (1955), and Knight (1929) show that the average angle of inclination of cross-beds is 17 to 19° in aqueous sediment and the maximum angle of inclination (angle of repose) is 36° (cf., Mann and Kanagy, 1990). Thus, average cross-bed dips in excess of 19° and maximum inclination dips in excess of 36° would indicate tectonic rotation of strata. Sites SFTE and SFSR have sufficient cross-beds to allow such an analysis to be made. The original dip of bedding at these sites was determined by the following procedure: (1) bedding, rotation axes of local faults, and more than 30 cross-beds were measured at each site and plotted on an equal angle net; (2) the cross-beds were rotated about the rotation axis maxima, in the direction that reduces dip (clockwise looking north for these two sites), until their average dip was 19°; (3) bedding was then rotated about the rotation axis maxima, the same number of degrees and in the same direction that the average dip rotated. The final bedding attitude is assumed to be its original attitude.

SFTE beds were rotated about a fault rotation axis maxima of 18°, S28°W. Rotating the cross-beds from an average dip of 27 to 19° changed the bedding from an attitude of N25°E, 15°N to N2°W, 8°W. SFSR cross-beds were rotated about a fault rotation axis of 0°, N28°W from an average dip of 28 to 19° and bedding from an attitude of N23°W, 12°S to N9°W, 3°W. These results indicate that the assumption of original horizontality is reasonable.

Magnetic inclination can be flattened by compaction of sediments and tectonic rotations. It has been recognized in metamorphic rocks that the grain-scale deformation associated with metamorphism physically rotates magnetic grains in the rock and disrupts the magnetic direction, possibly to a point where the original direction can not be recovered (e.g., Lowrie et al., 1986). Recently this effect has also been documented in folded sedimentary rocks (Stamatakos and Kodama, 1989). However, the basin fill of the Rio Grande rift was deformed under low-temperature, low-pressure conditions in an extensional environment, not the tectonic conditions required for rotation of magnetic grains. Therefore, any flattening of the magnetic inclination is probably a depositional effect.

A detailed sedimentological analysis was undertaken. From each site at least 20 samples were chosen at random from the stable samples and sieved to separate the different grain sizes. From site SFCA, with only ten stable samples, all ten were sieved. The sieve sizes were chosen at the standard rock texture divisions: 2 mm, 1 mm, 0.5 mm, 0.25 mm, 0.125 mm, and 0.063 mm. Because the samples were poorly indurated to begin with, only a minor amount of force was necessary to completely disaggregate the grains bonded by the weak cement. In a few cases where large amounts of clay were present, the thermal demagnetization samples fused upon heating and stronger crushing was necessary with a mortar and pestle. In all cases, only moderate force was used so the samples were probably not completely disaggregated and reduced to their minimum grain size. This would bias the size counts away from the finest grain sizes (<0.063 mm) and enhance the larger-sized divisions. Because our summary statistic is the median grain size picked from the cumulative weight percent grain size and because no median falls between 0.063 mm and <0.063 mm, this bias will not affect analysis of most samples. There will be a slight interpolation error for those samples with a median between 0.125 and 0.063 mm.

Our data (Fig. 2) satisfy the predictions of Levi and Banerjee (1990). Their laboratory experiments demonstrate that magnetite disaggregated by exposure to ultrasound produces no inclination error while nondisaggregated magnetite does cause an error. In one of their sediment settling experiments they found that inclination error increased with sediment size. They interpreted this as being attributable to larger magnetite grains and aggregates of grains being influenced by gravitational torque in addition to the magnetic torque. Although Levi and Banerjee used lake sediments in their studies, their experiment in which sediment was introduced in depositional episodes probably parallels our fluvial depositional environment. In addition, the physical interpretation of their observations makes it valid for environments with larger grain sizes such as ours. Their results, which show the inclination error is directly related to the coarseness of the sediment in the sample, agree with the prediction of one of the original studies of inclination error (King, 1958). This suggests that the flattening in the Española Basin sediments is depositional and not tectonic in origin.

The estimation of the initial mean magnetic direction is more difficult to make. Ideally, the mean magnetic direction is the magnetic field direction for the interpolated paleopole at the time of deposition. However, shallowing of the magnetic inclination may affect the accuracy of the field-magnetic direction, although laboratory experiments show that when inclination error occurs, the declination is preserved (King, 1958; Payne and Verosub, 1982; Levi and Banerjee, 1990).

Comparison of vertical and net rotation axes results

Using the method of net rotation axes, a pattern of rotation axes orientations and rotation amounts emerges (Table 3, Figs. 5 and 6). At sites SFTE, SFAS, and SFSR the rotation axes plunge 7 to 34° northeast; and sites SFTE and SFAS have similar amounts of rotation (11° and 10°, respectively). Site SFSR has a higher rotation, 24°, twice the rotation observed at SFTE and SFAS, the same two to one ratio observed in the vertical axis rotation analysis.

At site SFCA relatively few of the samples showed linear decreasing demagnetization paths. However, the high vertical angle rotation axis (–28.5 ± 10.6°) is in good agreement with Brown and Golombek's (1985) results from the adjacent Cerros del Rio basalt field (17.1 ± 20.5°). Similar agreement is seen in the net rotation axes. The amount of tilting of the basalt is not known but is probably very small as the dips of the sedimentary strata in this part of the rift are low. Recalculating the Cerros del Rio (BGCDR) rotation with the regional tilt result in a net rotation axis plunging 81°NE with 26° of counterclockwise rotation. The agreement between site SFCA and BGCDR (Table 2) gives us confidence that, in spite of the small number of samples, the site yields consistent results.

TABLE 3. NET ROTATION AXES FOR VARIATIONS IN ORIGINAL ORIENTATION OF BEDDING*

Site	Horizontal (°)	Dip/-15° (°)	Dip/0° (°)	Dip/+15° (°)
SFTE	7/025/-11	39/214/7	15/202/5	16/010/-6
SFAS	31/032/-10	31/224/5	5/031/-4	37/018/-5
SFSR	34/013/-24	19/016/-12	28/012/-13	36/007/-14
SFCA	65/001/-18	82/047/-43	82/039/-35	NS
MFBB	62/041/-16	75/071/-16	75/055/-12	72/032/-10
BGCDR	76/015/-17	NS	84/094/-24	83/102/-17
BGSFB6	55/254/-35	57/262/-32	56/258/-30	55/253/-28
BGSFB54	20/002/54	16/005/48	24/350/52	30/353/57
BGSFB55	65/281/-49	65/288/-43	65/285/-41	65/281/-40
BGSFB56	72/320/-93	69/321/-77	69/320/-76	70/320/-77

*Rotations are positive-clockwise and negative-counterclockwise viewed down the plunge of the axis direction. Horizontal rotations assume originally horizontal beds. Dip/0° are rotations for original bedding angle discussed in text and strike parallel to the present strike. Dip/-15° and Dip/+15° are rotation axes for the strike -15° and +15° from the present strike. NS = not solvable given the assumptions.

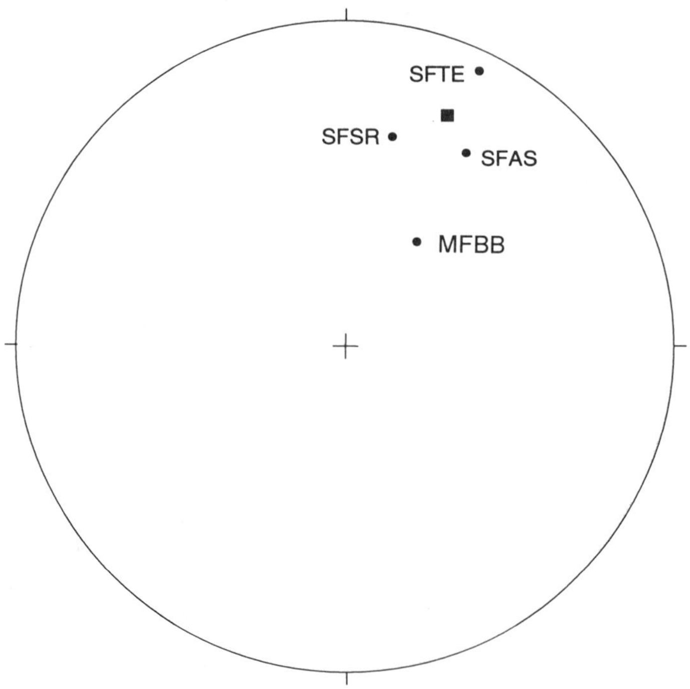

Figure 5. Equal-area plot showing net rotation axes for sites SFAS, SFTE, SFSR, and from the data of MacFadden (1977), MFBB. The square locates the mean direction of the first three axes.

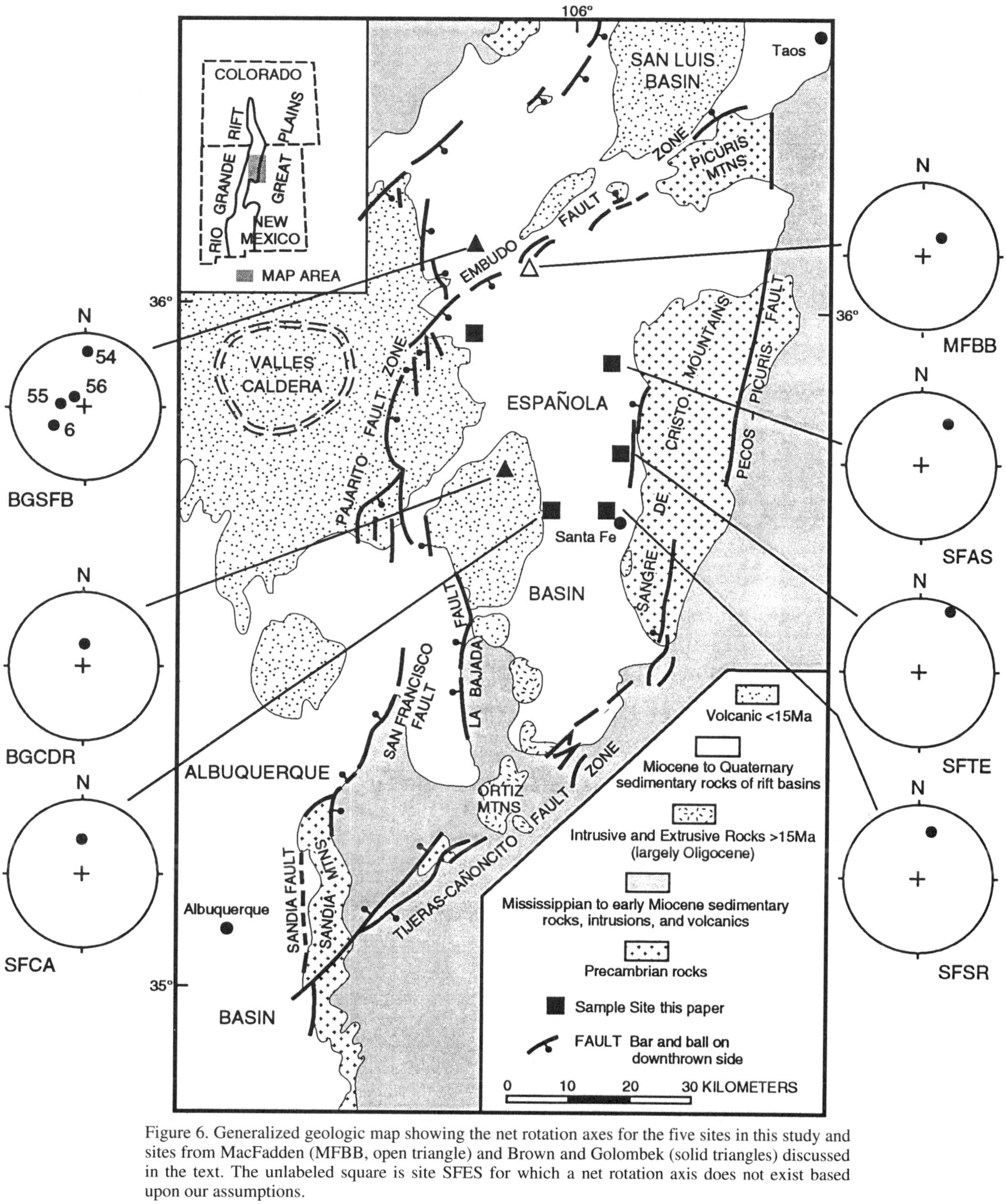

Figure 6. Generalized geologic map showing the net rotation axes for the five sites in this study and sites from MacFadden (MFBB, open triangle) and Brown and Golombek (solid triangles) discussed in the text. The unlabeled square is site SFES for which a net rotation axis does not exist based upon our assumptions.

Interestingly, the rotation at site SFES can not be explained by a single rotation axis, and yet Chasel's theorem (Whittaker, 1937) states that any series of rigid body rotations can be represented by a single rotation. There are a number of possible reasons for this including: (1) the site has not undergone rigid body rotation, (2) the mean paleomagnetic direction is not accurate, and/or (3) rotations have been both clockwise and counterclockwise during deposition. The high-vertical axis rotation of about 11° counterclockwise at the site SFES may be the result of movement along the western Embudo fault zone (Aldrich, 1986). Although the Embudo fault zone has been a left-lateral fault zone along its western end through much of its history, over the past 5 m.y. motion has been right lateral (Aldrich, 1986). Fault drag resulting from right-lateral motion would produce clockwise rotations. The section at SFES contains two polarity zones, a reversed zone in the upper 8 m and a normal zone in the remainder of the section. One interpretation of the slight eastward declination of the reversed samples would be their clockwise deformation in the right-lateral environment. To preserve an average counterclockwise rotation at the site would then require a larger counterclockwise rotation prior to the change in the sense of offset on the Embudo fault. This reversal of the sense of motion may, in part, explain our inability to calculate a net rotation axis. A more detailed study of this area is needed to resolve the problem.

As a substitute for site SFES, the nearby data of MacFadden were inverted for nonvertical axis rotation. Using the site mean direction 49°, N11°W reported by MacFadden (MacFadden, 1977, p. 781) and the bedding orientation from Kelley (1978), a northeastward, but steeply plunging, rotation axis is found. A second nearby location is the Santa Fe basalts (BGSFB) of Brown and Golombek (1985). Of these four sites, three have steep net rotation axes plunging to the northwest or southwest. The fourth (BGSFB-54) has a shallow inclination to the northeast. Because each of these sites is only one or a few individual flows, the magnetic field secular variation may not be properly averaged within each site. However, the agreement of three of the four flows probably gives a good general indication of the local tectonic movement.

The results of using various constraints on initial bedding orientation show that the orientation of the net rotation axis, and particularly the inclination, is a robust quantity (Table 3). The major difference is in the amount of rotation. The general orientation of the rotation axes are not significantly affected.

CONCLUSIONS

1. Sites SFTE, SFAS, and SFSR, which are located nearest the eastern boundary fault of the rift, all seem to have had similar kinematic histories. The beds at these sites rotated about low to moderately plunging (<34°) axes, and may reflect dominant extension across the rift with minor influence by rift-parallel, left-lateral shear. Sites SFTE and SFAS, located about 15 km apart, have very similar rotations and tilting, suggesting that they belong to the same fault-block domain (Table 3). But, the larger dissimilarity between SFAS and SFSR to the south indicates that SFSR belongs to a different domain.

2. Sites SFCA and BGCDR, separated by about 10 km and located near the center of the rift in the Cerros del Rio basalt field, have steeply plunging (>62°) rotation axes. These steep axes may reflect intrarift block rotations associated with left-lateral motion across the rift.

3. The steeply plunging axis at site MFBB (62°), located near the Embudo fault zone, may be the result of strike-slip displacement on this accommodation zone. Investigations by Aldrich (1986) show that the fault zone's net-slip has a significant strike-slip component in addition to its long-recognized dip-slip displacement.

4. Our paleomagnetic data do not support the rigid block rotation model for the north-central Rio Grande rift of Muehlberger (1979), in which the entire diamond-shaped crustal block bounded by the Embudo, Pajarito, San Francisco, Sandia, Tijeras-Cañoncito, and Pecos-Picuris fault zones (Fig. 1) is rotating counterclockwise. The large differences in the orientation of the rotation axes, near vertical to subhorizontal, and the amounts of rotation on these axes suggest that the rift is divided into a series of smaller, independently rotating fault-block domains. The significant difference between the fault-block domain that includes sites SFCA and BGCDR with steeply plunging axes and the fault-block domains that include sites SFTE, SFAS, and SFSR with gently plunging rotation axes supports this conclusion.

5. A combination of extension and left-lateral shearing across the rift, greater movement on the western boundary faults of the rift relative to the eastern boundary faults (resulting in an asymmetric rift block), dominant strike-slip deformation along the Embudo fault (accommodation) zone, and dominant dip-slip or strike slip on intrarift faults can collectively account for the observed paleomagnetic rotations in the north-central Rio Grande rift.

ACKNOWLEDGMENTS

We thank The Institute for Geophysics and Planetary Physics–Los Alamos for its support of this work, Dr. John Geissman for the generous use of the University of New Mexico paleomagnetics laboratory, and Dr. Craig Jones for a copy and support of his PaleoMag computer software. We also thank Laurie Brown and Michael J. Whitelaw for their careful reviews of the manuscript that helped clarify and improve it.

REFERENCES CITED

Aldrich, M. J., Jr., 1986, Tectonics of the Jemez lineament in the Jemez Mountains and Rio Grande rift: Journal of Geophysical Research, v. 91, p. 1753–1762.

Aldrich, M. J., and Dethier, D. P., 1990, Stratigraphic and tectonic evolution of the northern Española Basin, Rio Grande rift, New Mexico: Geological Society of America Bulletin, v. 102, p. 1695–1705.

Barghoorn, S., 1981, Magnetic-polarity stratigraphy of the Miocene type Tesuque Formation, Santa Fe Group, in the Española valley, New Mexico: Geological Society of America Bulletin, v. 92, p. 1027–1041.

Biehler, S., and 12 others, 1991, A geophysical model of the Española Basin, Rio Grande rift, New Mexico: Geophysics, p. 340–353.

Brown, L. L., and Golombek, M. P., 1985, Tectonic rotations within the Rio Grande rift: Evidence from paleomagnetic studies: Journal of Geophysical Research, v. 90, p. 790–802.

Brown, L. L., and Golombek, M. P., 1986, Block rotations in the Rio Grande rift, New Mexico: Tectonics,v. 5, p. 423–438.

Demarest, H. H., Jr., 1983, Error analysis for the determination of tectonic rotation from paleomagnetic data: Journal of Geophysical Research, v. 88, p. 4321–4328.

Ekas, L. M., Ingersoll, R. V., Baldridge, W. S., and Shafiqullah, M., 1984, The Chama–el Rito Formation, Española Basin, New Mexico: New Mexico Geological Society, 35th Field Conference, Guidebook, p. 137–143.

Galusha, T., and Blick, J. C., 1971, Stratigraphy of the Santa Fe Group, New Mexico: American Museum of Natural History Bulletin, v. 144, 127 p.

Golombek, M. P., McGill, G. E., and Brown, L., 1983, Tectonic and geologic evolution of the Española Basin, Rio Grande rift: Structure, rate of extension, and relation to the state of stress in the western United States: Tectonophysics, v. 94, p. 483–507.

Harrison, C.G.A., and Lindh, T., 1982, A polar wandering curve for North America during the Mesozoic and Cenozoic: Journal of Geophysical Research, v. 87, p. 1903–1920.

Irving, E., and Irving, G. A., 1982, Apparent polar wander paths Carboniferous through Cenozoic and the assembly of Gondwana: Geophysical Surveys, v. 5, p. 141–188.

Kelley, V. C., 1977, Geology of Albuquerque Basin, New Mexico: New Mexico Bureau of Mines and Mineral Resources Memoir 33, 60 p.

Kelley, V. C., 1978, Geology of Española Basin, New Mexico: New Mexico Bureau of Mines and Mineral Resources Geologic Map, v. 48, scale 1:125,000.

Kelley, V. C., 1979, Tectonics, middle Rio Grande rift, New Mexico, *in* Riecker, R. E., Rio Grande rift: Tectonics and magmatism: Washington, D.C., American Geophysical Union, p. 57–70.

King, R. F., 1958, Remanent magnetism of artificially deposited sediments: Monthly Notices of the Royal Astronomical Society, Geophysical Supplement 7, p. 115–134.

King, R. F., and Rees, A. I., 1966, Detrital magnetism in sediments: An examination of some theoretical models: Journal of Geophysical Research, v. 71, p. 561–571.

Kirschvink, J. L., 1980, The least-squares line and plane and the analysis of paleomagnetic data: Geophysical Journal of the Royal Astronomical Society, v. 62, p. 699–718.

Knight, S. H., 1929, The Fountain and Casper Formations of the Laramie Basin:University of Wyoming Public Science, Geology 1, no. 1, p. 1–82.

Levi, S., and Banerjee, S., 1990, On the origin of inclination shallowing in redeposited sediments: Journal of Geophysical Research, v. 95, p. 4383–4389.

Lowrie, W., Hirt, A. M., and Klizfield, R., 1986, Effects of tectonic deformation on the remanent magnetization or rocks: Tectonics, v. 5, p. 713–722.

MacDonald, W. D., 1980, Net tectonic rotation, apparent tectonic rotation, and the structural tilt correction in paleomagnetic studies: Journal of Geophysical Research, v. 85, p. 3659–3669.

MacFadden, B. V., 1977, Magnetic polarity stratigraphy of the Chamita Formation stratotype (Miocene-Pliocene) of north-central New Mexico: American Journal of Science, v. 277, p. 769–800.

Manley, K., 1979, Stratigraphy and structure of the Española Basin, Rio Grande rift, New Mexico, *in* Riecker, R. E., ed., Rio Grande rift: Tectonics and magmatism: Washington, D.C., American Geophysical Union, p. 71–86.

Mann, C. J., and Kanagy, S. P., II, 1990, Angles of repose that exceed modern angles: Geology, v. 18, p. 358–361.

May, J. S., 1984, Miocene stratigraphic relations and problems between Abiuquiu, Los Piños, and Tesuque Formations near Ojo Caliente, northern Española Basin: New Mexico Geological Society, 35th Field Conference Guidebook, p. 129–135.

McKenzie, D., and Jackson, J., 1986, A block model of distributed deformation by faulting: Journal of the Geological Society of London, v. 143, p. 349–353.

Muehlberger, W. R., 1979, The Embudo fault between Pilar and Arroyo Hondo, New Mexico: An active intracontinental transform fault: New Mexico Geological Society, Field Conference, Guidebook, v. 30, p. 77–82.

Payne, M. A., and Verosub, K. L., 1982, The acquisition of post-depositional detrital remanent magnetization in a variety of natural sediments: Geophysical Journal of the Royal Astronomical Society, v. 68, p. 625–642.

Potter, P., 1955, The petrology and origin of the Lafayette gravel: Part I. Mineralogy and petrology: Journal of Geology, v. 63, p. 1–38.

Potter, P. E., and Olson, J., 1954, Variance components of crossbedding in some basal Pennsylvanian sandstones of the Eastern Interior Basin: Geological applications: Journal of Geology, v. 62, p. 26–49.

Spell, T. L., Harrison, T. M., and Wolff, J. A., 1989, $^{40}Ar/^{39}Ar$ dating of the Bandelier tuffs and associated ignimbrites: Constraints on the evolution of the Bandelier magma system [abs.]: EOS Transactions of the American Geophysical Union, v. 70, p. 1413.

Sternberg, R. S., 1982, Archaeomagnetic secular variation of direction and paleointensity in the American Southwest [Ph.D. thesis]: University of Arizona, 305 p.

Stamatakos, J., and Kodama, K. P., 1989, The nature of the Bloomsburg Pole: Strain versus remagnetization [abs]: EOS Transactions of the American Geophysical Union, v. 70, p. 1068.

Verosub, K., 1977, Depositional and post-depositional processes in the magnetization of sediments: Review of Geophysics and Space Physics, v. 15, p. 129–143.

Whittaker, E. T., 1937, A treatise on the analytical dynamics of particles and rigid bodies; with an introduction to the problem of three bodies (fourth edition): Cambridge, Cambridge University Press, 456 p.

Manuscript Accepted by the Society October 13, 1993

Printed in U.S.A.

Geological Society of America
Special Paper 291
1994

Cenozoic stratigraphy, sandstone petrology, and depositional history of the Albuquerque Basin, central New Mexico

Richard P. Lozinsky*
New Mexico Bureau of Mines and Mineral Resources, Campus Station, Socorro, New Mexico 87801

ABSTRACT

The Albuquerque Basin is one of the largest and deepest basins of the Rio Grande rift. The latest Oligocene to middle Pleistocene Santa Fe Group is the major basin fill unit. Paleogene deposits underlie the Santa Fe and indicate that a depositional center predated the Albuquerque Basin. Pre–Santa Fe Tertiary deposits are subdivided into a lower unit that is at least partly correlative with the Eocene Galisteo and Baca Formations and an upper unit, the unit of Isleta #2, that is equivalent to the Datil Group and the overlying sequence of Oligocene volcanic rocks. Thickness of the Santa Fe Group ranges from 1,000 to 2,000 m along the basin margins to as much as 4,407 m in the basin center. Galisteo and Baca thicknesses are as much as 454 m; the unit of Isleta #2 is up to 2,185 m in the Shell West Mesa well.

Galisteo-Baca sediments were deposited in basins that predated the Albuquerque Basin. These depocenters continued to receive sediments of the unit of Isleta #2 during early to middle Oligocene time. Lower Santa Fe sediments (30 to 5 Ma) were deposited into two internally drained basins. After 10 Ma, the Santa Fe had filled the basins to the point where a single, internally drained Albuquerque Basin was formed. At about 5 Ma, the basin drainage became through-flowing with the development of the ancestral Rio Grande. The first major episode of Rio Grande Valley entrenchment at about 1.0 Ma ended Santa Fe deposition. Sedimentation rates ranged from 20 to 30 m/m.y. during early and late phases of Santa Fe deposition to as much as 600 m/m.y. during middle Miocene deposition.

INTRODUCTION

The Rio Grande rift is one of the major structural features of the southwestern United States. It consists of a series of roughly north trending, en echelon basins that extend from central Colorado to at least southern New Mexico, a distance of about 1,000 km (Chapin, 1971, 1979; Morgan et al., 1986). One of the largest and deepest basins of the rift is the Albuquerque Basin of central New Mexico (Fig. 1). This basin, covering an area of about 5,600 km^2, contains some of the best exposures of basin fill along its margins and has the greatest concentration of oil test wells in the entire rift. Utilizing outcrop and oil test well data, this paper discusses the stratigraphy, sandstone petrology, and depositional history of the basin fill. This study is one of the first to combine surface and subsurface data from deep oil test wells to analyze the stratigraphy and interpret the depositional history of the Albuquerque Basin.

*Present address: Division of Natural Sciences, Fullerton College, 321 E. Chapman Avenue, Fullerton, California 92634.

Lozinsky, R. P., 1994, Cenozoic stratigraphy, sandstone petrology, and depositional history of the Albuquerque Basin, central New Mexico, *in* Keller, G. R., and Cather, S. M., eds., Basins of the Rio Grande Rift: Structure, Stratigraphy, and Tectonic Setting: Boulder, Colorado, Geological Society of America Special Paper 291.

GEOLOGIC SETTING

As defined here, the Albuquerque Basin extends from the San Felipe fault belt in the north to the Joyita uplift in the south, a distance of about 112 km (Fig. 1). Basin width varies from about 20 km in the north to about 60 km in the central region. Although the Albuquerque Basin is a single topographic feature, seismic reflection data from Shell Oil Company (Russell and Snelson, 1990, and this volume) has revealed that it is composed of a northern, eastward-tilted half graben and southern, westward-tilted half graben separated by an accommodation zone along the southwest extension of the Tijeras fault (Fig. 1). Basin structure generally consists of a deep, inner graben that is flanked by relatively shallow benches (such as the Hubbell bench) leading up to the margin areas (Fig. 2). Many faults are listric (Russell and Snelson, 1990). The structural relief on the Precambrian basement is well over 10,000 m between the eastern margin uplifts and the deepest part of the basin.

The eastward-tilted Sandia–Manzano–Los Pinos uplift forms the prominent, relatively linear eastern border. This uplift consists primarily of Precambrian plutonic and metamor-

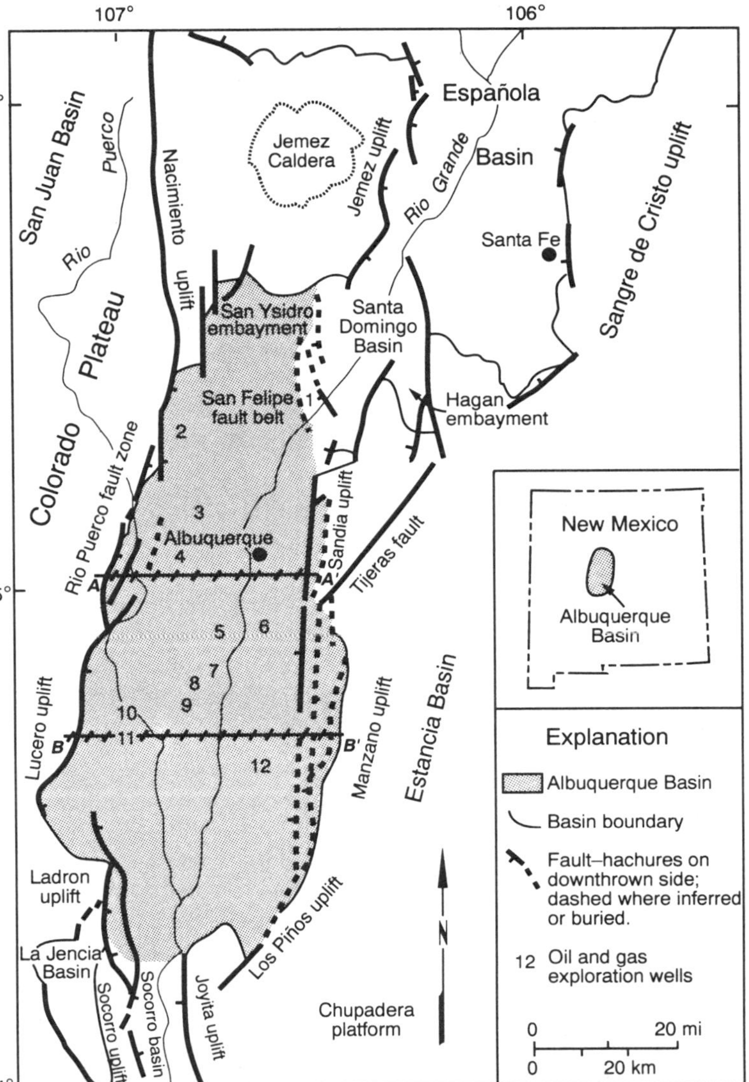

Well names: (1) Shell SFP No. 1, (2) Shell SFP No. 3, (3) Shell West Mesa Federal No. 1, (4) Carpenter Atrisco No. 1, (5) Shell Isleta No. 2, (6) TransOcean Isleta No. 1, (7) Shell Isleta Central No. 1, (8) Joiner San Clemente No. 1, (9) Long Dalies No. 1, (10) Humble Oil SFP No. 1, (11) Shell SFP No. 2, (12) Grober Fugua No. 1.

Figure 1. Tectonic map of Albuquerque Basin as defined in this study. Numbers refer to locations of oil test wells listed on Table 1. Hatched lines show the locations of cross sections in Figure 2. Modified from Anonymous (1961) and Woodward et al. (1978).

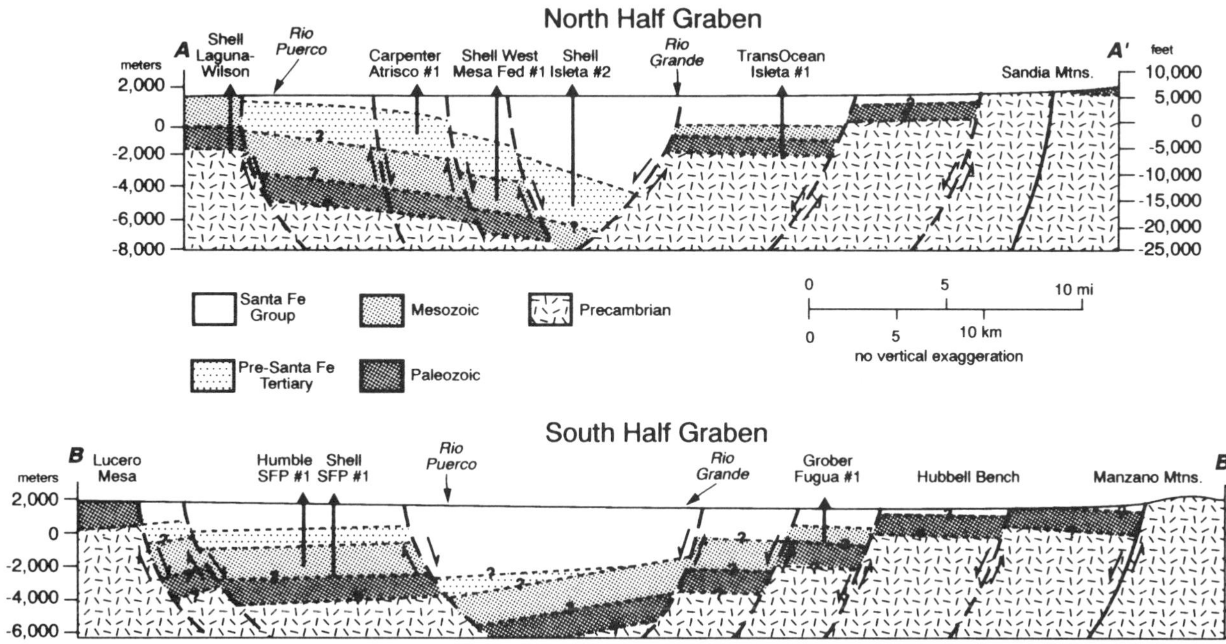

Figure 2. Diagrammatic cross sections of the Albuquerque Basin. Cross section locations shown on Figure 1.

phic rocks unconformably overlain by upper Paleozoic limestone, shale, and sandstone. The Ladron Mountains and Lucero uplift form the southwestern basin border. The Ladron Mountains consist mainly of Precambrian plutonic and metamorphic rocks with minor upper Paleozoic units. The gently west dipping Lucero uplift contains upper Paleozoic limestone, shale, and sandstone capped by upper Cenozoic basalt flows.

North of the Lucero uplift, the basin boundary is marked by the topographically subdued Rio Puerco fault zone (Slack and Campbell, 1976). Rocks exposed directly west of the fault zone include Cretaceous sandstone and shale, and, to a lesser extent, Jurassic clastic units. The Nacimiento uplift and Jemez Mountains border the basin to the northwest and north, respectively. Precambrian plutonic and metamorphic rocks overlain by Paleozoic and Mesozoic units crop out in the Nacimiento uplift (Woodward, 1987). The Jemez Mountains contain upper Miocene to Pleistocene mafic to silicic volcanic rocks (Smith et al., 1970; Gardner et al., 1986).

METHODS

Surface and subsurface data were utilized in analyzing the sedimentary fill of the Albuquerque Basin. Selected outcrops were mapped, described in measured sections, and sampled for petrography. Paleocurrent directions were measured from imbricated clasts, cross-beds, and parting lineations. Clast compositions were determined by counting 50 pebble- to cobble-size clasts in a given exposure. Sedimentation rates are based on the age and thickness of the unit and do not take into consideration compaction.

Cuttings and some cores from 12 deep, oil test wells (Table 1) were analyzed to determine stratigraphic units and thicknesses. Cuttings were initially examined with a binocular microscope at approximate 6-m intervals in order to construct provisional stratigraphic columns for each well and to determine sample intervals for petrographic work. Borehole geophysical logs (primarily resistivity and spontaneous potential) aided the cutting analysis.

A total of 148 thin sections from outcrops, well cuttings, and cores were point counted, using 400 points per section and following the method of Dickinson (1970). In the well cuttings, sand samples were collected at approximately 300-m intervals for thin-section work. Shorter sample intervals were used where more detail was needed. Detailed descriptions of the above methods are given in Lozinsky (1988).

STRATIGRAPHIC UNITS

The stratigraphic units of the Albuquerque Basin consist mostly of siliciclastic, continental deposits with minor volcanic rocks that can be subdivided into three units (Fig. 3): (1) Galisteo-Baca Formation deposits, (2) the unit of Isleta #2 well, and (3) the Santa Fe Group. The pre–Santa Fe Tertiary deposits crop out only in limited exposures within the basin and were examined mostly using cuttings and cores from the oil test wells. The Santa Fe Group was studied both in outcrop and in cuttings and cores from oil test wells. In well cuttings, Pre–Santa Fe Tertiary deposits can usually be distinguished from the Santa Fe Group by better induration and a distinct color change from reddish orange in the Santa Fe to purplish red in the pre–Santa Fe Tertiary deposits.

TABLE 1. STRATIGRAPHIC DATA FOR 12 OIL TEST WELLS STUDIED IN THE ALBUQUERQUE BASIN*

Well No.	Name; Location	Surface Elevation (m)	Total Depth (m)	Tops Tvs (m)	Tops Tg (m)	Tops K-Tr (m)	Thickness Qtsf (m)	Thickness Tvs (m)	Thickness Tg (m)
1	Shell Santa Fe Pacific #1; 18-13N-3E	1,747	3,366	NR	905	1,110	905	NR	205
2	Shell Santa Fe Pacific #3; 28-13N-1E	1,917	3,132	NR	1,218	1,240	1,218	NR	22
3	Shell West Mesa Federal #1; 24-11N-1E	5,774	5,906	2,603	4,788	5,126	2,603	2,185	338+
4	Carpenter Atrisco #1 28-10N-1E	1,766	2,028	1,006	NP	NP	1,006	1,202+	NP
5	Shell Isleta #2 16-8N-2E	1,524	6,482	4,407	6,194	NP	4,407	1,787	288+
6	Transocean Isleta #1; 8-8N-3E	1,909	3,163	NR	NR	1,536	1,536	NR	NR
7	Shell Isleta #1; 7-7N-2E	1,537	4,982	2,679	3,216	3,670	2,679	537	454
8	Joiner San Clemente #1; 23-7N-1E	?	1,709	NP	NP	NP	1,709+	NP	NP
9	Long Dalies #1; 32-7N-1E	1,620	1,856	NP	NP	NP	1,856+	NP	NP
10	Humble Santa Fe Pacific #1; 18-6N-1W	1,552	3,868	1,494	(2,164?)	2,591	1,494	670	(427?)
11	Shell Santa Fe Pacific #2; 29-6N-1W	1,583	4,360	1,460	(2,316?)	2,511	1,460	856	(195?)
12	Grober Fuqua #1; 19-5N-3E	1,531	1,920	NR	NR	1,384	1,384	NR	NR

*QTsf = Santa Fe Group; Tvs = unit of Isleta #2; Tg = Galisteo-Baca Formations; K-Tr = Cretaceous or Triassic; NP = not penetrated; NR = not recognized. From Lozinsky (1988).

Galisteo-Baca Formations

Description. In well cuttings and core samples, these deposits consist of purplish red to gray, well indurated, poorly to moderately sorted, fine- to coarse-grained sandstone with claystone and silty sandstone interbeds. Core samples exhibit cross-bedding and graded bedding. The contact with the underlying Upper Cretaceous section is marked by an abrupt downward change to gray shale and sandstone with coal interbeds. Thickness of the Eocene section is generally less than 500 m.

These deposits are generally restricted to the western and central portions of the basin and are not recognized in the eastern margin area (Fig. 4a). This is based on data from the Grober and Transocean wells where the Santa Fe Group rests directly on Mesozoic strata (Lozinsky, 1988). No definitive evidence for Eocene rocks in the southern basin was found, although well control in this area is limited. The Galisteo-Baca deposits are thicker in the central basin and thin toward the north.

Due to limited well control in the northeastern and southern basin area, it is not known how far the deposits extend. They probably continue northward into the Hagan embayment and correlate with the Galisteo Formation in that region (Fig. 1). This prerift basin is believed to be part of the Galisteo–El Rito basin of Gorham and Ingersoll (1979). To the south, these deposits may correlate with the Baca Formation exposed in the Joyita uplift. The Baca in the Joyita uplift was deposited in the Carthage–La Joya basin of Cather and Johnson (1984).

Petrology. Sandstones consist primarily of monocrystalline quartz and small amounts of plagioclase, K-feldspar, and chert (Fig. 5a). Many of the monocrystalline grains are clear, anomalously well rounded, and have overgrowth rims. Plagioclase grains are usually twinned, and occur in about equal proportions as K-feldspar. K-feldspar grains consist of microcline and orthoclase with minor perthite. Lithic fragments are chiefly sedimentary and include silt, clay, and carbonate. Samples from the West Mesa and Atrisco wells contain low volcanic-lithic percentages.

Provenance. The high quartz percentages, overgrowth

Unit			Thickness	Age
Santa Fe Group	upper		100-200m	Plio-Pleistocene
Santa Fe Group	Popatosa Fm	Zia Fm	0-4,207m	late Miocene to latest Oligocene
unit of Isleta #2			0-2,185m	Oligocene to Late Eocene
Galisteo-Baca Fms.			0-<500m	Eocene

Figure 3. Cenozoic stratigraphic chart for the Albuquerque Basin.

rims on quartz grains, sedimentary lithic fragments, and chert clearly show that these deposits were derived mainly from a sedimentary source area. However, the presence of microcline, plagioclase, orthoclase, and perthite suggests at least limited exposure of a granitic-metamorphic source (or recycled therefrom). Low volcanic-lithic percentages rule out major contributions from a volcanic terrane. It should be noted that the Lv percentage (49%) shown on Table 2 is misleading because of the low total lithic fragments for these deposits.

Age. Shell Oil (unpublished staff communication, December 1986) reported Eocene pollen from these beds in the Isleta #2 well. Thus, these deposits are at least partly correlative with the Galisteo and Baca Formations that are exposed to the northeast and south of the Albuquerque Basin, respectively.

Unit of Isleta #2

Description. This unit is informally referred to as the "unit of Isleta #2" because it was first recognized within the Isleta #2 well (Lozinsky, 1988). Cuttings are similar to those from the Galisteo-Baca deposits and consist of purplish red to gray, well indurated, poorly to moderately sorted, fine- to coarse-grained sandstone with claystone interbeds. However, the Isleta #2 also includes interbeds of silicic to intermediate volcanic flows and tuffs.

The "unit of Isleta #2 well" is significantly thicker than the Galisteo-Baca deposits. In the West Mesa well, it is up to 2,185

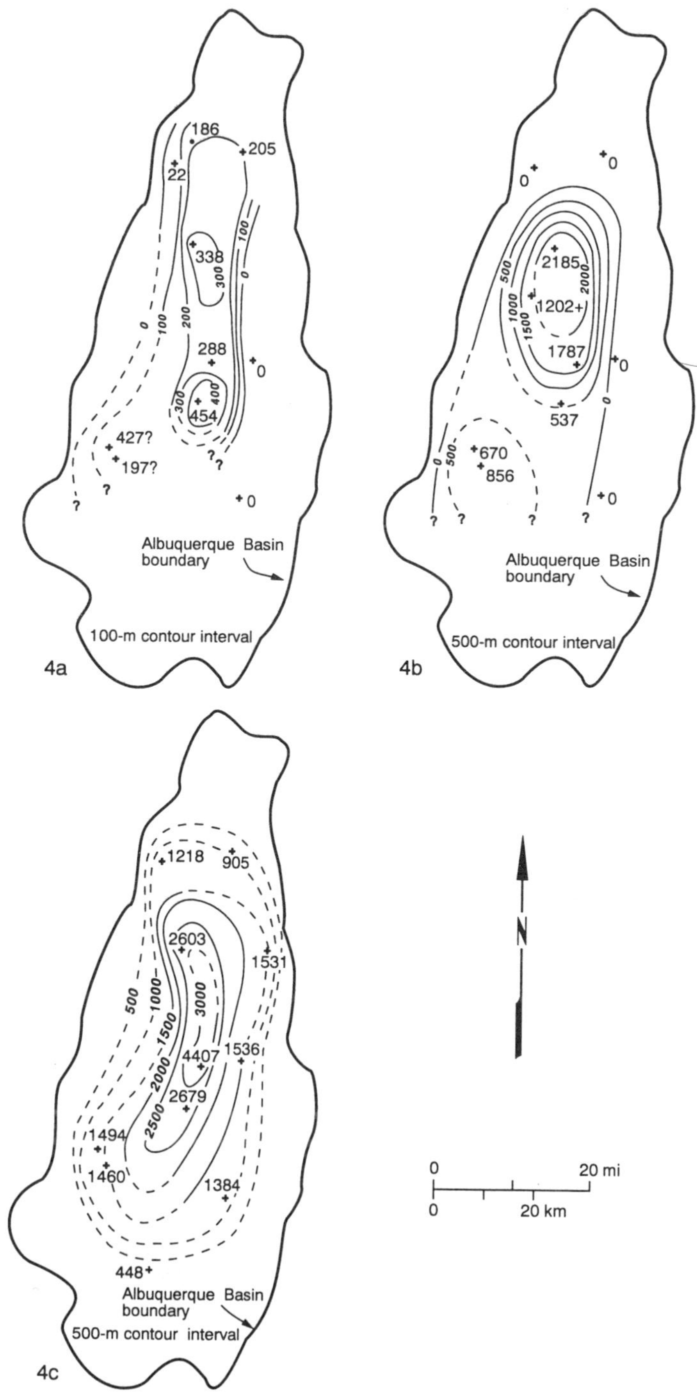

Figure 4. Isopach map of a, Galisteo-Baca Formations, the 186-m-thick location is an outcrop measurement by Gorham (1979); b, unit of Isleta #2 well; and c, Santa Fe Group, the 1,531- and 448-m-thick locations are from driller's logs of the Norins Realty and Central New Mexico Brown-Livingston oil test wells, respectively. Dashed lines are inferred contours.

m thick (Table 1). As with the Galisteo-Baca Formations, this unit is restricted to the western and central portions of the Albuquerque Basin and is not recognized along the eastern margin (Fig. 4b). Cuttings from the two northern wells (SFP Nos. 1 and 3) show that the unit of Isleta #2 does not extend into the northernmost part of the basin. The southern extent of these deposits is unclear. They may correlate with outcrops in the Black Butte area and Joyita uplift (see Fig. 1).

Petrology. Sandstones of the unit of Isleta #2 are mostly dominated by monocrystalline quartz, plagioclase, and volcanic-lithic fragments. Plagioclase grains are twinned, occasionally show oscillatory zoning, and are more abundant than K-feldspar. Orthoclase and sanidine are the major K-feldspars. In thin section, these deposits typically contain more volcanic-lithic fragments and less quartz than the Galisteo-Baca deposits (Fig. 5a).

Exceptions occur in the West Mesa and Atrisco wells where the unit of Isleta #2 contains very low volcanic-lithic percentages. However, abundant intermediate volcanic flows (?) occur in this interval suggesting that they are equivalent to the unit of Isleta #2.

Provenance. Volcanic-lithic fragments and twinned pla-

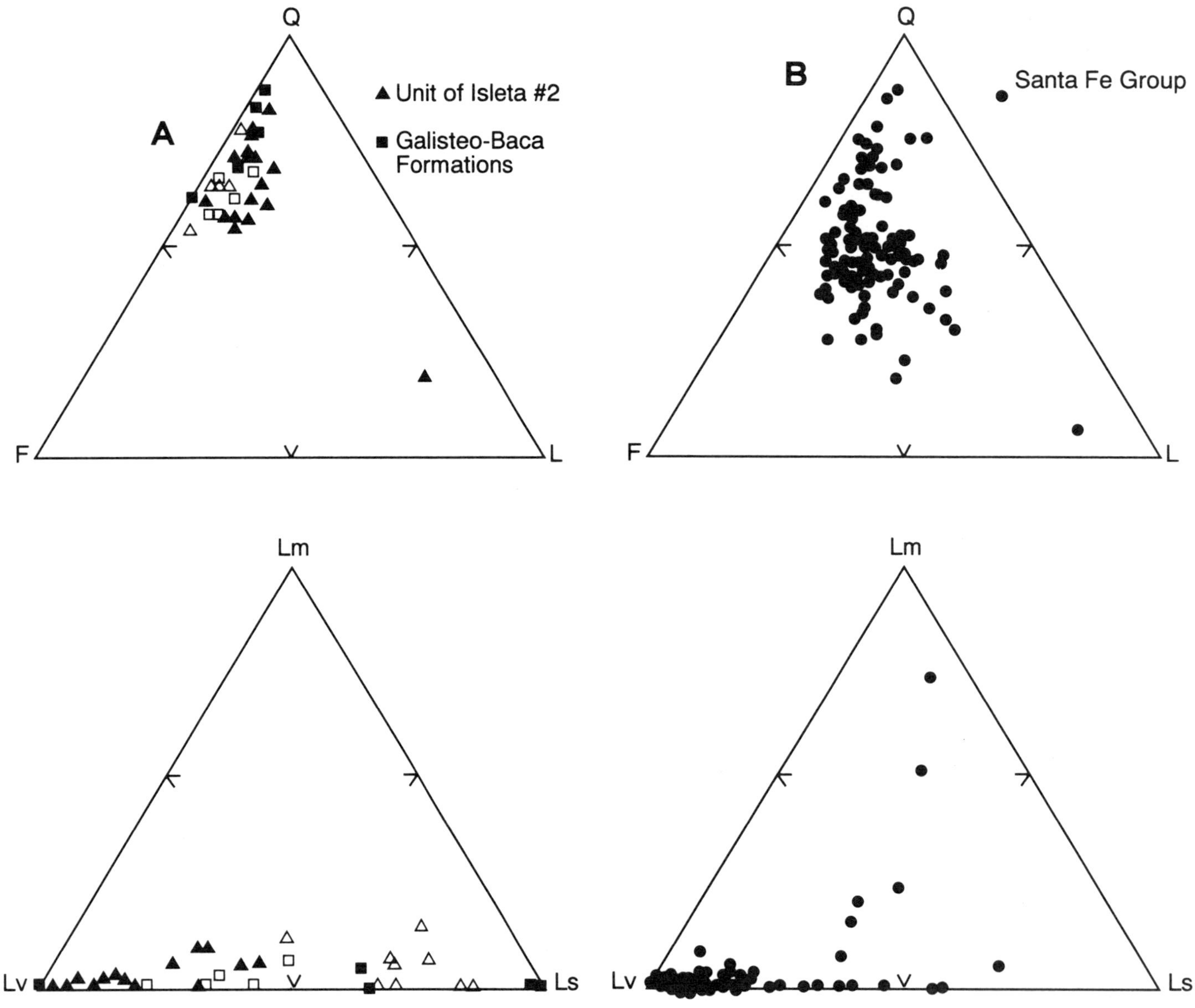

Figure 5. QFL and LmLvLs compositional diagrams for a, pre–Santa Fe Tertiary; and b, Santa Fe Group sandstones. See Table 2 for definition of letters. In Figure 5a, open symbols are samples from the West Mesa and Atrisco wells. In Figure 5b, samples closest to the Q pole on QFL diagram are from the northwestern basin and samples closest to the Lm pole on the LmLvLs diagram are from the southeastern basin.

TABLE 2. DETRITAL MODES OF CENOZOIC SANDSTONES IN THE ALBUQUERQUE BASIN*

	Q	F	L	P/F	Qp/Q	Lv	Ls	Lm
Galisteo-Baca Formations n = 11								
x	69.2	26.9	3.9	0.64	0.10	49.5	49.0	1.5
s	9.9	9.2	2.7	0.21	0.03	30.4	30.7	2.7
Isleta #2 Unit n = 24								
x	67.9	26.0	6.1	0.66	0.09	56.8	39.2	4.0
s	8.6	7.6	5.9	0.17	0.04	29.3	28.0	5.5
Santa Fe Group:								
Central basin n = 39								
x	45.7	33.9	20.4	0.63	0.08	90.8	8.3	0.9
s	10.7	7.2	13.4	0.10	0.03	6.2	5.9	1.8
Southwest basin n = 30								
x	46.7	35.7	17.6	0.60	0.08	89.8	10.0	0.2
s	3.5	6.9	6.2	0.16	0.03	6.3	6.1	0.6
Southeast basin n = 12								
x	56.3	27.5	16.2	0.60	0.12	63.3	20.5	16.2
s	18.0	6.2	12.9	0.19	0.06	32.2	16.5	24.1
North basin n = 32								
x	57.2	28.6	14.2	0.62	0.10	85.0	14.3	0.7
s	16.5	11.0	9.1	0.12	0.05	18.8	18.0	1.8

Detrital modes based on 400 counts per sample, exclusive of cement and matrix. n = total number of samples; x = mean; s = standard deviation. Q = total quartz, including chert; F = total feldspar; L = total lithic fragments; P = plagioclase; Qp = polycrystalline quartz, including chert; Lv = volcanic-lithic fragments; Ls = sedimentary-lithic fragments, including carbonates; Lm = metamorphic-lithic fragments.

gioclase in the unit of Isleta #2 clearly indicate a volcanic source that was compositionally diverse, ranging from silicic to mafic. A sedimentary source terrane is also strongly implied by the abundance of clear, monocrystalline quartz (some exhibiting overgrowth rims) with lesser amounts of sedimentary-lithic fragments and chert, especially in the northwestern part of the basin.

Age. The Shell Isleta #2 well penetrated an ash-flow tuff that was K-Ar dated at 36.3 ± 1.8 Ma (May and Russell, this volume). Therefore, this unit is believed to correlate with the mid-Tertiary volcanic and volcaniclastic units exposed to the southwest of the Albuquerque Basin (Osburn and Chapin, 1983). If true, this unit ranges in age from late Eocene to late Oligocene. Some of the lavas in the north basin may also be related to the Espinaso Formation.

Santa Fe Group

Description. The Santa Fe Group is subdivided into lower and upper units based on lithologies resulting from different depositional environments (Fig. 3). Lower Santa Fe basin fill is dominated by intertonguing piedmont, eolian, and fine-grained basin-floor deposits. Fan and coalescent-fan alluvium characterizes the piedmont facies, whereas playa sediments and playa margin alluvium are the major components of the basin-floor facies. The lower Santa Fe Group contains interbeds of light reddish brown to reddish orange, poorly to well sorted, fine- to very coarse grained sandstone, siltstone, conglomerate, and claystone. Mafic flows (or dikes) and ash beds are scattered throughout the entire Santa Fe section. In the southern basin area, the lower Santa Fe is called the Popotosa Formation (Machette, 1978a, b) and, in the northern part, it is generally referred to as the Zia Formation (Tedford, 1982).

Upper Santa Fe Group basin fill is characterized by intertonguing piedmont (alluvial-fan) and axial-river (fluvial) deposits. The fluvial deposits, located in the central basin area, include light gray to buff, moderately to well sorted, cross-stratified, medium- to coarse-grained sandstone and conglomerate. Beds are generally poorly consolidated, but are locally well cemented. The upper Santa Fe generally includes the Ceja Member of Kelley (1977) in the north basin and the Sierra Ladrones Formation of (Machette, 1987a, b) in the south basin.

Santa Fe Group thickness varies greatly within the Albuquerque Basin, ranging from 1,000 to 2,000 m along the margin areas to as much as 4,407 m in the deep central part of the basin (Fig. 4c). The thickest section was penetrated in the Shell Isleta #2 test well. The bottom of the Santa Fe was picked on a lithologic and petrologic change (as described below) in the cuttings. Based on seismic data, May and Russell (this volume) picked the bottom of the Santa Fe Group at 4,938 m in the Isleta #2 well. Either pick represents the thickest documented section of the Santa Fe Group in the entire Rio Grande rift. The upper Santa Fe Group is only 100 to 200 m thick.

Petrology. Detailed thin section analyses show that the Santa Fe Group typically contains less quartz and more lithic fragments than the pre–Santa Fe Tertiary deposits, especially volcanic-lithic fragments (Fig. 5b, Table 2). Monocrystalline quartz and plagioclase are usually the dominant detrital grains. Volcanic-lithic fragments are by far the most numerous lithic fragment (up to 90%).

On closer examination, however, there are some significant variations in the composition of the Santa Fe both geographically and stratigraphically (Table 2). Total quartz percentages generally increase with depth, whereas K-feldspar percentages usually increase upsection. Volcanic-lithic grains are mostly intermediate to silicic in the lower section, but mafic grains become dominant in the upper section. An exception occurs in the eastern basin-margin area where metamorphic- and plutonic-lithic grains increase upsection. The most apparent change is in the northwestern basin area, where most samples are highly quartzose (up to 60%) with lesser lithic fragments (14%), particularly volcanic-lithic grains.

Provenance. Variations in the Santa Fe Group can be directly attributed to changing source areas. The lower Santa Fe Group was derived mainly from volcanic and sedimentary source areas, as evidenced by abundant volcanic-lithic fragments, twinned plagioclase, chert, and quartz. However, in the northwestern basin, low volcanic-lithic percentages show that these deposits were derived mostly from a sedimentary source, probably Upper Cretaceous sandstones exposed west of the Rio Puerco fault zone. The low percentages of plutonic- and metamorphic-lithic fragments indicate that these source areas were minor. This is very apparent in lower Santa Fe outcrops in the southeastern basin where Precambrian detritus is absent in the lower Santa Fe only a few hundred meters west of the Manzano Mountains (Lozinsky, 1988). Thus, the Precambrian core of the Manzano Mountains was not exposed during most of lower Santa Fe time. Volcanic source areas consisted mainly of silicic to intermediate material with only minor mafic volcanic rocks. Ingersoll et al. (1987) has identified the San Juan Mountains as a possible volcanic source area for the volcaniclastics in lower Santa Fe Group in the northern part of the basin.

Sedimentary and volcanic terranes continued to be major source areas during the latter part of lower Santa Fe deposition. Mafic volcanic-lithic grains increase in the upper part of the lower Santa Fe. Plutonic and metamorphic fragments also become more abundant upsection in samples from the eastern margin of the basin, indicating that the Precambrian core of the eastern uplift had become exposed near the end of lower Santa Fe deposition.

Upper Santa Fe piedmont deposits were derived from the surrounding uplifts; however, the axial river deposits were derived from areas outside the basin to the north and west.

Age. The lower Santa Fe Group is believed to range in age from about 30 to 5 Ma based on fossil and radiometric dates (Lozinsky, 1988). The age of the upper Santa Fe Group is based on dates from within and outside the basin. These dates indicate that the upper Santa Fe Group ranges from about 5 to 1 Ma (Machette, 1978a, b; Lozinsky and Hawley, 1986; Lozinsky et al., 1991; Cather et al., 1991).

DEPOSITIONAL HISTORY

During the Eocene, the Albuquerque Basin area received sediments from nearby highlands that included the Nacimiento, Sandia, Lucero, and Sierra uplifts (Kelley, 1977; Gorham and Ingersoll, 1979; Cather and Johnson, 1984). These uplifts shed mainly sedimentary and Precambrian detritus into possible extensions of the Galisteo–El Rito basin of Gorham and Ingersoll (1979) in the north and the Carthage–La Joya basin of Cather and Johnson (1984) in the south.

These basins continued to receive sediments, the unit of Isleta #2, into the Oligocene. However, the uplifts to the west and southwest were by then providing mainly silicic to intermediate volcanic detritus into the basin. At least one ash-flow tuff was deposited as far north as the Isleta #2 well. The volcanic rocks were derived from the Mogollon-Datil volcanic field. Some may also have been derived from the Espinaso Formation to the northeast.

Lower Santa Fe Group deposition began about 30 Ma with initial development of the Rio Grande rift. It is speculated that at this time, the Albuquerque Basin may have consisted of two subbasins that corresponded to the two half grabens (Lozinsky, 1988). These internally drained subbasins were receiving mainly silicic to intermediate volcanic and sedimentary debris eroded from the surrounding uplifts. Based on outcrop data from the Zia Formation, sedimentation rates during early Santa Fe deposition may have ranged from 24 to 74 m/m.y. (Lozinsky, 1988). This would suggest relatively low tectonism for this time interval. By the end of lower Santa Fe deposition, the Precambrian core of the eastern uplifts had become exposed.

The most active tectonic episode of the Rio Grande rift is believed to be middle to late Miocene (Chapin and Seager, 1975; Morgan et al., 1986; Cather et al., this volume). This is supported in the Albuquerque Basin by data from the Gabaldon badlands where a sedimentation rate of 204 m/m.y. was calculated. Located in the southwestern Albuquerque Basin, the Gabaldon exposures range in age from 7 to 9 Ma (Lozinsky and Tedford, 1991). In the deeper portions of the basin, sedimentation rates may have been as much as 600 m/m.y. (Lozinsky, 1988).

During this time, drainage was still internal and basin fill was derived from the nearby uplifts. However, two major fluvial systems, one flowing from the northwest and the other from northeast, may have also transported detritus from more distance sources outside the Albuquerque Basin. These fluvial systems probably terminated in the southern basin area. Following the increase in sedimentation rates after about 10 Ma, these subbasins may have filled to the point where deposits coalesced to from a single aggradational basin. The bulk of the Santa Fe was probably deposited between 16 and 10 Ma when tectonism was most active in the rift (Cather et al., this volume).

Development of through-flowing drainage in the Albuquerque Basin and integration of an ancestral upper Rio Grande system occurred at about 5 Ma. Two major ancestral tributaries, the Rio San Jose from the west and the Rio Puerco from the northwest, joined the ancestral Rio Grande to form a large plain of fluvial aggradation in the central basin area. Upper Santa Fe sedimentation rates ranged from 22 to 33 m/m.y. Santa Fe deposition ended about 1.0 Ma when the first of several incision episodes began. Since the initial incision, the basin has been undergoing net degradation as several more cycles of incision have formed the modern inner valley of the Rio Grande.

ACKNOWLEDGMENTS

This work is part of a doctoral dissertation completed at the New Mexico Institute of Mining and Technology. Funding was provided by the New Mexico Bureau of Mines and Mineral Resources and by the New Mexico Geological Society. I am grateful to Shell Oil Company for loaning well cuttings and cores from several of their oil test wells and for allowing me to examine some of their extensive seismic data in the basin. This paper benefitted greatly from reviews by John Hawley, Greg Mack, Steve Cather, and Jud May.

REFERENCES CITED

Anonymous, 1961, Structural problems of the Rio Grande trough in Albuquerque country: New Mexico Geological Society, 12th Annual Field Conference, Guidebook, p. 144–147.

Cather, S. M., and Johnson, B. D., 1984, Eocene tectonics and depositional setting of west-central New Mexico and eastern Arizona: New Mexico Bureau of Mines and Mineral Resources Circular 192, 33 p.

Cather, S. M., Chapin, C. E., Chamberlin, R. M., and McIntosh, W. C., 1991, Stratigraphy and age of the Santa Fe Group near Socorro, New Mexico: Geological Society of America Abstracts with Programs, v. 23, p. 10.

Chapin, C. E., 1971, The Rio Grande rift, Part I—Modifications and additions: New Mexico Geological Society, 22nd Annual Field Conference, Guidebook, p. 191–201.

Chapin, C. E., 1979, Evolution of the Rio Grande rift—A summary, *in* Riecker, R. E., ed., Rio Grande rift: Tectonics and magmatism: Washington, D.C., Washington, D.C. American Geophysical Union, p. 1–5.

Chapin, C. E., and Seager, W. R., 1975, Evolution of the Rio Grande rift in the Socorro and Las Cruces area: New Mexico Geological Society, 26th Field Conference, Guidebook, p. 297–321.

Dickinson, W. R., 1970, Interpreting detrital modes of graywacke and arkose: Journal of Sedimentary Geology, v. 40, p. 695–707.

Gardner, J. N., Goff, F., Garcia, S., and Hagan, R. C., 1986, Stratigraphic relations and lithologic variations in the Jemez volcanic field, New Mexico: Journal of Geophysical Research, v. 91, p. 1763–1778.

Gorham, T. M., 1979, Geology of the Galisteo Formation, Hagan Basin, New Mexico [M.S. thesis]: Albuquerque, New Mexico, University of New Mexico, 136 p.

Gorham, T. M., and Ingersoll, R. V., 1979, Evolution of the Eocene Galisteo basin, north-central New Mexico: New Mexico Geological Society, 30th Annual Field Conference, Guidebook, p. 219–224.

Ingersoll, R. V., Cavazza, W., Baldridge, W. S., and Shafiqullah, M., 1987, Oligocene-Miocene volcaniclastic petrofacies and basin evolution in northern New Mexico: Implications for initiation of Rio Grande rift: Geologic Society of America Abstracts with Programs, v. 19, p. 712.

Kelley, V. C., 1977, Geology of Albuquerque Basin, New Mexico: New Mexico Bureau of Mines and Mineral Resources Memoir 33, 59 p.

Lozinsky, R. P., 1988, Stratigraphy, sedimentology, and sand petrology of the Santa Fe Group and pre–Santa Fe Tertiary deposits in the Albuquerque Basin, central New Mexico [Ph.D. thesis]: Socorro, New Mexico, New Mexico Institute of Mining and Technology, 298 p.

Lozinsky, R. P., and Hawley, J. W., 1986, Upper Cenozoic Palomas Formation of south-central New Mexico: New Mexico Geological Society, 37th Annual Field Conference, Guidebook, p. 239–247.

Lozinsky, R. P., and Tedford, R. H., 1991, Geology and paleontology of the Santa Fe Group, southwestern Albuquerque Basin, Valencia County, New Mexico: New Mexico Bureau of Mines and Mineral Resources Bulletin 132, 35 p.

Lozinsky, R. P., Hawley, J. W., and Love, D. W., 1991, Geologic overview and Pliocene-Quaternary history of the Albuquerque Basin, central New Mexico, *in* Julian, B., and Zidek, J., eds., Field guide to geologic excursions in New Mexico and adjacent areas of Texas and Colorado: New Mexico Bureau of Mines and Mineral Resources Bulletin 137, p. 157–162.

Machette, M. N., 1978a, Geologic map of the San Acacia Quadrangle, Socorro County, New Mexico: U.S. Geological Survey, Geologic Quadrangle Map GQ-1415, scale 1:24,000.

Machette, M. N., 1978b, Preliminary geologic map of the Socorro 1 × 2 Quadrangle, central New Mexico: U.S. Geological Survey Open-File Report 78–607, scale 1:250,000.

Morgan, P., Seager, W. R., and Golombek, M. P., 1986, Cenozoic thermal, mechanical and tectonic evolution of the Rio Grande rift: Journal of Geophysical Research, v. 91, p. 6263–6276.

Osburn, G. R., and Chapin, C. E., 1983, Nomenclature for Cenozoic rocks of northeast Mogollon-Datil volcanic field, New Mexico: New Mexico Bureau of Mines and Mineral Resources Stratigraphic Chart 1.

Russell, L. R., and Snelson, S., 1990, Structural style and tectonic evolution of the Albuquerque Basin segment of the Rio Grande rift, *in* Pinet, B., and Bois, C., eds., The potential of deep seismic profiling for hydrocarbon exploration, French Petroleum Institute Research Conference Proceedings: Paris, Editions Technip, p. 175–207.

Slack, P. B., and Campbell, J. A., 1976, Structural geology of the Rio Puerco fault zone and its relationship to central New Mexico tectonics: New Mexico Geological Society, Special Publication no. 6, p. 46–52.

Smith, R. L., Bailey, R. A., and Ross, C. S., 1970, Geologic map of the Jemez Mountains, New Mexico: U.S. Geological Survey Miscellaneous Investigations Series, Map I-571, scale 1:250,000.

Tedford, R. H., 1982, Neogene stratigraphy of the northwestern Albuquerque Basin: New Mexico Geological Society, 33rd Annual Field Conference, Guidebook, p. 273–278.

Woodward, L. A., 1987, Geology and mineral resources of Sierra Nacimiento and vicinity, New Mexico: New Mexico Bureau of Mines and Mineral Resources Memoir 42, 84 p.

Woodward, L. A., Callender, J. F., Seager, W. R., Gries, J. C., Shaffer, W. L., and Zilinski, R. E., 1978, Tectonic map of the Rio Grande rift region in New Mexico, Chihuahua, and Texas, *in* Hawley, J. W., compiler, Guidebook to the Rio Grande rift in New Mexico and Colorado: New Mexico Bureau of Mines and Mineral Resources Circular 163, Sheet 2, scale 1:1,000,000.

Manuscript Accepted by the Society October 13, 1993

Geological Society of America
Special Paper 291
1994

Structure and tectonics of the Albuquerque Basin segment of the Rio Grande rift: Insights from reflection seismic data

L. R. Russell* and S. Snelson*
Shell Development Company, Bellaire Research Center, Houston, Texas 77001

ABSTRACT

The Albuquerque Basin, located in the central portion of the Rio Grande rift, is filled by 7,350 m of Tertiary clastic sediments deposited on a "basement" of Mesozoic and Paleozoic sedimentary rocks and Precambrian crystalline rocks. The basin has been a center of Cenozoic volcanic activity over the last 37 m.y.

Detailed examination of regional seismic reflection data, supplemented by well control and field work, has demonstrated that the Albuquerque Basin is asymmetric and structurally complex, consisting of two subbasins downdropped along low-angle to listric-normal faults of opposing structural polarity, some of which flatten at depths of about 10 km. The northern subbasin has been downdropped along a major west-dipping, listric-normal fault system whereas the southern subbasin is bordered by a system of major east-dipping, low-angle normal faults. Palinspastic restorations show that the amount of extension in the Albuquerque Basin ranges from 17% in the north basin to at least 28 to 30% in the south basin. The differential amount and polarity of extension between the two subbasins is taken up across a complex midbasin transverse structural zone/accommodation zone that seismically displays characteristics of a transtensional, negative-flower structure.

General agreement on overall structural style exists relative to COCORP interpretations of the structure of the south basin. However, an alternative interpretation of the COCORP data is presented that advocates fold geometries on major basin-bounding normal faults in this portion of the basin.

Seismic and outcrop evidence suggest that preexisting structures in the Precambrian basement rocks may in part have controlled the geometry of Tertiary structures in the region.

INTRODUCTION

This study of the Albuquerque Basin was conducted between 1978 and 1980, and was undertaken in order to improve our understanding of the rift stage of passive continental margin development. As a young, active continental rift with abundant subsurface and surface control, it was hoped that a study of this segment of the Rio Grande rift would provide answers to some of the fundamental questions of structural style and geometry of rifting during initial stages of continental breakup. Of interest were questions such as:

1. What is the geometry of rift-controlling normal faults at depth? Are they classic high-angle faults or low angle and/or listric (downward flattening) in geometry?
2. If these rift-controlling faults are indeed low angle or listric, what controls the depth of flattening?
3. Is continental breakup a symmetrical or asymmetrical phenomenon in terms of the structural configuration of the resultant conjugate margins?

How much crustal stretching or attenuation occurs during rifting and breakup? How variable is it? What are the implications for palinspastic plate-tectonic reconstructions and fits?

*Present address: Russell, ARCO Exploration and Production Technology, 2300 West Plano Parkway, Plano, Texas 75075; Snelson, P.O. Box 1175, Eastsound, Washington 98254.

Russell, L. R., and Snelson, S., 1994, Structure and tectonics of the Albuquerque Basin segment of the Rio Grande rift: Insights from reflection seismic data, *in* Keller, G. R., and Cather, S. M., eds., Basins of the Rio Grande Rift: Structure, Stratigraphy, and Tectonic Setting: Boulder, Colorado, Geological Society of America Special Paper 291.

5. To what extent is rifting controlled by preexisting basement structures both in terms of the overall geographic location of rift development and the detailed geometry of rift-zone structures?

In the years following the inception of this work in 1978, many of the above questions, as well as questions on mechanisms of rifting have been variously addressed by numerous workers. These include McKenzie (1978) and Wernicke (1981, 1985, 1986), and many others who have discussed the pure shear versus whole lithosphere simple shear models of rifting; and Bally (1982), and Lister et al. (1986) whose discussions focus on the symmetry versus asymmetry of the rift process and its implications for passive continental margin evolution. The uniqueness and value of this study of the Albuquerque Basin is in the completeness of the data set and the clarity with which the geometry of the rift zone is expressed on seismic reflection lines, in deep exploration wells, and on the outcrop. It is hoped that the interpretations presented here will further add to our understanding of rift phenomena and hopefully stimulate more research into the still highly debated question of rift mechanisms. A previous version of this paper (Russell and Snelson, 1990) has been published that is far less comprehensive in its discussion of recent work in the area and its implications.

REGIONAL TECTONIC FRAMEWORK OF THE RIO GRANDE RIFT

Geophysical characteristics of the Rio Grande rift

The Rio Grande rift, like the greater Basin and Range province, is quite distinctive geophysically. It is characterized by thin crust (25 to 45 km), an anomalously low velocity (7.4 to 7.9 km/sec) and hence low-rigidity upper mantle, regionally high levels of seismicity at shallow focal depths (10 to 15 km), and a zone of seismic velocity inversion in the upper crust (10 to 12 km), possibly indicative of the brittle-ductile transition (Smith, 1978; Keller et al., 1975, 1979; Braile et al., 1974; McCullar and Smithson, 1977; Olsen et al., 1979; Sanford et al., 1979). The region is also typified by high heat flow (>4.5 h.f.u. locally), a large negative Bouguer gravity anomaly, and a zone of high electrical conductivity in the upper (10 to 20 km) to mid (20 to 30 km) crust (Reiter et al., 1975, 1978, 1979; Cordell, 1978; Cordell et al., 1978; Hermance and Pederson, 1980; Jiracek et al., 1979).

Major tectonic elements of the Rio Grande rift

Extending from just north of Leadville, Colorado, southward at least 1,000 km to the El Paso, Texas, area, the Rio Grande rift is the zone of Cenozoic extension resulting from the separation of the Colorado Plateau block from the Great Plains of the North American craton (Fig. 1). As such, the rift represents a northerly trending extension of the southern Basin and Range province into the southern Rocky Mountain region. The geographic location of the Rio Grande rift very closely follows the distribution of Laramide foreland basement block uplifts in the southern Rocky Mountain region (Stewart, 1978; Eaton, 1979). Compressional deformation in the Laramide foreland possibly resulted in a weakened crust that Tertiary extensional stresses exploited during rifting.

The northern segment of the Rio Grande rift extends from 20 km north of Leadville, Colorado, to the Taos, New Mexico, area and includes the San Luis Basin and Upper Arkansas graben (Figs. 2 and 3). It is characterized by north-northwest structural trends, roughly paralleling the late Paleozoic and Laramide structural grain. The southern boundary of this northern rift segment is defined by a northeasterly trending zone of volcanism and contemporary seismicity extending through the Taos and Jemez Mountains area.

The central segment of the rift runs from Taos, New Mexico, southward to the Socorro, New Mexico, area and is comprised of the Espanola and Albuquerque Basins and the smaller Santo Domingo Basin (Figs. 2 and 3). This segment is distinguished by a north-northeast structural orientation, again parallel to the Laramide tectonic grain, and is bounded on the south by a northeasterly trending zone of volcanism and con-

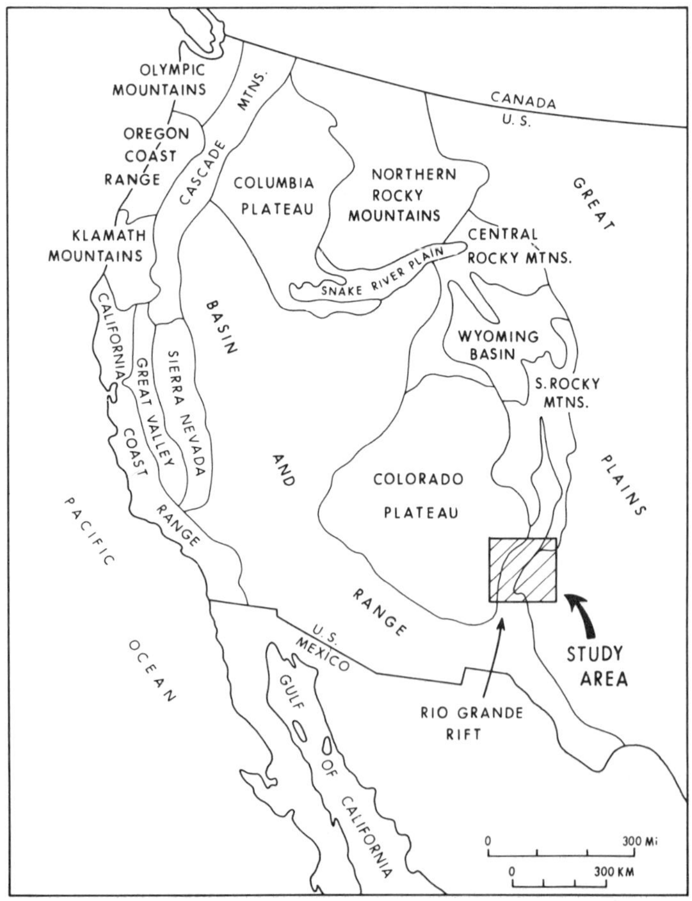

Figure 1. General map of the provinces of the western United States showing the location of the study area.

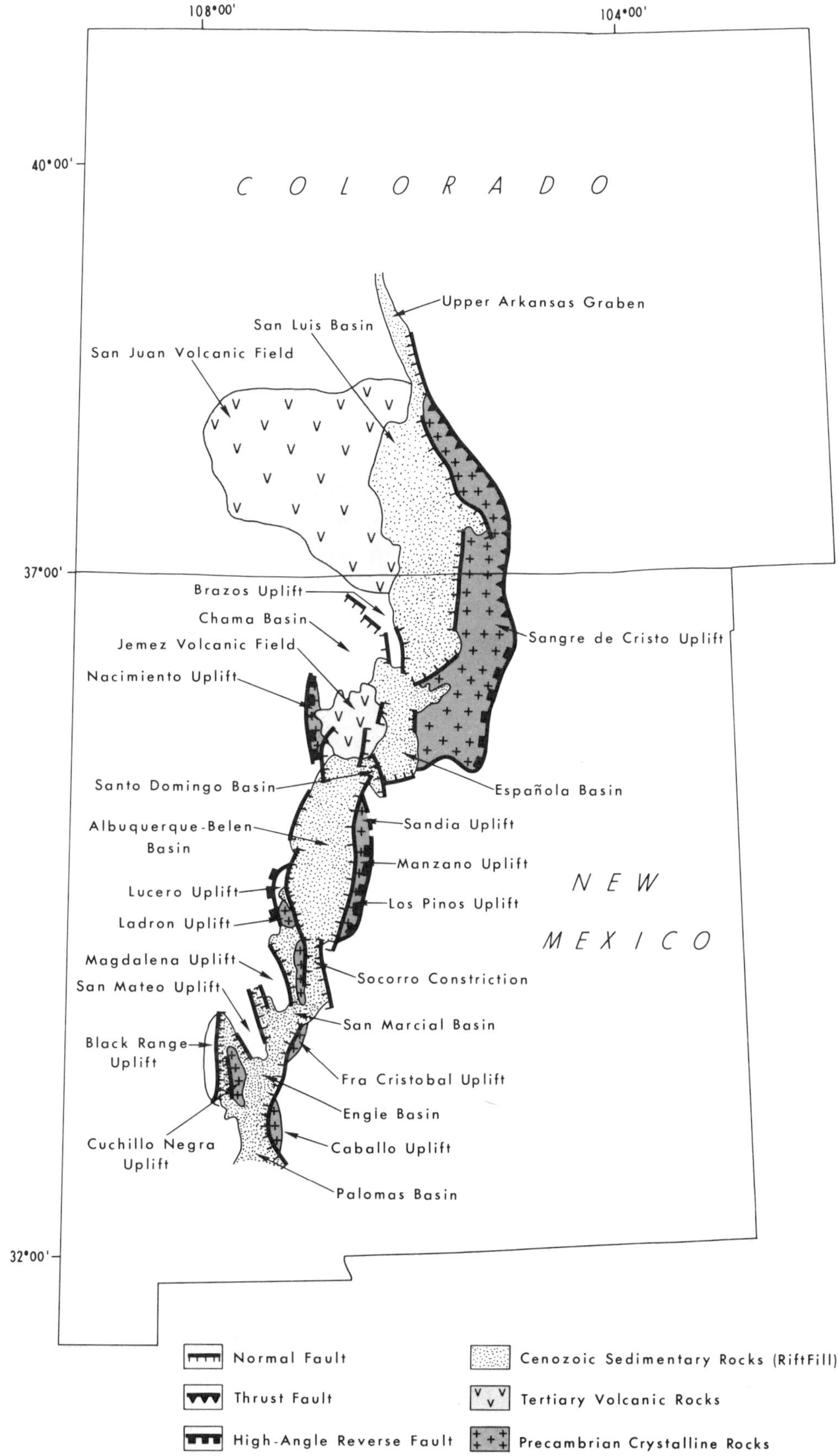

Figure 2. Index map of basins along the Rio Grande rift.

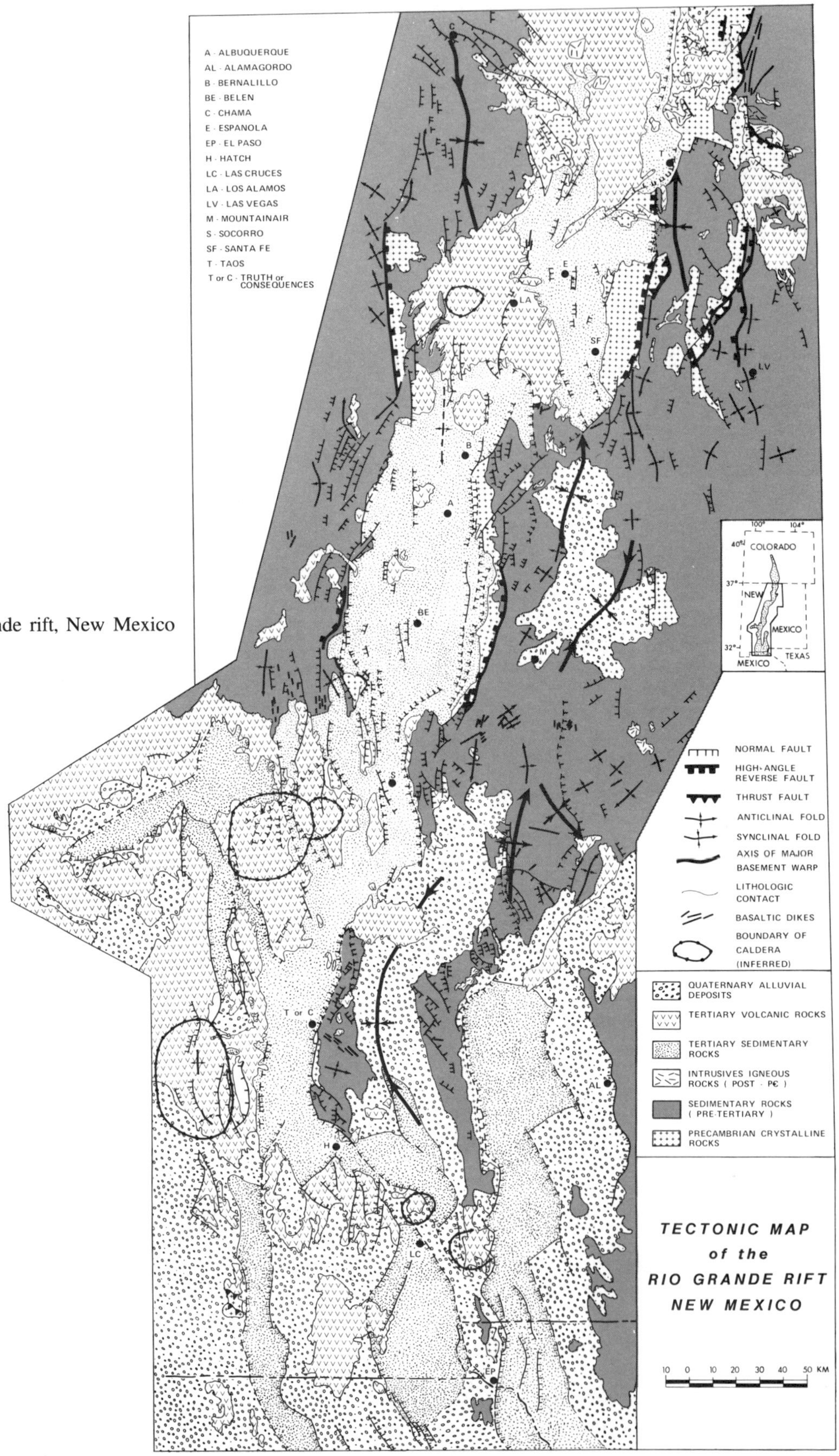

Figure 3. Tectonic map of the Rio Grande rift, New Mexico (modified after Woodward et al., 1978).

temporary seismicity which trends across the rift through the Socorro area.

The southern rift segment extends from Socorro, New Mexico, to the El Paso, Texas, area (Fig. 3) and is typified by a wide zone of northerly trending basins and ranges that contrasts sharply with the topographically distinct and well-defined northern and central segments of the rift. The more diffuse character of the southern rift is likely a consequence of the fact that it has probably undergone a greater amount of crustal attenuation (crust is approximately 26 km thick; McCullar and Smithson, 1977), and also that it merges with or overprints northwest-trending Paleogene structures of the southern Basin and Range province (Fig. 1).

GEOLOGIC FRAMEWORK OF THE ALBUQUERQUE BASIN

Location and physiography

With approximate dimensions of 160 km north-south and 64 km east-west, the Albuquerque Basin is one of the major sedimentary basins along the length of the Rio Grande rift. To the east, the basin is abruptly bounded by the Sandia, Manzanita, Manzano, and Los Pinos Mountains (Figs. 2, 3, and 4), all of which are uplifted blocks of Precambrian crystalline basement rocks. Sandia Crest marks the culmination of the eastern-bounding uplifts at 3255 m. To the west lies the Colorado Plateau, capped by the Mt. Taylor Cenozoic volcanic center and a cover of Paleozoic-Mesozoic sedimentary rocks (Fig. 4). Rocks of the plateau dip gently eastward into the basin making the western margin of the rift essentially a monoclinal ramp. Thus, the western border of the Albuquerque Basin is topographically subdued when compared with the impressive uplifts to the east. Two exceptions are the Ladron Mountains and Nacimiento uplift, which are uplifted blocks of Precambrian basement that respectively form the southwest and northwest corners of the basin (Figs. 2 and 4). Both of these ranges attain elevations of greater than 2750 m.

North of the Albuquerque Basin lies the Jemez Mountains, a major Cenozoic volcanic center, the southern Sangre de Cristo Range of the Rocky Mountains, and the intervening Espanola Basin of the Rio Grande rift (Fig. 2). South of the Albuquerque Basin, the rift becomes less distinct topographically and is typified by a regional pattern of alternating basins and ranges (Fig. 3).

Stratigraphy

Seismic-reflection data (Fig. 4) and exploration-well data (Fig. 5) reveal that the Albuquerque Basin is filled with as much as or greater than 7050 m of Tertiary continental clastic sedimentary rocks. Neogene rift fill makes up at least 4900 m of this and is composed largely of fanglomerates, alluvial, eolian, and playa deposits of the Santa Fe Formation. Underlying Eocene rocks of the Galisteo and Baca Formations were deposited in shallow pre-rift basins and are dominantly fluvial in nature, although some organic-rich lacustrine facies are observed at places. The Eocene section is typically no greater than 900 m in thickness but locally may reach 2150 to 2450 m in thickness in the deep basin. Detailed discussion of the Tertiary stratigraphy of the basin may be found in Lozinsky (1988, 1989).

As much as 3350 m of Paleozoic and Mesozoic rocks make up the pre-rift sedimentary basement on which the Tertiary fill was deposited. Paleozoic rocks range in age from Mississippian to Permian and vary from marine limestones to near coastal continent clastics. Exposures of the Paleozoic section occur along the entire uplifted eastern margin of the basin and on the southwestern basin flank (Fig. 4). Mesozoic rocks are Triassic–Upper Cretaceous in age and occur largely along the northern and western flanks of the basin (Fig. 4). Triassic formations are continental in nature and dominated by fluvial sandstones and mudstones. Jurassic rocks were also deposited in continental environments and vary from fluvial and eolian sandstones to lacustrine limestones and associated evaporites. Upper Cretaceous units of the Albuquerque Basin are dominated by regressive coastal marine sandstones intertonguing with open-marine shales. In addition, the Cretaceous is also characterized by coastal-plain fluvial and paludal deposits that are rich in coal (Molenaar, 1983).

The Precambrian basement is broadly divisible into 1,300-Ma granitic rocks and 1,700-Ma metamorphic rocks ranging from pelitic schists and phyllites, to quartzites, amphibolites, granitic gneisses, metarhyolites, and greenstones (Condie, 1976; Kelley, 1977; White, 1979).

Cenozoic volcanism

Based upon the work of various workers (Kudo et al., 1977; Bachman and Mehnert, 1978; Kelley and Kudo, 1978; Baldridge, 1979; White, 1979; Kudo, 1982), the Cenozoic volcanic history of the Albuquerque Basin region can be broken down into three broad stages. Pre-rift volcanism occurred in the mid-Oligocene (32 to 37 Ma) and was dominated by rhyolitic ash flows and shallow-level felsic plutons. Incipient rifting in the late Oligocene (28 to 21 Ma) is marked by basaltic-andesite flows and shallow intrusives. Much of this activity took place in the southernmost portion of the basin. Volcanism associated with the main stage of rifting in Miocene through recent time (19 Ma to present) is predominantly basaltic (alkaline) and is concentrated primarily in the northern and central portions of the basin. Eruptive activity has continued almost up to the present day, with the youngest basalt flows being those of the Albuquerque and Cat Mesa fields at 0.19 and 0.14 Ma, respectively.

Pre-rift deformational history

Major reverse faulting during the Laramide orogeny was the most significant episode in the pre-rift deformational history of the Albuquerque Basin region. Reverse faults that have

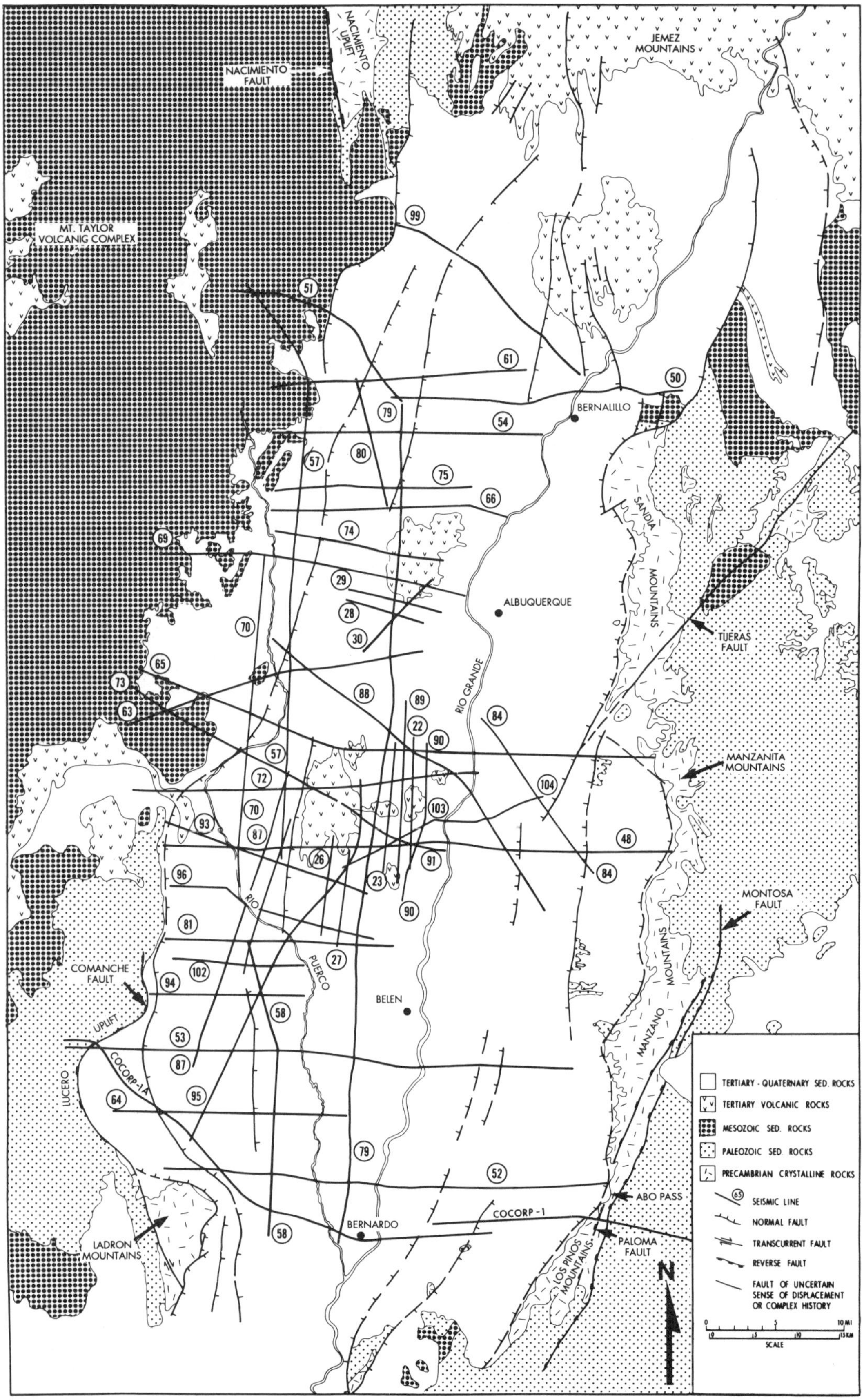

Figure 4. Map showing general outcrop geology of the Albuquerque Basin and the reflection seismic control used in this study.

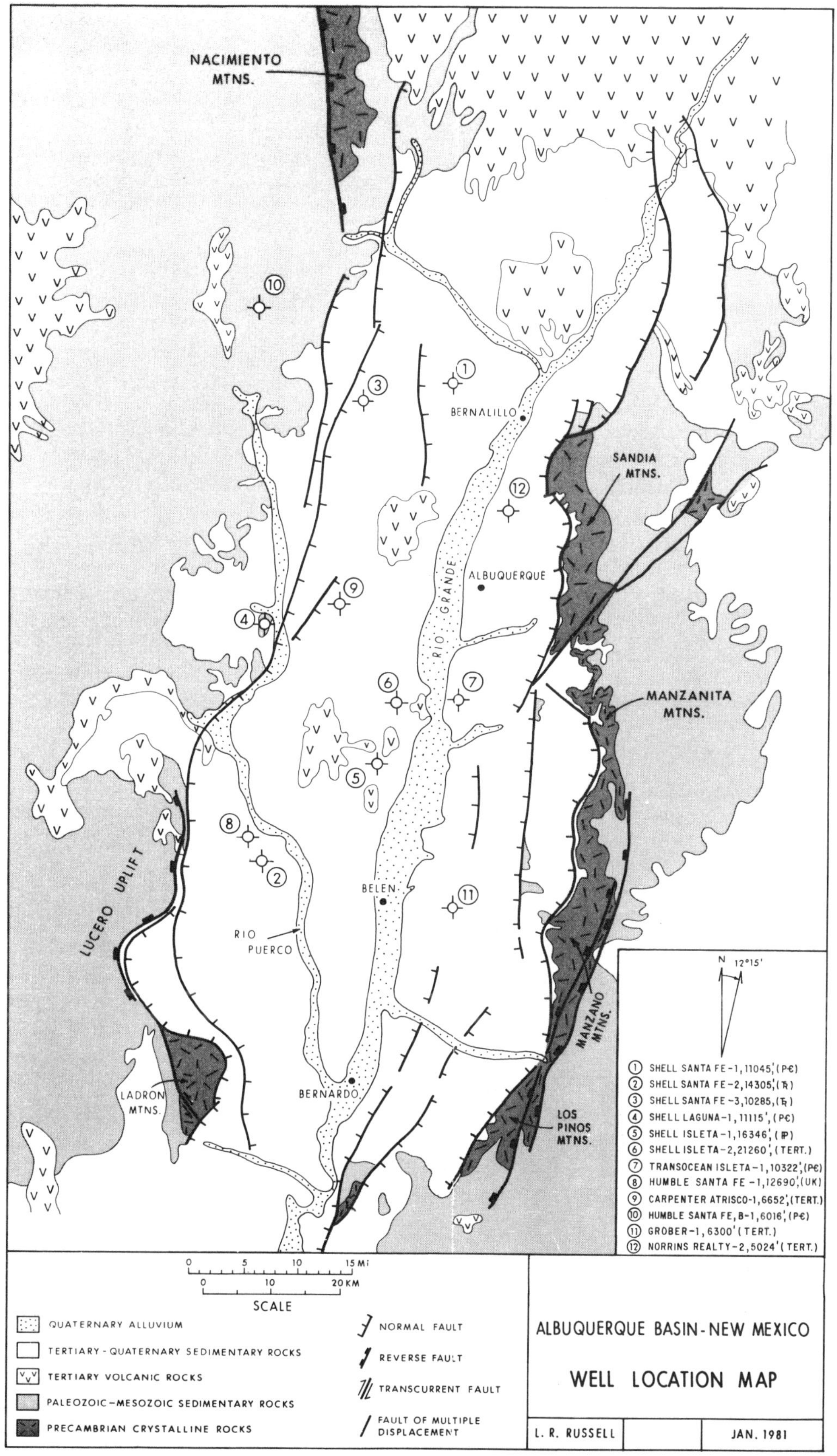

Figure 5. Map showing key exploration well control used in this study.

resulted in the local overturning of Paleozoic and Mesozoic formations are exposed in the Manzano and Los Pinos Mountains and in the Nacimiento uplift (Figs. 2, 3, 4). Upper Cretaceous rocks have been folded along the Tijeras fault zone at the southern end of the Sandia Mountains (Fig. 4), deformation that likely is of Laramide age. This is supported by the fact that the Sandia uplift broadly has the configuration of a north-plunging antiform with Eocene rocks of the Galisteo Formation resting in angular unconformity on Upper Cretaceous rocks in the northern foothills of the range (Kelley, 1977) suggesting some uplift of probable Laramide age. In the southwest corner of the Albuquerque Basin is the modest Lucero uplift (Fig. 4) whose uplift is related to motion along the Comanche fault, a reverse fault that induced monoclinal upwarp and gravity sliding in Laramide time (Callender and Zilinski, 1976; Kelley, 1977).

Another school of thought exists that relates these basin margin uplifts, and structures associated with them, to dextral wrench faulting of either Laramide or Tertiary (rift-related) age (Chapin, 1983; Hayden, 1991; Hayden and Mawer, 1991). Work by Molzer and Erslev (1991) farther to the north in the Laramide foreland (Owl Creek Range, Wyoming) describes structural patterns very similar to those in the uplifts flanking the Albuquerque Basin, and attributes them to Laramide thrusting with a modest component of oblique slip. Lewis and Baldridge (this volume), as well as May et al. (this volume), present arguments that relate the uplift and structures along the Lucero front as largely due to footwall uplift during Tertiary rifting. Considerable room for discussion and debate clearly still exists relative to this topic.

Some late Paleozoic deformation in the region related to the development of the ancestral Rocky Mountains is likely (Chapin and Seager, 1975), although its character is not well defined. Precambrian basement rocks exposed in the basin-flanking uplifts were isoclinally folded and metamorphosed to high grade at approximately 1,700 Ma and later were intruded by major granitic plutons at 1,300 Ma (Kelley, 1977; Condie, 1976). In addition, there is strong evidence of significant Precambrian strike-slip motion (post-metamorphic) along the Tijeras fault system (J. F. Callender, personal communication, 1979) that correlates well with similar motion along the Pecos-Picuris fault in the southern Sangre de Cristo Range to the north (Miller et al., 1963). Furthermore, there is growing evidence for probable reverse movement along these faults at some point during the Precambrian (J. F. Callender, personal communication, 1979).

STRUCTURAL FRAMEWORK OF THE ALBUQUERQUE BASIN

The structural framework map shown in Figure 6 was constructed using regional reflection seismic data (Fig. 4), supplemented by exploration well control (Fig. 5). This map depicts the structural complexity of the pre-rift "basement" (top Upper Cretaceous largely) and illustrates that the Albuquerque Basin is made up of two subbasins (north and south graben blocks) downdropped along major low-angle normal faults of opposing polarity. The north graben is controlled by the west-dipping Rio Grande fault, whereas the south graben is principally controlled by the east-dipping Santa Fe–Coyote fault (Fig. 6).

The structural centers of the two subbasins are quite narrow and reach maximum depths of approximately 7350 m in the north graben and 7050 m in the south graben. Flanking the narrow basin deeps to the east and west are a variety of structural benches and detached blocks. To the east, both the north and south grabens are bounded by a broad structural bench extending the entire length of the basin (north and south Albuquerque bench; Fig. 6). The north graben is bounded to the west by a major east-dipping antithetic normal fault (West Mesa fault; Fig. 6), which separates the deep basin from a wide and shallow structural bench (Laguna bench; Fig. 6). In contrast, the south graben is flanked to the west by two smaller blocks or benches (Monte Largo bench and Rio Puerco block) produced by displacements along the Jeter and Santa Fe–Coyote faults (Fig. 6).

One conspicuous aspect of basin geometry is that the physiographic margins of the basin do not correspond with the location of the major basin-controlling normal faults (Fig. 6). The main structural depressions occur much farther into the basin with structural relief at the topographic margins generally being comparatively subdued. Rigid and relatively undeformed basement terraces flank the basin to the east and west (High Plains block and Colorado Plateau block; Fig. 6).

The northern and southern terminations of the basin have different structural characteristics. The north basin floor rises gradually northward, essentially in ramplike fashion out of the structural deep. In the south, the pattern is more complex and obscure, with the major characteristic being a distinct narrowing and shallowing of the basin. The southern terminus has been called the Socorro Constriction (Chapin, 1979), south of which the rift is more basin and rangelike in its topographic expression.

Major basin-controlling normal faults/northern Albuquerque Basin

The Rio Grande fault, the major basin-controlling structural element of the northern Albuquerque Basin, is a west-dipping listric-normal fault having a throw of between 4600 and 6100 m (Figs. 6, 7, 8). It is expressed seismically by strong fault plane reflections and pre-rift (Cretaceous and older) basement reflections that distinctly tilt into the fault plane (Figs. 7, 8). Dips range from greater than 60° on the upper portion of the fault to as low as 17° along its lower reaches. This master fault appears to be flattening at approximately 10 km when the seismic data are depth corrected utilizing an average velocity-depth function for the basin extrapolated from exploration well velocity surveys.

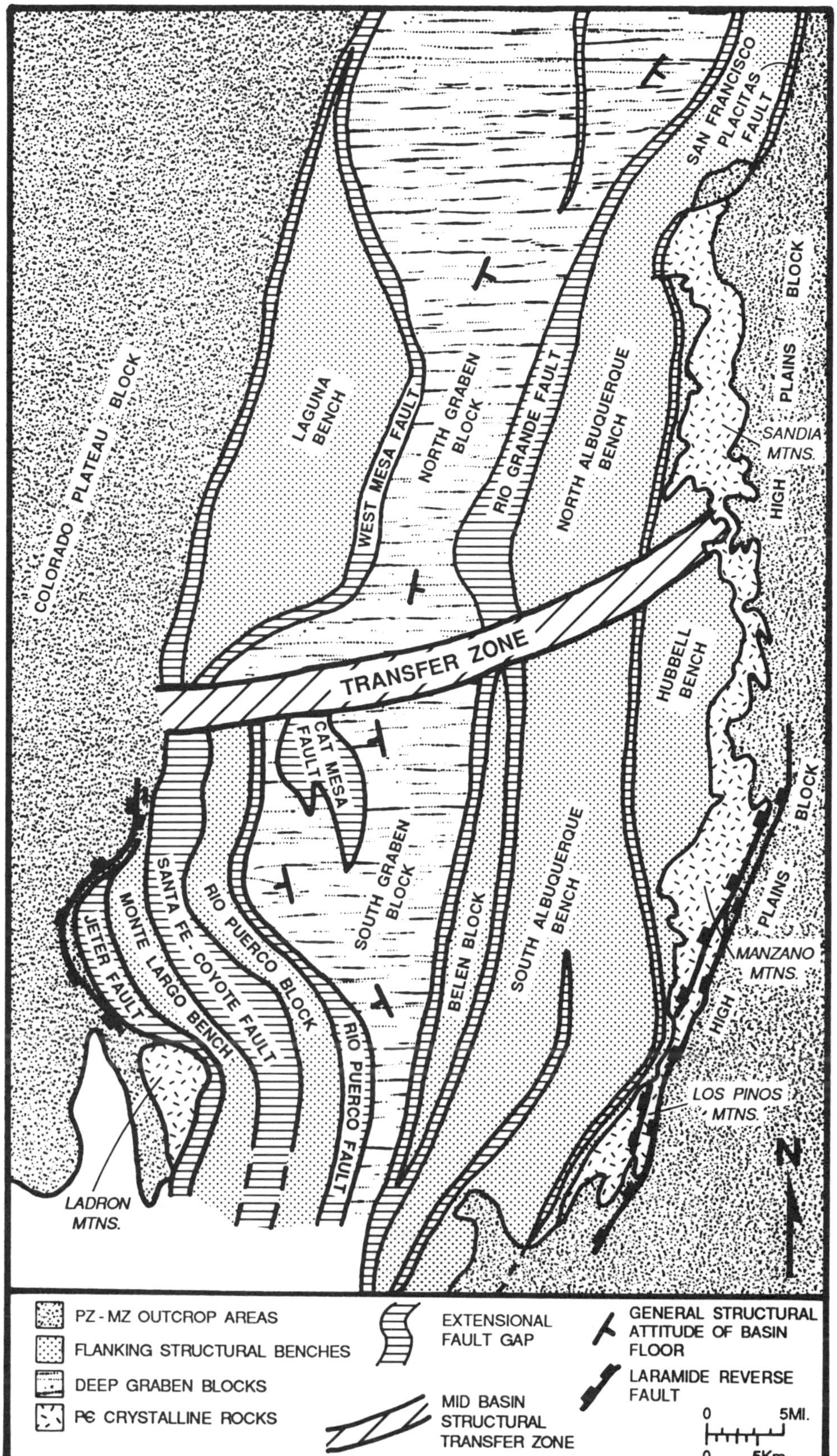

Figure 6. Generalized structural framework map of the Albuquerque Basin depicting the structural configuration of the pre-Tertiary "basement" (mostly top Upper Cretaceous).

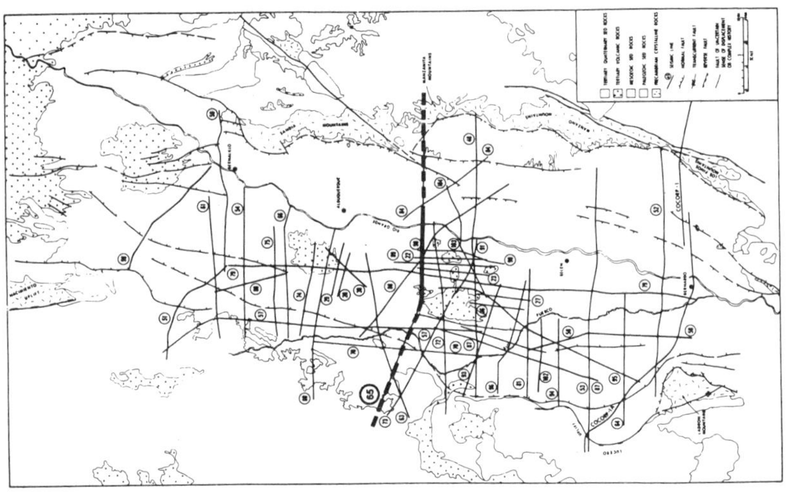

A

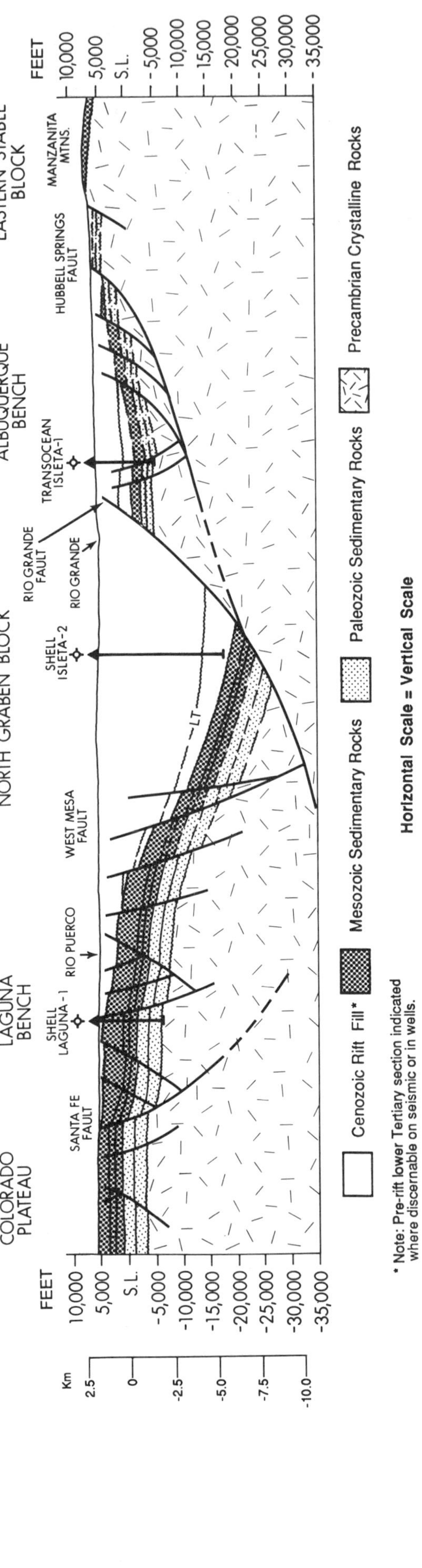
W
E
FEET
10,000
5,000
S.L.
-5,000
-10,000
-15,000
-20,000
-25,000
-30,000
-35,000
Km
2.5
0
-2.5
-5.0
-7.5
-10.0
COLORADO PLATEAU
LAGUNA BENCH
NORTH GRABEN BLOCK
ALBUQUERQUE BENCH
EASTERN STABLE BLOCK
SANTA FE FAULT
SHELL LAGUNA-1
RIO PUERCO
WEST MESA FAULT
SHELL ISLETA-2
RIO GRANDE
RIO GRANDE FAULT
TRANSOCEAN ISLETA-1
HUBBELL SPRINGS FAULT
MANZANITA MTNS.
LT
Cenozoic Rift Fill*
Mesozoic Sedimentary Rocks
Paleozoic Sedimentary Rocks
Precambrian Crystalline Rocks
Horizontal Scale = Vertical Scale
* Note: Pre-rift lower Tertiary section indicated where discernable on seismic or in wells.

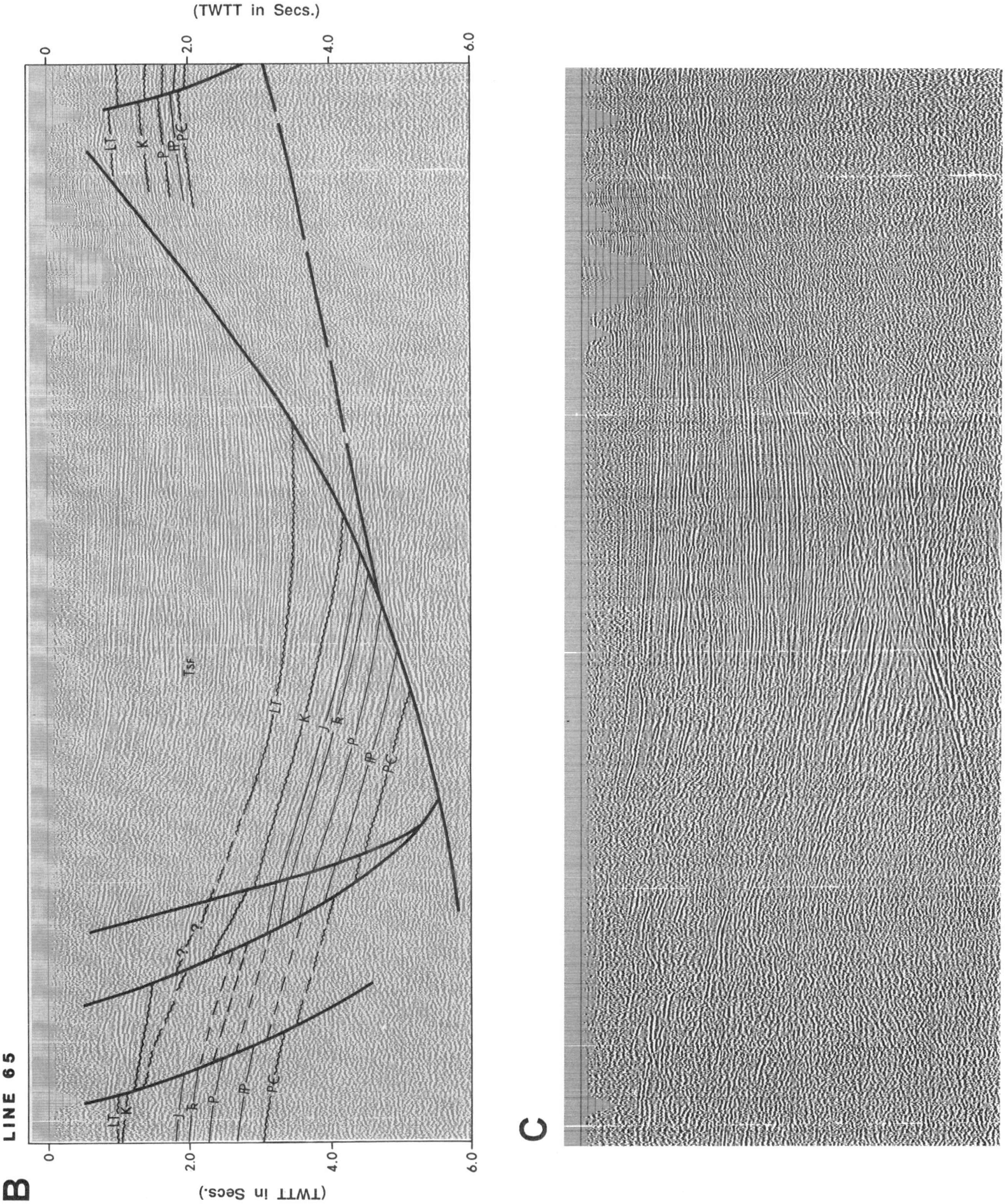

Figure 7. Geologic (A) and seismic (B, interpreted; C, uninterpreted) sections illustrating the structural configuration of the southern portion of the northern Albuquerque Basin. Location of geologic section shown by dashed line on index map; position of seismic line indicated by solid line.

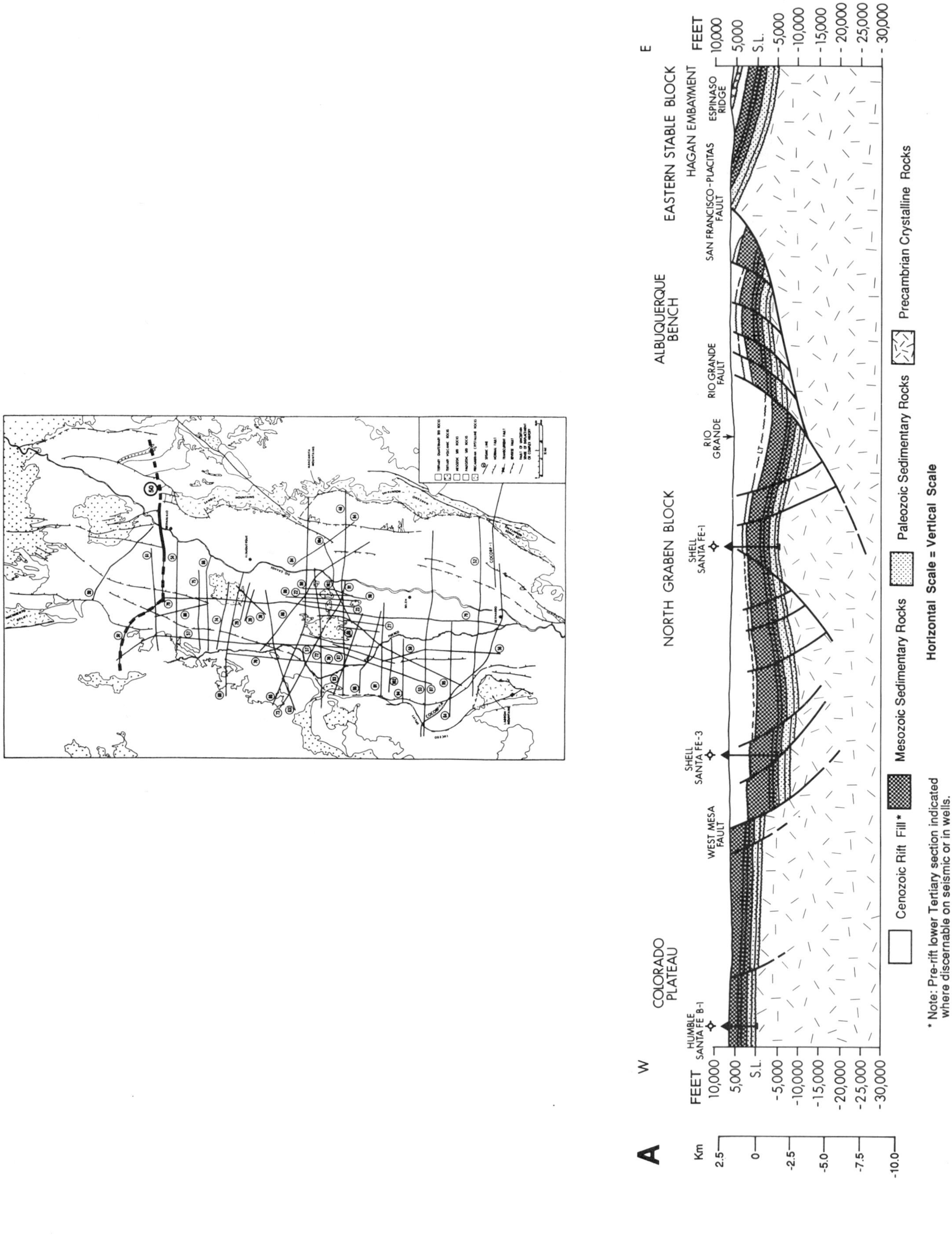
A
Km
2.5
0
-2.5
-5.0
-7.5
-10.0
FEET
10,000
5,000
S.L.
-5,000
-10,000
-15,000
-20,000
-25,000
-30,000
W
E
HUMBLE SANTA FE B-1
COLORADO PLATEAU
WEST MESA FAULT
SHELL SANTA FE-3
NORTH GRABEN BLOCK
SHELL SANTA FE-1
RIO GRANDE
RIO GRANDE FAULT
LT
ALBUQUERQUE BENCH
SAN FRANCISCO-PLACITAS FAULT
EASTERN STABLE BLOCK
HAGAN EMBAYMENT
ESPINASO RIDGE
Cenozoic Rift Fill *
Mesozoic Sedimentary Rocks
Paleozoic Sedimentary Rocks
Precambrian Crystalline Rocks
Horizontal Scale = Vertical Scale
* Note: Pre-rift lower Tertiary section indicated where discernable on seismic or in wells.
50

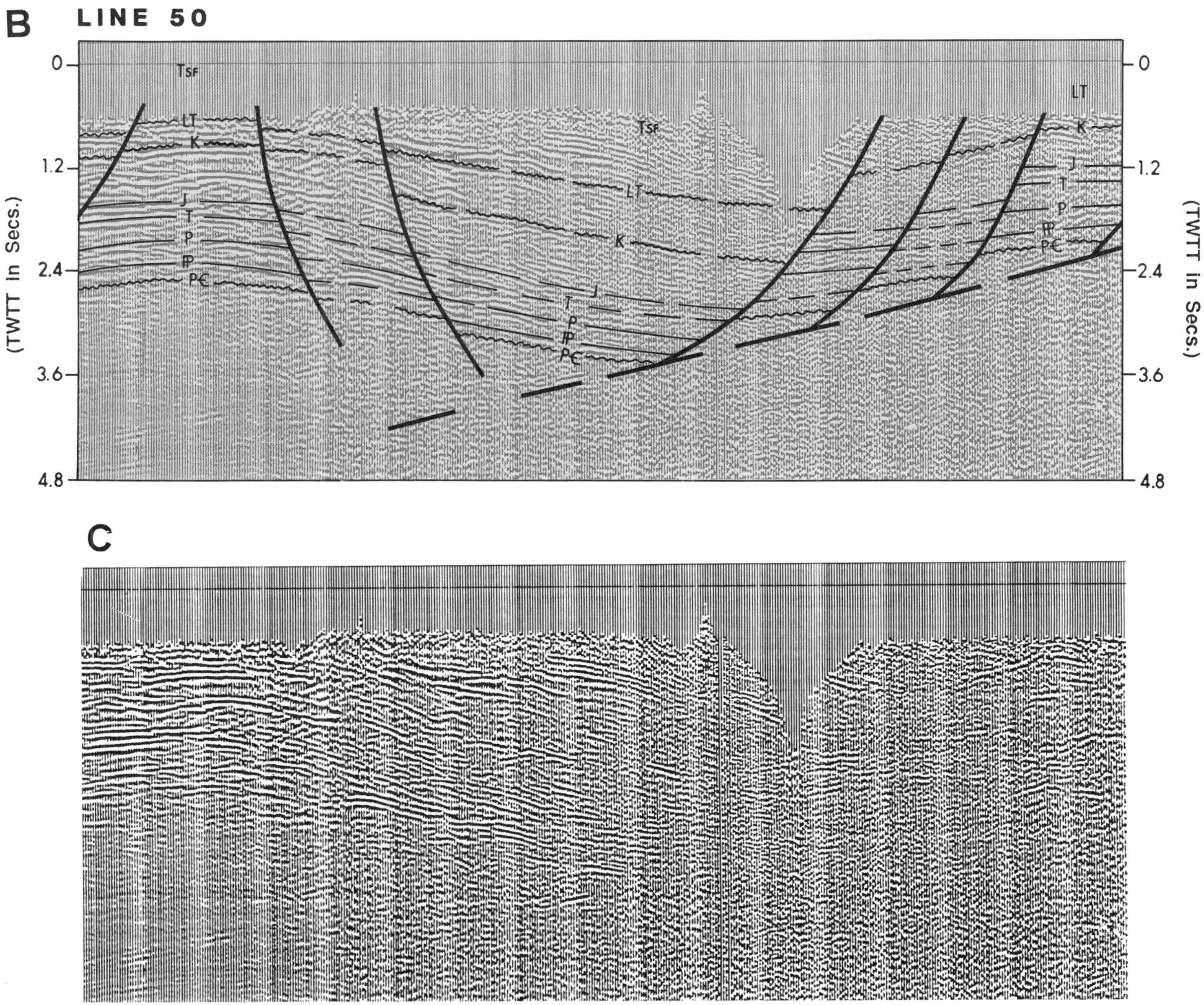

Figure 8 (on this page spread and following page). Geologic (A) and seismic (B, interpreted; C, uninterpreted) sections, as well as outcrop relationships (D) at the northern end of the Sandia Mountains, illustrating the structural configuration of the northern portion of the northern Albuquerque Basin. Location of geologic section shown by dashed line on index map; position of seismic line indicated by solid line.

QUATERNARY SEDIMENTARY ROCKS
UPPER TERTIARY SEDIMENTARY ROCKS (SANTA FE FORMATION)
LOWER TERTIARY SEDIMENTARY ROCKS (GALISTEO FORMATION)
UPPER CRETACEOUS SED. ROCKS
MESA VERDE GROUP
MANCOS SHALE
DAKOTA SANDSTONE
JURASSIC SEDIMENTARY ROCKS
MORRISON FM.
TODILTO FM.
TRIASSIC SEDIMENTARY ROCKS
CHINLE FM.
SHINARUMP FM.
PERMIAN SEDIMENTARY ROCKS
SAN ANDRES & GLORIETTA FM.
YESO FM
ABO FM
PENNSYLVANIAN SEDIMENTARY ROCKS
MADERA FM.
SANDIA FM.
MISSISSIPPIAN SEDIMENTARY ROCKS (ARROYO PENASCO FM.)
PRECAMBRIAN GRANITIC ROCKS
PRECAMBRIAN METAMORPHIC ROCKS
LITHOLOGIC CONTACT
NORMAL FAULT
STRIKE & DIP OF BEDDING
FOLD AXIS TREND & PLUNGE
STRIKE & DIP OF FOLIATIONS
TRUE NORTH
MAGNETIC NORTH
12°15'
SCALE
PLACITAS
SAN FRANCISCO-PLACITAS FAULT
D

A related low-angle normal fault along the northeast flank of the Albuquerque Basin is the San Francisco–Placitas fault (Fig. 6). To the west, this low-angle fault merges with or is truncated at depth by the higher angle Rio Grande fault (Fig. 8). The low-angle character of the San Francisco–Placitas fault is suggested by faint fault plane reflections on seismic data (Fig. 8) and by viewing geologic map relationships of the northern Sandia Mountains (Fig. 8) using the down-structure method of Mackin (1950). As discussed previously, the Sandia uplift is a northerly plunging antiformal structure. This is illustrated on the geologic map (Fig. 8) by dips in the Pennsylvanian strata south of the arcuate surface trace of the fault. Dips change from due east, to northeast, and then northerly to slightly northwesterly at the northern end of the range. If this geologic map (Fig. 8) is viewed in a northerly direction, then an approximate cross-sectional view of the area is produced. This graphically supports the interpretation that the San Francisco–Placitas fault is low angle and listric in geometry, as suggested on seismic, and that the fault was uplifted and exposed as a result of the late Tertiary uplift of the Sandia range. The subsidiary normal faults seen on the geologic map (Fig. 8) merge with the master San Francisco–Placitas fault in such a way that, when considered in cross-sectional perspective, suggests they are synthetic normal faults to the master normal fault. In addition, geologic map relationships suggest rollover or reverse drag of the Mesozoic section into the San Francisco–Placitas fault, geometries common to listric fault systems.

This joint use of both seismic data and surface geologic map relationships allows for the more confident completion of the geologic section (Fig. 8). The fact that the San Francisco–Placitas fault merges with or is truncated by the Rio Grande fault suggests that extension initially began on the San Francisco–Placitas fault to the east, and that with time, the main focus of extension shifted westerly to the Rio Grande fault. This relationship is supported by the existence of older rift fill sediments on the structural bench above the San Francisco–Placitas fault, with continued deposition and structural tilting of younger rift fill section on the downthrown side of the Rio Grande fault (May et al., this volume).

The overall listric, downward-flattening geometry of the Rio Grande fault system is compatible stylewise with the pure shear model of extension (McKenzie, 1978). It appears that the depth of flattening of the fault is in the range of 10 to 12 km, which correlates well with the base of earthquake activity in the basin (Sanford et al., 1979). This suggests the fault system is decoupling at the top of the brittle-ductile transition in the crust.

Major basin-controlling normal faults/southern Albuquerque Basin

In the southern Albuquerque Basin, the east-dipping Santa Fe–Coyote, Cat Mesa, and Jeter faults that form the key structural elements of basin extension have virtually the same geometric relationship to one another as the Rio Grande and San Francisco–Placitas faults in the northern basin (Figs. 6, 9, 10). Although the comparable stratigraphic evidence is lacking, by analogy with the Rio Grande/San Francisco–Placitas fault system, the Santa Fe–Coyote fault system appears to be older than the Cat Mesa fault, which like the Rio Grande fault in the northern portion of the basin continued to move throughout the later stages of rifting. The Santa Fe–Coyote fault displays more than 3050 m of throw at top Cretaceous level, while the Cat Mesa fault has an additional 4,600 to 5,200 m of throw at top Cretaceous level. Both fault systems are delineated seismically by strong fault-plane reflections, synthetic faults rooting into the master fault, and prominent tilt of pre-rift reflections into the fault planes (Fig. 9). In outcrop, surface dips on the Santa Fe fault are about 65°, whereas dips as low as 15° are suggested on seismic at depths of about 10 km.

In the southern part of the basin the Coyote fault is the younger, dominant structure and it truncates the earlier, lower-angle Jeter fault (Fig. 10). Seismic reflections from the Coyote fault suggest it dips between 15 and 17° at the depth of maximum seismic resolution.

The Jeter fault is well exposed along the northeastern and eastern fronts of the Ladron Mountains (Fig. 6) where it dips basinward at angles as low as 26°. Features such as strongly slickensided fault planes, sheared Precambrian basement rocks in the footwall, and slivers or horses of lower Tertiary, Permian, and Triassic strata (Black, 1964) collectively support the seismic interpretation that the major basin-controlling faults, and the Jeter fault specifically, penetrate the basement.

In contrast to the apparently pure shear geometry of the north subbasin where the listric Rio Grande fault is interpreted to sole at the base of the brittle crust, the low-angle, ramplike, basin-controlling faults in the south subbasin (Jeter, Santa Fe–Coyote, and Cat Mesa faults) are strongly analogous to the simple shear model of full lithosphere extension proposed by Wernicke (1985, 1986). The basin-bounding faults of the south Albuquerque Basin show no signs of detaching or soleing at midcrustal depths. Instead, they appear to be continuing to perhaps full crustal or lithospheric depths when seismic resolution is lost.

PALINSPASTIC ANALYSIS OF THE BASIN

In order to test our structural interpretations of the seismic data, an attempt was made to volumetrically balance and restore our cross sections palinspastically.

The reconstructions were accomplished utilizing the computer program PREP, discussed in detail in Worrall and Snelson (1989). Reconstructions were done at the Precambrian basement surface on each section (Fig. 11). As a result of the fact that each reconstruction was accomplished essentially through rigid body rotations, some slight gaps and overlaps between blocks were produced (Fig. 11). If ductile deformation in the basement rocks were to be built into the reconstruc-

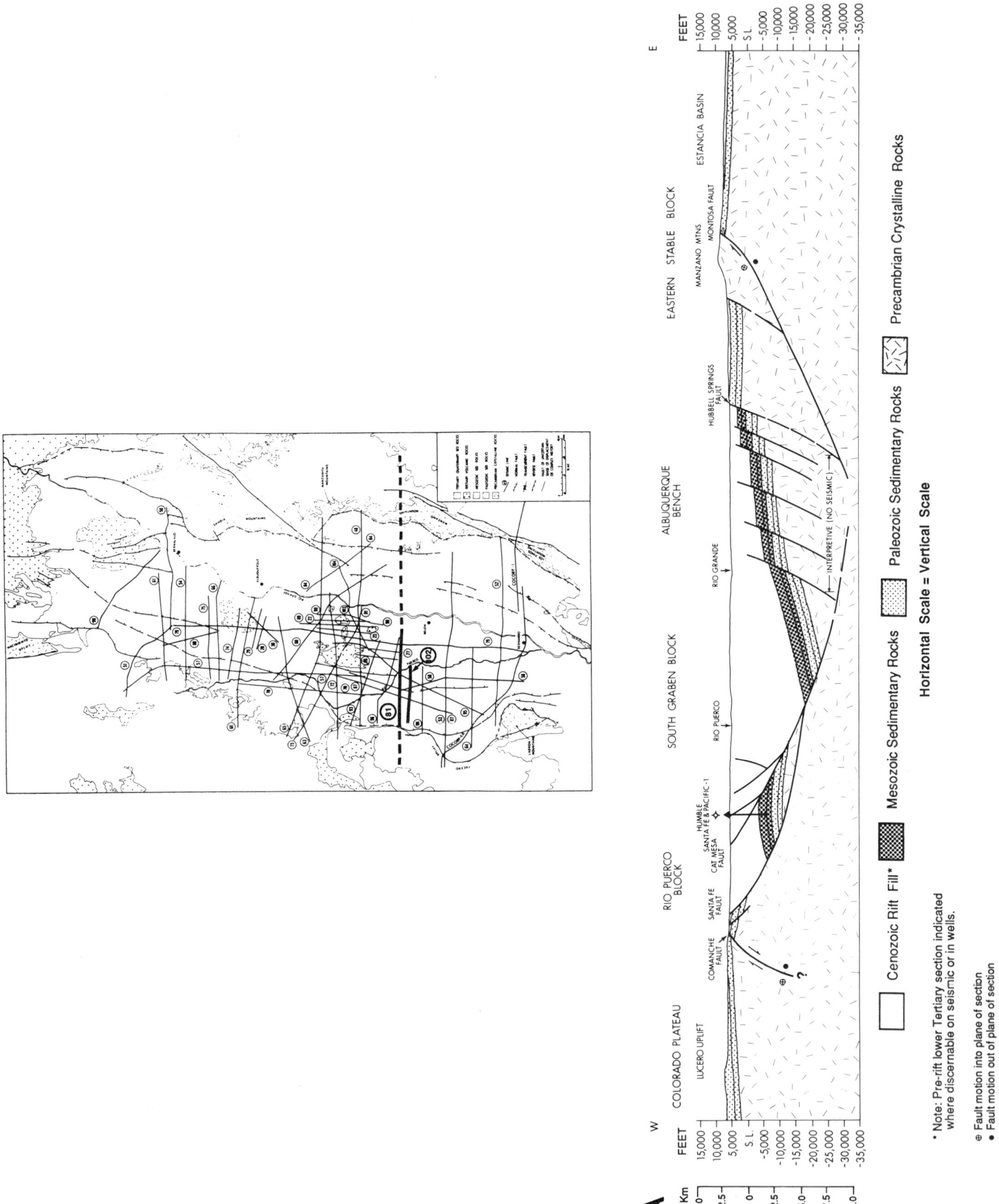
W
E
FEET
Km
COLORADO PLATEAU
LUCERO UPLIFT
COMANCHE FAULT
SANTA FE FAULT
RIO PUERCO BLOCK
CAT MESA FAULT
HUMBLE SANTA FE & PACIFIC-1
SOUTH GRABEN BLOCK
RIO PUERCO
RIO GRANDE
ALBUQUERQUE BENCH
INTERPRETIVE (NO SEISMIC)
HUBBELL SPRINGS FAULT
EASTERN STABLE BLOCK
MANZANO MTNS
MONTOSA FAULT
ESTANCIA BASIN
Cenozoic Rift Fill*
Mesozoic Sedimentary Rocks
Paleozoic Sedimentary Rocks
Precambrian Crystalline Rocks
Horizontal Scale = Vertical Scale
* Note: Pre-rift lower Tertiary section indicated where discernable on seismic or in wells.
⊕ Fault motion into plane of section
● Fault motion out of plane of section

A

B

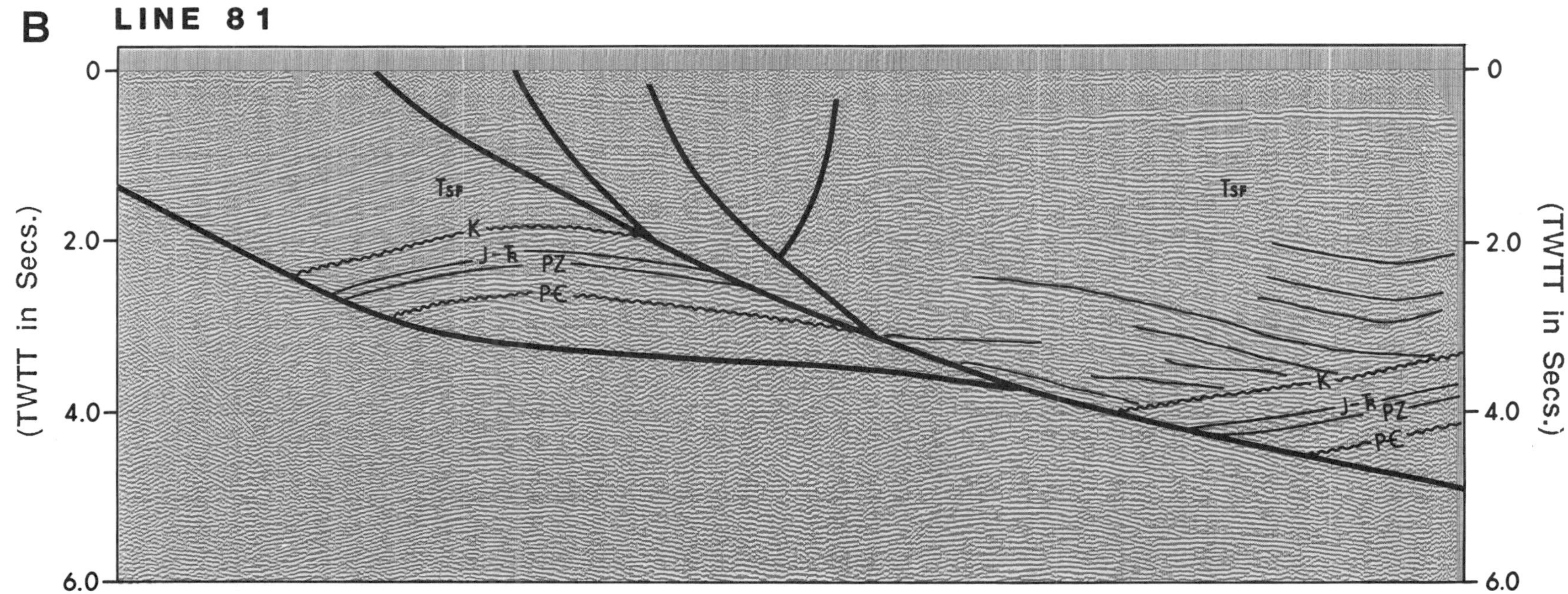

C

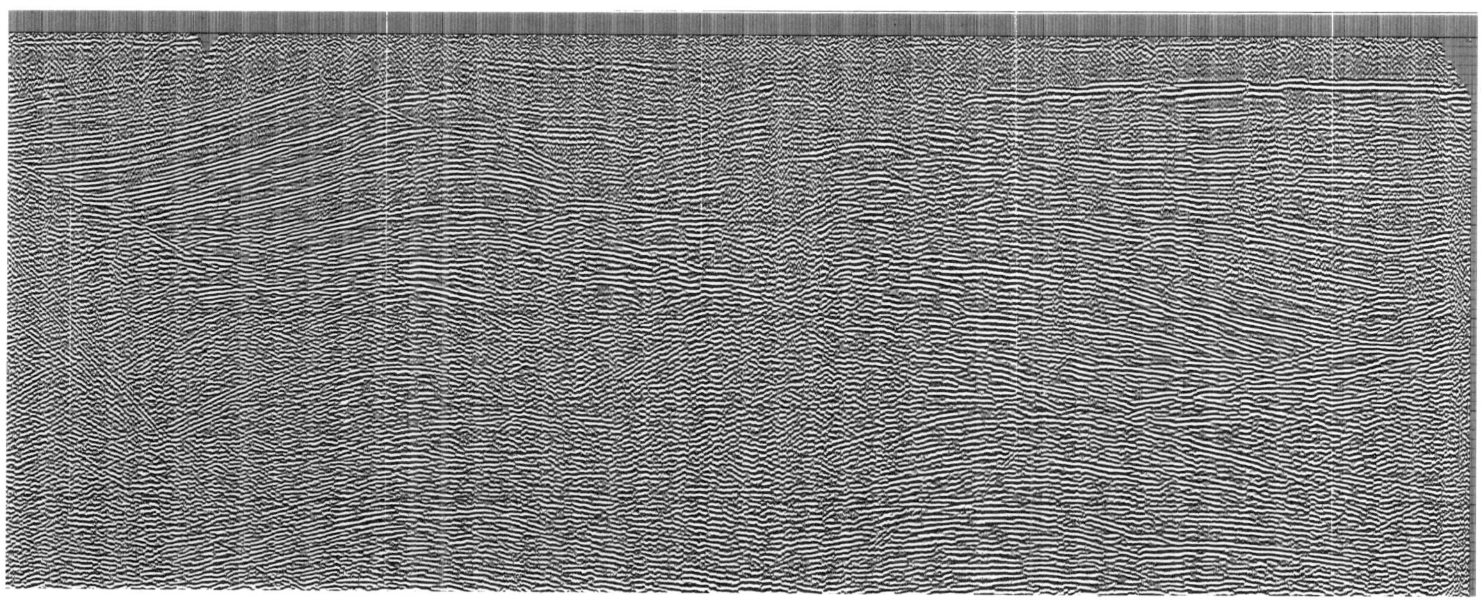

Figure 9 (on this page spread and following page). Geologic (A) and seismic sections (B and D, interpreted; C and E, uninterpreted) illustrating the structural configuration of the north portion of the southern Albuquerque Basin. Location of geologic section shown by dashed line on index map; position of seismic lines indicated by solid lines.

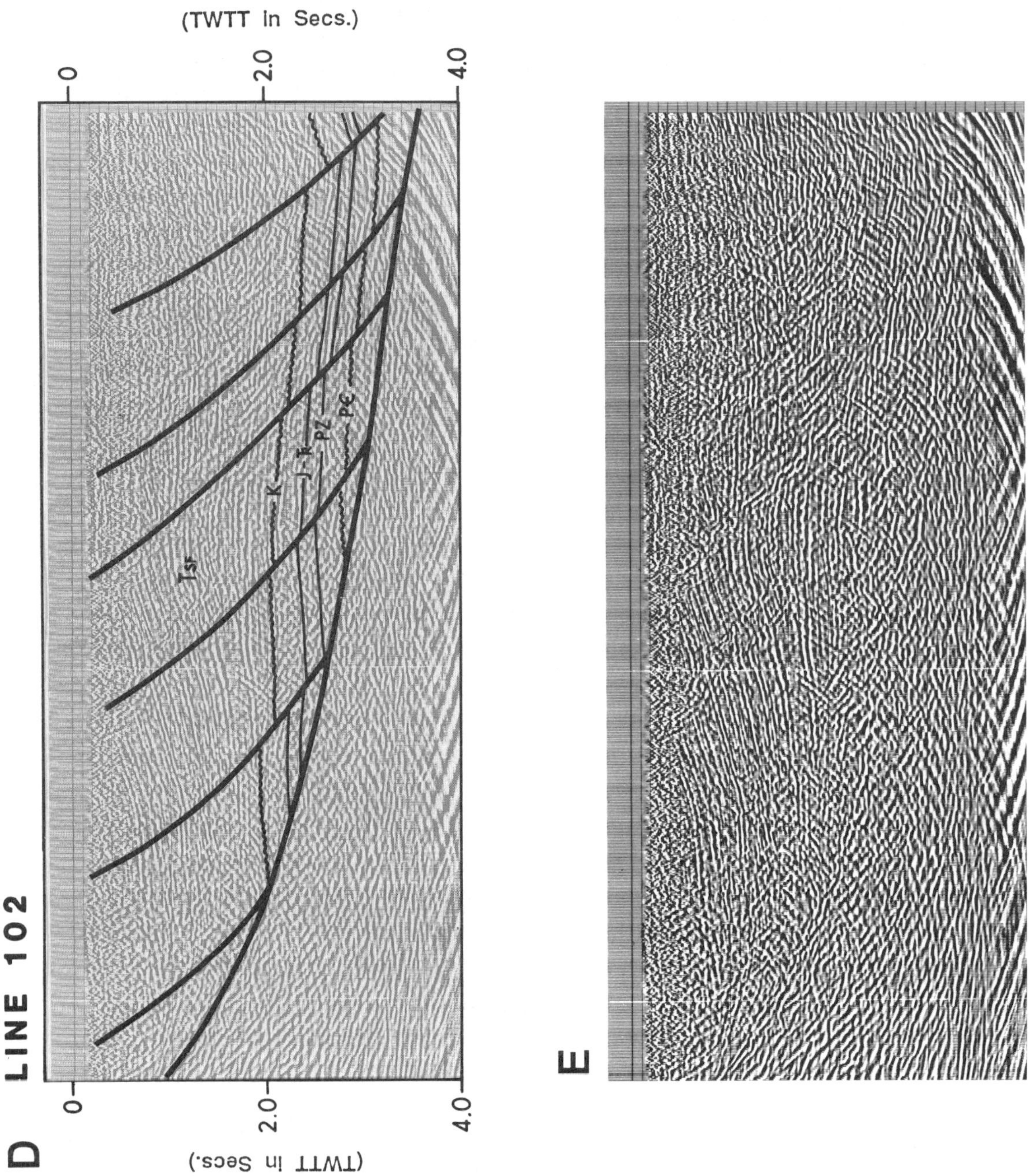
D
LINE 102
(TWTT in Secs.)
0
2.0
4.0
K
PZ
E

tion process, which in nature would result in flow and change in shape of blocks, these apparent inaccuracies, we feel, would be removed.

Palinspastic restorations across the northern and southern halves of the Albuquerque Basin (Fig. 11) have revealed a distinct variation in the amount of Tertiary extension. The northern Albuquerque Basin has undergone a minimum of 17% extension whereas the southern part of the basin has been stretched by at least 28%. These minimum extension values apply to reconstruction to a preextension basement assumed to be horizontal. It is suspected, however, that after Laramide thrusting, the basement underlying the southern Albuquerque Basin was structurally higher than the basement in the Manzano and Los Pinos Mountains to the east as a result of reverse motion along the Montosa and Paloma faults. In this likely scenario, a larger, yet unconstrained, amount of extension would have been necessary in order for the basement to reach present-day depths beneath the rift in the south basin, thus making the difference in extension between subbasins potentially even greater.

MIDBASIN ZONE OF STRUCTURAL COMPENSATION/ACCOMMODATION ZONE

The differential amount and polarity of extension between the two subbasins necessitates the presence of a transfer zone or zone of structural accommodation in the basin center (Figs. 6, 12). This accommodation zone between the northern and southern subbasins is a relatively narrow and enigmatic zone. Unfortunately, the seismic quality at this critical point in the basin is, in general, quite poor because of the effects of the extensive surface volcanic cover localized in this portion of the basin. However, a recently reprocessed key dip line across this midbasin zone interpreted by L. G. Garmezy (Fig. 12) reveals geometries similar to those in negative flower structures associated with zones of transcurrent tectonics (Harding, 1985). This interpretation is structurally compatible with the zone being the locus of differential lateral motion accommodating the opposing extensional regimes of the northern and southern subbasins.

The location of this accommodation zone seems to correspond roughly with the basinward extension of the Tijeras fault zone at the southern end of the Sandia Mountains (Fig. 4). The Tijeras system continues to the northeast of the basin into the High Plains, and perhaps to the southwest onto the Colorado Plateau. This allows for the geometrically required transfering of the differential extension within the basin to structures that extend out of the basin boundaries, with the displacements dying out to the northeast and southwest.

CONTROL OF TERTIARY EXTENSIONAL GEOMETRIES BY OLDER STRUCTURES

The question as to whether the geometry of Tertiary extension was influenced by preexisting zones of structural weakness in the crust can be addressed somewhat by the data presented here. As discussed earlier in the paper, the Rio Grande rift is located along the zone of Laramide basement uplifts of the southern Rocky Mountains arguing for some element of spatial control on the location of the rift. More detailed relationships can be seen when examining the Montosa and Paloma faults, which are west-dipping Laramide reverse faults of the southern Manzano and Los Pinos Mountains (Kelley, 1977; Figs. 2, 3, 4, 6, and 10). Compressional deformation associated with these two faults is best expressed in the Abo Pass area where overturned bedding in Pennsylvanian rocks is observed.

Shell seismic data shot in the vicinity of Abo Pass display two west-dipping zones of faint basement reflections, which, when projected a short distance updip, correlate well with the outcrop location of these two reverse faults (Fig. 10). West-dipping Tertiary normal faults in the basin apparently are rooting into the older reverse faults. Thus, the pattern of Tertiary extension seems to have been controlled by the preexisting structural grain in the basement. In this case, the roots of Laramide reverse faults have provided a zone of weakness or structural anisotropy into which the Tertiary normal faults sole. This interpretation has also been proposed by the COCORP group in reexamining the data from the Rio Grande rift (de Voogd and others, 1988).

Another probable example of basement fabric controlling younger extensional faulting occurs in the Ladron Mountains on the southwest flank of the basin. The Tertiary-aged Jeter fault is well exposed along the eastern and northeastern fronts of the Ladron Mountains and dips basinward at angles of between 26 and 30°. Precambrian metamorphic units exposed in the range in close proximity to the Jeter fault are largely well foliated, and locally mylonitic, quartzites, schists, gneisses, and granitic gneisses with easterly dips as low as 20 to 30°. Seismic data shot immediately basinward of the exposures of the Jeter fault show it to be distinctly low angle in nature with dips between 20 and 30° (Fig. 10). These relationships suggest the possibility that the geometry of Tertiary extension was controlled, at least in part, by the preexisting structural fabric in the basement.

In addition, the position of the midbasin transfer or accommodation zone relative to the pre-rift Tijeras fault zone (Figs. 6, 12) also argues for tectonic inheritance of older structural trends. At the very least, the presence of zones such as the Tijeras fault zone provided a convenient weakness or anisotropy in the crust that was utilized during rifting.

IMPLICATIONS OF THIS STUDY FOR PREVIOUS INTERPRETATIONS OF COCORP DATA

During the 1980s a number of interpretations of the COCORP deep seismic profiles in this area were published that differ to varying degrees from the interpretations proposed in this paper. Early interpretations of the COCORP lines in the southern Albuquerque Basin suggested that a highly faulted

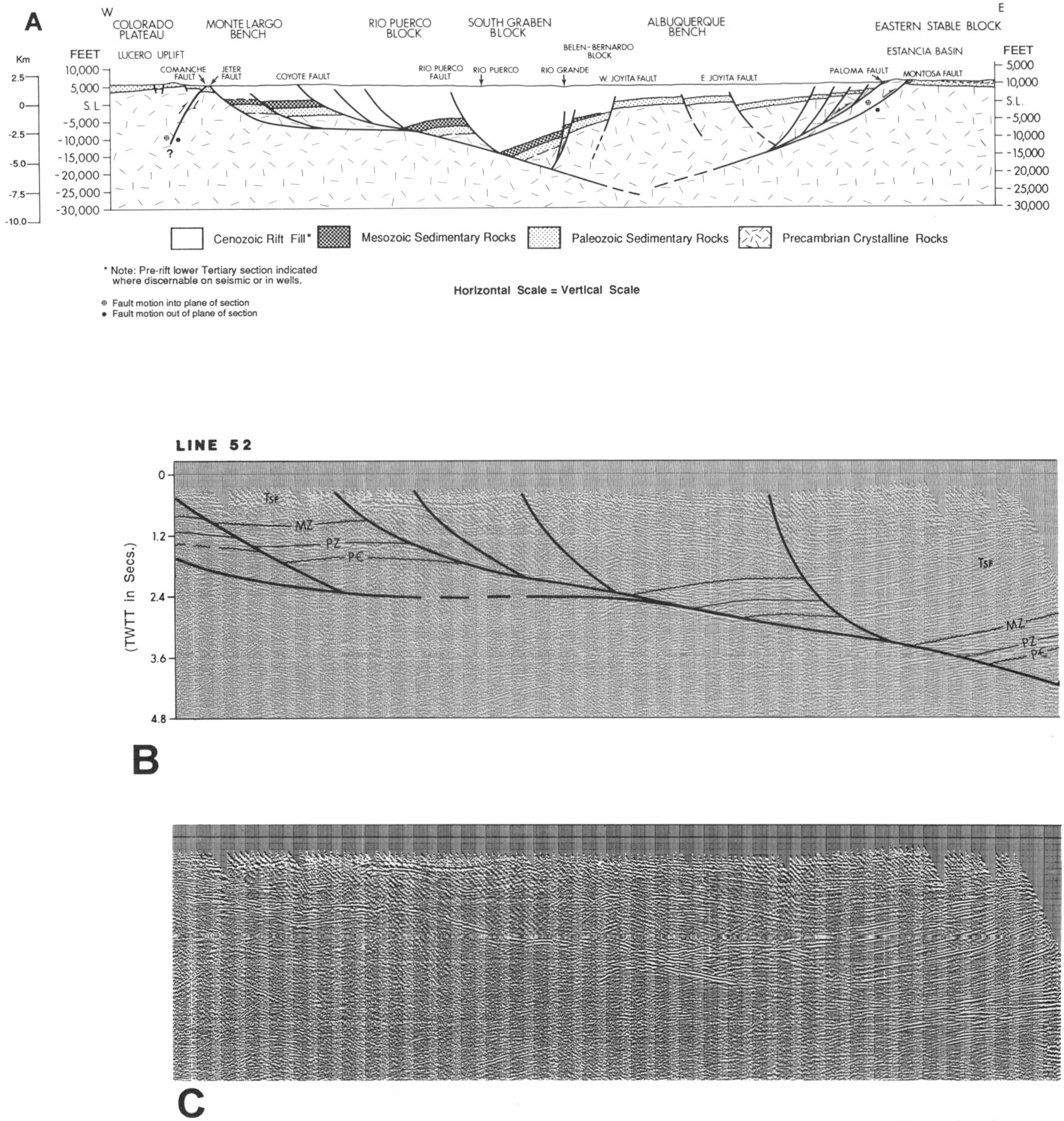

Figure 10. Geologic (A) and seismic (B, interpreted; C, uninterpreted) sections illustrating the structural configuration of the south portion of the southern Albuquerque Basin. Location of geologic section shown by dashed line on index map; position of seismic line indicated by solid line.

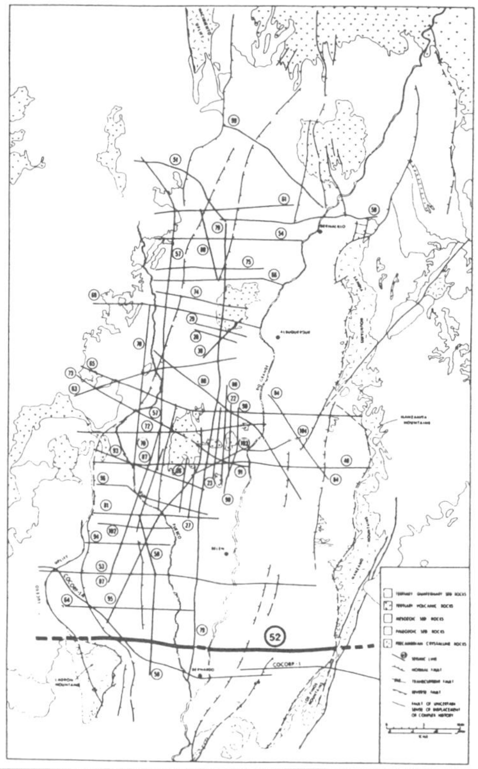
52
COCORP-1

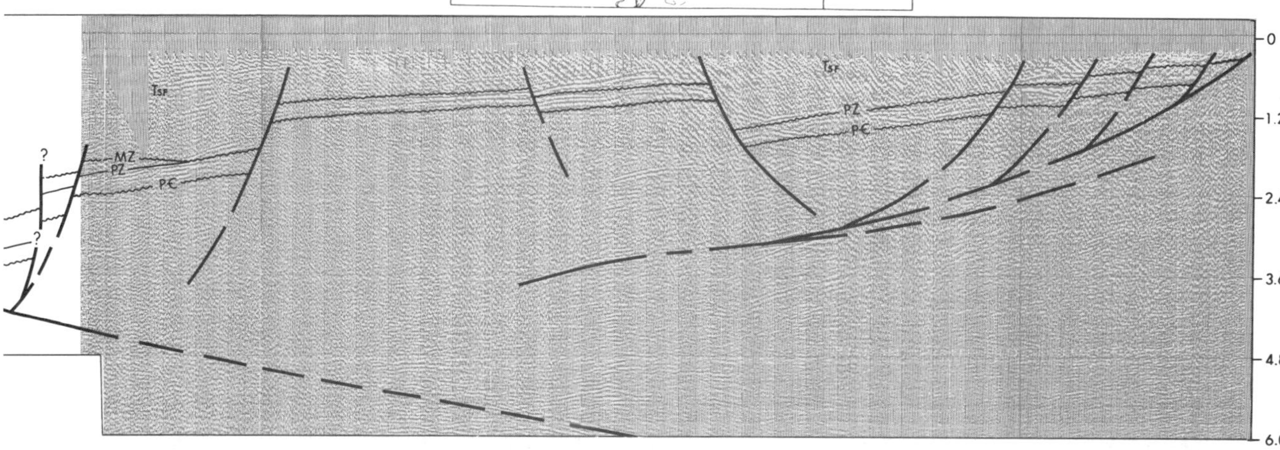
Tsp
MZ
PZ
PЄ
?
?
Tsp
PZ
PЄ
0
1.2
2.4
3.6
4.8
6.0
(TWTT in Secs.)

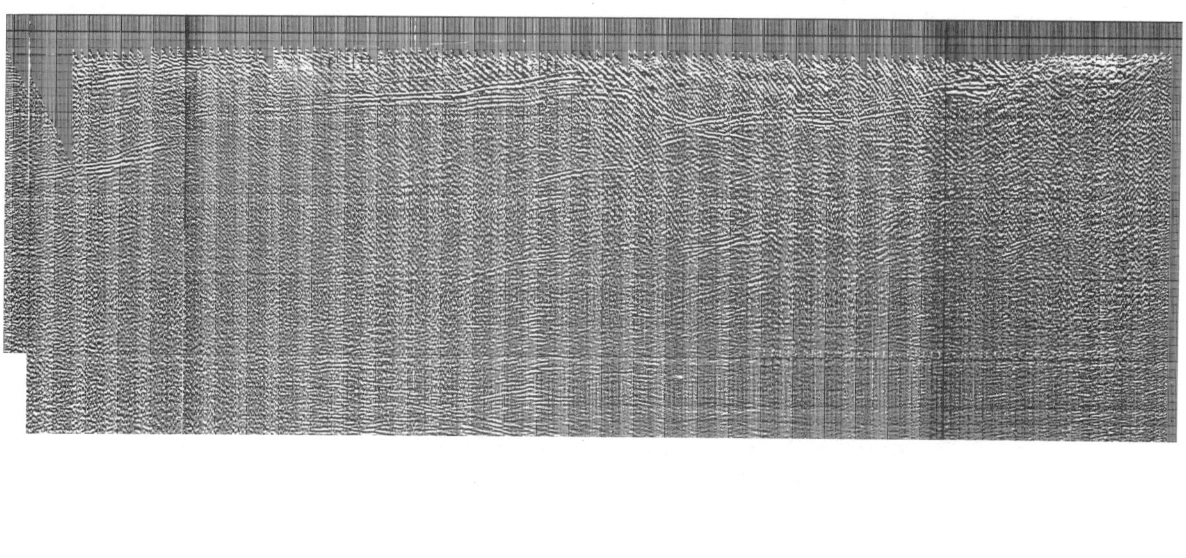

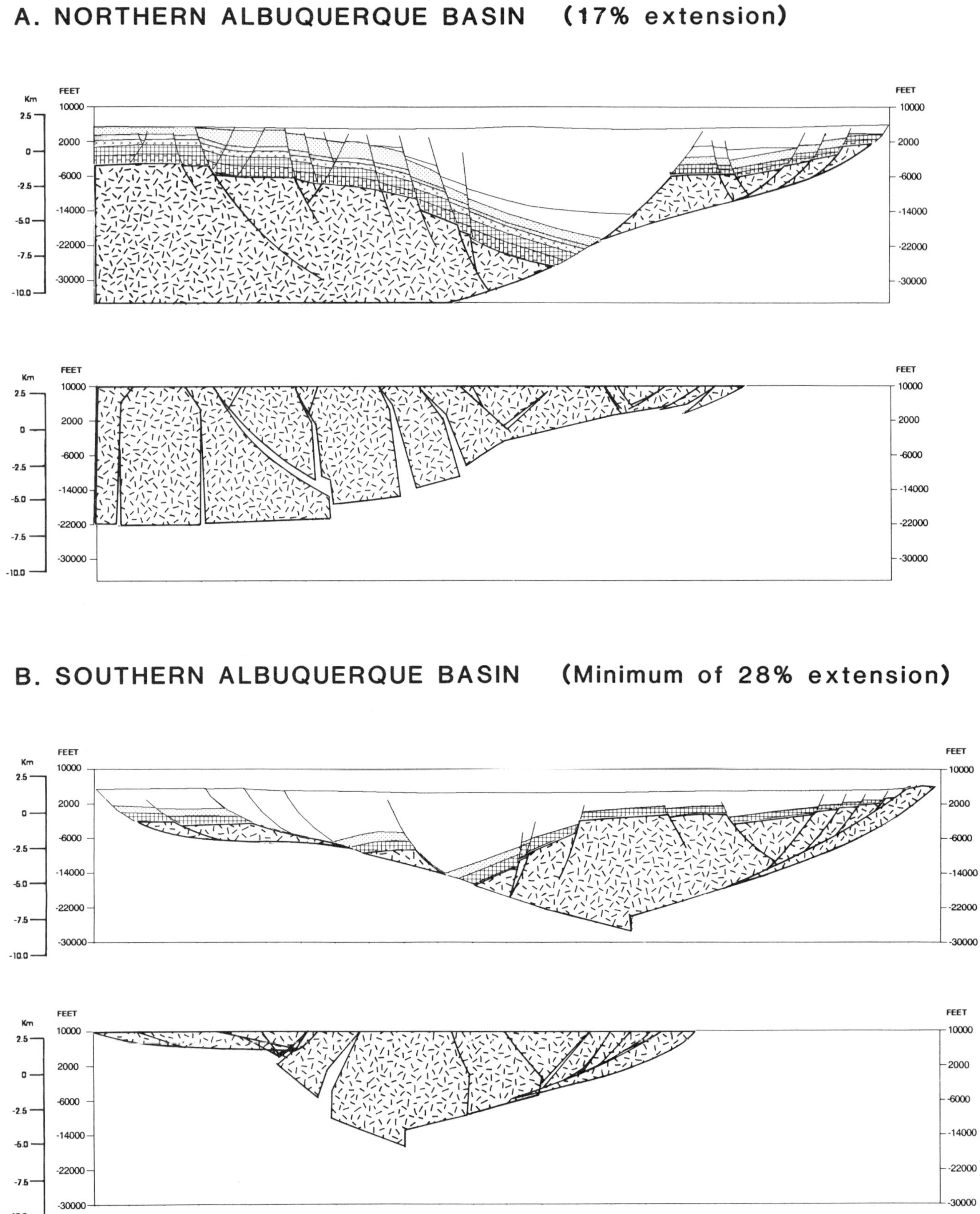

Figure 11. Palinspastic reconstructions of cross sections illustrating the variation in amount of extension between the northern (A) and southern (B) Albuquerque Basin (horizontal scale equals vertical scale). Sections reconstructed on top Precambrian basement.

and fractured, uplifted basement horst existed beneath the central portion of the basin. This horst block was attributed to be the down-plunge, basinward expression of the Ladron uplift (Brown et al., 1980; Fig. 13D). In later interpretations of reprocessed versions of the same data, better seismic resolution allowed for the role of listric normal faults to be discussed including the postulate that the Ladron horst block is detached above a low-angle normal fault (Cape et al., 1983; de Voogd et al., 1988; Fig. 13C).

Several factors need to be addressed in reconciling the differing interpretations of this area. The first of these is the orientation of the seismic lines themselves. Shell line 52 is a nearly true dip section across the basin. On the other hand, COCORP line 1-A is largely a strike line relative to the major basin controlling structures and only becomes a dip section in its easterly third (Fig. 13B). Previous interpretations of these data apparently did not account for this fact. Consequently, whereas the apparent midbasin horst block is in the middle of the line, it actually lies very near the western margin of the basin.

Another key relationship important to the interpretation of these data is the arcuate trace of the Jeter and Santa Fe–Coyote faults, the two major western boundary faults in this portion of the basin. The trace of these faults, broadly arcuate in a concave sense regionally, becomes tightly, and convexly, arcuate in the vicinity of the Ladron Mountains and conforms closely to the northern and eastern flanks of the range (Fig. 6). This relationship suggests that the faults were probably folded late in their history during the uplift of the Ladron structural block. This uplift is constrained at 10 to 15 Ma by apatite fission track cooling ages (Kelley et al., 1992). The folded nature of these faults is not nearly as pronounced on seismic dip transects across the basin as on strike lines. Thus, Shell line 52 reveals a mildly folded geometry (Fig. 13A). COCORP line 1-A, however, closely follows the strike of these faults. Our interpretation of these data (Fig. 13B) shows a distinctly folded nature of the structures. One inherent distortion is created by this strike section: the footwall cutoff, in the Paleozoic-Mesozoic section of the Santa Fe–Coyote fault on the eastern edge of the Monte Largo Bench (Fig. 13B) is much longer in length than in the hanging wall. This results because the footwall is imaged on that portion of the COCORP line that is strike parallel to the fault and the hanging wall is imaged on the dip-oriented section of the line.

In spite of the above differences, the interpretations of all these authors share some fundamental and important similarities. All show the basin to be quite asymmetric and also interpret the western boundary faults of the rift to be low angle to listric in character. Finally, all interpretations of basin margin faults show basement involvement.

CONCLUSIONS AND BROADER TECTONIC IMPLICATIONS

The Albuquerque Basin segment of the Rio Grande rift has been shown by this study to be two subbasins of opposing structural polarity and differing amounts of Tertiary extension. The northern subbasin is controlled by a major west-dipping, listric-normal fault system that appears to flatten, or detach, at the base of the brittle crust. In contrast, the southern subbasin is controlled by a system of major east-dipping, low-angle, ramp-like, and largely planar normal faults that seem to penetrate to deeper crustal levels. Separating these two subbasins is an accommodation zone that allows for the differential polarity and degrees of stretching observed between the two subbasins.

The observations and conclusions derived from this work on the Rio Grande rift have major implications as applied to the problems of passive continental margin evolution. In the past, the gaps and overlaps produced by plate reconstructions have been major reasons for modifications and rearrangements in the fits between continental plates. Palinspastic analysis of Albuquerque Basin cross sections suggests that variations in extension along a rifted continental margin could be quite significant. Accounting for these variations might reduce problems with gaps and overlaps. This study also suggests that passive continental margins, as is the case with their ancestral rift zones, are likely to be segmented into tectonic domains of varied amounts of extension and reversed structural polarity that are separated by transcurrent structural zones or accommodation zones. Thus, when analyzing a margin for variations of this sort, particular attention should be paid to evidence of transcurrent features. These transverse zones could connect with true oceanic transforms and might then be used as tentative flow lines to guide the restoration of the margin to its original position and structural configuration.

In addition, the geometric lessons acquired when studying a continental rift basin such as the Albuquerque Basin, lead one to the conclusion that conjugate passive margins that evolve after rifting progresses to sea-floor spreading, will not be mirror images of one another. On the contrary, they will be asymmetric, just as the initial rift was, and display very different structural geometries, subsidence histories, thermal structures, and perhaps hydrocarbon potential. The ideas presented in this paper are not unlike those of other workers who in recent years have discussed the asymmetric nature of rifting, continental breakup, and the role of low-angle detachment tectonics along passive continental margins (Bally, 1982; Lister et al., 1986). Further, the concepts related to accommodation zones have been abundantly illustrated by Rosendahl (1987) in his studies of variations along the East African rift lakes.

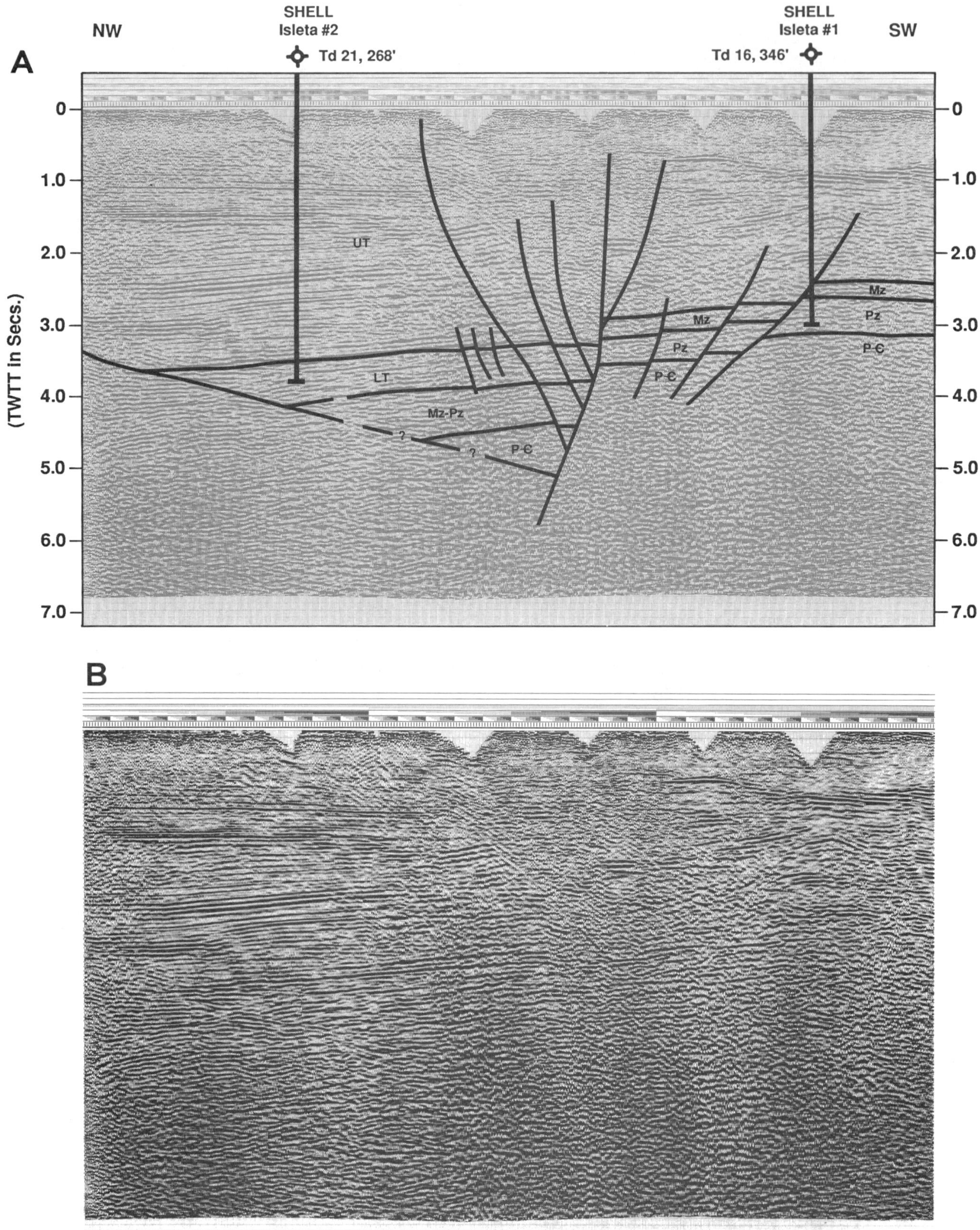

Figure 12. Seismic section (A, interpreted; B, uninterpreted) across the midbasin structural transfer zone (unpublished interpretation after L. Garmezy). Location of seismic line shown on index map (C).

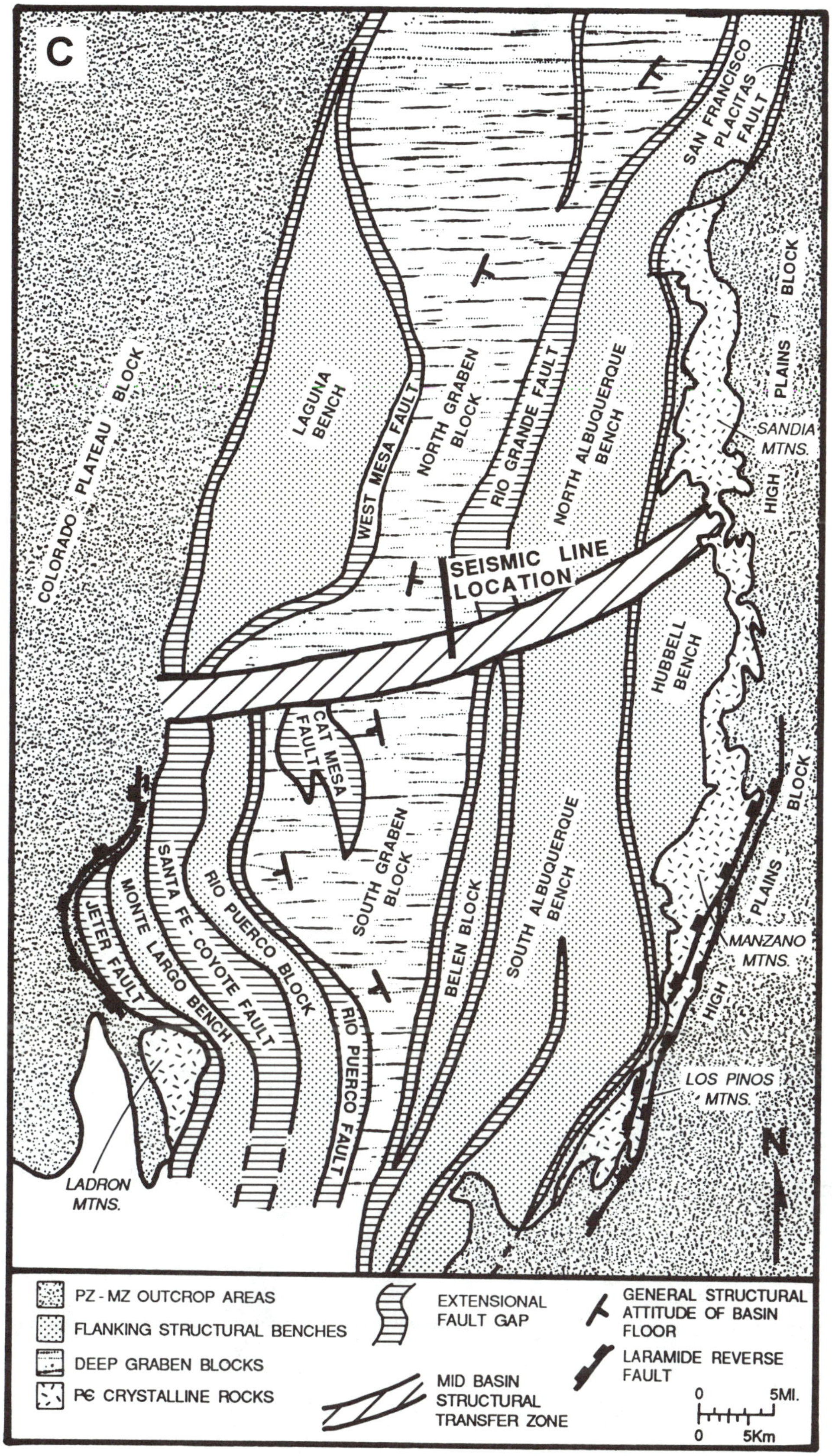
C
SAN FRANCISCO PLACITAS FAULT
COLORADO PLATEAU BLOCK
LAGUNA BENCH
WEST MESA FAULT
NORTH GRABEN BLOCK
RIO GRANDE FAULT
NORTH ALBUQUERQUE BENCH
HIGH PLAINS BLOCK
SANDIA MTNS.
SEISMIC LINE LOCATION
HUBBELL BENCH
CAT MESA FAULT
SOUTH GRABEN BLOCK
BELEN BLOCK
SOUTH ALBUQUERQUE BENCH
JETER FAULT
MONTE LARGO BENCH
SANTA FE-COYOTE FAULT
RIO PUERCO BLOCK
RIO PUERCO FAULT
MANZANO MTNS.
LOS PINOS MTNS.
LADRON MTNS.
N
PZ-MZ OUTCROP AREAS
FLANKING STRUCTURAL BENCHES
DEEP GRABEN BLOCKS
P€ CRYSTALLINE ROCKS
EXTENSIONAL FAULT GAP
MID BASIN STRUCTURAL TRANSFER ZONE
GENERAL STRUCTURAL ATTITUDE OF BASIN FLOOR
LARAMIDE REVERSE FAULT
0 5MI.
0 5Km

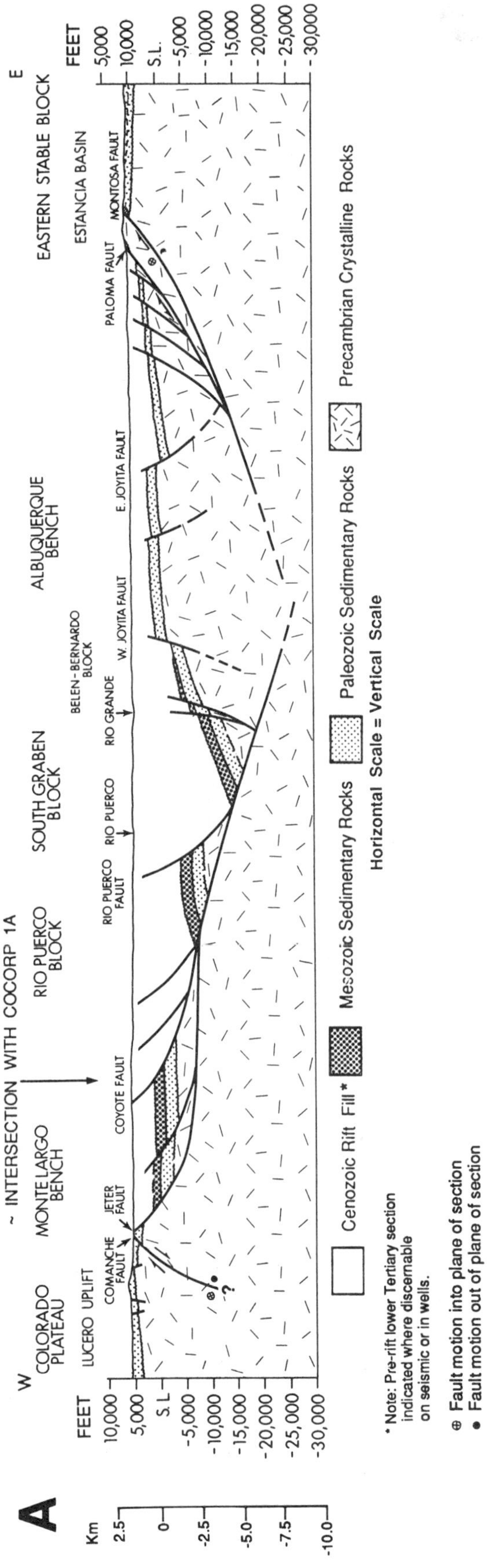
A
W
E
COLORADO PLATEAU
~ INTERSECTION WITH COCORP 1A
MONTE LARGO BENCH
RIO PUERCO BLOCK
SOUTH GRABEN BLOCK
BELEN-BERNARDO BLOCK
ALBUQUERQUE BENCH
EASTERN STABLE BLOCK
LUCERO UPLIFT
COMANCHE FAULT
JETER FAULT
COYOTE FAULT
RIO PUERCO FAULT
RIO PUERCO
RIO GRANDE
W. JOYITA FAULT
E. JOYITA FAULT
PALOMA FAULT
MONTOSA FAULT
ESTANCIA BASIN
Km
FEET
10,000
5,000
S.L.
-5,000
-10,000
-15,000
-20,000
-25,000
-30,000
2.5
0
-2.5
-5.0
-7.5
-10.0
Cenozoic Rift Fill *
Mesozoic Sedimentary Rocks
Paleozoic Sedimentary Rocks
Precambrian Crystalline Rocks
Horizontal Scale = Vertical Scale
* Note: Pre-rift lower Tertiary section indicated where discernable on seismic or in wells.
⊕ Fault motion into plane of section
● Fault motion out of plane of section

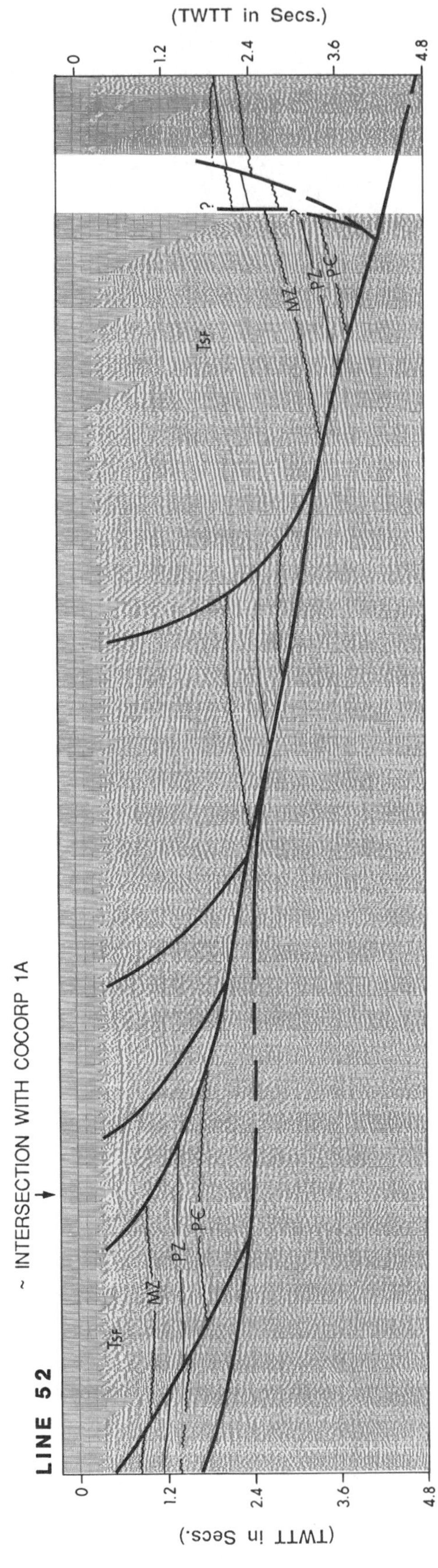
LINE 52
~ INTERSECTION WITH COCORP 1A
(TWTT in Secs.)
0
1.2
2.4
3.6
4.8
Tsf
MZ
PZ
PC

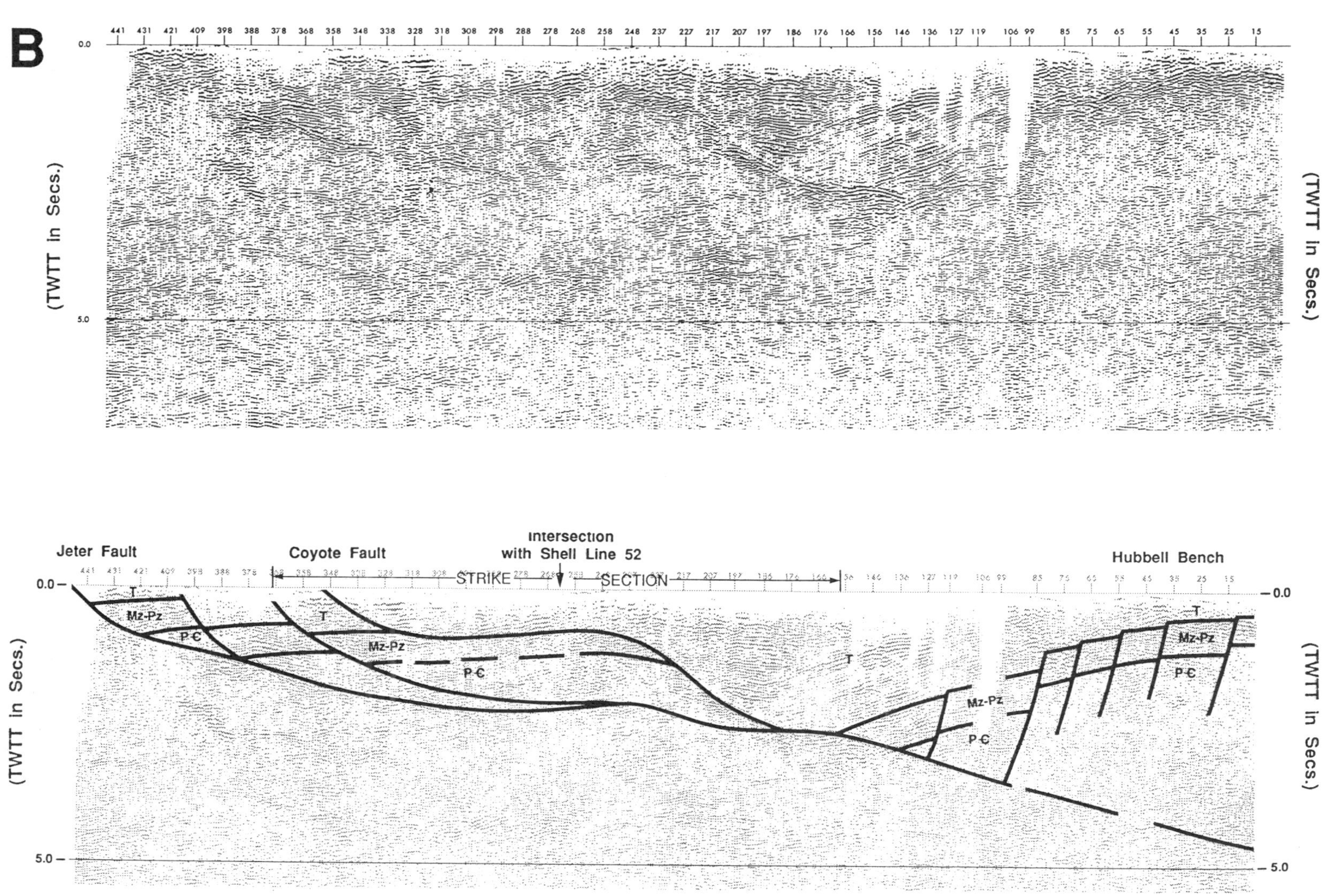

Figure 13 (on this page spread and following page). A. Geologic and seismic sections illustrating the structural interpretation from this study in the vicinity of COCORP Line 1A. Location of geologic section shown by dashed line number 52 on index map; position of seismic lines indicated by solid lines. B. COCORP Line 1A uninterpreted and an interpretation based on this study. Note from the index map that the western portion of the COCORP line is largely a strike line relative to basin-controlling structures. C. A contrasting interpretation of COCORP Line 1A by COCORP workers. D. An early interpretation of COCORP Line 1A by COCORP workers.

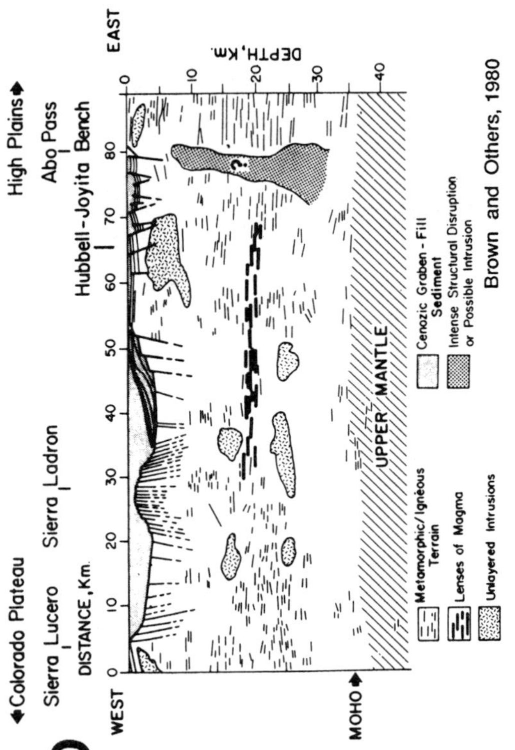
D
◆Colorado Plateau
Sierra Lucero
Sierra Ladron
Hubbell-Joyita Bench
Abo Pass
High Plains◆
WEST
EAST
DISTANCE, Km.
DEPTH, Km.
MOHO◆
UPPER MANTLE
Metamorphic/Igneous Terrain
Lenses of Magma
Unlayered Intrusions
Cenozoic Graben-Fill Sediment
Intense Structural Disruption or Possible Intrusion
Brown and Others, 1980

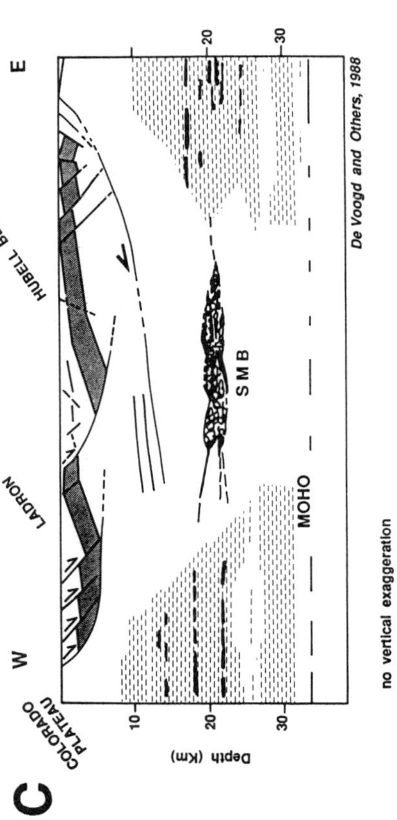
C
COLORADO PLATEAU
W
LADRON
HUBELL BENCH
E
Depth (Km)
SMB
MOHO
no vertical exaggeration
De Voogd and Others, 1988

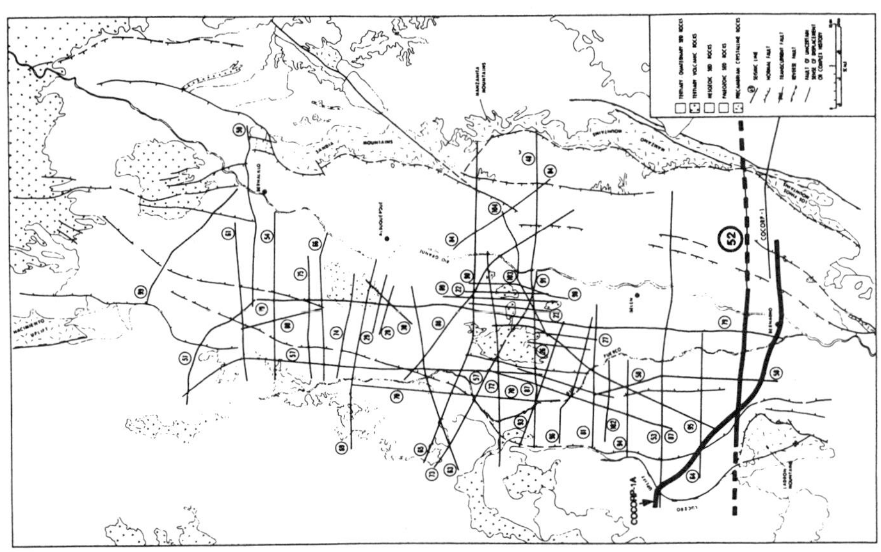
52
COCORP-1A
COCORP-1

ACKNOWLEDGMENTS

The authors wish to thank Shell Oil Company for release of the seismic data pivotal to this paper, and to acknowledge the contributions and insights provided by our many colleagues. A. W. Bally, in particular, was instrumental in the conception of the study and provided much encouragement along the way. In addition, the comments and suggestions of Geological Society of America reviewers E. W. Collins and C. F. Kluth measurably improved the manuscript. Their contributions are gratefully acknowledged.

REFERENCES CITED

Bachman, G. O., and Mehnert, H. H., 1978, New K-Ar dates and the late Pliocene to Holocene geomorphic history of the central Rio Grande region, New Mexico: Geological Society of America Bulletin, v. 89, p. 283–292.

Baldridge, W. S., 1979, Petrology and petrogenesis of Plio-Pleistocene basaltic rocks from the central Rio Grande rift, New Mexico, and their relation to rift structure, *in* Riecker, R. E., ed., Rio Grande rift, tectonics and magmatism: Washington, D.C., American Geophysical Union, p. 323–353.

Bally, A. W., 1982, Musings over sedimentary basin evolution: Royal Society of London Philosophical Transactions, A305, p. 325–328.

Black, B. A., 1964, The geology of the northern and eastern parts of the Ladron Mountains, Socorro County, New Mexico [M.S. thesis]: Albuquerque, University of New Mexico, 117 p.

Braile, L. W., Smith, R. B., Keller, G. R., and Welch, R. M., 1974, Crustal structure across the Wasatch Front from detailed seismic refraction studies: Journal of Geophysical Research, 79, p. 2669–2677.

Brown, L. D., Chapin, C. E., Sanford, A. R., Kaufman, S., and Oliver, J., 1980, Deep structure of the Rio Grande rift from seismic reflection profiling: Journal of Geophysical Research, v. 85, p. 4773–4800.

Callender, J. F., and Zilinski, R. E., 1976, Kinematics of Tertiary and Quaternary deformation along the eastern edge of the Lucero uplift, central New Mexico, *in* Woodward, L. A., and Northrop, S. A., eds., Tectonics and mineral resources of southwestern North America: New Mexico Geological Society Special Publication no. 6, p. 53–61.

Cape, C. D., McGeary, S., and Thompson, G. A., 1983, Cenozoic normal faulting and the shallow structure of the Rio Grande rift near Socorro, New Mexico: Geological Society of America Bulletin, 94, p. 3–14.

Chapin, C. E., 1979, Evolution of the Rio Grande rift, a summary, *in* Riecker, R. E., ed., Rio Grande rift, tectonics and magmatism: Washington, D.C., American Geophysical Union, p. 1–5.

Chapin, C. E., 1983, An overview of Laramide wrench faulting in the southern Rocky Mountains with emphasis on petroleum exploration: Rocky Mountain Association of Geologists, p. 169–179.

Chapin, C. E., and Seager, W. R., 1975, Evolution of the Rio Grande rift in the Socorro and Las Cruces areas: New Mexico Geological Society, 26th Field Conference, Guidebook, p. 297–321.

Condie, K. C., 1976, Precambrian rocks of the Ladron Mountains, Socorro County, New Mexico: New Mexico Bureau of Mines and Mineral Resources Geological Map 38, scale 1:24,000.

Cordell, L., 1978, Regional geophysical setting of the Rio Grande rift: Geological Society of America Bulletin, 89, no. 7, p. 1073–1090.

Cordell, L., Keller, G. R., and Hildenbrand, T. G., 1978, Complete Bouguer gravity anomaly map of the Rio Grande rift: U.S. Geological Survey Open-File Map, scale 1:1,000,000.

de Voogd, B., Serpa, L., and Brown, L. D., 1988, Crustal extension and magmatic processes, COCORP profiles from Death Valley and the Rio Grande rift: Geological Society of America Bulletin 100, p. 1550–1567.

Eaton, G. P., 1979, A plate tectonic model for late Cenozoic crustal spreading in the western United States, *in* Riecker, R. E., ed., Rio Grande rift, tectonics and magmatism: Washington, D.C., American Geophysical Union, p. 7–32.

Harding, T. P., 1985, Seismic characteristics and identification of negative flower structures, positive flower structures, and positive structural inversion: American Association of Petroleum Geologists Bulletin 69, p. 582–600.

Hayden, S., 1991, Dextral oblique slip deformation along the Montosa fault zone at Abo Pass, Valencia and Socorro Counties, New Mexico: New Mexico Geology, v. 13, p. 64.

Hayden, S., and Mawer, C. K., 1991, Dextral transtensional deformation along the western margin of the Rio Grande rift, central New Mexico: Geological Society of America Abstracts with Programs, v. 23, no. 4, p. 31.

Hermance, J. F., and Pedersen, J., 1980, Deep structure of the Rio Grande rift—A magnetotelluric interpretation: Journal of Geophysical Research, v. 85, p. 3899–3912.

Jiracek, G. R., Ander, M. E., and Holcombe, H. T., 1979, Magnetotelluric soundings of crustal conductive zones in major continental rifts, *in* Riecker, R. E., ed., Rio Grande rift, tectonics and magmatism: Washington, D.C., American Geophysical Union, p. 209–222.

Keller, G. R., Smith, R. B., and Braile, L. W., 1975, Crustal structure along the Great Basin–Colorado Plateau transition from seismic refraction studies: Journal of Geophysical Research 80, no. 8, p. 1093–1098.

Keller, G. R., Braile, L. W., and Morgan, P., 1979, Crustal structure, geophysical models and contemporary tectonism of the Colorado Plateau: Tectonophysics, v. 61, p. 131–147.

Kelley, S. A., Chapin, C. E., and Corrigan, J. D., 1992, Late Mesozoic to Cenozoic cooling histories of the flanks of the northern and central Rio Grande rift: New Mexico Bureau of Mines and Mineral Resources Bulletin 145, 40 p.

Kelley, V. C., 1977, Geology of the Albuquerque Basin: New Mexico Bureau of Mines and Mineral Resources Memoir 33, 60 p.

Kelley, V. C., and Kudo, A. M., 1978, Volcanoes and related basalts of the Albuquerque Basin, New Mexico: New Mexico Bureau of Mines and Mineral Resources Circular 156, 30 p.

Kudo, A. M., 1982, Rift volcanics of the Albuquerque Basin: Overview with some new data: New Mexico Geological Society, 33rd Field Conference, Guidebook, p. 285–289.

Kudo, A. M., Kelley, V. C., Damon, P. E., and Shafiquillah, M., 1977, K-Ar ages of basalt flows at Canjilon Hill, Isleta volcano, and Cat Hills volcanic field, Albuquerque Basin, central New Mexico: Isochron/West, no. 18, p. 15–16.

Lister, G. S., Etheridge, M. A., and Symonds, P. A., 1986, Detachment faulting and the evolution of passive continental margins: Geology, v. 14, p. 246–250.

Lozinsky, P. R., 1988, Stratigraphy, sedimentation, and sand petrology of the Santa Fe Group and pre–Santa Fe Tertiary deposits in the Albuquerque Basin, central New Mexico [Ph.D. thesis]: Socorro, New Mexico Institute of Mining and Technology, 298 p.

Lozinsky, R. P., 1989, Cenozoic basin-fill stratigraphy and depositional history of the Albuquerque Basin, central New Mexico: New Mexico Geological Society 40th Field Conference, Guidebook: p. 269–272.

Mackin, J. H., 1950, The down-structure method of viewing geologic maps: Journal of Geology, v. 58, p. 55–72.

McCullar, D. B., and Smithson, S. B., 1977, Unreversed seismic crustal refraction profile across the Southern Rio Grande rift [abs.]: EOS Transactions of the American Geophysical Union, v. 58, p. 1184.

McKenzie, D. P., 1978, Some remarks on the development of sedimentary basins: Earth and Planetary Sciences Letters, v. 40, p. 25–32.

Miller, J. P., Montgomery, A., and Sutherland, P. K., 1963, Geology of part of the southern Sangre de Cristo Mountains, New Mexico: New Mexico Bureau of Mines and Mineral Resources, Memoir 11, 106 p.

Molenaar, C. M., 1983, Major depositional cycles and regional correlations of

Upper Cretaceous rocks, southern Colorado Plateau and adjacent areas, *in* Reynolds, M. W., and Dolly, E. E., eds., Mesozoic paleogeography of the west-central United States: Society of Economic Geologists and Mineralogists, Rocky Mountain Section, Rocky Mountain Paleogeography Symposium 2, p. 201–224.

Molzer, P. C., and Erslev, E. A., 1991, Oblique thrusting in east-west Laramide foreland uplifts: Geological Society of America Abstracts with Programs, v. 23, no. 4, p. 50.

Olsen, K. H., Keller, G. R., and Stewart, J. N., 1979, Crustal structure along the Rio Grande rift from seismic refraction profiles, *in* Riecker, R. E., ed., Rio Grande rift, tectonics and magmatism: Washington, D. C., American Geophysical Union, p. 127–143.

Reiter, M., Edwards, C. L., Hartman, H., and Weidman, C., 1975, Terrestrial heat flow along the Rio Grande rift, New Mexico and Colorado: Geological Society of America Bulletin, v. 86, p. 811–818.

Reiter, M., Shearer, C., and Edwards, C. L., 1978, Geothermal anomalies along the Rio Grande rift: Geology, v. 6, p. 85–88.

Reiter, M., Mansure, A. J., and Shearer, C., 1979, Geothermal characteristics of the Rio Grande rift with the Southern Rocky Mountains complex, *in* Riecker, R. E., ed., Rio Grande rift, tectonics and magmatism: Washington, D.C., American Geophysical Union, p. 253–267.

Rosendahl, B. R., 1987, Architecture of continental rifts with special reference to east Africa: Annual Review of Earth and Planetary Sciences, v. 15, p. 445–503.

Russell, L. R., and Snelson, S., 1990, Structural style and tectonic evolution of the Albuquerque Basin segment of the Rio Grande rift, *in* Pinet, B., and Bois, C., eds., The potential of deep seismic profiling for hydrocarbon exploration: Paris, Editions Technip, p. 175–207.

Sanford, A. R., Olsen, K. H., and Jaksha, L. H., 1979, Seismicity of the Rio Grande rift, *in* Riecker, R. E., ed., Rio Grande rift, tectonics and magmatism: Washington, D.C., American Geophysical Union, p. 145–168.

Smith, R. B., 1978, Seismicity, crustal structure, and intraplate tectonics of the interior of the Western Cordillera, *in* Smith, R. B., and Eaton, G. P., eds., Cenozoic tectonics and regional geophysics of the Western Cordillera: Geological Society of America Memoir 152, p. 111–144.

Stewart, J. H., 1978, Basin and Range structure in western North America: A review, *in* Smith, R. B., and Eaton, G. P., eds., Cenozoic tectonics and regional geophysics of the Western Cordillera: Geological Society of America Memoir 152, p. 1–31.

Wernicke, B., 1981, Low-angle normal faults in the Basin and Range province: Nappe tectonics in an extending orogen: Nature, v. 291, p. 645–647.

Wernicke, B., 1985, Uniform-sense normal simple shear of the continental lithosphere: Canadian Journal of Earth Sciences, v. 22, p. 108–125.

Wernicke, B., 1986, Whole-lithosphere normal simple shear: An interpretation of deep reflection seismic profiles in Great Britain, *in* Barazangi, M., and Brown, L., eds., Reflection seismology: The continental crust: Washington, D.C., American Geophysical Union Geodynamics Series, v. 14, p. 331–339.

White, D. L., 1979, Rb-Sr age and isotopic data of some Precambrian plutons in central New Mexico: Mountain Geologist, v. 16, p. 131–137.

Woodward, L. A., and 6 others, 1978, Tectonic map of the Rio Grande rift region in New Mexico, Chihuahua, and Texas, *in* Hawley, J. W., ed., Guidebook to the Rio Grande rift in New Mexico and Colorado: New Mexico Bureau of Mines and Mineral Resources Circular 163, 241 p.

Worrall, D. M., and Snelson, S., 1989, Evolution of the northern Gulf of Mexico, with emphasis on growth faulting and the role of salt, *in* Bally, A. W., and Palmer, A. R., eds., The Geology of North America—An Overview: Boulder, Colorado, Geological Society of America, The Geology of North America, v. A, p. 97–138.

Manuscript Accepted by the Society October 13, 1993

Printed in U.S.A.

Geological Society of America
Special Paper 291
1994

Thickness of the syn-rift Santa Fe Group in the Albuquerque Basin and its relation to structural style

S. Judson May and Lee R. Russell
ARCO Exploration and Production Technology, 2300 West Plano Parkway, Plano, Texas 75075

ABSTRACT

The Santa Fe Group is a succession of Neogene sandstone, siltstone, conglomerate, and intercalated volcanics that fills the basins of the central Rio Grande rift. Data from twelve deep wells and recently released seismic profiles in the Albuquerque rift basin document a dramatic basinward thickening of the Santa Fe Group across the major basinal faults. Abbreviated thicknesses of less than 1,220 m (4,000 ft) occur on the structurally higher, outer benches, but the section thickens to more than 4,270 m (14,000 ft) in the basin center, on the downthrown sides of the basin's seismically defined master normal faults.

Thickness of the Santa Fe mimics the basin's structural asymmetry. In the north, it forms an eastward-thickening clastic wedge in response to eastward tilting of the north part of the basin toward the westward-flattening master faults (Rio Grande, Sandia, and San Francisco–Placitas faults). In the south, it thickens westward toward the Santa Fe–Coyote master faults along the southwestern margin of the basin. As a result, depositional rates of the Santa Fe were rapid on the structurally deep, master-fault sides of the basin in the northeast and southwest and relatively slow on the basin's hinge sides in the northwest and southeast. At the northeast-trending Tijeras accommodation zone, which divides the north and south halves of the basin, the section thickens abruptly to the north in the basin center, indicating a major dip-slip as well as probable strike-slip component of motion between the two subbasins.

Much of the thickening in the Santa Fe Group seems to occur in the upper Miocene part of the section, possibly implying a phase of accelerated crustal extension at that time.

INTRODUCTION

Tectonic and stratigraphic models of the syn-rift Santa Fe Group (upper Oligocene to Pleistocene) in the Rio Grande rift have historically been based on data from surface exposures and, to a lesser extent, from analyses of cuttings from wells drilled in the basins. Seismic reflection data in the Albuquerque Basin, released recently be Shell Oil Company (Russell and Snelson, 1990, and this volume), provide new constraints on how extensional structural styles control the thickness and depositional processes of the Santa Fe Group by providing additional control on subsurface dips, thicknesses, and seismic character that can be correlated with petrologic descriptions of well cuttings and surface exposures.

Major Cenozoic stratigraphic units in the Albuquerque Basin are shown in Figure 1. The pre-rift Galisteo and Baca

May, S. J., and Russell, L. R., 1994, Thickness of the syn-rift Santa Fe Group in the Albuquerque Basin and its relation to structural style, *in* Keller, G. R., and Cather, S. M., eds., Basins of the Rio Grande Rift: Structure, Stratigraphy, and Tectonic Setting: Boulder, Colorado, Geological Society of America Special Paper 291.

Formations (Eocene) contain petrologically mature, nonvolcanic sandstones and lacustrine mudstones reflecting late-stage filling of Laramide basins (Gorham and Ingersoll, 1979; Lozinsky, 1988). The overlying "Unit of Isleta #2" is primarily an Oligocene section, named informally by Lozinsky (1988) for andesitic volcaniclastic debris preserved in the basin center. It is thought to be the approximate volcaniclastic equivalent of the Datil Group and overlying volcanic section, up through the La Jara Peak Basaltic Andesite, exposed southwest of the Albuquerque Basin (Osburn and Chapin, 1983; Lozinsky, 1988), and the Espinaso Formation, exposed to the northeast (Kautz et al., 1981). This unit appears to have been deposited prior to and during the earliest stages of rifting in the Albuquerque Basin.

The overlying Santa Fe Group (Fig. 1), the primary rift-fill unit, is composed of petrologically immature arkosic and volcaniclastic debris derived locally from rift shoulders surrounding the basin and from rift-related volcanic centers. In the north, the lower to mid-Miocene section includes the quartz-rich Zia Formation (largely eolian) and volcaniclastic Abiquiu Formation. These units are approximate age equivalents of the lower part of the Popotosa Formation in the south. Presence of intermediate and silicic volcanic debris in these lower beds of the Santa Fe implies some depositional recycling of the underlying Oligocene volcaniclastics. Minor sandstone clasts derived from the Mesozoic section on surrounding uplifts increase upward in the lower Santa Fe Group, indicating early, minor uplift of the rift shoulders (Lozinsky, 1988).

The middle unnamed unit of the Santa Fe Group is an arkosic alluvial section equivalent to the middle part of the formerly used "Middle Red Member" of the Santa Fe. Although the definition of this unit is subject to controversy, it is the approximate age equivalent of the arkosic upper part of the Popotosa Formation in the south. The upper part of this interval contains abundant K-feldspar and clasts of granite and quartzite, primarily on the east side of the basin, indicating that the rift shoulders on the northeast margin of the basin had risen dramatically and were eroded to the Precambrian basement in many areas by the late Miocene (Lozinsky, 1988).

The Sierra Ladrones Formation forms an unconformably overlying veneer of mildly deformed, Plio-Pleistocene fluvial sand and gravel deposited by the ancestral Rio Grande and associated piedmonts in the last 4 to 5 m.y. (Lozinsky and others, 1991). The unconformity at the base of this unit is exposed locally on the structurally higher, outer benches of the basin because of Quaternary incision by the Rio Grande and its tributaries.

The relationship between structural style and thickness of the syn-rift section in the Albuquerque Basin is illustrated by comparing a highly simplified block diagram of the basin's structure (Fig. 2) and an isochore map of the Santa Fe Group (Fig. 3). In the north, the basin dips gently east and the Santa Fe Group thickens to the east as a result of eastward tilting of the northern part of the basin toward the Rio Grande and Sandia master faults. The Sandia uplift forms the eastern rift shoulder of the basin at this latitude. The west side of the basin rolls down gently to the east in a hingelike fashion with smaller down-to-the-east normal faults forming the western fault margin. The southern half of the basin is an approximate mirror image of the north in which hanging-wall dips and stratigraphic thickening of the Santa Fe Group are toward the west-southwest, reflecting large-scale rotation of the southern subbasin toward the Jeter–Santa Fe–Coyote master faults. The west-dipping Lucero and Ladron uplifts form the western rift shoulders in the southwest. In the southeast, the Manzano–Los Pinos uplift and adjacent Albuquerque Basin form a broadly antiformal (hingelike) structure, the western half of which is down dropped along the basin-bounding Manzano and Los Pinos faults. The Manzano and Los Pinos rift shoulders form the eastern uplifted limb of this antiform at the surface.

Structurally intermediate benches in the basin are classified as hinge benches where the structure rolls over in the northwest and southeast, and as master fault benches in the northeast and southwest. This distinction between bench styles is useful for explaining stratigraphic variations in the Santa Fe.

NORTHERN ALBUQUERQUE BASIN

In the northern Albuquerque Basin, the Shell seismic data illustrate the eastward thickening of the Santa Fe alluvial

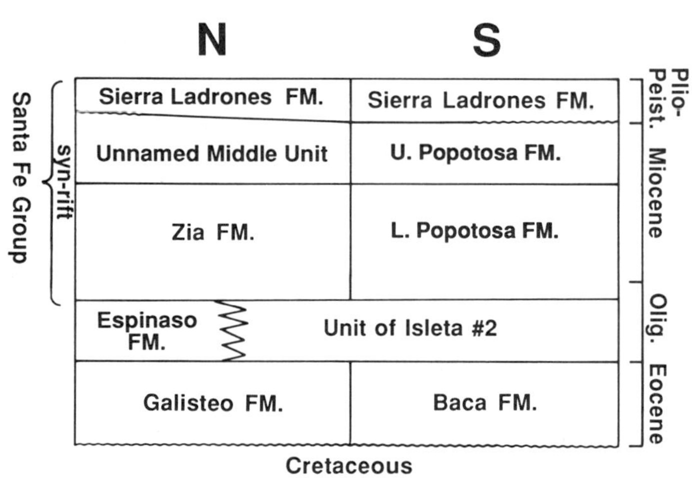

Figure 1. Simplified Cenozoic stratigraphic chart of the Albuquerque Basin. Variable uplift rates and structurally evolving source terranes in the rift shoulders have led to intertonguing of alluvial-fan systems, lateral facies changes, and complex stratigraphic terminology in the Santa Fe Group. Formation contacts are typically gradational and time transgressive in most areas. Contacts are smoothed here to simplify discussion of the rift fill.

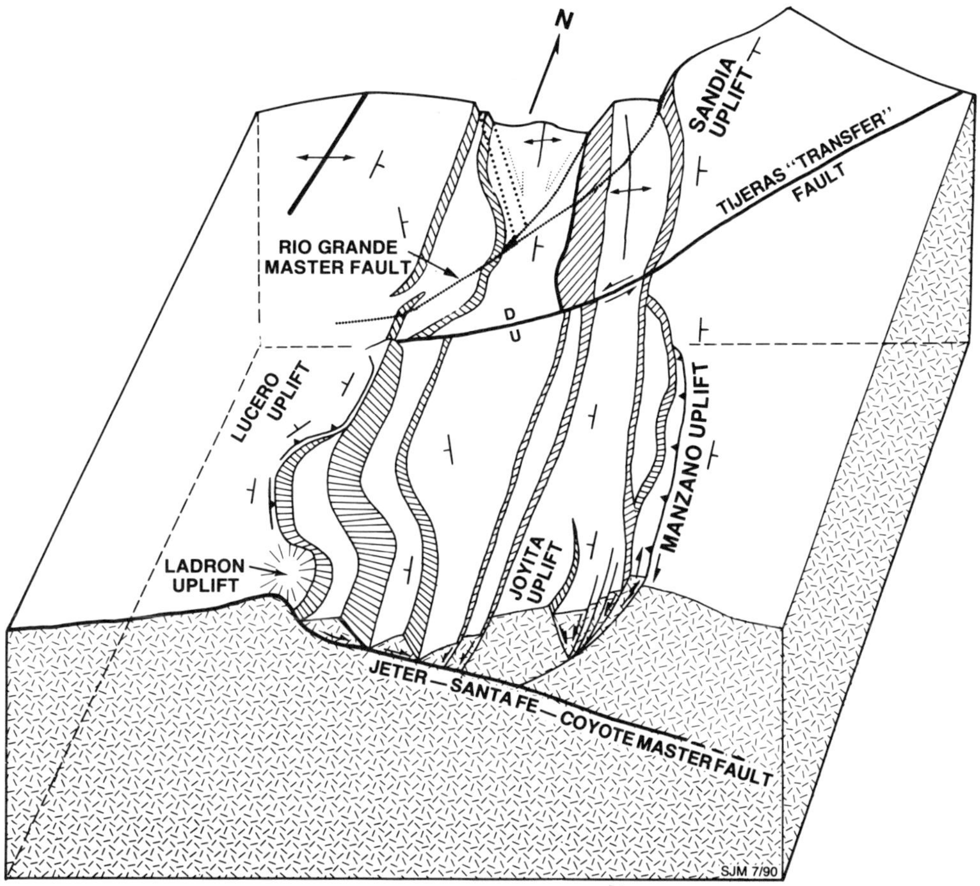

Figure 2. Highly generalized structure model of Albuquerque Basin showing opposing structural asymmetry of north and south halves of the basin and controlling master normal faults. Neogene syn-rift Santa Fe Group has been removed. Extreme north and south ends of basin are not shown. The city of Albuquerque is located above the large scarp of the Rio Grande master fault. Arcuate dotted line in the northern part of the figure represents intersection of the Rio Grande fault with the north face of the block diagram. Model is based on seismic data from Russell and Snelson (1990), surface geology, and unpublished ARCO isostatic residual gravity map by S. Rowley at ARCO. Some of the structural details in this model may conflict with unpublished Shell seismic data not available for this study.

wedge toward the Rio Grande master fault (Fig. 4). The listric geometry of this westward-flattening fault and others to the east, such as the San Francisco–Placitas and Sandia faults, appear to accommodate the large-scale, eastward tilting of the northern basin. The low-angle fault east of the Rio Grande fault is the root of the San Francisco–Placitas fault system. This system has been uplifted and exposed at a deeper structural level at the north end of the Sandia uplift in a down-plunge perspective (Russell and Snelson, 1990, and this volume).

The north-trending Ziana anticline (Fig. 4) is the surface expression of a high-standing block in the hanging wall of the Rio Grande fault and can be recognized at the surface (Fig. 5). The Shell Santa Fe #1 penetrated an abbreviated section of the Santa Fe Group (905 m; 2,970 ft; Lozinsky, 1988) on the crest of the structure (Figs. 3 and 4). This midbasin ridge may be related to complex interaction between the east-dipping northern Albuquerque Basin and west-dipping southern Espanola basin along the very diffuse Santa Ana accommodation zone (Fig. 3). The anticline is a low-relief accommodation zone in the sense of Rosendahl (1987) and is approximately similar to a midbasin ridge in the Morley et al. (1990, Fig. 2d) model of a conjugate, convergent, overlapping transfer zone.

Figure 4 includes an interpreted depth section crossing the entire width of the north basin, parallel to Shell line 50. The

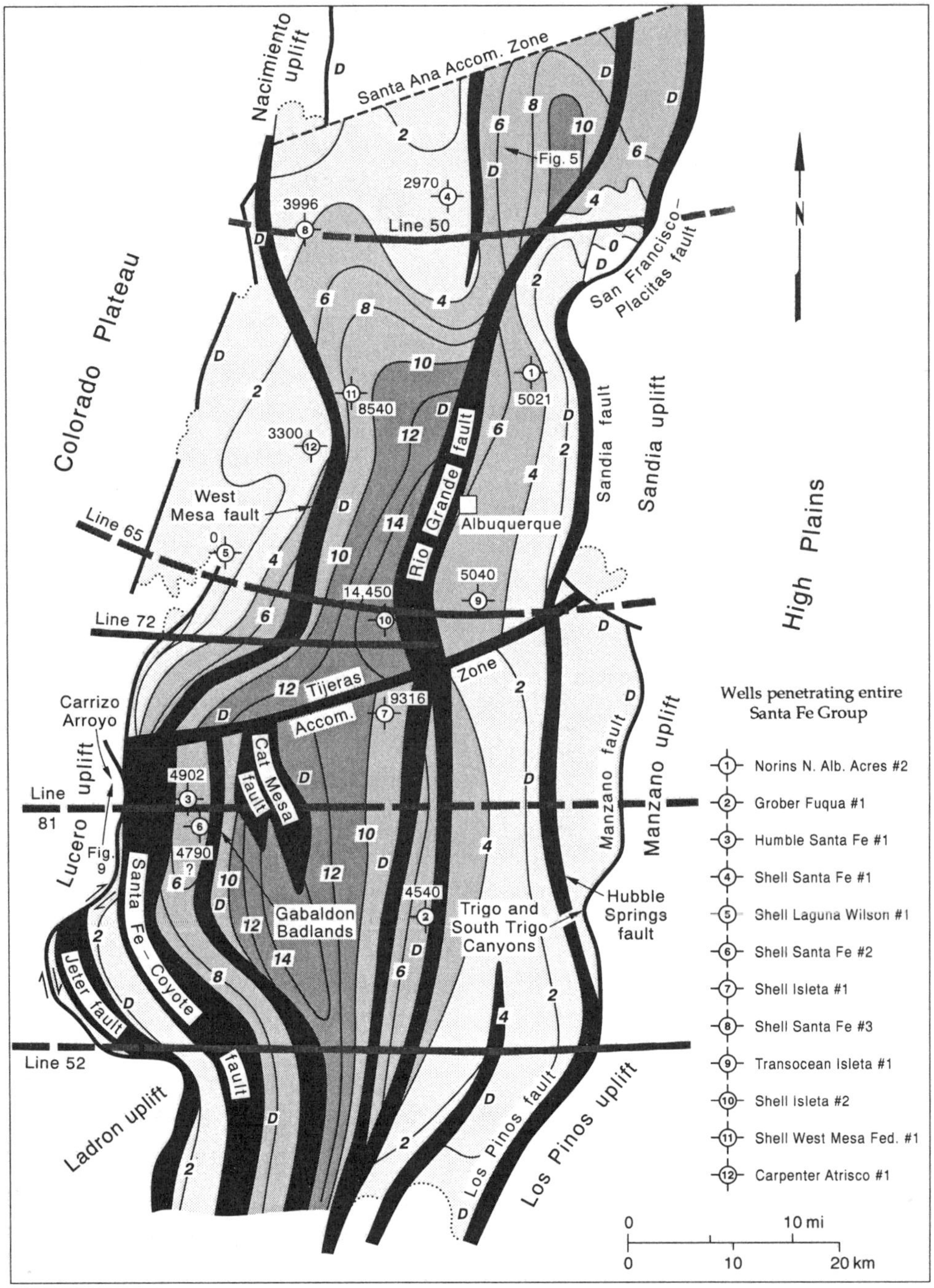

Figure 3. Generalized isochore map of Albuquerque Basin showing approximate thickness of syn-rift Santa Fe Group in thousands of feet. Light gray shading indicates Santa Fe Group thicknesses of 0 to 1,220 m (0 to 4,000 ft) medium gray 1,220 to 3,050 m (4,000 to 10,000 ft), and dark gray, greater than 3,050 m (10,000 ft). Contours are based on data from Russell and Snelson (1990; from Shell seismic lines), Lozinsky (1988; from wells), an ARCO isostatic residual gravity map by S. Rowley, and surface geology. Numbers next to well symbols indicate measured thickness of Santa Fe Group (in feet). Only wells that penetrate the entire syn-rift section are shown. Seismic profiles are shown by numbered lines. Isochore contours more than 5 km (3 mi) from wells and seismic lines are loosely constrained by gravity data and in a few areas may conflict with unpublished Shell seismic data not available for this study. D indicates downthrown side of faults.

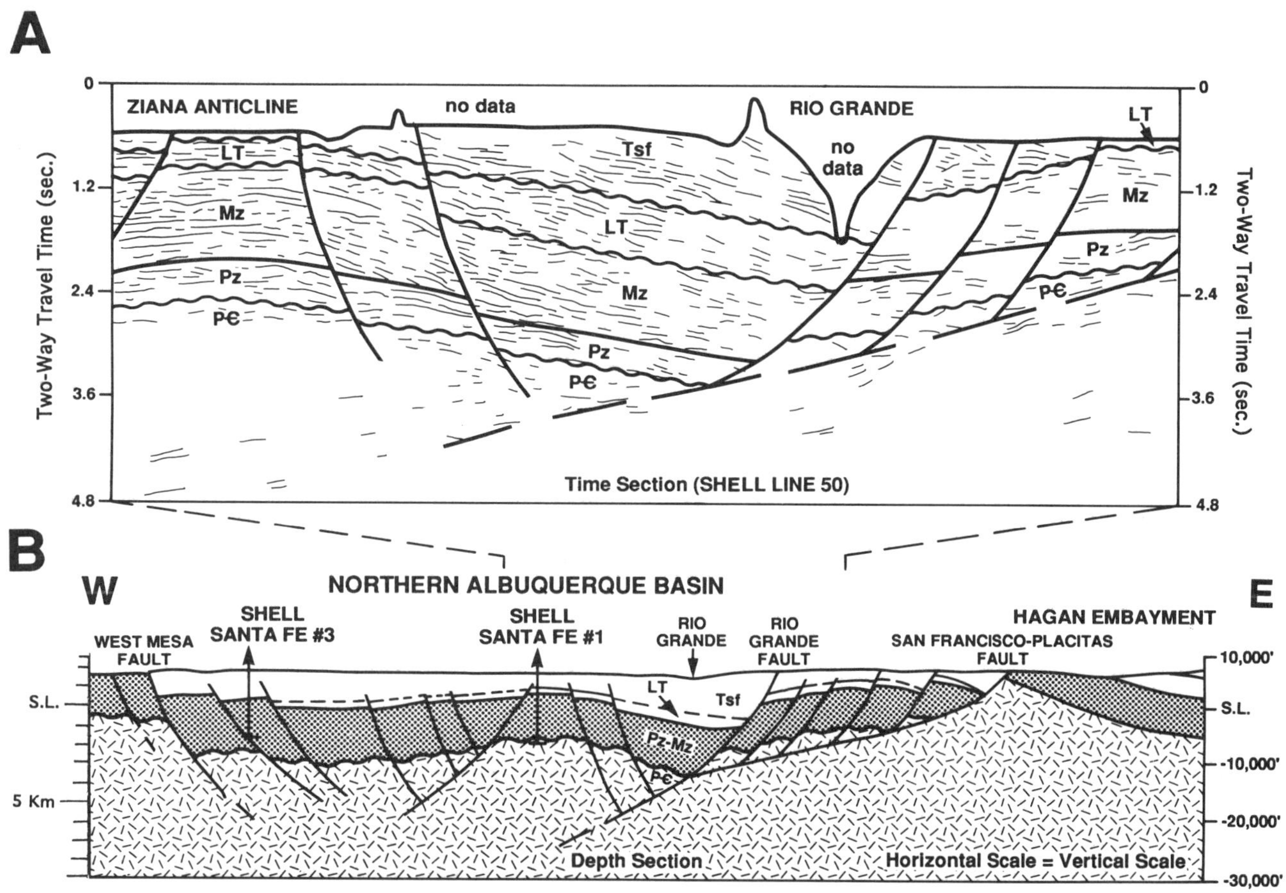

Figure 4. A, Line drawing of Shell seismic line 50 in central part of northern Albuquerque Basin. B, Depth section of northern Albuquerque Basin at latitude of Shell line 50. Tsf, Santa Fe Group; LT, lower Tertiary (Eocene Galisteo and Oligocene Espinaso Formations); Mz, Mesozoic; Pz, Paleozoic (Pennsylvanian and Permian only); PC, Precambrian basement. Modified from Russell and Snelson (1990). See Figure 3 for location of section.

Figure 5. View north from Highway 44 in the northern Albuquerque Basin along eastern flank of the Ziana anticline. In the foreground, the unnamed middle unit of the Santa Fe Group dips 5 to 10° southeast (right) toward the Rio Grande master fault, which is out of view to the right. Dip of beds is highlighted by dashed lines. Stratigraphically higher beds in the Santa Fe are exposed beneath the 2.5-Ma basalt flows capping Santa Ana Mesa on the skyline. These higher beds and overlying flows are gently faulted and dip only 1 to 4° east, reflecting continued east tilting in the hanging wall of the Rio Grande fault in the last few million years. See Figure 3 for location of photograph.

Shell Santa Fe #3 was drilled about 14 km west of the Santa Fe #1 and encountered only 1,218 m (3,996 ft) of the Santa Fe Group (Lozinsky, 1988). Both wells were drilled on the west (hinge) side of the basin. Lozinsky (1988) reported abbreviated thicknesses of only approximately 457 m (1,500 ft) for the unnamed middle unit of the Santa Fe Group and 610 m (2,000 ft) for the underlying Zia Formation in these two wells. The Zia is widely exposed at the surface on this high-standing bench on the northwestern side of the basin and data from Wright (1946), Galusha (1966), Manley (1978), Gawne (1981), and Tedford (1982) indicate local unconformities in the Santa Fe Group. Therefore, it seems likely that depositional rates in the syn-rift section must have been relatively slow in this hinge region and were punctuated by episodes of erosion.

Much of the eastward thickening of the Santa Fe Group in the north basin seems to occur in the unnamed middle unit, which makes up most of the syn-rift section on the master fault benches in the east. This thickening is also indicated about 18 km south of the Santa Fe #1 and #3 wells, where total Santa Fe thickness reaches 2,604 m (8,540 ft) in the Shell West Mesa Federal well (Fig. 3). Here, the unnamed middle unit is tripled in thickness to 1,647 m (5,400 ft; Lozinsky, 1988), possibly implying a phase of accelerated, basinwide extension and motion on the Rio Grande–Sandia master faults in the late Miocene (Lozinsky, 1988; May et al., 1991).

Figure 6 is Shell seismic line 72 in the southern part of the northern basin, which further documents the amount of post–mid-Miocene sedimentation and motion along the Rio Grande fault. The Shell Isleta #2 well bottomed at 6,486 m (21,266 ft), above the top of the Cretaceous and recovered Eocene pollen near the bottom of the hole from beds equivalent to the Galisteo and Baca Formations. The contact between the Cretaceous and Eocene appears to be an unconformity with the Eocene beds showing an apparent onlap relationship at their base. At 5,779 m (18,950 ft), the well penetrated an ash-flow tuff that yielded a K-Ar age date of 36.3 ± 1.8 Ma. This tuff is

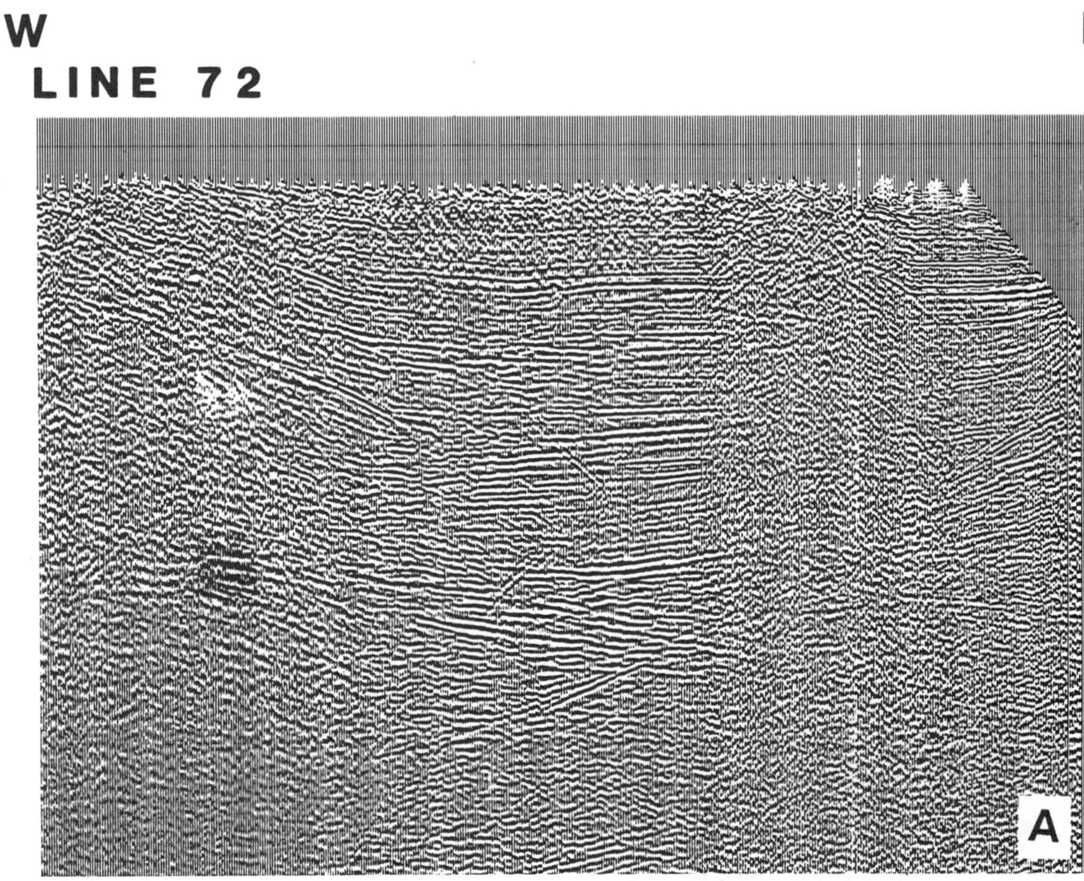

Figure 6 (on this and facing page). A, Uninterpreted version of Shell seismic line 72, in southern part of the northern Albuquerque Basin. This line is approximately 5 km (3 mi) south of line 65 on Figure 3. B, Interpreted version of Shell line 72. At right (east) is the westward-flattening Rio Grande master fault. The Shell Isleta #2 well has been tied to the reflectors using synthetic seismograms. Tsf, Santa Fe Group; LT, lower Tertiary (Oligocene Unit of Isleta #2 and Eocene Galisteo Formation); K, Cretaceous; J, Jurassic. Two-way seismic travel times are shown in seconds. 16.1-Ma age for reflector at 2.4 seconds is based on seismic correlation with basalt flow(?) in Shell Isleta #1 well dated at 16.1 Ma. Interpretation by L. R. Russell and S. Snelson.

in Lozinsky's (1988) Oligocene "Unit of Isleta #2." At 4,941 m (16,200 ft), the well penetrated another unconformity exhibiting an onlap relationship. This appears to be the base of the syn-rift section, based on the seismic data, but occurs about 533 m (1,750 ft) below the depth (4,407 m; 14,450 ft) where Lozinsky (1988) noted a change from typical Santa Fe lithologies to a more indurated section of purplish red–brown sand and clay that he assigned to the "Unit of Isleta #2." Regardless of which of these two horizons represents the base of the syn-rift fill, this is the thickest known section of the Santa Fe Group in the Rio Grande rift (Lozinsky, 1988). The reflector at 2.4 seconds in the Shell Isleta #2 well appears to correlate seismically with a 16.1-Ma basalt flow(?) in the nearby Shell Isleta #1, although this correlation occurs across the structurally complex Tijeras accommodation zone. If the basalt is a flow and the seismic correlation is correct, there has been much post–16-Ma deposition. Lozinsky (1988) estimated an average depositional rate of 600 m (2,000 ft)/m.y. in the Isleta #2 well. This implies a large amount of motion on the Rio Grande master fault sometime in the last 16 m.y. On the basis of evidence here and elsewhere in the basin, much of this motion probably occurred in the late Miocene (May et al., 1991; Lozinsky, 1988) during deposition of the unnamed middle unit of the Santa Fe Group.

Interpretation of the underlying Unit of Isleta #2 is more problematical. That portion of the Unit of Isleta #2 beneath the unconformity at 4,941 m (16,200 ft) depth (approximately 1,220 m; 4,000 ft thick) probably represents pre-rift deposition in a remnant Laramide basin centered approximately beneath the Neogene Albuquerque rift basin or (less likely) a previously undocumented early phase of basinal rifting that began in the middle Oligocene. We conclude that rifting probably started shortly before the end of Unit of Isleta #2 deposition in the basin center, leading to the unconformity at 4,941 m (16,200 ft). This occurred at a time when this unit (Espinaso Formation in the north) blanketed much of the region, possibly including the lowermost slopes of the early rift shoulders.

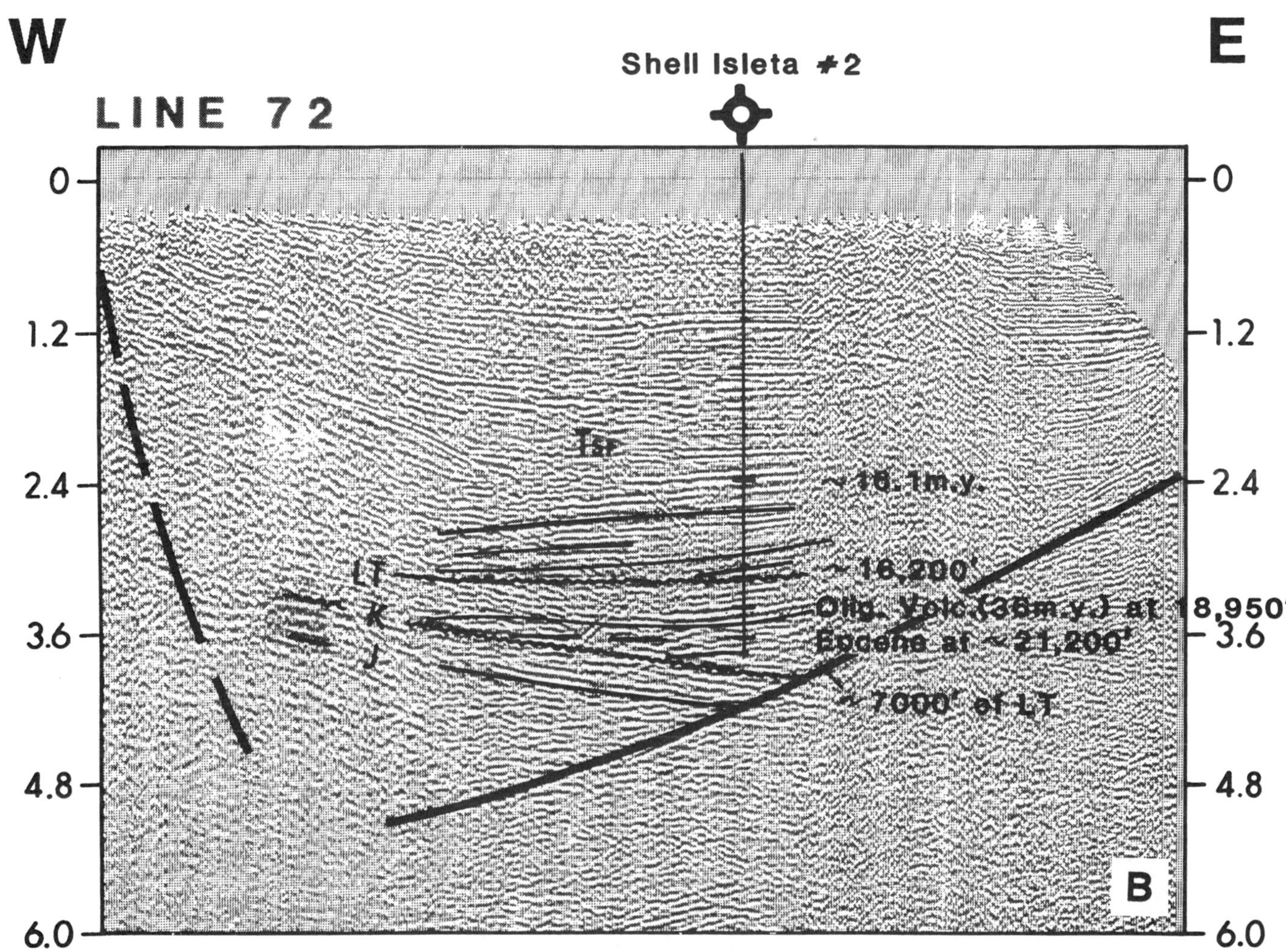

Shortly after rifting began, strata beneath the Unit of Isleta #2 in the surrounding source areas were reexposed and deposition of the more arkosic beds of the Santa Fe Group began.

Figure 7 includes an interpreted depth section across the basin along Shell line 65, immediately north of Figure 6. This cross section illustrates the geometric distinction between the detached master fault benches of the basin, shown by the Albuquerque Bench, and the downward-flexed hinge benches on the opposing side. On the Albuquerque Bench, the Transocean Isleta #1 well penetrated 1,536 m (5,040 ft) of the Santa Fe Group, primarily the upper part of the Popotosa Formation (Lozinsky, 1988). On the hinge side of the basin, the Shell Laguna Wilson well spudded in the Cretaceous at the surface and this area was probably never buried very deeply by the Santa Fe.

SOUTHERN ALBUQUERQUE BASIN

Figure 8 includes an east-west cross section of the southern Albuquerque Basin along Shell line 81, illustrating the basin's westward dip, south of the Tijeras accommodation zone. In the east, a wide hinge bench is flexed downward toward the basin-controlling faults in the west, with a narrow master fault bench in a structurally intermediate position, west of Rio Puerco. The Santa Fe fault is the western border fault of the basin at this latitude and its steep, upper part is well exposed along the eastern flank of the Lucero uplift near Carrizo Arroyo (Figs. 3 and 9). However, the fault flattens eastward beneath the master fault bench as indicated by the seismic data. The Cat Mesa fault forms the eastern side of the bench and appears to merge with or truncate the downward flattening Santa Fe fault on the seismic data. The Cat Mesa fault exhibits approximately 10 km (6 mi) of horizontal separation of the Cretaceous section between its hanging-wall and footwall cutoffs. This fault appears to be located west of the Cat Mesa fault as mapped at the surface by Kelley (1977) and may correlate with Kelley's Gabaldon fault.

The Humble Santa Fe #1 and Shell Santa Fe #2 drilled approximately 1,463 to 1,494 m (4,800 to 4,900 ft) of Santa Fe Group on the master fault bench between the Santa Fe and Cat Mesa faults (Lozinsky, 1988; Fig. 3). The upper two-thirds of the Santa Fe in these wells appears to be the upper part of the Popotosa Formation, based on petrologic descriptions of cuttings (Lozinsky, 1988). Santa Fe strata in this area show westward hanging-wall rotation into the Santa Fe fault and the dips gradually decrease upward in the section on the seismic data.

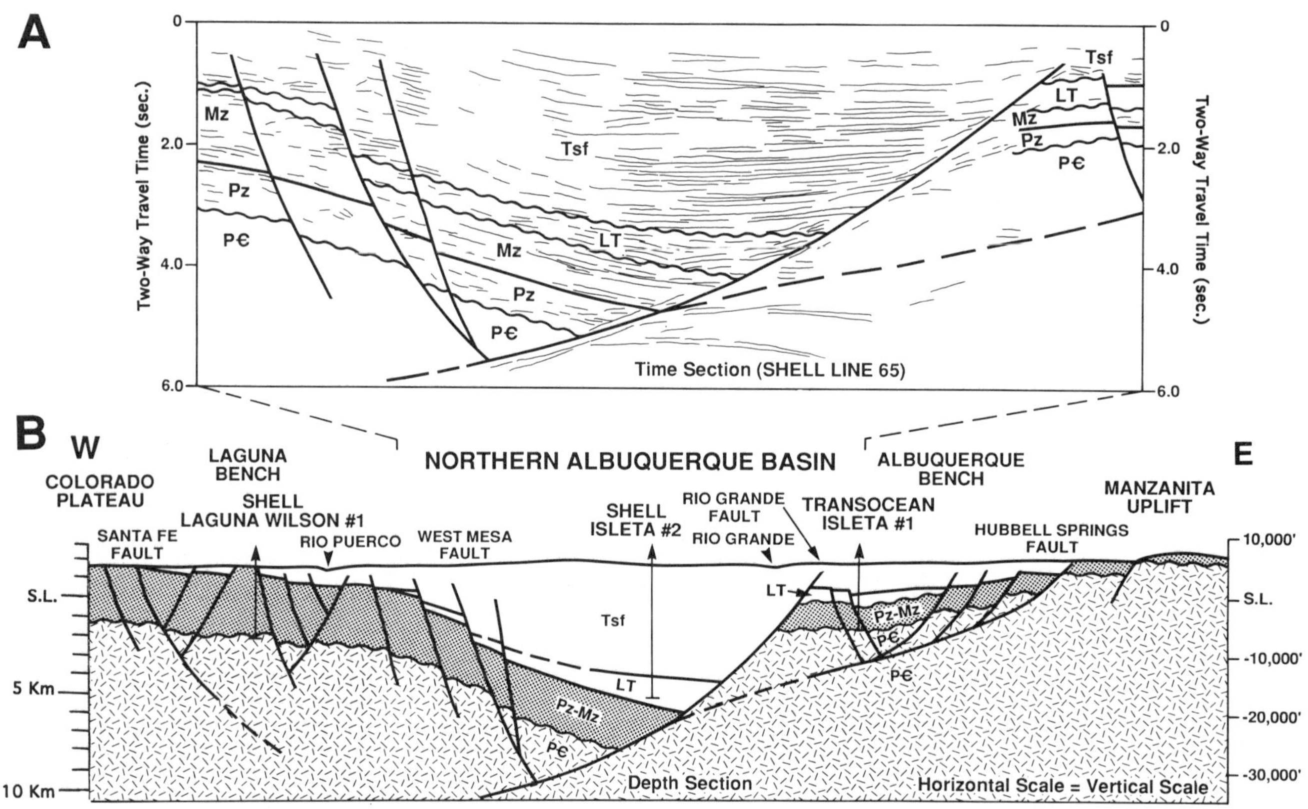

Figure 7. A, Line drawing of Shell seismic line 65 in southern part of northern Albuquerque Basin. B, Depth section of northern Albuquerque Basin at latitude of Shell line 65. Tsf, Santa Fe Group; LT, lower Tertiary (Oligocene Unit of Isleta #2 and Eocene Galisteo Formation); Pz-Mz, Paleozoic and Mesozoic; PC, Precambrian basement. Modified from Russell and Snelson (1990). See Figure 3 for location of section.

This upward decrease can also be seen at the surface in the 7 to 9 Ma part of the upper Popotosa in the Gabaldon badlands area, southeast of Carrizo Arroyo, as illustrated by Lozinsky and Tedford's (1991) mapping of that area. This probably reflects syndepositional hanging-wall rotation into the Santa Fe fault. Lozinsky (1988) calculated relatively rapid depositional rates of about 200 m (650 ft)/m.y. for the upper Popotosa in this area, as might be expected. This implies moderately rapid extension on the Santa Fe master fault in the late Miocene. However, uppermost Santa Fe relationships in the Sierra Ladrones Formation suggest that motion of the Santa Fe fault may have slowed considerably in the last few million years (May et al., 1991). In addition, Lozinsky and Tedford (1991) mapped a major fault that offsets the upper Popotosa in this area, but dies out in the overlying Sierra Ladrones Formation. If the roots of the Santa Fe fault are truncated basinward by the Cat Mesa fault, as implied on the seismic data, Santa Fe relationships at the surface near Carrizo Arroyo may provide additional evidence that extension is being transferred basinward to the Cat Mesa fault, which roots at a lower level and may have started to tectonically strand the Santa Fe fault in the last few million years. This inward migration of faulting appears to be a common, but very subtle, theme in the basin and is discussed further by May et al. (this volume).

In spite of hanging-wall rotation into the Santa Fe fault in the west, farther east there is a somewhat perplexing easterly dip in the Santa Fe Group on the hanging wall of the Cat Mesa fault (Fig. 8). The Santa Fe Group shows a possible downlap or onlap relationship at its base with the Cretaceous or lower Tertiary section, which has rotated in the opposite direction, back toward the fault. These relationships are also seen elsewhere in the basin, indicating that there is still more work to be done to understand the timing and kinematics of both rift-related and earlier Laramide tectonic events in the basin. This problem is complicated further by the possibility of dextral and/or sinistral components of wrench faulting along the Rio Grande rift during the Laramide and Neogene (Kelley, 1977; Chapin, 1983; Hammond, 1987; Beck, 1991; Hayden and Mawer, 1991; S. Baldridge, personal communication, 1991).

Stratigraphic relations on the southeastern (hinge) side of the basin are poorly exposed but appear different than those in the southwest. The lower Santa Fe Group preserved below the lower Pliocene, intra–Santa Fe unconformity (Fig. 1) is much older and thinner in the southeastern part of the basin, according to data from Lozinsky (1988). South of the area shown on Figure 8, the lower part of the Santa Fe Group is exposed west of the Manzano rift shoulder in the Trigo and South Trigo Canyon area (Fig. 3). Here, Lozinsky (1988) reported white to gray pebbly sand equivalent to the lower part of the Popotosa Formation. These beds appear to contain a basalt flow dated at 21 Ma by Bachman and Mehnert (1978), confirming an early Miocene age, although limited exposures and presence of faults

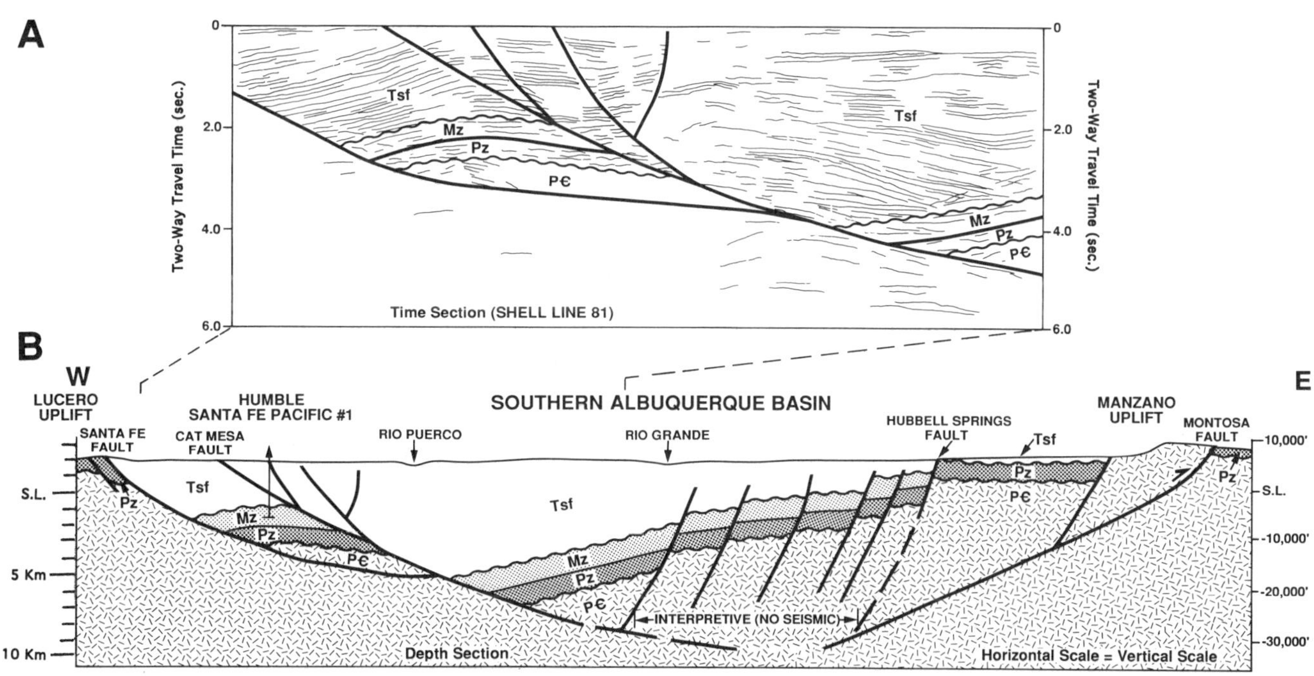

Figure 8. A, Line drawing of Shell seismic line 81 in southern Albuquerque Basin. B, Depth section of southern Albuquerque Basin at latitude of Shell line 81. Tsf, Santa Fe Group and an uncertain thickness of underlying lower Tertiary strata (Oligocene Unit of Isleta #2 and Eocene Galisteo Formation); Mz, Mesozoic; Pz, Paleozoic (Pennsylvanian and Permian only); PC, Precambrian basement. Modified from Russell and Snelson (1990). See Figure 3 for location of section.

Figure 9. View north toward Santa Fe fault near Carrizo Arroyo at western margin of the Albuquerque Basin (Fig. 3). The fault dips to right (east) beneath the basin. East-dipping upper Popotosa beds of the Santa Fe Group (Tsf) are down dropped against east-dipping pre-rift section of the Lucero rift shoulder. Km, Cretaceous Mancos Shale; Trc, Triassic Chinle Formation; Ps, Permian San Andres Formation.

in the area make it difficult to demonstrate the precise stratigraphic position of the basalt (R. Lozinsky, personal communication, 1992). These beds are overlain unconformably by a 46-m-thick (150 ft) section of bouldery conglomerate equivalent to the Plio-Pleistocene Sierra Ladrones Formation of the upper Santa Fe (Lozinsky, 1988). The intervening mid- to upper Miocene part of the Santa Fe Group (upper Popotosa) appears to be missing below the unconformity.

The Grober Fuqua #1 well is located about 12 km west (down dip) of Trigo and South Trigo Canyons, on the hinge side of the basin (Fig. 3) and penetrated a thicker (1,384 m; 4,540 ft) section of Santa Fe (Lozinsky, 1988). This well projects north into Figure 8 a short distance east of the Rio Grande. Lozinsky reported a lower Popotosa section, overlain by a thick section of upper Popotosa, apparently preserved beneath the lower Pliocene unconformity. As on the northwestern hinge side of the basin, basinward thickening of the Santa Fe Group in the southeast results mainly from thickening of the upper Miocene (upper Popotosa) part of the section, implying accelerated tilting toward the western master-fault side of the basin at that time.

CONCLUSIONS

Recently released Shell seismic data have enhanced our understanding of how structural styles in the Albuquerque Basin have influenced deposition of the syn-rift Santa Fe Group. Two examples of these enhancements are:

1. An improved picture of the stratigraphic asymmetry in the basin that results from the basin's structural asymmetry. In the north basin, the Santa Fe thickens eastward in the subsurface, reflecting eastward rotation into the Rio Grande and related master faults. In the southern basin, the Santa Fe thickens to the west, toward the Santa Fe–Coyote fault system.

2. The concept of distinguishing benches on the master fault and hinge sides of the basin provides a structural framework for explaining stratigraphic variations in the Santa Fe. This distinction is seen in the isochore map of the Santa Fe Group (Fig. 3). The hinge benches in the northwest and southeast are relatively wide and shallow. Here, mean depositional rates of the Santa Fe appear to have been relatively slow. Erosion and deposition have probably been closer to a state of dynamic equilibrium. Erosion at the lower Pliocene unconformity beneath the upper Santa Fe was very effective in removing the relatively thin accumulations of the unnamed middle unit of the Santa Fe in the northwest and upper Popotosa beds in the southeast. As a result, these hinge benches contain the stratigraphically deepest exposures of the Santa Fe at the surface.

In contrast, the master fault benches in the northeast and southwest are somewhat narrower and structurally deeper.

Significant thicknesses of the unnamed middle unit and upper Popotosa are well preserved in these areas and depositional rates were more rapid. While the hinge benches remained structurally neutral, crustal extension in the Albuquerque Basin (probably most rapid during the late Miocene) caused the master-fault benches in the northeast and southwest to subside rapidly and provide locally greater accommodation space for deposition of the Santa Fe Group.

ACKNOWLEDGMENTS

We thank S. M. Cather, C. E. Chapin, J. W. Hawley, R. V. Ingersoll, R. P. Lozinsky, and L. A. Woodward for insightful discussions and critical reviews of the manuscript. We also thank Shell Oil Company for releasing some of their seismic data in the basin, James Duke and Rebecca Titus for drafting of figures, and ARCO Oil and Gas Company for general support of the authors' efforts.

REFERENCES CITED

Bachman, G. O., and Mehnert, H. H., 1978, New K-Ar dates and the late Pliocene to Holocene geomorphic history of the central Rio Grande region, New Mexico: Geological Society of America Bulletin, v. 89, p. 283–292.

Beck, W. C., 1991, Definition of Laramide and mid-Tertiary stress orientations: Structural data from the Joyita Hills, Socorro County, New Mexico: Geological Society of America, Rocky Mountain and South-Central Section Meeting, Abstracts with Program, p. 6.

Chapin, C. E., 1983, An overview of Laramide wrench faulting in the southern Rocky Mountains with emphasis on petroleum exploration: Rocky Mountain Association of Geologists, p. 169–179.

Galusha, T., 1966, The Zia Sand Formation, new early to medial Miocene beds in New Mexico: American Museum Novitates 2271, 12 p.

Gawne, C. E., 1981, Sedimentology and stratigraphy of the Miocene Zia Sand of New Mexico: Geological Society of America Bulletin, v. 92, part I (summary), p. 999–1007.

Gorham, T. W., and Ingersoll, R. V., 1979, Evolution of the Eocene Galisteo Basin, north-central New Mexico: New Mexico Geological Society Guidebook, 30th Field Conference, Santa Fe Country, p. 219–224.

Hammond, C. M., 1987, Geology of the Navajo Gap area between the Ladron Mountains and Mesa Sarca, Socorro County, New Mexico: A structural analysis [M.S. thesis]: Socorro, New Mexico, New Mexico Institute of Mining and Technology, 212 p.

Hayden, S. N.,and Mawer, C. K.,1991, Dextral transtensional deformation along the western margin of the Rio Grande rift, central New Mexico: Geological Society of America, Rocky Mountain and South-Central Section Meeting, Abstracts with Program, p. 31.

Kautz, P. F., Ingersoll, R. V., Baldridge, W. S., Damon, P. E., and Shafiqullah, M., 1981, Geology of the Espinaso Formation (Oligocene), north-central New Mexico: Geological Society of America Bulletin, pt. II, v. 92, p. 2318–2400.

Kelley, V. C., 1977, Geology of the Albuquerque Basin: New Mexico Bureau of Mines and Mineral Resources Memoir 33, 60 p.

Lozinsky, R. P., 1988, Stratigraphy, sedimentology, and sand petrology of the Santa Fe Group and pre–Santa Fe Tertiary deposits in the Albuquerque Basin, central New Mexico [Ph.D. thesis]: Socorro, New Mexico, New Mexico Institute of Mining and Technology, 298 p.

Lozinsky, R. P., and Tedford, R. H., 1991, Geology and paleontology of the Santa Fe Group, southwestern Albuquerque Basin, Valencia County, New Mexico: New Mexico Bureau of Mines and Mineral Resources Bulletin 132, 35 p.

Lozinsky, R. P., Hawley, J. W., and Love, D. W., 1991, Geologic overview and Pliocene-Quaternary history of the Albuquerque Basin, central New Mexico, *in* Julian, B., and Zidek, J., eds., Field guide to geologic excursions in New Mexico and adjacent areas of Texas and Colorado: New Mexico Bureau of Mines and Mineral Resources Bulletin 137, p. 157–162.

Manley, K., 1978, Geologic map of Bernalillo NW Quadrangle, Sandoval County, New Mexico: U.S. Geological Survey Geologic Quadrangle Map GQ-1445, scale 1:24,000.

May, S. J., Russell, L. R., Dula, W. F., and Kilsdonk, B., 1991, Field Guide to the Rio Grande rift (Albuquerque Basin): (unpublished guidebook) ARCO Oil and Gas Co. Extensional Structural Styles School,141 p.

Morley, C. K., Nelson, R. A., Patton, T. L., and Munn, S. G., 1990, Transfer zones in the East African rift system and their relevance to hydrocarbon exploration in rifts: American Association of Petroleum Geologists Bulletin, v. 74, no. 8, p. 1234–1253.

Osburn, G. R., and Chapin, C. E., 1983, Nomenclature for Cenozoic rocks of northeast Mogollon-Datil volcanic field, New Mexico: New Mexico Bureau of Mines and Mineral Resources, Stratigraphic Chart 1.

Rosendahl, B. R., 1987, Architecture of continental rifts with special reference to East Africa: Annual Review of Earth and Planetary Sciences, v. 15, p. 445–503.

Russell, L. R., and Snelson, S., 1990, Structural style and tectonic evolution of the Albuquerque Basin segment of the Rio Grande rift, *in* Pinet, B., and Bois, C., eds., The potential of deep seismic profiling for hydrocarbon exploration: Institut Francais Petrole Research Conference Proceedings: Paris, Editions Technip, p. 175–207.

Tedford, R. H., 1982, Neogene stratigraphy of the northwestern Albuquerque Basin: New Mexico Geological Society, Guidebook, 33rd Field Conference, Albuquerque Country II, p. 273–278.

Wright, H. E., Jr., 1946, Tertiary and Quaternary geology of the lower Rio Puerco area, New Mexico: Geological Society of America Bulletin, v. 57, p. 383–456.

Manuscript Accepted by the Society October 13, 1993

Printed in U.S.A.

Geological Society of America
Special Paper 291
1994

Footwall unloading and rift shoulder uplifts in the Albuquerque Basin: Their relation to syn-rift fanglomerates and apatite fission-track ages

S. Judson May
ARCO Exploration and Production Technology, 2300 West Plano Parkway, Plano, Texas 75075
Shari A. Kelley
Department of Geological Sciences, Southern Methodist University, Dallas, Texas 75275-0395
Lee R. Russell
ARCO Exploration and Production Technology, 2300 West Plano Parkway, Plano, Texas 75075

ABSTRACT

A systematic relationship exists between seismically defined, master normal faults of the Albuquerque Basin and the structural style of rift shoulders on the basin margins. Rift shoulders occupying the footwalls of these master faults have at least three characteristics distinguishing them from the other shoulders of the basin:

1. They are the source areas for the basin's major fanglomerates in the syn-rift Santa Fe Group.

2. They display distinctly Neogene apatite fission-track ages.

3. They have undergone the greatest amount of rift-related uplift in the Neogene.

These relationships imply that uplift of these rift shoulders may be in part, the result of isostatic rebound of the lithosphere, accompanying tectonic unloading by the seismically defined master faults of the basin.

INTRODUCTION

Recent models by Buck et al. (1988), Wernicke and Axen (1988), and Egan (1992) illustrate the dynamic relationship between upper crustal thinning by major listric normal faults and the uplift of basin-bounding range blocks in their footwalls as the flexural isostatic response of the lithosphere to tectonic unloading. Kinematic histories of these basin-controlling faults and the resultant uplift and unroofing of adjacent rift shoulders can be constrained by the study of basin-margin unconformities, fanglomerates in the syn-rift basin fill, and apatite fission-track ages in the rift shoulders. The Albuquerque Basin is one of the best rift basins to illustrate the relations between these features for several reasons:

1. Late Quaternary incision of the Rio Grande provides locally good exposures of the syn-rift basin fill (Neogene Santa Fe Group) on both the master fault and hinge sides of the Albuquerque Basin, thereby displaying the distribution of the major fanglomerates.

2. Good subsurface definition of the basin's master normal faults has been achieved by the study of more than 800 km of seismic-reflection data shot by Shell Oil Company in the 1970s. Some of these data have been released (Russell and Snelson, 1990, this volume).

3. Formation tops in a suite of eight deep Shell wells in the basin have been correlated to the seismic data with synthetic seismograms and substantiated by petrographic examination of well cuttings by Lozinsky (1988).

May, S. J., Kelley, S. A., and Russell, L. R., 1994, Footwall unloading and rift shoulder uplifts in the Albuquerque Basin: Their relation to syn-rift fanglomerates and apatite fission-track ages, *in* Keller, G. R., and Cather, S. M., eds., Basins of the Rio Grande Rift: Structure, Stratigraphy, and Tectonic Setting: Boulder, Colorado, Geological Society of America Special Paper 291.

4. Apatite fission-track apparent ages have been determined for rock samples from uplifts surrounding the basin, providing additional constraints on uplift and cooling histories of the basin's rift shoulders.

RIFT SHOULDERS AND MASTER NORMAL FAULTS

Figure 1 is a map of the basin showing the relationship between the master faults of the Albuquerque Basin, as defined by the Shell seismic data, and geometry of the basin's rift shoulders. In the north, the Rio Grande fault is the basin-controlling fault. It trends north-northeast beneath the Rio Grande and is a westward-flattening, listric-normal fault, based on the Shell data. The north half of the basin tilts east as a result of hanging-wall rotation into this and other major faults (Sandia and San Francisco–Placitas faults) along the eastern margin of the basin. The southern half of the basin is tilted west toward the Jeter–Santa Fe–Coyote master faults along the southwestern basin margin.

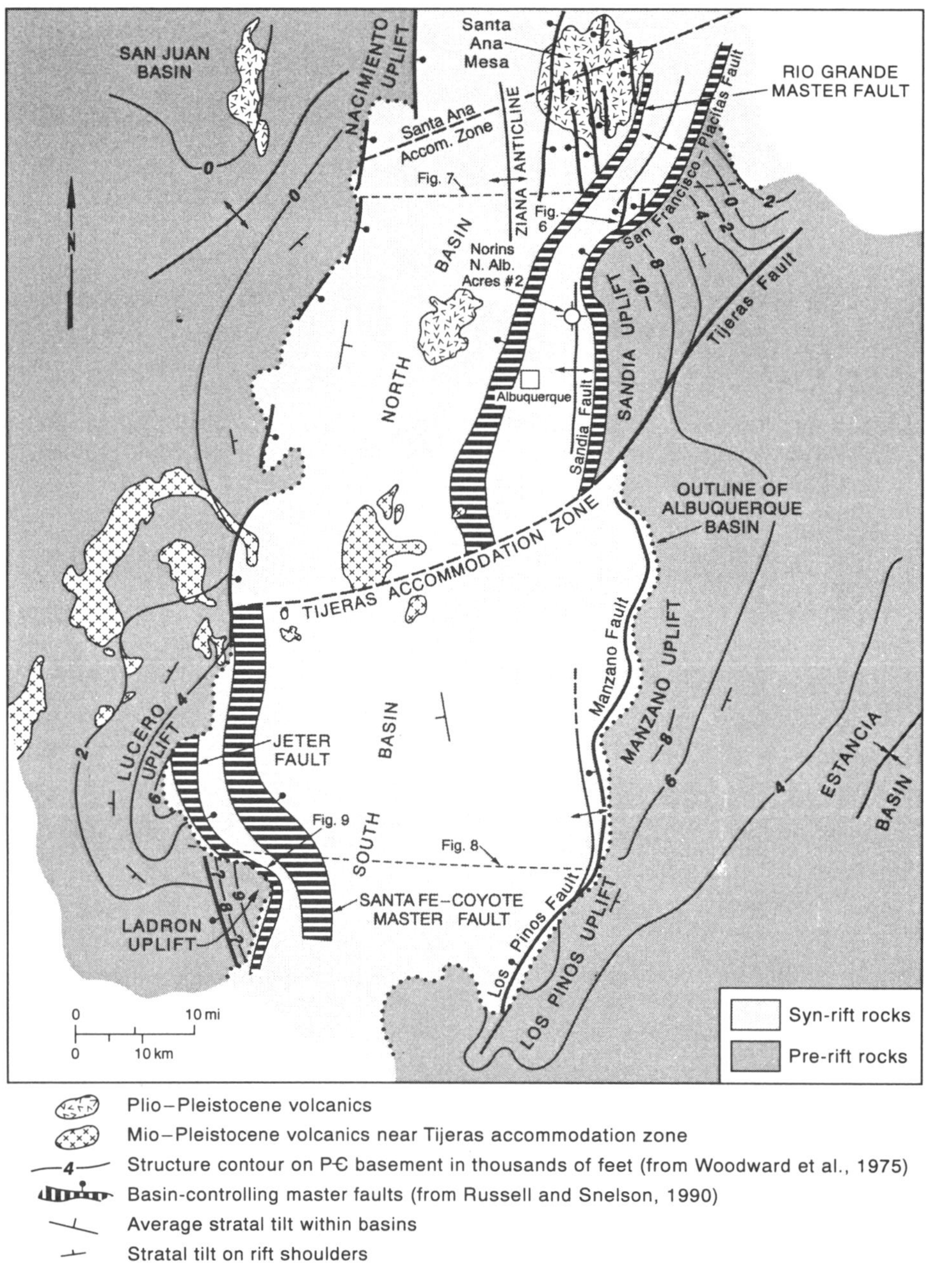

Figure 1. Map of Albuquerque Basin emphasizing the relationship between seismically defined master normal faults and the geometry of surrounding rift shoulder uplifts.

In the north basin, structure contours on top of the Proterozoic basement indicate that the Sandia uplift is the structurally highest and most strongly east tilted of the rift shoulders and occupies the uplifted footwall of the Rio Grande and Sandia master faults. These faults and the steep easterly dips of the Sandia block terminate to the south near the Tijeras fault and Tijeras accommodation zone. This relationship implies a mechanical connection between motion on the Rio Grande—Sandia fault system and the unusually high elevation and eastward tilting of the Sandia uplift. The northwest side of the basin lacks a prominent rift shoulder and master fault. Instead, a broad, northeast-trending antiform separates the northern Albuquerque and southeastern San Juan Basins.

In the south basin, the rift shoulders and master faults exhibit a mirror image relationship to those in the north and are separated from the north basin by the Tijeras accommodation zone. In the south, the Jeter–Santa Fe–Coyote faults dip to the east, unloading the east flank of the Lucero and Ladron uplifts. Structure contours on the Lucero uplift show that this uplift terminates to the north near the westward projection of the Tijeras accommodation zone. The north ends of the seismically defined master faults also terminate at this same latitude, implying a mechanical connection between uplift of the Lucero-Ladron block and motion on these faults.

The simplistic relationship just described between master faults and rift shoulders of the basin is complicated slightly by several factors:

1. The presence of other, deep-seated processes affecting rift shoulder uplift and flexural rigidity of the lithosphere, such as asthenospheric diapirism beneath the rift, ductile thinning of the lower crust, and density changes in the lithosphere related to variable heat flow.

2. The superposition of Neogene rift shoulders on preexisting Laramide and mid-Tertiary structures. Some of the basin's rift shoulders were the approximate sites of Laramide (latest Cretaceous? to Eocene) uplifts, indicating that part of their present structural relief may have originated in the early Tertiary. For example, the Laramide or mid-Tertiary Lucero monocline on the eastern flank of the Lucero uplift implies some pre-rift structural elevation of that rift shoulder.

On the southeastern (hinge) side of the basin, the Manzano–Los Pinos uplift is actually higher than the Lucero uplift, yet lacks an adjacent master fault to explain its relatively high structural elevation. However, Laramide compressional and transpressional faults on the east flank of the Manzano–Los Pinos Mountains (Chapin, 1983; Hayden, 1991) imply a Laramide (pre-rift) ancestry for the uplift. A deeply eroded Mesozoic section beneath the Santa Fe Group in the southeastern part of the basin also implies that part of the Manzano uplift's relief is Laramide. In general, however, petrologic data in the Santa Fe Group and underlying Paleogene section from Lozinsky (1988) imply that much uplift of these rift shoulders relative to the basin occurred during Neogene extension.

APATITE FISSION-TRACK AGES

Figure 2 shows the distribution of apatite fission-track apparent ages with respect to the rift shoulders and seismically defined master faults of the basin. A more detailed description of the fission-track data can be found in Kelley et al. (1992). The round dots show the location of fission-track sampling traverses that extend up the basinward slopes of the rift shoulders. Note that distinctly Neogene fission-track ages of less than 35 Ma (open and "double circle" dots) are restricted to the Sandia uplift in the northeast and Ladron uplift in the southwest. These young dates are consistent with particularly rapid elevation of these range blocks since mid-Miocene time, due to footwall unloading of the master faults. On the opposing hinge sides of the basin, the fission-track ages are older (early to mid-Tertiary), shown by the filled and partly filled dots, and are probably more related to Laramide cooling at this level in the crust. Although the shoulders on the hinge sides of the basin were also uplifted along range-marginal faults during rifting, the magnitude of their uplift was apparently not enough to cause as much erosional removal of the uppermost rocks in these ranges.

Figure 3 is a plot of apatite fission-track apparent age versus elevation in the Ladron uplift in the southwest and the Manzano uplift in the southeast. The data from the Ladron uplift are taken from the east side of the range, which is upthrown along the Jeter fault. The ages range from 14 to 9 Ma and show no correlation with elevation, indicating rapid cooling. Two samples ("double circle" dot on Ladron uplift, Fig. 2) from the northwest side of the Ladron block and from Ladron Peak yield ages of 32 ± 1.2 Ma and 29 ± Ma and are thought to reflect westward tilting of the range.

Fission-track ages from the Manzano uplift, on the hinge side of the basin, are generally older, ranging from 42 to 22 Ma and show relatively good correlation with elevation, suggesting slower cooling. The older ages at the top of the Manzano block reflect cooling during Eocene erosion, following Laramide uplift of the range. The open square near the bottom of the figure shows the age of a detrital apatite from the Yeso Formation exposed on the Hubbell Bench, adjacent to the west side of the Manzano uplift. It is anomalously old relative to its elevation. This sample age is also related to Eocene erosion on the top of the range but the Hubbell Bench (western) portion of the range has been subsequently down dropped to a structurally intermediate position during rifting.

Figure 4 shows the fission-track data for representative

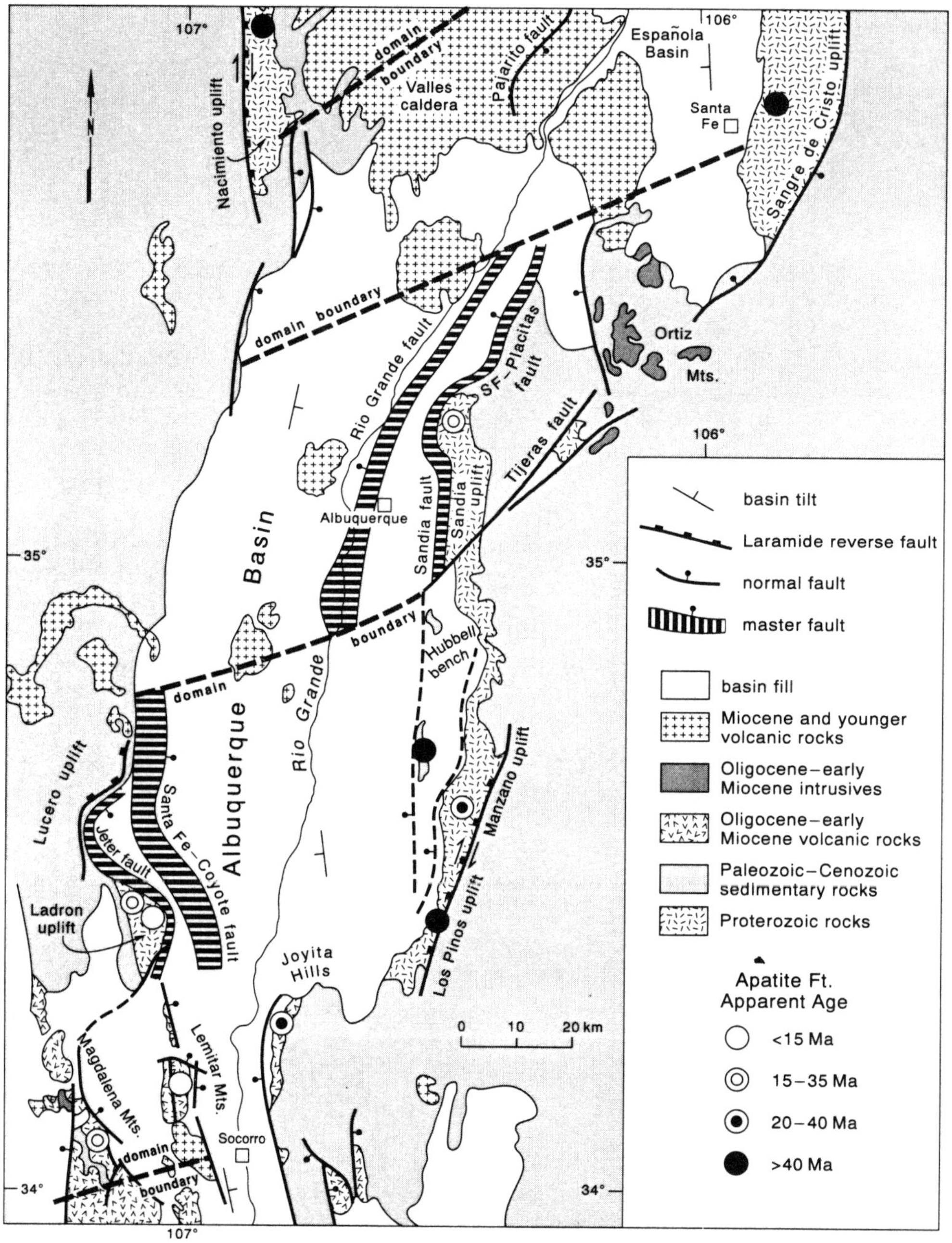

Figure 2. Map of Albuquerque Basin showing relationship between apatite fission-track apparent ages in rift shoulders and the location of seismically defined master normal faults of the basin. The large strike and dip symbols show average dip of Santa Fe strata toward the master faults. Note that the basins dip alternately to the west and east as one moves north along the axis of the rift. The approximate boundaries between these domains are accommodation zones shown by the northeast-trending dashed lines.

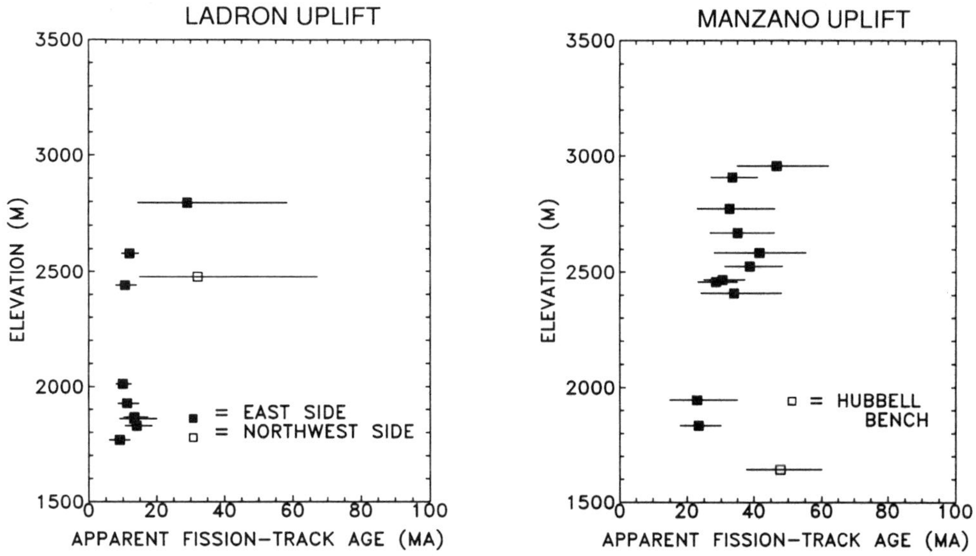

Figure 3. Plots of apatite fission-track age versus elevation in the Ladron and Manzano uplifts along the southern margins of the Albuquerque Basin.

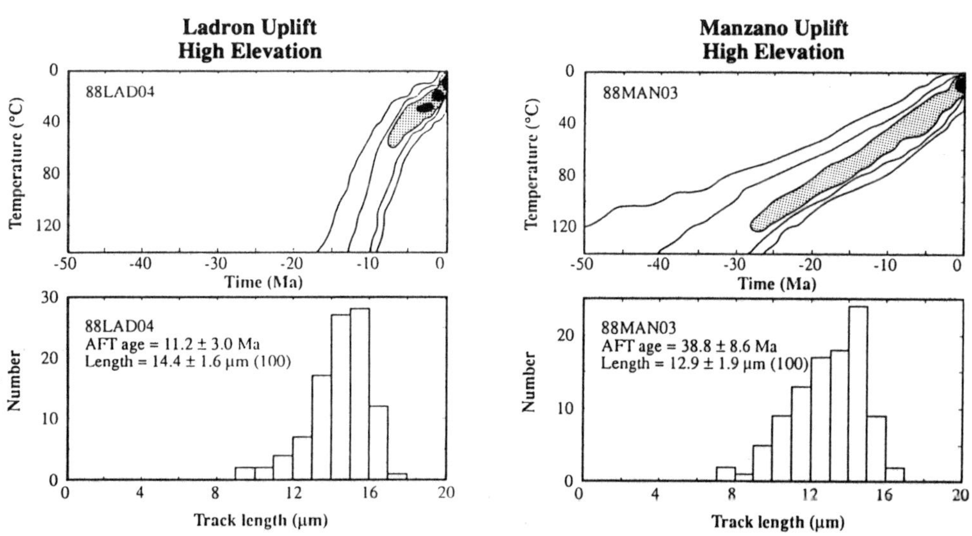

Figure 4. Time-temperature histories and track-length distributions for apatite fission-track (AFT) apparent ages in the Ladron and Manzano uplifts. Time-temperature contour plots are used to describe the thermal history. 50% of the solutions lie in the dark gray areas and 75% lie in the light gray regions. Outer envelopes enclose 90 and 99% of the solutions, respectively. Numbers in parentheses following the track-length measurements correspond to the number of confined tracks measured. (From Kelley et al., 1992.)

samples from the Ladron and Manzano rift shoulders. The time/temperature plots at the top of the figure are a quantitative evaluation of the thermal history based on a Monte Carlo inversion scheme developed by Corrigan (1991). This model is based on laboratory studies of the annealing behavior of tracks in apatite. The model uses the fission-track age and track-length distribution to calculate time-temperature paths that satisfy the data. This model gives a reasonably realistic view of how well the thermal history is known. The Ladron sample from the master fault side of the basin (Fig. 4) is young and the track length distribution and mean are characteristic of a sample with a rapid cooling history, as might be expected with rapid uplift of the range. Cooling rates for the Ladrons are about 15 to 20°C/m.y. These rapid cooling rates are probably related to tectonic denudation beneath the low-angle Jeter fault, uplift and erosion resulting from that denudation, and westward tilting of the Ladron block. In contrast, the Manzano sample from the hinge side of the basin is older and the wider distribution in track lengths is characteristic of slower cooling. Cooling rates are only 3 to 4°C/m.y. for the time interval 10 to 28 Ma. This early slow cooling event is probably related to erosion following Laramide uplift of the Manzano block that continued through the early stages of rifting. Only in the past 10 m.y. or so has there been significant uplift related to rifting.

SYN-RIFT FANGLOMERATES

The structural and fission-track asymmetry between the north and south halves of the basin is also reflected in fanglomerate distribution in the Santa Fe Group (upper Oligocene to Pleistocene). Stratigraphy of the syn-rift Santa Fe Group is described by Lozinsky (this volume) and briefly by May and Russell (this volume). Figure 5 shows a model for depositional features of an internally drained continental rift

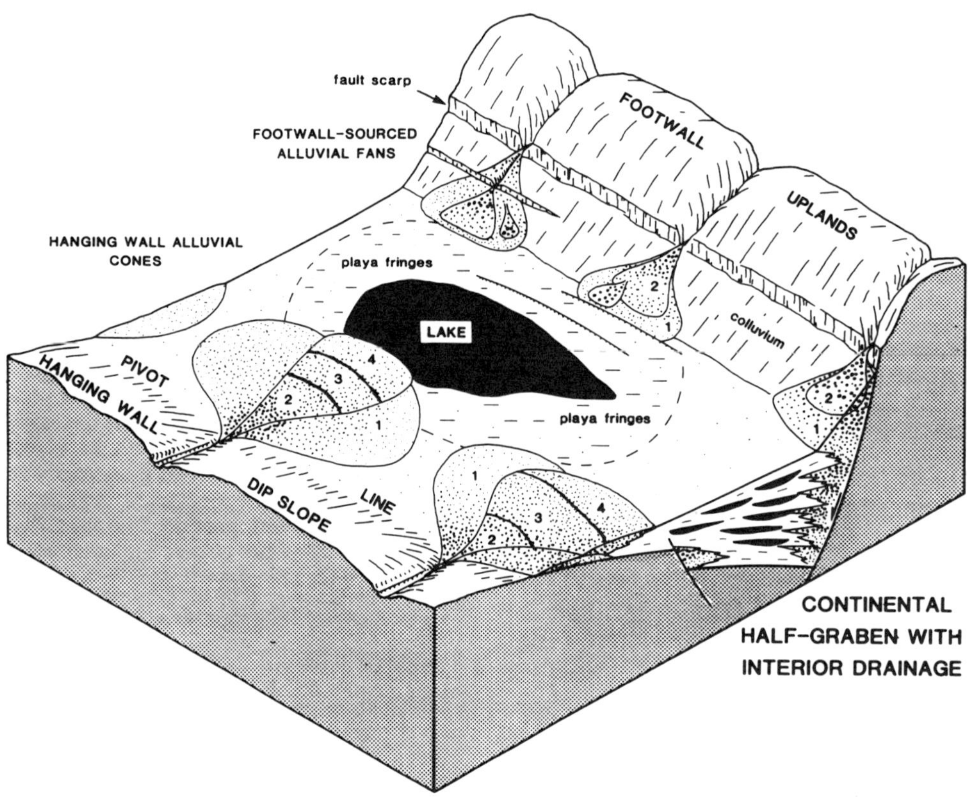

Figure 5. Facies model of syn-rift deposits in a continental half graben (Leeder and Gawthorpe, 1987). Note stratigraphic asymmetry between hinge and master fault sides of the basin, in particular, the position of the vertically stacked fanglomerates against the basin's master fault. In this diagram, the basin is drained internally like the Albuquerque Basin was during middle to late Miocene deposition of the upper Popotosa part of the Santa Fe Group.

basin undergoing hanging-wall rotation into a large listric normal fault. Coarse, aggrading fanglomerates, like those that are stacked vertically against the fault, are generally best developed in the Santa Fe Group along the northeast and southwest margins of the Albuquerque Basin, where Neogene fault throws are the greatest.

In the northeast part of the basin, evidence of major fanglomerates occurs on the down-thrown sides of the Sandia, Rincon, and San Francisco–Placitas faults. The Norins Northern Albuquerque Acres #2 well (Fig. 1) was drilled northeast of Albuquerque, a short distance west of the Sandia fault. According to drillers' logs, the upper 655 m (2,150 ft) contained coarse bouldery gravels and granite wash, the bulk of which was assigned to the upper Popotosa (mid- to upper Miocene) part of the Santa Fe Group by Lozinsky (1988).

Figure 6 shows an outcrop of upper Miocene(?) fanglomerates along the north-trending Ranchos fault at the north end of the Sandia uplift (Fig. 1). The fanglomerates dip gently east toward the upthrown Cretaceous section in the footwall of the fault. The fanglomerates contain angular, locally derived clasts of Precambrian through Cretaceous lithologies that are probably derived from the north end of the Sandia uplift. The Ranchos fault is one of several faults that in map view, merge to the south into the San Francisco–Placitas fault. These faults, based on their cross-sectional view, are thought to be hanging-wall imbricates of the westward-flattening San Francisco fault. In this area the San Francisco–Placitas fault system has been uplifted and eroded down to the pre-rift section at the north-plunging northern nose of the Sandia Range.

Figure 7 shows an east-west seismic section of the north basin, a short distance north of and down-plunge from the outcrops seen in Figure 6. The seismic data appear to image a series of linked, westward-flattening listric-normal faults, such as the Ranchos fault. These faults merge into a more gently west dipping feature that is thought to be the low-angle root of the San Francisco–Placitas fault.

Though the Rio Grande and San Francisco–Placitas faults have operated synchronously, in the last few million years some of the extension may have been transferred basinward to the Rio Grande fault, which seems to control the modern position of the Rio Grande. In map view, throw on the downward-flattening imbricate faults above the San Francisco fault appears to diminish rapidly to the north in the upper Miocene–lower Pliocene part of the Santa Fe, suggesting that extension on the San Francisco fault system, as a whole, may have slowed in the last few million years. The presence of east tilt and young faults offsetting the 2.5-Ma basalt flows of Santa Ana Mesa in the basin center (Fig. 1), may represent current deformation focused in the hanging wall of the Rio Grande fault. In map view, some of these young faults are

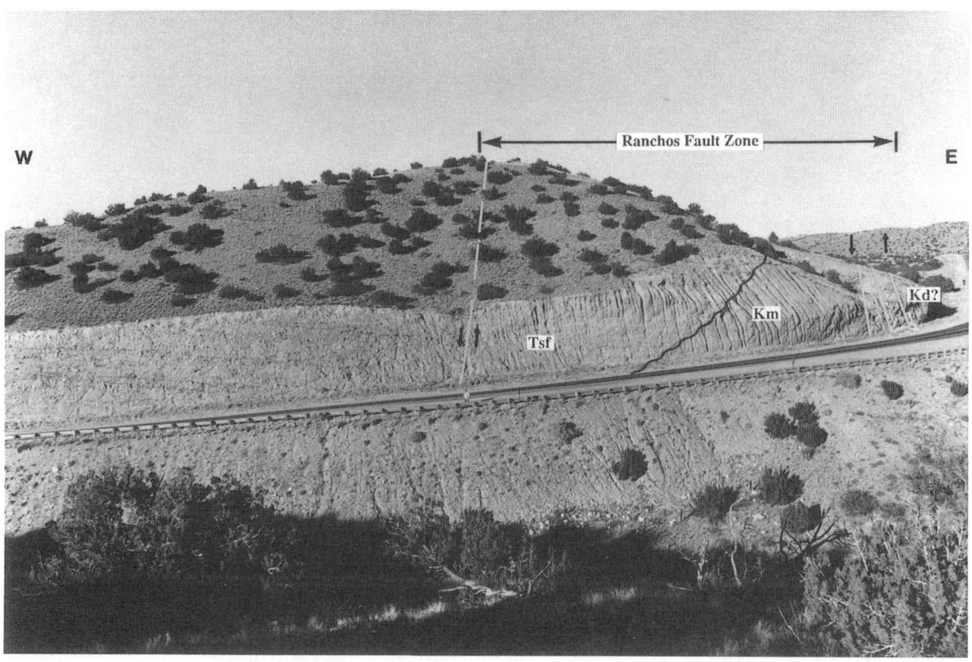

Figure 6. View north toward Ranchos (Rincon) fault zone exposed along Highway 44 at north end of Sandia uplift. Coarse-grained, upper(?) Miocene fanglomerates in the Santa Fe Group (Tsf) dip gently east (right) toward Cretaceous Mancos and Dakota Formations (Km and Kd?) on upthrown side of the fault zone. Clasts in fanglomerates are derived locally from uplifted Precambrian to Cretaceous rocks in the Sandia rift shoulder. See Figure 1 for location of photograph.

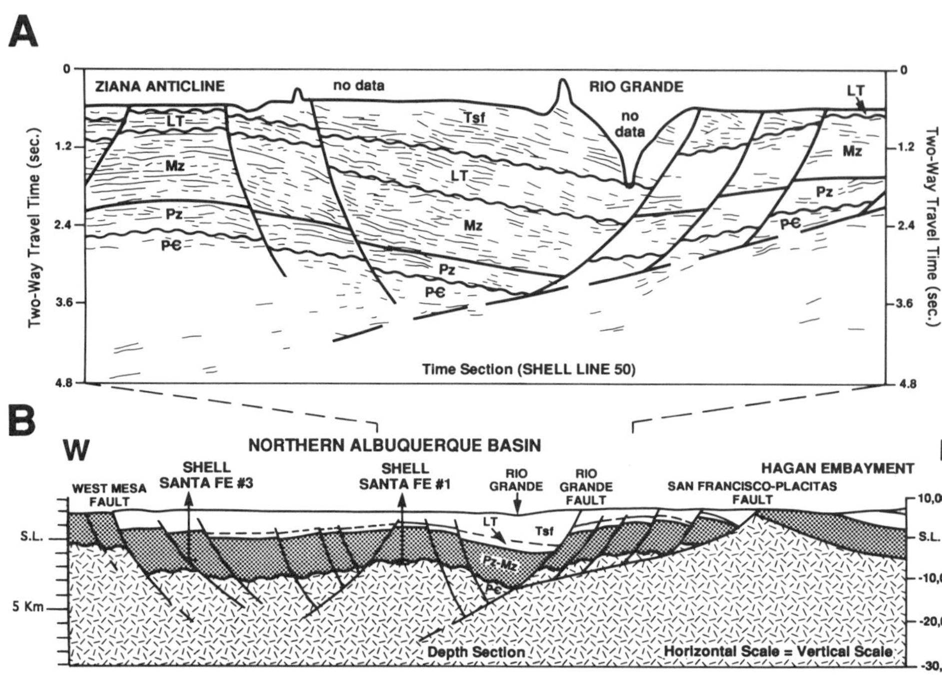

Figure 7. A, Line drawing of Shell seismic line 50 in central part of northern Albuquerque Basin. B, Depth section of northern Albuquerque Basin at latitude of Shell line 50. Tsf, Santa Fe Group; LT, lower Tertiary (Eocene Galisteo Formation and Oligocene Unit of Isleta #2); Mz, Mesozoic; Pz, Paleozoic (Pennsylvanian and Permian only); PC, Precambrian basement. Modified from Russell and Snelson (1990). See Figure 1 for location of section.

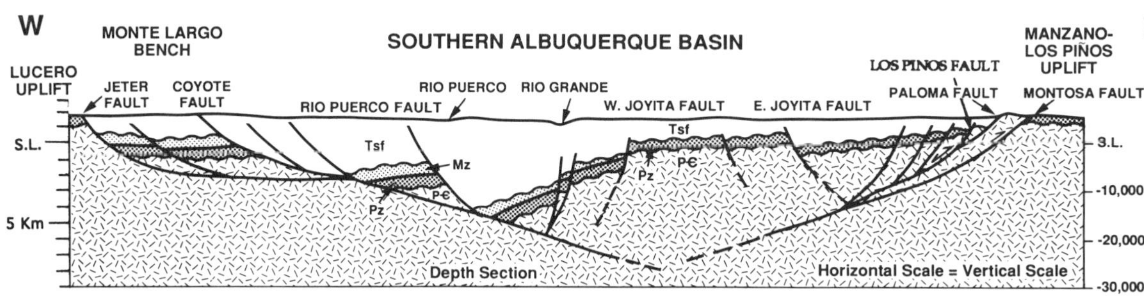

Figure 8. Cross section of southern Albuquerque Basin along Shell seismic line 52. Tsf, Santa Fe Group and an uncertain thickness of underlying lower Tertiary strata (Eocene Baca Formation and Oligocene Unit of Isleta #2); Mz, Mesozoic; Pz, Paleozoic (Pennsylvanian and Permian only); PC, Precambrian basement. Modified from Russell and Snelson (1990). See Figure 1 for location.

thought to terminate southward against the Rio Grande fault.

An approximate mirror image of these relationships exists in the southern Albuquerque Basin, also implying a subtle inward migration of faulting with time, concurrent with uplift, the early stages of tectonic stranding, and deeper erosion into basin-margin fault systems like the San Francisco–Placitas fault. Figure 8 is an east-west cross section of the southern basin, based on Shell line 52, from Abo Pass on the Manzano uplift to an area just north of the Ladron uplift. Note how the westernmost fault, the low-angle Jeter fault, either ramps down to the east or is truncated eastward by the more deeply rooted Coyote fault. If truncated, this implies basinward transfer of extension during the later stages of rifting and may be consistent with data on fault-scarp morphology in the southern Albuquerque Basin from Machette (1982), which seem to imply a very subtle basinward progression of faulting, at least during the late Pliocene and Quaternary. The arcuate traces of the Jeter and San Francisco–Placitas faults around the Ladron and Sandia uplifts, respectively, imply gently dipping faults that have been uplifted, folded, and partially eroded down to their low-angle roots. The nearly flat dip on the roots of the Jeter fault (Fig. 8) and unusually high elevation of the Ladron uplift appear to be consistent with flexural, isostatic uplift due to tectonic unloading by the Jeter and Santa Fe–Coyote faults, as also noted by Wernicke and Axen (1988) and Lewis (1989), and are an approximate analog of the San Francisco and Rio Grande faults in the north basin.

Figure 9 is a south-facing view of the Jeter fault (Fig. 1) at the Jeter mine. The uplifted Ladron block is out of view to the right (west). The fault dips 25° to the left (east) into the basin and separates highly sheared Precambrian basement in the footwall from the syn-rift fanglomerate of Ladron Peak on the hanging wall. Clasts in the fanglomerate contain Precambrian lithologies derived locally from the adjacent Ladron block (R. Chamberlin, personal communication, 1990). Clast orientations in the fanglomerates imply a large amount of hanging-wall rotation of the fanglomerates into the fault. Considering the relatively low dip and planar geometry of the fault at this erosional level and the relatively great amount of antithetic dip in the hanging wall, large amounts of tectonic transport, uplift, and erosion have probably occurred in this area. Although the precise age of these gravels is not yet well constrained, we believe that these gravels are an approximate stratigraphic and tectonic analog of the upper Miocene(?) fanglomerates along the northeast margin of the basin.

CONCLUSION

Integration of seismic data and surface geology indicates the presence of large-displacement, downward-flattening normal faults along the northeast and southwest margins of the Albuquerque Basin. The footwalls of these master faults contain the most strongly uplifted and outwardly rotated rift shoulders of the basin. Fission-track data in surrounding uplifts and approximately defined age control on the basin's fanglomerates indicate synchronous, rapid uplift and cooling of these master-fault rift shoulders and rapid deposition of the basin's major fanglomerates, especially in the late Miocene. These relationships imply both temporal and mechanical connections between motion on the seismically defined master faults of the Albuquerque Basin and uplift of the rift shoulders in the footwalls of these faults.

ACKNOWLEDGMENTS

We thank S. M. Cather, C. E. Chapin, J. W. Hawley, R. V. Ingersoll, R. P. Lozinsky, and L. A. Woodward for helpful discussions and critical reviews of the manuscript. We also thank Shell Oil Company for releasing some of their seismic data in the basin, James Duke and Rebecca Titus for drafting of figures, and ARCO Oil and Gas Company for general support of the authors' efforts. Fission-track work of S. Kelley was supported by NSF grant EAR-8804203 and a grant from the New Mexico Bureau of Mines and Mineral Resources.

REFERENCES CITED

Buck, W. R., Martinez, F., Steckler, M. S., and Cochran, J. R., 1988, Thermal consequences of lithospheric extension: Pure and simple: Tectonics, v. 7, p. 213–234.

Chapin, C. E., 1983, An overview of Laramide wrench faulting in the

Figure 9. View southeast toward Jeter fault at Jeter mine. Fault dips 25° to the left (east) and can be traced beneath southwestern Albuquerque Basin on Shell seismic line 52 (Fig. 8). West-dipping upper Miocene(?) fanglomerates in hanging wall are in fault contact with highly sheared Precambrian basement in footwall. Ladron uplift is out of view to the right (west). See Figure 1 for location of photograph.

Southern Rocky Mountains with emphasis on petroleum exploration: Rocky Mountain Association of Geologists, p. 169–179.

Corrigan, J. D., 1991, Deriving thermal history information from apatite fission-track data: Journal of Geophysical Research, v. 96, p. 10347–10360.

Egan, S. S., 1992, The flexural isostatic response of the lithosphere to extensional tectonics: Tectonophysics, v. 202, p. 291–308.

Hayden, S., 1991, Dextral oblique-slip deformation along the Montosa fault zone at Abo Pass, Valencia and Socorro Counties, New Mexico [abs.]: New Mexico Geology, v. 13, p. 64.

Kelley, S. A., Chapin, C. E., and Corrigan, J., 1992, Late Mesozoic to Cenozoic cooling histories of the flanks of the northern and central Rio Grande rift, Colorado and New Mexico: New Mexico Bureau of Mines and Mineral Resources, Bulletin 145, 40 p.

Leeder, M. R., and Gawthorpe, R. L., 1987, Sedimentary models for extensional tilt block/half-graben basins, *in* Coward, M. P., Dewey, J. F., and Hancock, P. L., eds., Continental extensional tectonics: Geological Society of London Special Publication no. 28, p. 139–152.

Lewis, C. J., 1989, Crustal extension and shoulder uplift in the Rio Grande rift, New Mexico: Half grabens and accommodation zones in the Ladron Peak area [M.S. thesis]: Los Angeles, California State University, 79 p.

Lozinsky, R. P., 1988, Stratigraphy, sedimentology, and sand petrology of the Santa Fe Group and pre–Santa Fe Tertiary deposits in the Albuquerque Basin, central New Mexico [Ph.D. thesis]: Socorro, New Mexico, New Mexico Institute of Mining and Technology, 298 p.

Machette, M. N., 1982, Quaternary and Pliocene faults in the La Jencia and southern part of the Albuquerque-Belen Basins, New Mexico: Evidence of fault history from fault-scarp morphology and Quaternary geology: New Mexico Geological Society Guidebook, 33rd Field Conference, Albuquerque Country, II, p. 161–169.

Russell, L. R., and Snelson, S., 1990, Structural style and tectonic evolution of the Albuquerque Basin segment of the Rio Grande rift, *in* Pinet, B., and Bois, C., eds., The potential of deep seismic profiling for hydrocarbon exploration: Institut Francais Petrole Research Conference Proceedings, Editions Technip, Paris, p. 175–207.

Wernicke, B., and Axen, G. J., 1988, On the role of isostasy in the evolution of normal fault systems: Geology, v. 16, p. 848–851.

Woodward, L. A., Callender, J. F., and Zilinski, R. E., 1975, Tectonic map of the Rio Grande rift, New Mexico: Geological Society of America, Map and Chart Series MC-11, scale 1:500,000.

Manuscript Accepted by the Society October 13, 1993

Printed in U.S.A.

Geological Society of America
Special Paper 291
1994

Crustal extension in the Rio Grande rift, New Mexico: Half-grabens, accommodation zones, and shoulder uplifts in the Ladron Peak–Sierra Lucero area

Claudia J. Lewis*
Department of Earth and Planetary Sciences, Harvard University, Cambridge, Massachusetts 02138
W. Scott Baldridge
Earth and Environmental Sciences Division, MS D462, Los Alamos National Laboratory, Los Alamos, New Mexico 87545

ABSTRACT

The geometry of major basin-bounding faults in the central Rio Grande rift indicates that extension is characterized by overlapping and nested half grabens linked together by accommodation zones and bounded by tilted block uplifts. The Albuquerque-Belen Basin is composed of two separate, opposite-facing, structurally complex basins linked by an accommodation zone. South of the Albuquerque-Belen Basin, the rift broadens southwestward around the Colorado Plateau. A zone of distributed sinistral shear at the margin of the plateau accommodated differential extension between the little-extended plateau and the half grabens (Milligan Gulch, La Jencia, and Socorro) of the rift. Extension in the latter accommodation zone is expressed in the sedimentary cover rocks by north-striking, high-angle normal faults and a segmented, northeast-striking, sinistral fault.

The Ladron horst and the Lucero uplift border the rift on the western side of the Albuquerque-Belen Basin and were isostatically uplifted as a result of tectonic denudation along the bounding faults of adjacent half grabens. Footwall structures include folded Paleozoic and Mesozoic rocks, intrabed thrust faults, and high-angle shear zones.

We present new models for the Neogene evolution of the central Rio Grande rift and integrate contractional and extensional structures previously attributed to different stress regimes. Many of the structures in the central Rio Grande rift previously attributed to Laramide compression can be adequately explained by late Tertiary extension.

INTRODUCTION

In the last decade, field and laboratory structural studies of extensional terranes have yielded a clearer understanding of the geometry of extensional faulting and the kinematics of lithospheric thinning. In regions of low to moderate extension, much recent work on extensional faulting has elucidated the three-dimensional geometry of fault systems bounding rift basins. Many basins in continental rift systems are asymmetric, hinged half grabens in which one side is bounded by a major fault system and the other side, characterized by monoclines, steps, and flexures, ramps up to a platform (e.g., King, 1978; Smith and Bruhn, 1984; Bosworth, 1985; Rosendahl, 1987; Ebinger, 1989). These features have led to a simple geometric model for individual half grabens in which basin-

*Present address: 2111-D 34th Street, Los Alamos, New Mexico 87544.

Lewis, C. J., and Baldridge, W. S., 1994, Crustal extension in the Rio Grande rift, New Mexico: Half-grabens, accommodation zones, and shoulder uplifts in the Ladron Peak–Sierra Lucero area, *in* Keller, G. R., and Cather, S. M., eds., Basins of the Rio Grande Rift: Structure, Stratigraphy, and Tectonic Setting: Boulder, Colorado, Geological Society of America Special Paper 291.

bounding faults are listric in cross section and arcuate in plan view (e.g., Rosendahl, 1987). However, numerous geologic and seismic studies have shown that although some map-view curved faults may be listric, more appear to be planar in cross section (e.g., Morley, 1988, 1989; Russell, and Snelson, 1990, this volume).

Rift geometry exhibits marked changes along strike (e.g., Crossley and Crow, 1980; Bosworth, 1985; Rosendahl et al., 1986). Half grabens alternate polarity along the axis of rifts (e.g., Rosendahl, 1987; Morley et al., 1990) and link together in a variety of ways, including overlapping and nonoverlapping, similar and opposing polarity half grabens. Interferences caused by linking produce regions of great structural complexity, called "zones of compensation" (Anderson, 1973), "transform structures" (Davis and Burchfiel, 1973), "transfer faults" (Gibbs, 1984), "accommodation zones" (e.g., Bosworth, 1985; Rosendahl, 1987), or "transfer zones" (Morley et al., 1990). These structural boundaries serve to subdivide rifts into discrete basins and allow for variations in the style and scale of extensional structures, the direction of upper-plate transport, and tilt directions in adjacent basins. Accommodation zones may be expressed by single, through-going faults (e.g., Davis and Burchfiel, 1973), narrow regions of distributed shear (e.g., Rosendahl, 1987), or broad areas of overlapping faults that gradually transfer displacement (Morley et al., 1990).

Recent work in regions of high extension has produced many new ideas on the accommodation of extension at mid- to lower-crustal depths. A breakthrough in our understanding of continental extension was the recognition that large-displacement, low-angle normal faults, termed "detachment faults," serve as boundary faults to mosaics of both rotational and nonrotational extension faults, (e.g., Davis, 1979; Davis et al., 1980; Dokka, 1981; Gibbs, 1984; Wernicke and Burchfiel, 1982; Bosworth et al., 1986). Results of the Consortium for Continental Reflection Profiling (COCORP) have been pivotal to an understanding of extensional structures by imaging horizontal reflectors in the upper to middle crust that are interpreted as detachment faults. COCORP data from the Sevier Desert area demonstrate that large, fault-block ranges are an integral part of a large-scale system of down-to-the-west shear on low-angle normal faults (Allmendinger et al., 1987). In some locations, low-angle reflectors interpreted as extensional detachments can be traced to surface faults (e.g., northern Snake Range décollement, Allmendinger et al., 1983; McCarthy, 1986). Large, nonrotational, low-angle normal faults may dominate regions of large-magnitude extension (Wernicke and Burchfiel, 1982). Listric- and planar-normal faults are common components of their hanging walls. Listric-normal faults relieve space problems between families of planar fault blocks but are not the principal contributors to extension in highly extended terranes.

Although many geologists have stressed the importance of low-angle normal faults to Cenozoic extension in the Basin and Range province (e.g., Anderson, 1971; Armstrong, 1972; Wernicke, 1981), controversy exists concerning the initial geometry of these structures. In some localities, extension in the uppermost crust is accommodated by faults active at low angles. For example, palinspastic reconstruction of Panamint Valley, southeast California, indicates that the valley was formed by movement on a shallow crustal, low-angle normal fault of 0 to 15° dip (Burchfiel et al., 1987). In other areas, normal faults may have originated as deeply penetrating high-angle faults that were seismically active with steep dips, but reached the surface by rotation to low dips (e.g., Buck, 1988; Hamilton, 1988; Wernicke and Axen, 1988). Rotation of moderate- to high-angle normal faults to low dips may occur in response to isostatic forces caused by tectonic denudation of the footwall (e.g., Spencer, 1984; Buck, 1988; Hamilton, 1988; Wernicke and Axen, 1988; Weissel and Karner, 1989; Kusznir and Egan, 1990; Kusznir et al., 1991; Fig. 1A). In general, all these models assume that relatively small deformations associated with individual earthquakes accumulate over time to create geological structures. Fault motion associated with each earthquake results in immediate deformation followed by a long period of readjustment as stresses relax in the viscous lower crust and isostatic equilibrium is restored.

Current controversy on the origin of footwall uplift centers on whether the geometry of isostatic rebound in the crust is controlled by elastic processes (e.g., Buck, 1988; Weissel and Karner, 1989; Kusznir and Egan, 1990; Bartley and Fletcher, 1990; Kusznir et al., 1991) or nonelastic processes (e.g., Zandt and Owens, 1980; Wernicke and Axen, 1988; Kruse et al., 1990; Axen and Wernicke, 1991; Wdowinski and Axen, 1992). In elastic models, footwall uplift is primarily controlled by forces that arise from plate bending (e.g., Buck, 1988; Block and Royden, 1990; Fig. 1). The elastic, flexural-failure model of Buck (1988) predicts that compressional fiber stresses in the upper part of the concave-upward footwall flexure lead to cataclastic failure in the upper crust (Figs. 1B, 2A). Cataclastic shortening decreases the radius of curvature of the flexure, producing large bending stresses and significantly reducing the rigidity of the upper crust. This mechanism has been cited as an explanation for the short wavelength of topographic response to normal fault loads (Buck, 1988).

Flexural models achieve good agreement with the geometry of extensional basins and metamorphic core complexes (e.g., Weissel and Karner, 1989; Block and Royden, 1990; Kusznir and Egan, 1990; Kusznir et al., 1991). However, these models have not been able to match geological and geophysical observations that bound continental crustal strength. Low values of effective elastic thickness obtained by flexural modeling of extension are substantially less than the thickness of the seismogenic upper crust (e.g., Buck, 1988; Block and Royden, 1990; Kusznir et al., 1991). In addition, significant elastic strength is required to support the locally uncompensated topographic loads of the Basin and Range (Eaton et al., 1978). Studies of the postglacial rebound of Lake Bonneville, Utah, suggest an elastic plate thickness of about 20 to 30 km despite

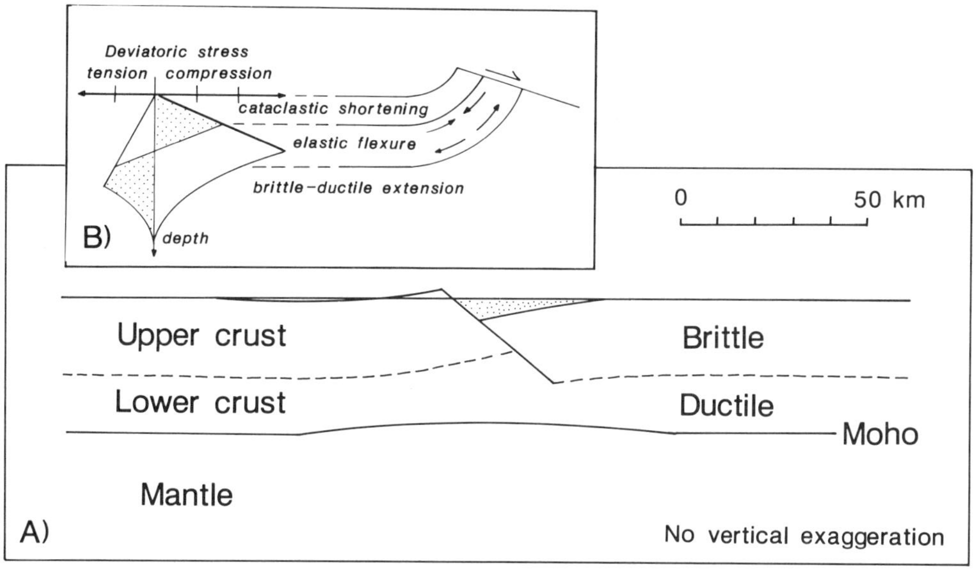

Figure 1. Flexural-failure model for footwall uplift. A) Schematic representation of lithospheric extension and footwall uplift. Upper crust (depth <15 km) is assumed to be brittle and deforms by fracture on planar faults; lower crust and mantle are assumed to behave ductilely. Modified from Kusznir and Egan (1990). B) Shaded area on stress-depth curve represents predicted stress distribution in the crust due to elastic flexure and flexurally driven failure. Contractional strain predicted in the concave- upward footwall flexure. Modified from Bartley and Fletcher (1990).

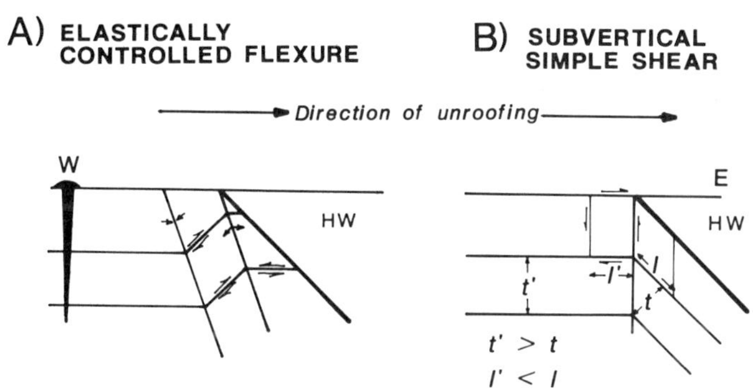

Figure 2. Comparison of end-member models for development of footwall folds. Footwall sustains all deformation. A) Elastic model predicts either elongation followed by contraction or layer-parallel shear strain as anticlinal axial surface migrates eastward with continued extension. Footwall pin indicates no layer-parallel slip. B) Subvertical simple shear model predicts layer-parallel shear strain, layer shortening and thickening, and steep faults in the footwall. This model does not predict shortening and thickening of layers that have never resided beneath the normal fault. HW, hanging wall; W, west; E, east; t, initial thickness; t′, final thickness; l, initial length; l′, final length. Modified from Axen and Wernicke (1991).

very high heat flow in that part of the Basin and Range province (Walcott, 1970; Nakiboglu and Lambeck, 1983; Bilis and May, 1987). In the East African rift, the effective elastic plate thickness is estimated to be 25 to 30 km (Forsyth, 1985). Thus, for loads of small amplitude and for time periods of about 10^4 yr, the continental lithosphere behaves elastically. However, for longer term loads the lithosphere responds viscously.

Other recent modeling efforts suggest that the elastic response of the crust may not control the observed pattern of strain in footwalls of normal fault systems (Spencer, 1984; Axen and Wernicke, 1991; Wdowinski and Axen, 1992). Instead, nonelastic responses may dominate deformation (Fig. 2B). The moment applied to the footwall crust by isostatic forces may just break the plate as well as bend it (Wernicke and Axen, 1988). In the combined elastic-nonelastic model of Axen and Wernicke (1991), elastic strain accumulates until the load exceeds the elastic limit of the lithosphere, causing the plate to break. In this scenario, shear along subvertical planes and layer-parallel thickening are important modes of footwall failure (Fig. 2B). It is important to note that in both elastic and nonelastic models, footwall folds and reverse faults form during crustal extension. In some cases, these structures have been misinterpreted as older than normal faulting strictly because of their contractional nature (e.g., the Virgin anticline; Seager, 1970; Moore, 1972; Hintze, 1986).

Structural studies in the Rio Grande rift provide an opportunity to apply and further develop new ideas about the structural style and kinematics of extension. The western margin of the central Rio Grande rift (Fig. 3), was studied for two reasons. First, arcuate basin boundaries (Figs. 3, 4) and seismically imaged, low-angle normal faults east and south of Sierra Lucero and Ladron Peak (Fig. 4) prompted consideration of a new interpretation of the regional structural geometry. We were motivated by the availability of COCORP (e.g., Brown et al., 1980; Cape et al., 1983; De Voogd et al., 1988) and Shell (Russell and Snelson, 1990, this volume) seismic data in the Monte Largo embayment (Fig. 4) to map in detail the complex structure along the eastern edge of the Lucero uplift. Seismic data provide compelling evidence that structures in the Monte Largo embayment are related to late Tertiary extension.

Second, some previous investigators invoked Laramide crustal shortening to explain contractional structures and arcuate or sinuous fault geometry along the eastern front of the Lucero uplift (e.g., Kelley and Wood, 1946; Callendar and Zilinski, 1976; Hammond, 1987; Cabezas, 1989). For example, published cross sections perpendicular to the edge of the Monte Largo embayment (De Voogd et al., 1988; Kelley and

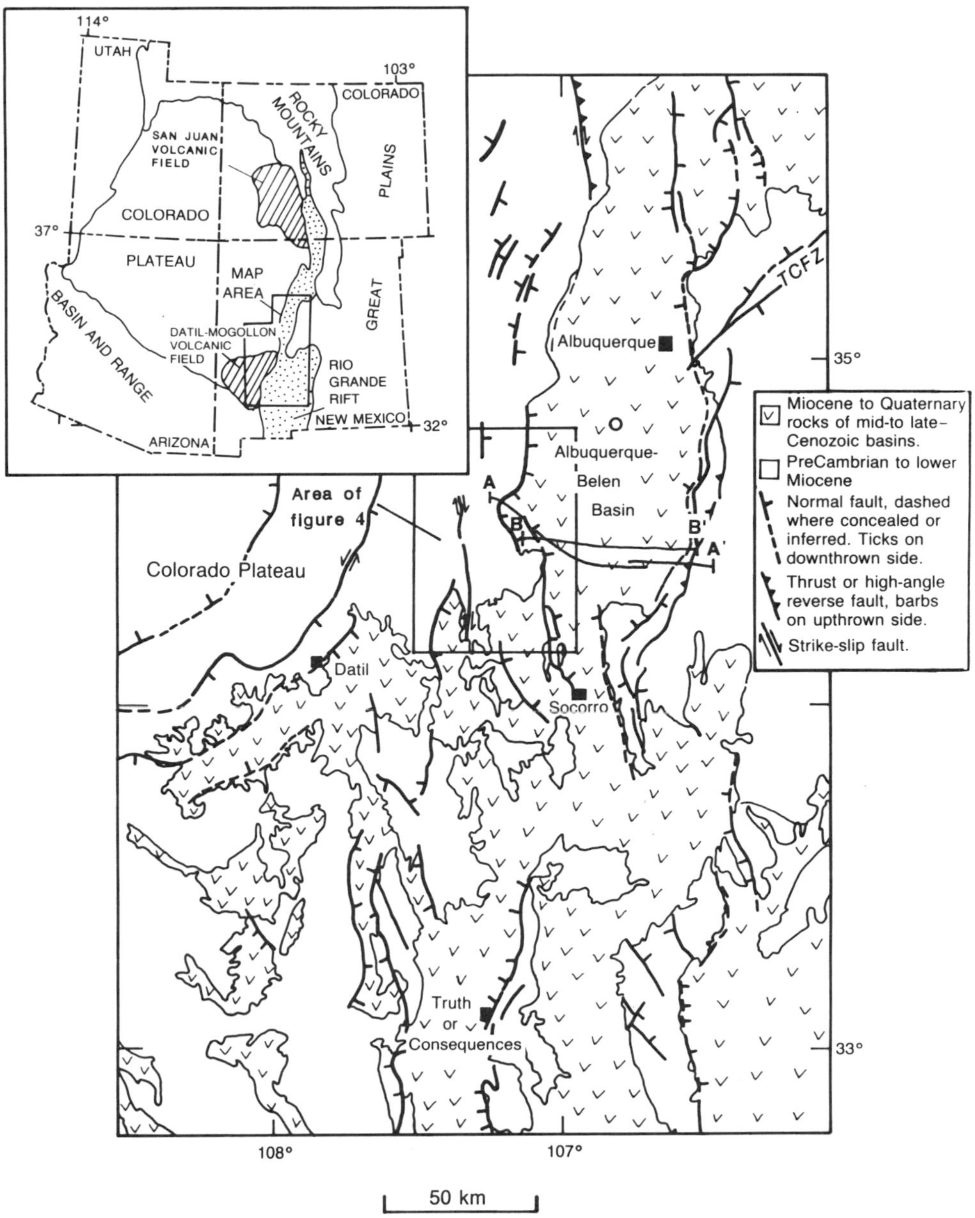

Figure 3. Location of Ladron Peak–Sierra Lucero area with reference to major physiographic features of the region and basins and uplifts of the Rio Grande rift. Faults west of rift on Colorado Plateau are in the plateau/rift transition zone. Modified from Baldridge et al. (1984). Lines of section: A–A′, Shell seismic profile (Russell and Snelson, 1990); B–B′, COCORP seismic profile (e.g., Brown et al., 1980; Cape et al., 1983; De Voogd et al., 1988). TCFZ, Tijeras-Cañoncito fault zone. Open circle in Albuquerque-Belen Basin indicates Shell Isleta #2 oil test well.

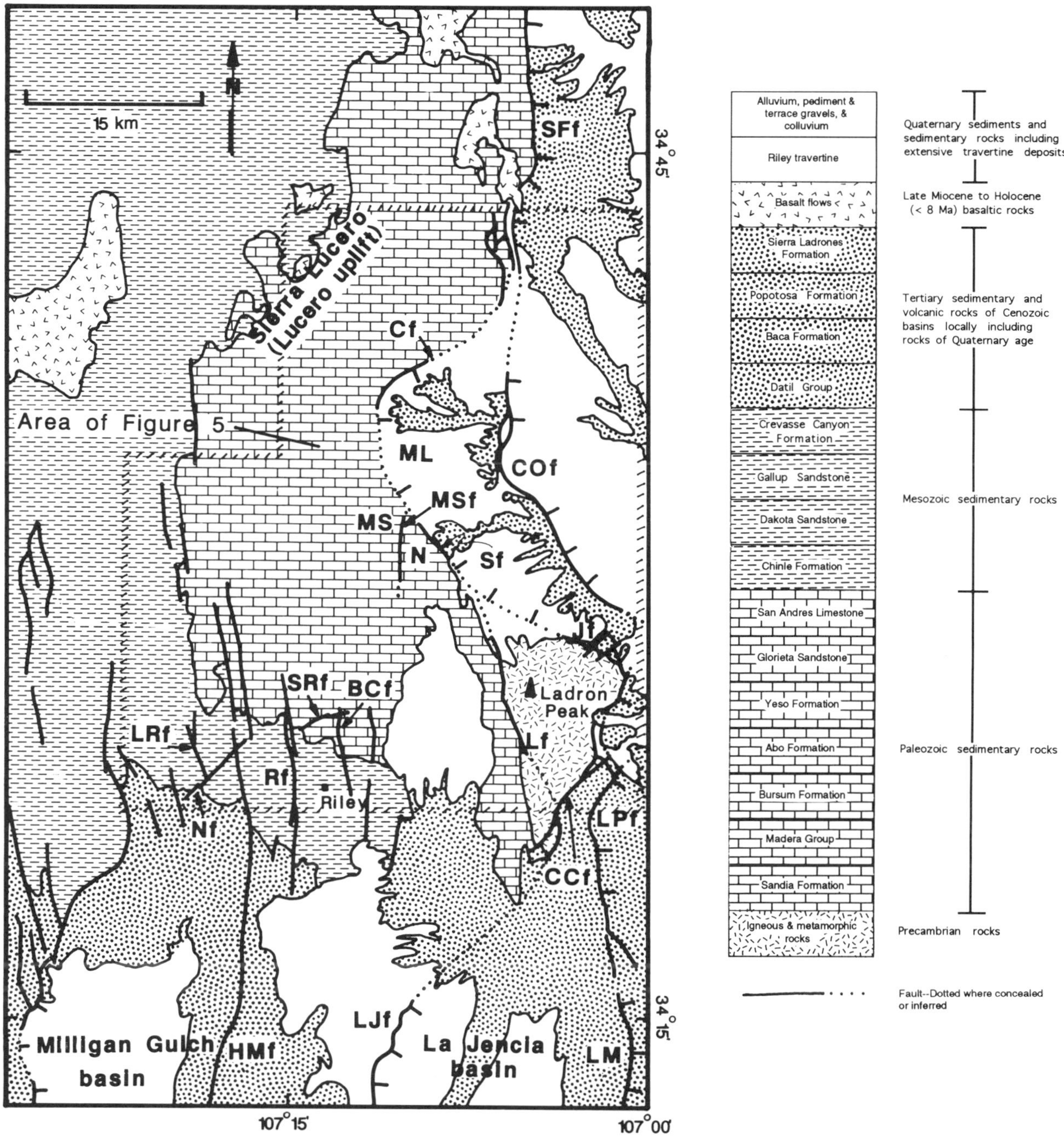

Figure 4. Generalized geology and principal geographic features of Ladron Peak–Sierra Lucero area. Compiled from Kelley and Wood (1946), Tonking (1957), Jicha (1958), Zilinski (1976), Kelley (1977), Massengill (1979), Chamberlin (1982), Osburn (1984), Hammond (1987), and Lewis (1989). Faults: BCf, Baca Canyon; Cf, Comanche; CCf, Cerro Colorado; COf, Coyote; HMf, Hell's Mesa; Jf, Jeter; LJf, La Jencia; LPf, Loma Pelada; LRf, La Rosa; MSf, Mesa Sarca; Nf, Negros; Rf, Riley; Sf, Saiz; SFf, Santa Fe; SRf, Santa Rita. Topographic features: LM, Lemitar Mountains; MS, Mesa Sarca uplift; ML, Monte Largo embayment; N, Navajo Gap. Faults with ticks are major, western-bounding faults of rift. Area of map indicated on Figure 3.

Wood, 1946; Russell and Snelson, 1990) depict a west-dipping thrust or high-angle reverse fault and an east-dipping normal fault either in near proximity or intersecting at the surface.

We find little or no evidence for such a west-dipping fault and suggest that basin geometry resulted from middle to late Tertiary extension. Typically, contractional strain has been assumed to occur only very locally in extensional terranes. In the Rio Grande rift, contraction has usually been attributed to Laramide crustal shortening. New ideas regarding contractional strain in extensional environments provide alternatives to the traditional interpretation of the structural evolution of the southern Lucero uplift. A major contribution of our work lies in integrating contractional and extensional structures previously attributed to different stress regimes.

Four major questions, arising out of new ideas about extensional structures, thus directed our investigation in the Rio Grande rift. (1) What are the major tectonic elements in the central Rio Grande rift, and how are they related geometrically? (2) How is the structural transition between the Rio Grande rift and the relatively unextended Colorado Plateau accomplished? (3) What was the driving force for uplift of the Ladron block? (4) What is the origin of contractional strain along the eastern front of the Lucero uplift?

GEOLOGIC SETTING

The Rio Grande rift is a north-northeast–trending extensional zone consisting of a series of asymmetrical basins and fault-block uplifts extending more than 1,000 km from central Colorado into northern Mexico (Fig. 3). Vertical structural relief of as much as 12 km occurs in the main rift basins (Chapin, 1988). Basin-fill deposits may reach 8 km in thickness (Lozinsky, 1988).

The brittle-ductile transition, as defined by maximum earthquake epicentral depths (Olsen et al., 1987), is at a depth of 13 to 15 km beneath the rift. In the central rift, crustal thickness is approximately 33 km (Olsen et al., 1987), indicating 25 to 35% crustal thinning compared to the thicker crust of the adjacent Great Plains (50 km) and Colorado Plateau (45 km) provinces.

Physiographically, the Colorado Plateau/Rio Grande rift boundary is marked by the major western bounding faults of the rift, a complex zone of faulting and folding (e.g., Callendar and Zilinski, 1976) that forms the eastern margin of the Lucero uplift (Figs. 4 and 5). The Ladron uplift, flanking part of the central Rio Grande rift, is located in a structurally complex region where the late Miocene to Holocene rift widens south of the Colorado Plateau. North of Ladron Peak is the 50-km-wide southern Albuqerque-Belen Basin. Directly south of Ladron Peak, the rift is a 70-km-wide zone of extension, composed of three parallel basins and intervening mountain ranges.

The Albuquerque-Belen Basin is composed of two distinct subbasins, a northern basin deeper on its east side and a southern basin deeper on its west side (Lozinsky, 1988; Russell and Snelson, 1990, this volume). The northern basin has subsided along a major, west-dipping, listric normal fault system on its east side, whereas the southern basin is bordered on the west by a system of major, east-dipping, listric normal faults (Russell and Snelson, 1990, this volume).

Previous investigators have suggested that the structural transition between these two basins may be a strike-slip accommodation zone (Lozinsky, 1988; Russell and Snelson, 1990, this volume). Seismic interpretations indicate substantial down-to-the-north, dip-slip displacement and flower structures in this zone (Russell and Snelson, 1990). Different thicknesses of Cenozoic basin fill, approximately 8 km in the northern half graben and 5 km in the southern half graben (Lozinsky, 1988, this volume), bolster the interpretations made from seismic data. This fault zone has also been interpreted from aeromagnetic and gravity data (Cordell, 1978).

At the surface, this zone may be linked, on the eastern side of the rift, with the Tijeras-Cañoncito fault system (Fig. 3), a major, northeast-trending crustal shear zone. This is a zone of en echelon and anastamosing high-angle faults that have experienced repeated and segmented fault displacement since Precambrian time (Lisenbee et al., 1979). Complex strike-slip offset in the most recent phase may result from the fault zone's role as a transfer fault; strike-slip block displacements may be accommodating greater extension on the south side of the fault zone relative to the north side (Russell and Snelson, 1990).

The Monte Largo embayment (Fig. 4) is a 10-km offset in the north-striking, western-bounding faults of the Albuquerque-Belen Basin. The bounding faults of the embayment, the Comanche and Saiz faults (Fig. 4), were previously assumed to be Laramide reverse faults reactivated as normal faults during mid- to late Tertiary extension (e.g., Kelley and Clinton, 1960; Callendar and Zilinski, 1976; Kelley, 1977; Hammond, 1987). Russell and Snelson (1990) distinguished two bounding faults, (1) the Comanche fault, assumed to be a Laramide reverse fault; and (2) the Jeter fault, a mid- to late-Tertiary normal fault.

Development of the Rio Grande rift

Two major episodes of extension are recognized in the Rio Grande rift: (1) a late Oligocene to early Miocene phase (beginning about 29 Ma and lasting 10 to 12 m.y.) characterized by low-angle faults and broad basins; and (2) a mid-Miocene to Quaternary phase (beginning about 17 Ma and continuing today; e.g., Chapin and Seager, 1975; Baldridge et al., 1980; Gardner and Goff, 1984; Seager et al., 1984) characterized by high-angle normal faults and narrow, deep basins, resulting in the modern physiography of the rift.

During the early phase of extension, the least principal stress direction was oriented west-southwest to east-northeast (Aldrich et al., 1986). Paleostress data from the Rio Grande rift (Aldrich et al., 1986) and the southwestern United States in general (Zoback et al., 1981) indicate that the direction of regional extension changed about 10 Ma to east-west or west-

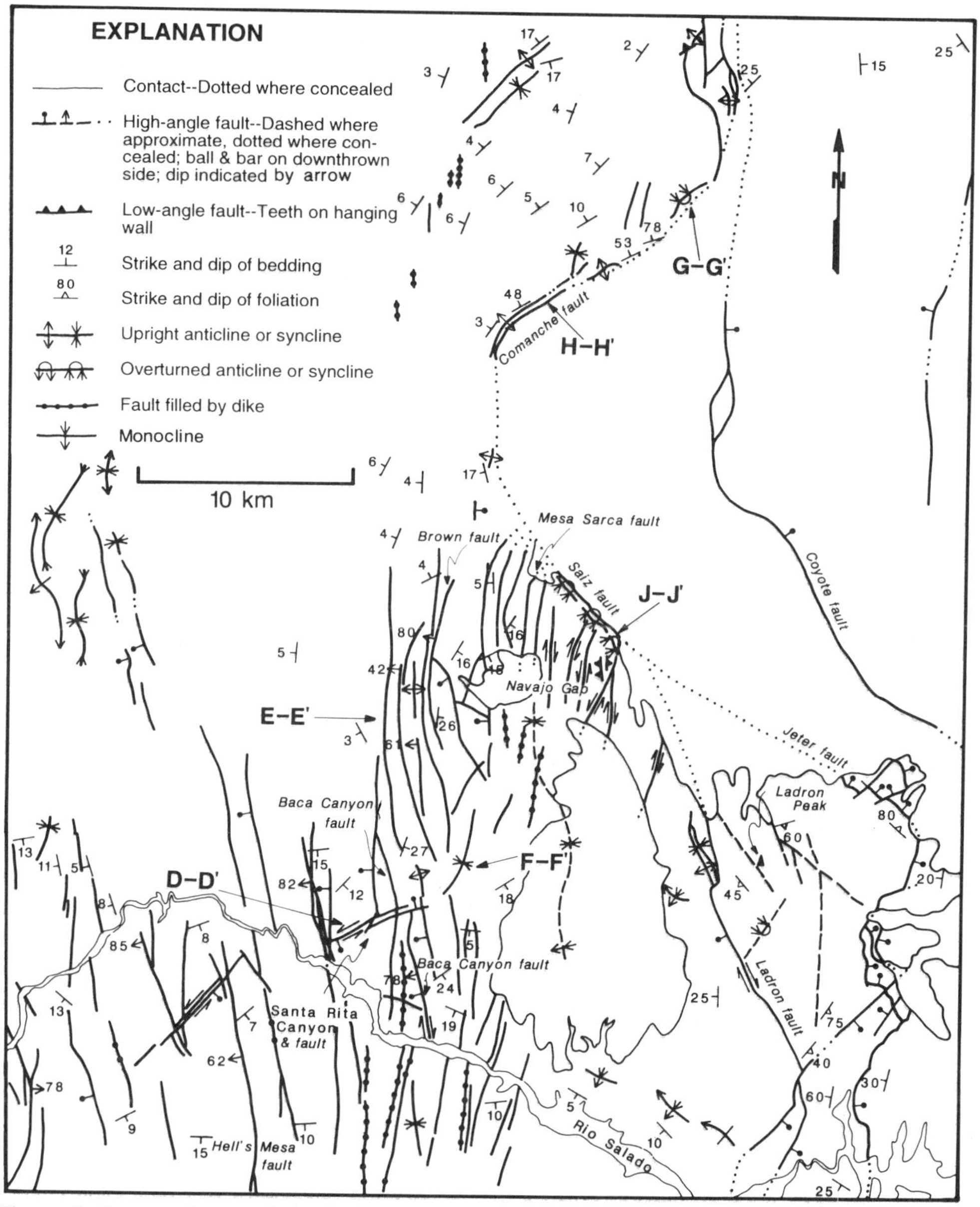

Figure 5. Structural map of the Ladron and southern Lucero uplifts. Compiled from Kelley and Wood (1946), Zilinski (1976), Massengill (1979), Chamberlin (1982), Osburn (1984), Hammond (1987), and Lewis (1989). Area of map indicated on Figure 4. See Figure 7 for cross sections D–D′, E–E′; and F–F′; see Figure 8 for cross sections G–G′, H–H′, and J–J′.

northwest to east-southeast. This 45° change in orientation of the least principal stress occurred as development of the San Andreas transform fault imposed dextral shear on the previous back-arc setting (e.g., Atwater, 1970; Snyder et al., 1976; Personius and Machette, 1984). Common to both episodes of rifting is an east-west component of extension.

Evidence of recent crustal extension in the rift includes fault scarps (Machette, 1982), seismicity (e.g., Sanford et al., 1979), high heat flow (Reiter et al., 1979), ongoing uplift as shown by geodetic measurements (Sanford et al., 1977), and the possible presence of a midcrustal magma body (e.g., Sanford et al., 1979).

Stratigraphy

The stratigraphic section in the Ladron Peak–Sierra Lucero area (Fig. 4), above the Precambrian crystalline rocks of the Ladron block, consists of upper Paleozoic and Mesozoic sedimentary strata typical of the eastern Colorado Plateau (e.g., Kelley and Wood, 1946; Tonking, 1957; Jicha, 1958; Massengill, 1979). Up to about 2.7 km of Eocene and Oligocene rocks underlie mid- to late Tertiary basin-fill sedimentary rocks (e.g., Cather, 1983; Lozinsky, 1988). The latter rocks are equivalent in age to the sedimentary Eocene Baca Formation and volcanic and volcaniclastic Oligocene Datil Group (e.g., Lozinsky,

1988). Miocene and Pliocene Santa Fe Group rocks (Popotosa and Sierra Ladrones Formations), deposited during the early and late phases of rifting, respectively, reach thicknesses of about 4.4 km in the Albuquerque-Belen Basin (Lozinsky, 1988). South of the Ladron uplift, rocks of the Sierra Ladrones Formation are separated from the older Popotosa Formation by an angular unconformity (e.g., Chamberlin, 1982). This unconformity is not seen in seismic sections of the Albuquerque-Belen Basin, suggesting a low level of angularity there (Russell, personal communication, 1992).

PREVIOUS STUDIES

Wright (1946) and Kelley and Wood (1946) first conducted geological studies in the Lucero uplift area. Both studies recognized the effects of middle to late Tertiary extensional deformation, but Kelley and Wood (1946) also documented features related to contractional deformation that they attributed to a prior regional crustal shortening event. In the absence of precise limits on ages of structures, the uplift, folds, reverse faults, and thrust faults were assigned an early Tertiary (Laramide) age, principally because of their contractional nature. Kelley's ideas were further developed in subsequent publications (e.g., Kelley, 1952, 1977). Duschatko (1953) conducted detailed fracture studies on the Lucero uplift, and further geologic mapping, including parts of the Lucero uplift, was accomplished by Tonking (1957) and Jicha (1958). Subsequent detailed mapping of the Lucero uplift–Rio Grande rift margin was conducted by Zilinski (1976), Callender and Zilinski (1976), Hammond (1987), and Cabezas (1989). With the exception of Duschatko (1953), these investigators have all accepted and further developed the two-stage regional compression-extension interpretation of Kelley and Wood (1946). In contrast to models of Laramide crustal shortening, Hayden and Mawer (1991) attributed compressional features to local crustal shortening by oblique slip during middle to late Tertiary dextral transtensional deformation.

Structures within the basins of the Rio Grande rift have been inferred from geologic mapping of the rift shoulders and from COCORP and Shell seismic profiles (Fig. 3; e.g., Brown et al., 1980; Cape et al., 1983; Wu, 1986; De Voogd et al., 1988; Russell and Snelson, 1990, this volume). Of particular interest to our investigation are (1) listric-normal faults in the upper crust (Cape et al., 1983), which sole into detachment faults at midcrustal depths (e.g., Wu, 1986; De Voogd et al., 1988); and (2) planar, ramplike normal faults in the southern Albuquerque-Belen Basin, which do not conspicuously sole. It is evident from both COCORP data (e.g., Brown et al., 1980) and Shell seismic data (Russell and Snelson, 1990, this volume) that the structural margin of the Albuquerque-Belen Basin is not at the Coyote fault (Fig. 4), as argued by Kelley (1977), but at the western edge of the Monte Largo embayment.

GEOLOGIC STRUCTURE OF THE SOUTHERN LUCERO AND LADRON UPLIFTS

Descriptions of major uplifts

Lucero uplift. The Lucero uplift is a half dome centered approximately on the Monte Largo embayment (Fig. 4). Rocks of the Pennsylvanian Madera Group (Fig. 4) form the core of the Lucero uplift and thus crop out around the perimeter of the Monte Largo embayment. The eastern front of the uplift is a complex zone of folding and faulting (Fig. 5). North of the Monte Largo embayment, the structure along the Lucero front is characterized by a pattern of anastamosing normal faults, Reidel shears, small pull-apart structures filled by 27.1-Ma intrusions, and décollements in the Permian section (Callendar and Zilinski, 1976; Hayden and Mawer, 1991). Madera Group rocks adjacent to the Santa Fe, Comanche, Saiz, and Ladron faults (Figs. 4, 5) are folded into sinuous, discontinuous anticlines subparallel to, and in some outcrops truncated by, the adjacent faults. Small thrust faults are locally present in the hinge zones of these folds.

The southern end of the uplift is faulted by closely spaced, north-striking normal faults and intruded by numerous dikes (Fig. 5) and sills. The dominant structure of the southern Lucero uplift consists of a set of closely spaced, north-striking, west-dipping, high-angle normal faults with down-to-the-west sense of offset (Figs. 5, 6A). Displacement on most of these faults was oblique in a dextral sense but dominantly dip-slip (Fig. 6B). Few of these faults are reverse or sinistral strike-slip faults. Southward-increasing separation on normal faults ramps Paleozoic and Mesozoic rocks down to the south into the Rio Grande rift.

The Permian Yeso and San Andres Formations (Fig. 4) played prominent roles in the structural development of the region, due to the ductility and solubility of interbedded gypsum. On the southern Lucero uplift, décollements are developed within the Yeso and San Andres Formations. Thrust faults and normal faults higher and lower in the stratigraphic section do not, in most cases, cut through gypsiferous members of the Yeso Formation. Instead, these rocks were penetratively deformed by cataclasis. Internally, beds are highly brecciated and recemented with calcium carbonate. Disharmonic folding is common. Locally, blocks of the very well lithified, and relatively impermeable, Permian Glorieta Sandstone (Figs. 4, 7) slid on gypsum and shale in the Yeso Formation, forming décollements between the two formations.

No precise age control is presently available on uplift of the Sierra Lucero. Nevertheless, some inferences can be drawn from the record of Tertiary sedimentary and volcanic rocks ruling out major uplift in Laramide time and suggesting most uplift occurred during middle to late Miocene time. Geological evidence from the Sierra Lucero suggests that most uplift oc-

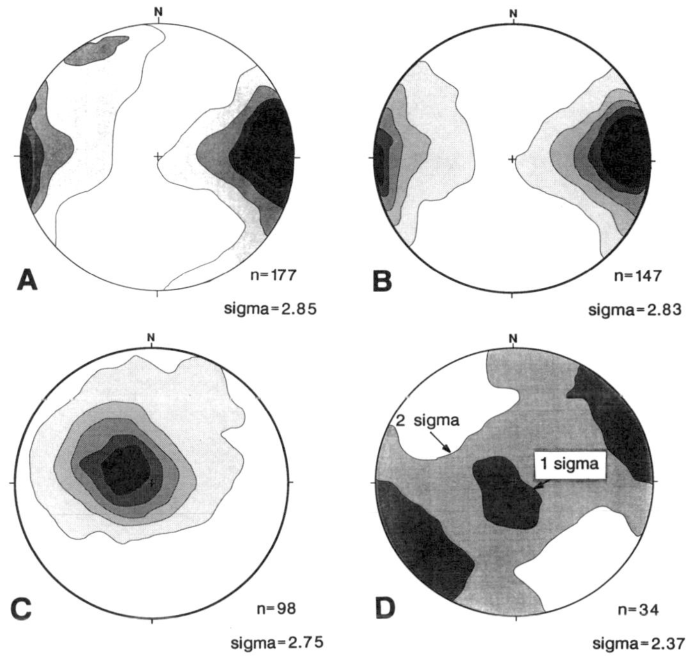

Figure 6. Contoured, lower-hemisphere, equal-area stereographic projections of structural data from the southern Lucero uplift. Contoured pole density projections of fault planes: A, North-striking faults are dominant throughout the area. Population of northeast-striking fault planes are from a single fault zone, the Santa Rita fault; B, Preferred northerly orientation of faults and predominance of high-angle, west-dipping faults emphasized by excluding data from Santa Rita fault. Contoured projection of trends and plunges of striae; C, Dip slip with a generally minor dextral component predominates on all faults (excludes data from Santa Rita fault); D, Bimodal distribution of data in dip-slip and strike-slip sets in plane of Santa Rita fault. Contoured by method of Kamb (1959), based on a variable counting area that is a function of the number of points to be contoured. Contour intervals are multiples of sigma, the standard deviation of the number of points that will fall within a given area under random sampling of the population. N is the number of fault attitude/slickenline data.

curred during middle to late Miocene time. First, basal pebble conglomerates of the Triassic Chinle Formation are exposed beneath late Miocene, mesa-capping basalts at the highest elevations of the Sierra Lucero (Fig. 4). Only about 1 km of the Paleozoic-Mesozoic section was removed prior to deposition of the basalts. Had the Sierra Lucero been uplifted to its present elevation (approximately 2,400 m) in Laramide time, it seems unlikely that half of the Paleozoic-Mesozoic section would have been preserved. Second, basalts in the Sierra Lucero place some bounds on the minimum age of uplift (Baldridge et al., 1987). Major uplift of the Sierra Lucero is older than 3.7 Ma because basalt flows of Mesa Carrizo, erupted from local vents on the Lucero uplift, cross the eastern boundary faults of the uplift without offset. Possibly, uplift began after eruption of 8.2 to 6.2-Ma basalts that cap the highest elevations of the Sierra Lucero. No remnants of these flows are preserved on the flanks of the uplift, as would be likely if these flows were emplaced at their high elevations after uplift. This suggests a minimum of 425 m of uplift of the 8.2 to 6.2-Ma basalts.

Ladron uplift. The Ladron uplift is a westward-tilted block of Paleozoic sedimentary rocks and Precambrian crystalline rocks, bounded on the west by the Ladron fault, on the north and east by the Jeter fault, and on the south by the Cerro Colorado fault (Fig. 4). Numerous north-northwest- and northeast-striking, conjugate(?) faults and shear zones cut the Precambrian rocks of the Ladron block (Fig. 4; Condie, 1976; Chamberlin, 1982) Some of these structures accommodated normal slip and some reverse slip.

Two lines of geologic evidence exist in the Tertiary sedimentary record to define the time of uplift of the Ladron block: (1) The basal Tertiary unconformity, averaged regionally, is only mildly angular across a 20-km traverse from Milligan Gulch basin (Fig. 4) to the east side of Ladron Peak. In Milligan Gulch basin, Eocene Baca Formation is deposited on Cretaceous Crevasse Canyon Formation. In isolated outcrops on the east side of the Jeter fault, Eocene Baca Formation is deposited on Permian Yeso Formation. About 1.0 km more of the Mesozoic section was preserved in the west relative to the east before deposition of Baca rocks. These relations suggest a regional angular discordance of 3°. Thus, the Ladron block underwent neither major uplift nor significant tilting before Eocene time. (2) The Ladron Mountains were repeatedly uplifted in Miocene time, acting as principal sources for the mid-Miocene Popotosa Formation and the late Miocene to Pliocene Sierra Ladrones Formation (Chapin and Seager, 1975). (3) Apatite fission-track analysis indicates a rapid cooling rate (5 to 10 °C/Ma) for the Ladron block in middle to late Miocene time (Kelley, 1990), requiring rapid uplift of the block. In summary, uplift of the Ladron block apparently coincides with early and late phases of rifting and is probably genetically related to rifting.

Descriptions of major faults

Santa Rita fault. The Santa Rita fault zone (new name; Figs. 5, 7A) is an east-northeast–striking, southeast-dipping, high-angle fault zone. Vertical displacement on one splay of the fault is 85 m down-to-the-south near the westernmost exposures of the Santa Rita fault and 120 m down-to-the-north on another splay near the easternmost exposures. Folds, décollements, and small thrust faults in the Permian section are locally associated with the Santa Rita fault. Anticlinal folds parallel the fault along much of its length. A set of north-striking faults intersects the Santa Rita fault. Cross-cutting relations indicate that some of these north-striking faults predate and some postdate the Santa Rita fault.

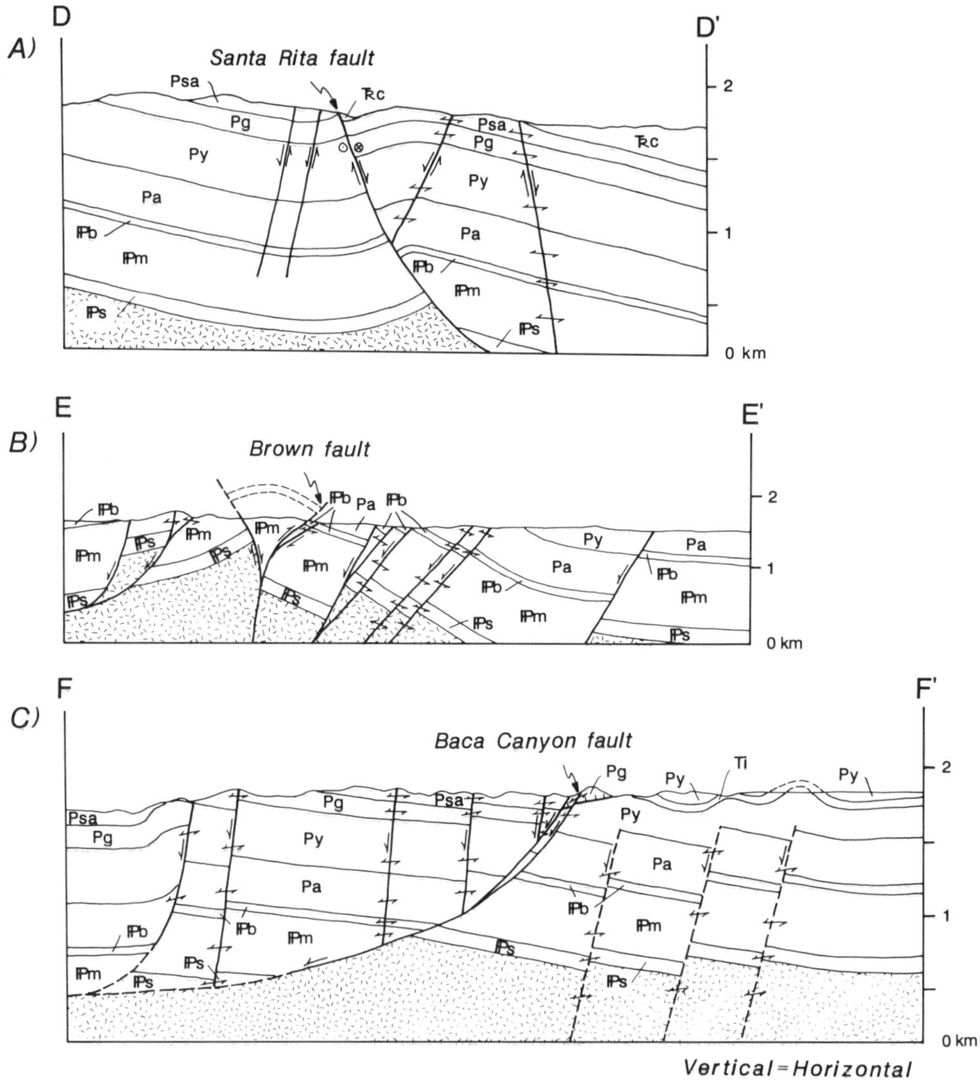

Figure 7. Cross sections of extensional and contractional structures on Lucero uplift. A, Santa Rita fault zone; B, Brown fault zone; C, Baca Canyon fault zone. Lines of section shown on Figure 5; all sections perpendicular to major structures. IPs, Sandia Formation; IPm, Madera Group; IPb, Bursum Formation; Pa, Abo Formation; Py, Yeso Formation; Pg, Glorieta sandstone; Psa, San Andres Formation; TRc, Chinle Formation; Ti, intrusion. Double-headed arrow is tie bar connecting fields of same lithologic unit. ʘ, motion toward; ⊗, motion away.

To the east, the Santa Rita fault is covered by the Riley travertine, a 1- to 10-m-thick Quaternary deposit (Barker, 1983; Fig. 5). To the west of Santa Rita Canyon, the course of the Rio Salado (Fig. 5) is locally controlled by the fault. Farther west along strike, the northeast-striking Negros fault (Fig. 4) underwent both sinistral and down-to-the-south relative offset (Massengill, 1979).

Striations on Santa Rita fault planes fall principally into two populations, dip slip and strike slip (Fig. 6D). Cross-cutting relations among superposed striations indicate that dip-slip motion preceded dextral strike slip, and dextral preceded sinistral strike slip.

North of the Santa Rita fault, rocks of the Yeso Formation and a prominent sill are folded into west-vergent folds (Figs. 5, 7A) whose axes curve westward into the trace of the fault. Vergence of the folds indicates that Yeso Formation was sheared between a lower block moving eastward and an upper block moving westward. Concavity of the folds implies they were dragged by dextral offset along the fault. The folded sill is almost certainly Oligocene in age, correlating with three dated igneous intrusions (Aldrich et al., 1986) located within a radius of about 15 km of the folded sill. No known Paleozoic or Mesozoic igneous rocks occur in this area. The involvement of Oligocene intrusions in folding is strong evidence for Tertiary contraction, at least locally.

Macroscopic evidence for sinistral strike-slip on the Santa

Rita fault includes offset of the contact between Permian San Andres Limestone and Triassic Chinle Formation at the western end of Santa Rita Canyon. The contact strikes west-northwest and dips moderately to the south (Lewis, 1989). Eighty-five meters of down-to-the-south dip slip would offset the contact at most a few tens of meters to the east. Another possible constraint is offset of the Baca Canyon fault (Fig. 5; see "Baca Canyon fault" below). Neither of these offsets can be accomplished by dip-slip displacement alone. Based on these two examples, 750 to 1,100 m of sinistral strike-slip offset has occurred.

Brown fault. The Brown fault (Fig. 5, 7B) is an east-vergent, moderate- to high-angle reverse fault on the Lucero uplift with a few tens of meters of displacement on Pennsylvanian rocks. The fault cuts the eastern limb of a doubly plunging anticline. The western limb of the anticline is cut by a west-vergent, dextral-oblique reverse fault. Slickenlines in the planes of the Brown fault and adjacent normal faults indicate dip slip with a dextral component. Two adjacent faults had strike-slip displacement. These features as a group may comprise a positive flower structure.

Mesa Sarca fault. The Mesa Sarca fault (Figs. 4, 5) is the western-bounding fault of Navajo Gap, a narrow, north-trending graben between the Mesa Sarca and Ladron uplifts. The fault is an east-dipping, down-to-the-east, high-angle normal fault with 840 m of dip-slip displacement (Hammond, 1987). The southward-projected trace of the Mesa Sarca fault coincides with a dike that dates 28.3 ± 1.1 Ma (Aldrich et al., 1986). Closely spaced normal and strike-slip faults, parallel in strike to the Mesa Sarca fault, shattered Madera Group rocks in Navajo Gap into numerous elongate blocks (Hammond, 1987). On the east side of the gap, beds of Madera Group rocks structurally ramp onto the flank of Ladron Peak. Faults in Navajo Gap die out to the south in a syncline filled with Tertiary/Quaternary Santa Fe Group sedimentary rocks and extensive deposits of Quaternary Riley travertine.

Baca Canyon fault. The Baca Canyon fault zone (new name) (Figs. 5, 7C) is a moderate- to high-angle, west-dipping normal fault and several splays. A cross section through the fault, indicating rollover of beds on the west side of the fault, requires that the fault be listric (Lewis, 1989). Slickenlines on fault planes indicate both dextral and sinistral oblique slip. Although the amount of displacement varies along strike, greatest throw (415 m) occurs between Santa Rita Canyon and the Rio Salado (Fig. 5).

The Baca Canyon fault offsets an Oligocene sill, indicating that normal faulting occurred after the emplacement of this intrusion. The fault also truncates a conspicuous décollement between Permian Glorieta sandstone and the underlying Cañas Member of the Permian Yeso Formation (Fig. 7C). The intersection of the décollement and the Baca Canyon fault provide a piercing point by which offset on the Santa Rita fault can be assessed. The trace of the Baca Canyon fault appears offset 1,100 m in a sinistral sense by the Santa Rita fault. Transverse faulting along the Santa Rita fault thus postdated the development of the north-striking Baca Canyon fault and may account for as much as 1,100 m of strike-slip offset.

Hell's Mesa fault. The Hell's Mesa fault (Figs. 4, 5) cuts 32- to 24-Ma volcanic and sedimentary rocks (Chapin and Seager, 1975). Its straight trace over 25 km of hilly terrain indicates an average dip that is steep. Sense of offset along the fault is down-to-the-west normal with a dextral component. Stratigraphic separation increases from 475 m in the north to 518 m in the south (Massengill, 1979).

Comanche fault. The Comanche fault zone forms the arcuate western edge of the Monte Largo embayment (Figs. 5, 8A and B), and as such bounds part of the Albuquerque-Belen Basin. No attitudes were obtained on faults plane due to poor exposure, although the faulted contact between Madera Group and Yeso Formation can be traced in isolated outcrops around the edge of the embayment. East-side-down throw on the fault, based on stratigraphic omission of the Bursum, the Abo, and part of the Yeso Formations, is at least 650 m and perhaps as much as 1,100 m. Seismic data suggests that Mesozoic rocks are downfaulted to a depth of about 1.2 km in the Monte Largo embayment (Russell and Snelson, 1990), suggesting that another major normal fault (perhaps the Jeter fault) is buried beneath extensive deposits of Tertiary sedimentary rocks and Quaternary travertine. Carbonate-rich springs are depositing travertine today where modern arroyos intersect the Comanche fault (Goff et al., 1983). Total throw in the Comanche-Jeter fault zone is probably 2 km or more down to the east.

Deformation of Madera Group rocks extends about 500 m into the footwall of the Comanche fault. Adjacent to the fault, rocks of the Madera Group are folded into a discontinuous anticline subparallel to, and in some outcrops truncated by, the fault (Fig. 5). Dips of anticlinal limbs are as great as 84° in some exposures. In one area, Pennsylvanian rocks are folded into an anticline and an overturned syncline (Fig. 8A). Outcrops of Yeso Formation in the hanging-wall block dip moderately eastward, away from the surface trace of the fault. A set of north-northeast–trending small folds in Madera Group rocks (Fig. 5) form an angle of about 60° with the trace of the Comanche fault. In addition, northeasterly striking normal faults of small offset cut the Comanche fault along the northern border of Monte Largo embayment. Numerous calcite-encrusted spring vents are exposed along small faults and fractures in Tertiary/Quaternary pediment gravels. The capping travertine is not faulted, however, indicating no recent activity on these faults.

Saiz fault. The Saiz fault (Figs. 5, 8C), which dips 20 to 50° northeastward, forms part of the bounding fault system of the Monte Largo embayment. Footwall Pennsylvanian rocks, striking nearly parallel to the fault, are deformed into anticlines and overturned synclines (Fig. 5). Within the hinge zones of the synclines are abundant, small-scale, asymmetric folds. Locally, north-striking, dextral- and sinistral-slip, high-

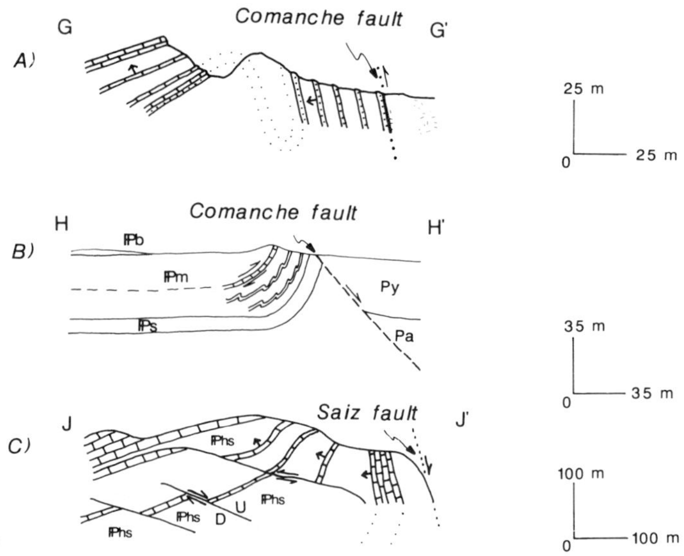

Figure 8. Diagrammatic cross sections of steeply west-dipping and overturned beds of Pennsylvanian rocks at eastern edge of Lucero uplift. A and B, Comanche fault zone; C, Saiz fault zone.Lines of section shown on Figure 5; all sections perpendicular to major structures. Patterns on marker beds: blocks for limestone; stipple for sandstone; cross hatch for crystalline rocks. Arrows on marker beds show stratigraphic up. IPhs (Pennsylvanian Story Formation within Madera Group) used as marker bed. See Figure 7 for symbols.

angle faults offset small-scale folds but do not cut the Saiz fault. Folding and strike-slip faulting thus appear to have occurred contemporaneously in a northeast-southwest compressive stress regime (Hammond, 1987).

On the east side of the Saiz fault, contractional and extensional structures are closely associated. Permian rocks of the hanging wall are gently to tightly folded, strongly faulted and sheared. Small-offset, northeast-dipping, listric normal faults in the hanging wall sole into the low-angle Saiz fault (Hammond, 1987).

Ladron fault. The Ladron fault (Figs. 4, 5) borders the west side of the Ladron uplift and juxtaposes Precambrian rocks of the Ladron block against Pennsylvanian rocks of the southern Lucero uplift. The fault is steeply east dipping to vertical, indicating reverse-sense offset. Rare striations in the plane of the fault indicate dextral oblique displacement. Pennsylvanian strata adjacent to the fault are moderately to steeply dipping and overturned. Contractional structures parallel to the Ladron fault in the downdropped western block include a west-dipping monocline, an anticline-overturned syncline pair, and intrabed thrust faults. Striations on the thrust faults are down dip, implying east-west compression (Hammond, 1987).

Jeter fault. The Jeter fault (Figs. 4, 5) is a low-angle (20 to 30°), normal fault bounding the Ladron uplift on its northern and eastern sides (Black, 1964; Kelley, 1977). The fault zone contains 2.5 to 3 m of cataclastic breccia and fault gouge, below which Black (1964) reports nearly a meter of considerably sheared Precambrian rocks of the footwall; many tens of meters of sheared rock have been reported for some outcrops (Russell, personal communication, 1992). The fault plane is convex towards the northeast, its hingeline plunging 20 to 30° towards N55°E. This direction of plunge is consistent with the northeasterly strike of the Ladron "horst" inferred from the associated Bouguer gravity anomaly by Cordell (1978) and Wu (1986).

The Jeter fault juxtaposes Precambrian crystalline rocks in the footwall against Paleozoic, Mesozoic, and Cenozoic rocks, including Oligocene volcanic rocks and Santa Fe Group rocks, in the hanging wall (Black, 1964). Rocks of the early to mid-Miocene Popotosa Formation and late Miocene to Pliocene Sierra Ladrones Formation have been tilted up to 25° toward the surface trace of the fault (Black, 1964). Thus, displacement on the Jeter fault continued after deposition of Sierra Ladrones Formation.

Cerro Colorado fault. The Cerro Colorado fault (Figs. 4, 5) is a high-angle, sinistral oblique-slip fault (Chamberlin, 1982) that bounds the south side of the Ladron uplift. To the southwest, the fault connects with the bounding fault of the La Jencia basin (Fig. 4). To the northeast, the Cerro Colorado fault truncates the Jeter fault.

KINEMATICS OF THE CENTRAL RIO GRANDE RIFT

Basin geometry of the central Rio Grande rift

We propose a geometric model (Fig. 9) for the central Rio Grande rift consisting of linked and nested half grabens bounded by accommodation zones and bordered by tilted block uplifts. Accommodation zones are inferred beneath Tertiary and younger basin fill based on geometric relationships between half grabens.

The Albuquerque-Belen Basin is underlain by two half grabens of opposite polarity, the Albuqerque half graben and the Belen half graben (Russell and Snelson, 1990). A zone of accommodation separates them in the subsurface (Lozinsky, 1988, this volume; Russell and Snelson, 1990). Numerous authors have noted that the northeast-striking northern edge of the Monte Largo embayment appears to be controlled by the Tijeras-Cañoncito fault system (e.g., Kelley, 1977; Ander, 1980; Lozinsky, 1988; Cabezas, 1989), a major fault zone that cuts the eastern-bounding faults of the Albuquerque-Belen Basin (Figs. 3, 9). However, the arcuate faulted margin of the embayment is bounded by a continuous arcuate fault; no discrete strike-slip fault continues beyond the margin of the embayment to the southwest. Russell and Snelson (1990) show the accommodation zone well to the north of the Monte Largo embayment (Fig. 9). The northern Lucero uplift is bounded by the Santa Fe fault (Fig. 4), which has accommodated approximately 3 km of stratigraphic separation on Mesozoic rocks (Russell and Snelson, 1990). The Santa Fe fault is a major Tertiary extensional fault unbroken by major transverse struc-

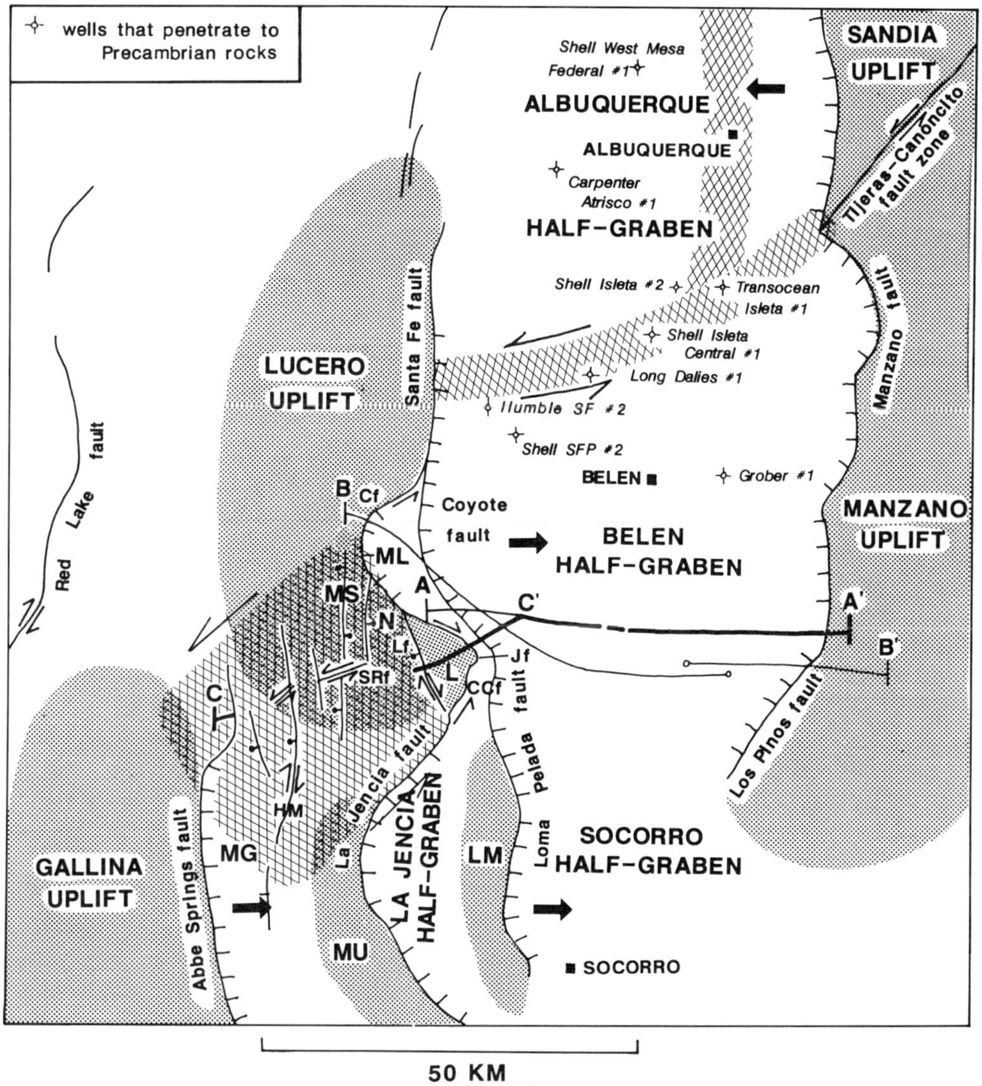

Figure 9. Geometry of linked and nested half grabens, tilted block uplifts, and accommodation zones of central Rio Grande rift. Hachured lines indicate bounding faults of half grabens; hachures on downthrown side. Shaded areas indicate uplifts. Cross-hatched pattern indicates accommodation zones. Widest cross-hatched area symbolizes shear couple created by rotating Colorado Plateau clockwise or extending crust beneath rift relative to stable plateau. Bold arrows indicate direction of dip of underlying detachment. Faults: Cf, Comanche; CCf, Cerro Colorado; HMf, Hell's Mesa; Jf, Jeter; Lf, Ladron; SRf, Santa Rita. Half grabens: MG, Milligan Gulch; ML, Monte Largo; N, Navajo Gap. Uplifts: L, Ladron; LM, Lemitar Mountains; MS, Mesa Sarca; MU, Magdalena. Lines of section from the following sources: A–A′, Shell (Russell and Snelson, 1990); B–B′, COCORP (e.g., Brown et al., 1980; Cape et al., 1983; De Voogd et al., 1988); C–C′, this study.

tures. Thus, the accommodation zone does not break the western boundary of the rift as a through-going structure. Although structures along the northern edge of the Monte Largo embayment indicate a northeast-striking structural grain (Fig. 5), nowhere on the western margin of the basin is this grain resolved into a discrete fault zone like the Tijeras-Cañoncito fault zone.

The cross-sectional geometry of the Belen half graben consists of an east-facing half graben, bounded by the Comanche, Jeter, and Santa Fe faults on the west side, and the secondary Manzano and Los Pinos faults on the east side (De Voogd et al., 1988; Russell and Snelson, 1990; Fig. 9). The Monte Largo embayment (Fig. 9) is underlain by a half graben (hereafter called "Monte Largo half graben") whose bounding fault, the Comanche fault, is the arcuate edge of the embayment. This interpretation differs from Russell and Snelson (1990) in that, in their model, the Jeter fault is the western-bounding fault of the Monte Largo half graben, and the

Comanche fault is a west-dipping Laramide thrust fault. We propose that the Jeter and Comanche faults are both normal faults and that the Comanche fault dips to the east.

The major offsets on the west side of the Belen half graben have occurred, however, not at the Comanche fault but near the axis of the rift on the Coyote fault (e.g., De Voogd et al., 1988; Russell and Snelson, 1990; Fig. 9). Thus, the Monte Largo half graben appears to ramp down to the east to a structurally deeper detachment at the Coyote fault, making the shoulder of the "Ladron horst," as imaged by COCORP (e.g., Brown et al., 1980; Chaimov, 1988), a fault-bend fold over an antilistric fault ramp. This interpretation assumes the crustal detachment is continuous beneath the "horst," as shown by interpretations of Shell seismic data (Russell and Snelson, 1990).

South of Ladron Peak, we infer that the Milligan Gulch, La Jencia, and Socorro basins (Fig. 9) are bounded by down-to-the-east listric normal faults that sole into one or more east-dipping crustal detachments (Fig. 10). The detachment(s) may be continuous with the 5- to 6-km-depth detachment underlying the west side of the Belen half graben, though this detachment has been folded by uplift of the Ladron block (Wu, 1986; Russell and Snelson, 1990).

Distributed shear along the Colorado Plateau/Rio Grande rift boundary

We interpret the set of north-striking, high-angle normal faults on the southern end of the Lucero uplift as a zone of distributed shear, related to a northeast-southwest–trending wrench zone that decouples the southeastern Colorado Plateau from the rift basins to the south. North of this zone, the central Lucero uplift is only slightly extended. South of this zone, in the Lemitar Mountains (Fig. 4), extension exceeded 100% (Chamberlin, 1983). Thus, this fault zone comprises an accommodation zone (Fig. 9). Structures in this zone form the transition between the relatively unextended Colorado Plateau and a highly extended part of the Rio Grande rift. This zone also links the Milligan Gulch and La Jencia half grabens.

The Santa Rita and Negros faults (Fig. 4), in this context, are parts of a segmented, synthetic, sinistral strike-slip fault that extends from the bounding fault of the Milligan Gulch half graben to Ladron Peak (Fig. 9). Most slip, however, was resolved as dip slip and oblique slip on normal faults oriented obliquely to the shear couple rather than as strike-slip motion on a through-going transfer fault. Striations in fault planes, indicating the most recent offset on faults, show a component of dextral slip on north-striking faults (Fig. 6). Field relations at the boundary between the rift and the plateau imply that a southwest-northeast–trending sinistral shear couple produced the resolved shear stresses observed in the study area. Shear was thus distributed by translation of fault blocks and tilting, or ramping, of beds down into the Milligan Gulch and La Jencia half grabens (Fig. 11).

Two kinematic models, actually just different perspectives on the same mechanics, might explain the stress regime. Clockwise rotation of the Colorado Plateau (Eardley, 1962), in an overall tensional regime, would have produced a sinistral shear couple along the boundary between the plateau and the rift. Alternatively, opening of the Milligan Gulch and La Jencia half grabens would have created a shear couple along the southern edge of the Colorado Plateau without rotation of the plateau. Depending on relative block displacements, the sense of shear resolved on individual fault planes could have been either dextral or sinistral. As discussed above (Santa Rita fault), both senses of shear are manifest at the Colorado Plateau/Rio Grande rift boundary.

An area of similar late Tertiary structural complexity en-

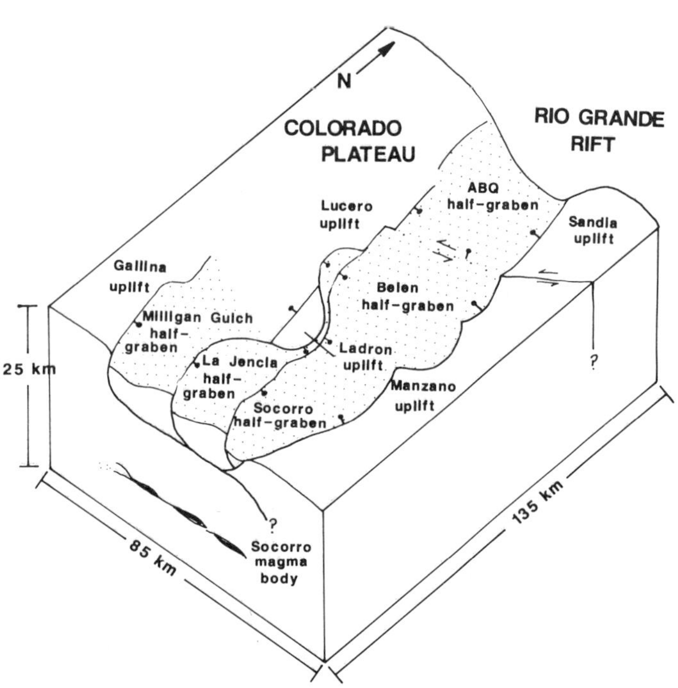

Figure 10. Schematic block diagram of linked and nested half grabens of central Rio Grande rift. Stippled area indicates Tertiary rift fill. Vertically exaggerated 2.5 times.

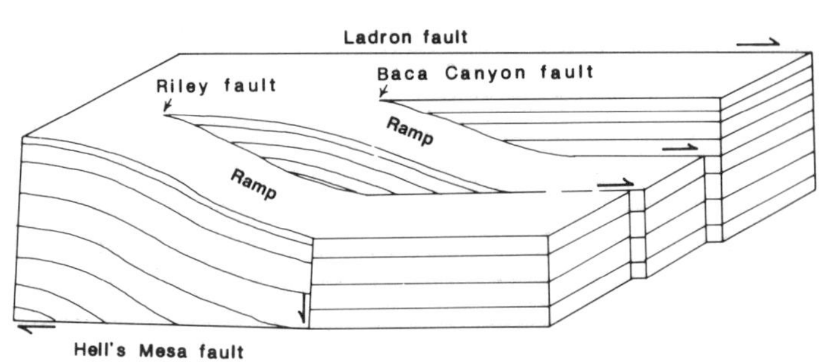

Figure 11. Schematic block diagram of left-stepping, en echelon relay faults and ramps at southern terminus of Lucero uplift. Arrows indicate second-order dextral shear couple parallel to the relay faults. Sense of displacement on faults is dextral oblique. Modified from Kelley (1979).

compasses the Muddy, North Muddy, and northern Black Mountains in Nevada where extensional and contractional structures developed synchronously (Bohannon, 1983). Fault blocks within this area have undergone significant (30°) counterclockwise rotations about vertical axes (Ron et al., 1986). This rotation was accommodated by slip on northwest-striking dextral faults conjugate to the principal sinistral wrench. This area may be analogous to the accommodation zone on the southern Lucero uplift. If so, we might expect paleomagnetic studies to reveal counterclockwise rotations of fault blocks bounded by north-striking faults in this area. Rotation about vertical axes may be an important mechanism of strain accommodation at the boundary between the Colorado Plateau and the rift.

Extension involving some of the north-striking structures in this accommodation zone nearly severed Ladron Peak from the Colorado Plateau. We postulate that Navajo Gap (Fig. 4), a narrow graben between Mesa Sarca and the Ladron Mountains, developed as an incipient half graben as the Mesa Sarca fault propagated southwards behind the Ladron block. The alignment of Navajo Gap with the synclinal basin west of Ladron Peak (Fig. 5) and with the La Jencia fault (Fig. 4) indicates the strike of these failed rift structures.

Isostatic uplift of the Ladron block

The Ladron block is bounded on the north and east by the low-angle, normal-sense Jeter fault and on the south by the strike-slip Cerro Colorado fault (Fig. 4). Thus, we interpret that the Ladron block was isostatically uplifted, as proposed for the Virgin-Beaver Dam breakaway zone along the west-central margin of the Colorado Plateau (Wernicke and Axen, 1988). The Ladron block rebounded as a result of unloading on three sides as the linked Monte Largo and La Jencia half grabens opened up (Figs. 9, 10).

We postulate that the low dip of the Jeter fault, and the steep westward dips of the pre-Cenozoic section, resulted from arching of an initially moderately dipping normal fault in response to buoyancy forces caused by tectonic denudation of the footwall (Fig. 12). Whereas present dips on the Jeter fault range from 20 to 30°, the initial dip was probably about 48° (Fig. 13). The hanging-wall allochthon has moved approximately 20 km east-northeastward along the Jeter fault, representing significant unloading of the footwall. Removal of a hanging wall approximately 6 km thick from the exposed level of the Jeter fault would requirc a similar amount of isostatic rebound of the footwall. Striations paralleling the dip of thrust planes in the footwall indicate east-west compression, consistent with westward tilting of the footwall of the Jeter fault and consequent layer-parallel shortening. Reverse displacement occurred on the Ladron fault in response to isostatic buoyancy forces. In addition, the numerous conjugate shear zones within the Ladron block (Fig. 5) may be associated with footwall isostatic uplift.

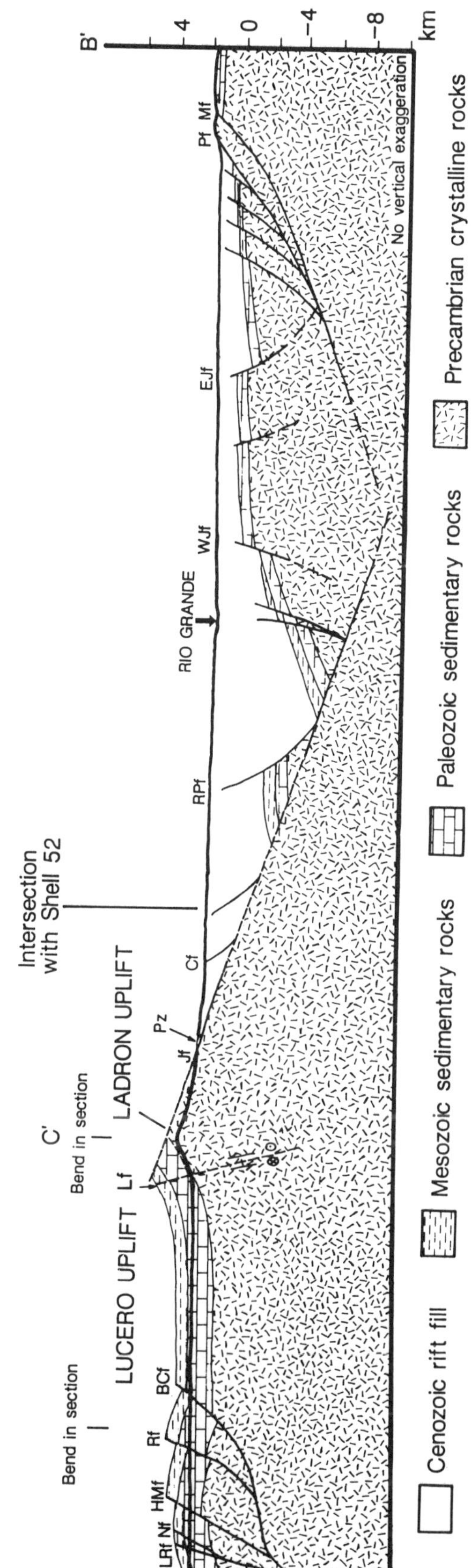

Figure 12. West-to-east cross section through Ladron Peak and southern Albuquerque-Belen Basin. Structure in basin adapted from Russell and Snelson (1990). Geology of Ladron Peak from Black (1964), Chamberlin (1982), Hammond (1987), and Lewis (1989). Faults: LRf, La Rosa; Nf, Negros; HMf, Hell's Mesa; Rf, Riley; BCf, Baca Canyon; Lf, Ladron; Jf, Jeter; Cf, Coyote; RPf, Rio Puerco; WJf, West Joyita; EJf, East Joyita; Pf, Paloma; Mf, Montosa; Pz, outcrop of Paleozoic sedimentary rocks. See Figure 9 for line of section. ⊙, motion toward; ⊗, motion away.

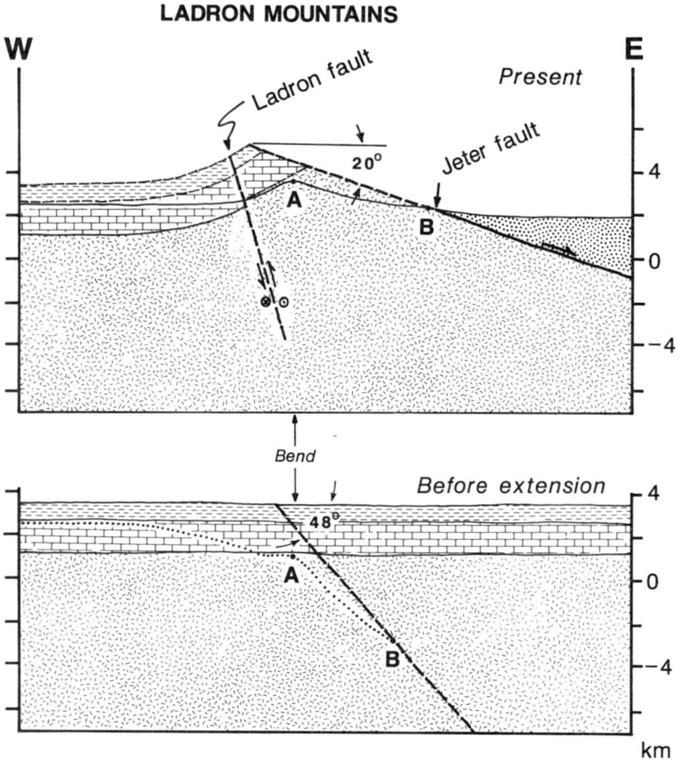

Figure 13. Balanced and reconstructed cross sections through Ladron Peak. See Figure 9 for line of section and Figure 4 for explanation. Restoration (bottom) indicates a moderate initial dip for the Jeter fault. Bend in section necessitated by linking with seismic section. A true west-east cross section would probably not differ enough to significantly affect interpretation. ⊙, motion toward; ⊗, motion away.

The amount of extension accommodated by the Jeter fault and hanging-wall faults to the east can be estimated from the cross section in Figure 9. The present width of the basin east of Ladron is about 65 km (counting the projected segment of the Jeter fault up to the footwall cutoff in the Paleozoic/Mesozoic section). The restored width of the basin is about 40 km. Thus, the southern part of the Belen half graben underwent about 60% extension. This magnitude of extension, accomplished primarily by one major fault, must have produced large buoyancy forces in the crust beneath the unloaded footwall and consequent rapid uplift of the Ladron block.

Our reconstruction of the Jeter fault suggests that it was originally a planar fault. The 48° hanging wall cutoff angle apparent in seismic profiles (DeVoogd et al., 1988; Russell and Snelson, 1990) constrains the footwall cutoff to be the same. The Jeter fault must project over the top of the Ladron block, through the Mesozoic section, without appreciable steepening. Much of the rollover of bedding in the hanging-wall fault blocks, evident in seismic sections (e.g., Russell and Snelson, this volume), appears related to soling of one of the hanging-wall faults, the Rio Puerco fault.

The relatively minor extension of the Jeter fault's wedge-shaped upper plate influenced the way that the footwall deformed. Spencer (1984) showed that the amount and distribution of attenuation and lateral displacement of the upper plate affect the final form of low-angle normal faults. Uniform internal extension and down-dip displacement of a wedge-shaped upper plate above a planar-normal fault results in a linear increase in the amount of tectonic denudation in the down-dip direction. An initially planar normal fault is therefore rotated to a more gentle dip but maintains its planar form. The result for the Jeter fault, where its dip and depth of penetration were greatest, was rotation from its initial 48° dip to its final 20° dip with little anticlinal folding of the fault plane.

We see evidence for both layer-parallel shortening and subvertical simple shear in the Ladron uplift. The major footwall syncline on the east side of the Ladron block, the reverse sense Ladron fault, and other contractional structures parallel to the fault are similar to footwall uplift structures in the Beaver Dam Mountains, where the timing of uplift is well known (Wernicke and Axen, 1988).

If the footwall flexure in the Ladron block is assumed to be elastic, the amplitude:wavelength ratio (about 0.15) implies low flexural rigidity of the lithosphere (Spencer, 1984). In the central Rio Grande rift, however, short wavelength, positive Bouguer gravity anomalies centered on the shoulder uplifts of the Albuquerque-Belen Basin (Ladron, Sandia, Manzano, and Lucero uplifts; Cordell, 1978), indicate these uplifts are locally uncompensated. The crust beneath the central rift thus has enough flexural rigidity to support the smaller uplifts of the rift shoulders despite high heat flow in this part of the rift (2.56 ± 0.65 HFU; Reiter et al., 1979). Following Zandt and Owens (1980) and Wernicke and Axen (1988), we suggest that footwall uplift in the Ladron block is a nonelastic response to tectonic denudation.

The magnitude of extension calculated for the vicinity of the Ladron uplift has an important consequence for our interpretation of the initial geometry of normal faults in this area. Russell and Snelson (1990) estimated about 28 to 30% extension from their seismic section through the Monte Largo embayment. Because we used their section from the Coyote fault eastward and estimate 60% extension east of the Ladron uplift, the main difference must be that more extension has been accommodated on the Jeter fault where it bounds the Ladron block than where it bounds the Monte Largo embayment. We suggest that the original geometry of the Jeter fault changed southward from listric and shallow to planar and deeply penetrating. This is probably one of several mechanisms by which differential extension was accommodated between the central and southern parts of the Rio Grande rift. Other mechanisms include sinistral strike-slip offset on the Cerro Colorado fault, in the direction of extension, and distributed shear at the southern edge of the Lucero uplift.

Origin of contractional strain on the Lucero uplift

Contractional strain on the Lucero uplift is restricted to a narrow margin along the eastern front and discrete zones in the southern part of the uplift (Brown and Santa Rita fault zones, Figs. 5, 7). The congruity of contractional structures and breakaway faults at the Lucero front, in the Saiz fault zone for example (Fig. 5), may represent flexural contraction in an uplifted footwall. Overturned folds at the edge of the Monte Largo embayment may be related to bedding-parallel shear and reverse faulting produced by a late change in the footwall strain field from extensional shear to hanging-wall-side-up shear. The Comanche and Saiz faults may have served as reverse faults, accommodating hanging-wall-side-up shear, during part of their movement history (Figs. 8A, B, and C). The magnitude of crustal shortening perpendicular to the Comanche and Saiz faults probably does not exceed 100 m. In both the Saiz and Comanche fault zones, rocks in the breakaway zone last experienced extension rather than contraction. This observation is consistent with subsequent, major normal displacement stepping structurally downward and closer to the axis of the rift on the Coyote–Loma Pelada faults (e.g., Russell and Snelson, 1990). Regional Laramide compression is thus not required for the origin of contractional strain on and along the edge of the Lucero uplift.

Along the northern border of the Monte Largo embayment, some local areas of very high dip, folds that intersect the Comanche fault at a 60° angle, and minor, northeasterly striking faults that cut the Comanche fault do not readily fit models of footwall isostatic uplift. However, the similarity in trend of folds and breakaway normal faults suggests a genetic association. Folding is probably not wrench related due to the lack of systematic en echelon arrays of folds and strike-slip faults.

In contrast to Ladron, structural relief between the Monte Largo embayment and the Lucero uplift is relatively subdued. The difference in the degree of footwall uplift and rotation may be due to variations along strike in thickness of the allochthon (Wdowinski and Axen, 1992). The Coyote-Jeter fault system presently penetrates to depths of 11 or more kilometers (Fig. 12), whereas the Comanche-Jeter fault system penetrates only to depths of 5 to 6 km. Our reconstruction of the Jeter fault suggests that it was originally planar and moderately dipping (48°), and it probably penetrated deeply into the middle crust. This contrasts with the Comanche/Jeter fault zone in the Monte Largo embayment, where the Jeter fault is listric in cross section and may never have penetrated more than 5 to 6 km. The variation we observe in the degree of footwall uplift agrees with flexural models that predict more footwall uplift for planar-normal faults than for listric-normal faults (Kusznir et al., 1991). Further investigations of the three-dimensional geometry of folded detachment systems may clarify these variations in footwall response to denudation.

Contractional strain on the uplift itself may be caused by wrenching in a combined normal/wrench setting (Fig. 14). With some minor exceptions, dextral slip occurred on north-northwest- to north-northeast–striking faults and sinistral slip on west-southwest–striking faults. Assuming these faults represent conjugate sets, the pattern of strain requires a west-northwest to east-southeast orientation for the least principal stress axis, which is the modern orientation determined for the Basin and Range province and Rio Grande rift (Zoback et al., 1981; Aldrich et al., 1986). This represents a rotation from the west-southwest to east-northeast extension direction determined for extension in the period from about 32 to 23 Ma (Aldrich et al., 1986) and apparent in pole density diagrams for the Lucero uplift (Fig. 6B). Thus, north-striking faults on the Lucero uplift, some of them intruded by Oligocene dikes, may have been produced by west-southwest to east-northeast extension and subsequently reactivated as dextral oblique-slip faults in the later west-northwest to east-southeast stress field.

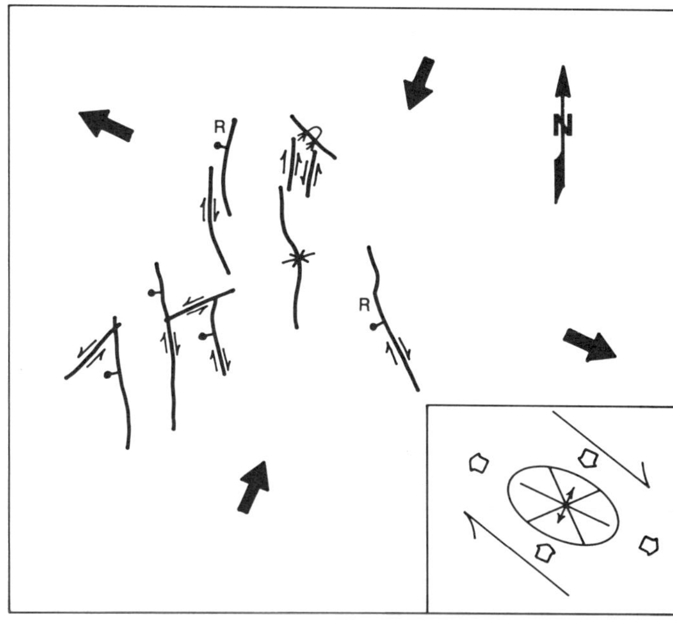

Figure 14. Major strike-slip and oblique-slip faults on southern Lucero uplift. Note dextral slip on north-northwest– to north-northeast–striking faults and sinistral slip on west-southwest–striking faults. Assuming these faults represent conjugate sets, the pattern of strain requires a west-northwest to east-southeast orientation for the least principal stress axis and north-northeast to south-southwest shortening. Inset shows first-order dextral shear couple. Ball and bar on downthrown side of fault; R indicates reverse sense of displacement.

The prevalence of oblique- and strike-slip offsets on the Lucero uplift suggests a constrictional finite-strain path (i.e., shortening along the intermediate principal strain axis), allowing for partial absorption of east-west extensional strain by north-south shortening rather than crustal thinning, as suggested by Wernicke et al. (1988) for the Basin and Range province near Las Vegas. A zone of concentrated microearthquake activity has been noted previously within the rift between Belen and Socorro (Sanford et al., 1979). Composite fault plane solutions suggest that both strike-slip and normal faulting are currently occurring in the rift; the composite T-axis for focal mechanisms is oriented west-northwest to east-southeast. Geodetic data from the Socorro area

(Fig. 3) show no ongoing east-west spreading, but they do indicate that the region may be undergoing horizontal compression with the intermediate stress axis oriented approximately north-south (Prescott et al., 1979).

Time-stratigraphic markers by which the timing of structural development can be bounded are scarce on the Lucero uplift. Some structures are clearly early; for example, the dike swarm near Riley (Fig. 4), intruded in Oligocene time (Aldrich et al., 1986). Some structures are clearly late; Quaternary offset has been documented on the Santa Fe, Coyote, Loma Pelada, and La Jencia faults (Fig. 4; Machette, 1982). Strong evidence for Oligocene or later contraction is the folded Oligocene sill in the vicinity of the Santa Rita fault. From available data, we do not know whether the Ladron Peak–Sierra Lucero area experienced a continuum of deformation or discrete events.

CONCLUSIONS

The geometry of major basin-bounding faults indicates that the style of extension in the central Rio Grande rift is characterized by overlapping and nested half grabens linked together by accommodation zones and bounded by tilted-block uplifts. The Albuquerque-Belen Basin is comprised of two subbasins separated by an accommodation zone confined between the north-striking bounding faults of the central rift.

Northeast-southwest–oriented shear between the relatively unextended Colorado Plateau and the more extended rift is taken up by translation and tilting of fault blocks along en echelon, north-striking normal faults. These faults, which decouple Ladron Peak from the plateau, constitute a zone of distributed strike-slip strain or accommodation between the plateau and rift.

The Ladron uplift is an isostatically uplifted footwall block of a major crustal detachment system. The low dip of the Jeter fault resulted from arching of the initially moderately dipping normal fault in response to buoyancy forces caused by tectonic denudation of the footwall.

Coincident contractional and extensional structures along the edge of the Monte Largo embayment imply that the Lucero uplift resulted from isostatic uplift of the footwall of the Santa Fe fault and the Comanche-Jeter and Saiz-Jeter fault systems. Thus, the Monte Largo embayment and the Lucero uplift are principally late Tertiary extensional features, although precise limits on the age of their formation have not been established.

An extensional or extensional/wrench origin for structures in the Ladron Peak–Sierra Lucero area seems mechanically simpler, and therefore more likely, than Laramide crustal shortening followed by late Tertiary extension. Despite the paucity of timing constraints, our model represents a plausible alternative to two-stage models invoking Laramide compression and middle to late Tertiary extension. Many of the structures in the central Rio Grande rift previously attributed to Laramide compression can be adequately explained by late Tertiary extension.

ACKNOWLEDGMENTS

The authors thank P. Cabezas, S. M. Cather, R. G. Chamberlin, C. E. Chapin, J. Hawley, D. Kraig, D. Love, and R. P. Lozinsky for enlightening discussions about Rio Grande rift geology, Laramide tectonism, and extensional processes. Topographic base maps were provided by M. Woolridge. Photographic reductions of field maps were prepared by R. C. Walther. We thank J. N. Gardner and J. M. Stock for reviews of early drafts of the manuscript. Discussions with G. J. Axen were very helpful in understanding the mechanics of footwall uplift. Reviews by C. Morley and L. Russell greatly improved the final manuscript. Lewis was supported by Associated Western Universities Thesis Parts and Laboratory Graduate Fellowships, Los Alamo National Laboratory, a California State University/Los Angeles Graduate Equity Fellowship, and National Science Foundation grants EAR 8904022 and EAR 9058217 awarded to J. M. Stock, California Institute of Technology. Baldridge was supported by the Los Alamos branch of the Institute of Geophysics and Planetary Physics of the University of California.

REFERENCES CITED

Aldrich, M. J., Chapin, C. E., and Laughlin, A. W., 1986, Stress history and tectonic development of the Rio Grande rift, New Mexico: Journal of Geophysical Research, v. 91, p. 6199–6211.

Allmendinger, R. W., and 6 others, 1983, Cenozoic and Mesozoic structure of the eastern Basin and Range province, Utah, from COCORP seismic-reflection data: Geology, v. 11, p. 532–536.

Allmendinger, R. W., and 7 others, 1987, Overview of the COCORP 40°N transect, western United States: The fabric of an orogenic belt: Geological Society of America Bulletin, v. 98, p. 308–319.

Ander, M. E., 1980, Geophysical study of the crust and upper mantle beneath the central Rio Grande rift and adjacent Great Plains and Colorado Plateau [Ph.D. thesis]: Albuquerque, University of New Mexico, 218 p.

Anderson, R. E., 1971, Thin-skin distension in Tertiary rocks of southeastern Nevada: Geological Society of America Bulletin, v. 82, p. 43–58.

Anderson, R. E., 1973, Large-magnitude late Tertiary strike-slip faulting north of Lake Mead, Nevada: U.S. Geological Survey Professional Paper 794, 18 p.

Armstrong, R. L., 1972, Low angle (denudation) faults, hinterland of the Sevier orogenic belt, eastern Nevada and western Utah: Geological Society of America Bulletin, v. 83, p. 1729–1754.

Atwater, T., 1970, Implications of plate tectonics for the Cenozoic tectonic evolution of western North America: Geological Society of America Bulletin, v.81, p. 3513–3536.

Axen, G. J., and Wernicke, B. P., 1991, Comment *on* 'Tertiary extension and contraction of lower-plate rocks in the central Mojave metamorphic core complex, southern California': Tectonics, v. 10, p. 1084–1086.

Baldridge, W. S., Damon, P. E., Shafiqullah, M., and Bridwell, R. J., 1980, Evolution of the central Rio Grande rift, New Mexico: New potassium-argon ages: Earth and Planetary Science Letters, v. 51, p. 309–321.

Baldridge, W. S., Olsen, K. H., and Callender, J. F., 1984, Rio Grande rift: Problems and perspectives: New Mexico Geological Society, 35th Field Conference, Guidebook, p. 1–12.

Baldridge, W. S., Perry, F. V., and Shafiqullah, M., 1987, Late Cenozoic volcanism of the southeastern Colorado Plateau: I. Volcanic geology of the Lucero area, New Mexico: Geological Society of America Bulletin, v. 99, p. 463–470.

Barker, J. M., 1983, Preliminary investigation of the origin of the Riley travertine, Socorro County, New Mexico: New Mexico Geological Society, 34th Field Conference, Guidebook, p. 269–276.

Bartley, J. M., and Fletcher, J. M., 1990, Tertiary extension and contraction of lower-plate rocks in the central Mojave metamorphic core complex, southern California: Tectonics, v. 9, p. 521–534.

Bills, B. G., and May, G. M., 1987, Lake Bonneville: Constraints on lithospheric thickness and upper mantle viscosity from isostatic warping of Bonneville, Provo, and Gilbert stage shorelines: Journal of Geophysical Research, v. 92, p. 11493–11508.

Black, B. A., 1964, The geology of the northern and northeastern parts of the Ladron Mountains, Socorro County, New Mexico [M.S. thesis]: Albuquerque, University of New Mexico, 117 p.

Block, L., and Royden, L. H., 1990, Core complex geometries and regional scale flow in the lower crust: Tectonics, v. 9, p. 557–567.

Bohannon, R. G., 1983, Mesozoic and Cenozoic tectonic development of the Muddy, North Muddy, and northern Black Mountains, Clark County, Nevada: Geological Society of America Memoir 157, p. 125–148.

Bosworth, W., 1985, Geometry of propagating continental rifts: Nature, v. 316, p. 625–627.

Bosworth, W., Lambiase, J., and Keislar, R., 1986, A new look at Gregory's rift: The structural style of continental rifting: EOS, Transactions of the American Geophysical Union, v 67, p. 577, 582, 583.

Brown, L. D., Chapin, C. E., Sanford, A. R., Kaufman, S., and Oliver, J., 1980, Deep structure of the Rio Grande rift from seismic reflection profiling: Journal of Geophysical Research, v. 85, p. 4773–4800.

Buck, W. R., 1988, Flexural rotation of normal faults: Tectonics, v. 7, p. 959–973.

Burchfiel, B. C., Hodges, K. V., and Royden, L. H., 1987, Geology of Panamint Valley–Saline Valley pull-apart system, California: Palinspastic evidence for low-angle geometry of a Neogene range-bounding fault: Journal of Geophysical Research, v. 92, p. 10422–10426.

Cabezas, P., 1989, Etude géologique d'un segment des Montagnes Rocheuses meridionales de l'ouest des Etats-Unis: Stratigraphie et tectonique des deux bordures du rift du Rio Grande dans la region de Socorro [Ph.D. thesis]: France, Université de Nice, 209 p.

Callendar, J. F., and Zilinski, R. E., Jr., 1976, Kinematics of Tertiary and Quaternary deformation along the eastern edge of the Lucero uplift, central New Mexico: New Mexico Geological Society Special Publication 6, p. 53–61.

Cape, C. D., McGeary, S., and Thompson, G. A., 1983, Cenozoic normal faulting and the shallow structure of the Rio Grande rift near Socorro, New Mexico: Geological Society of America, v. 94, p. 3–14.

Cather, S. M., 1983, Lacustrine deposits of the Eocene Baca Formation, western Socorro County, New Mexico: New Mexico Geological Society, 34th Field Conference, Guidebook, p. 179–185.

Chaimov, T. A., 1988, A 3-D seismic modeling study of the Ladron horst near Socorro, New Mexico: Geophysical Research Letters, v. 15, p. 1207–1210.

Chamberlin, R. M., 1982, Preliminary evaluation of the mineral resource potential of the Sierra Ladrones wilderness study area, Socorro County, New Mexico: New Mexico Bureau of Mines and Mineral Resources Open-File Report, 141 p.

Chamberlin, R. M., 1983, Cenozoic domino-style crustal extension in the Lemitar Mountains, New Mexico: A summary: New Mexico Geological Society, 34th Field Conference, Guidebook, p. 51–53.

Chapin, C. E., 1988, Axial basins of the northern and central Rio Grande rifts, *in* Sloss, L. L., ed., Sedimentary cover—North American craton: Geological Society of America, Decade of North American Geology, v. D-2, p. 165–170.

Chapin, C. E., and Seager, W. R., 1975, Evolution of the Rio Grande rift in the Socorro and Las Cruces areas: New Mexico Geological Society, 26th Field Conference, Guidebook, p. 297–321.

Condie, K. C., 1976, Precambrian rocks of Ladron Mountains, Socorro County, New Mexico: New Mexico Bureau of Mines and Mineral Resources, Geologic Map 38, scale 1:24,000.

Cordell, L., 1978, Regional geophysical setting of the Rio Grande rift: Geological Society of America Bulletin, v. 89, p. 1073–1090.

Crossley, R., and Crow, M. J., 1980, The Malawi rift, *in* Geodynamic evolution of the Afro-Arabian rift system: Academia Nazional dei Lincei, no. 47, p. 77–87.

Davis, G. A., 1979, Problems of intraplate extensional tectonics, western United States, with special emphasis on the Great Basin: Rocky Mountain Association of Geologists, Basin and Range Symposium, p. 41–54.

Davis, G. A., and Burchfiel, B. C., 1973, Garlock fault: An intracontinental transform structure, southern California: Geological Society of America Bulletin, v. 84, p. 1407–1422.

Davis, G. A., Anderson, J. L., and Frost, E. G., 1980, Mylonitization and detachment faulting in the Whipple-Buckskin–Rawhide Mountains terrane, southeastern California and western Arizona: Geological Society of America Memoir 153, p. 79–129.

De Voogd, B., Serpa, L., and Brown, L., 1988, Crustal extension and magmatic processes: COCORP profiles from Death Valley and the Rio Grande rift: Geological Society of America Bulletin, v. 100, p. 1150–1567.

Dokka, R. K., 1981, Thin-skinned extensional tectonics, central Mojave Desert, California: Geological Society of America Abstracts with Programs, v. 13, p. 52.

Duschatko, R. W., 1953, Fracture studies in the Lucero uplift, New Mexico: Oak Ridge, Tennessee, U.S. Atomic Energy Commission RME 3072, Technical Information Service, 49 p.

Eardley, A. J., 1962, Structural geology of North America: New York, Harper and Row, 743 p.

Eaton, G. P., Wahl, R. R., Protska, H. J., Mabey, D. R., and Kleinkopf, M. D., 1978, Regional gravity and tectonic patterns: Their relation to late Cenozoic epeirogeny and lateral spreading in the Western Cordillera, *in* Smith, R. B., and Eaton, G. P., eds., Cenozoic tectonics and regional geophysics of the Western Cordillera: Geological Society of America Memoir 152, p. 51–92.

Ebinger, C. J., 1989, Geometric and kinematic development of border faults and accommodation zones, Kivu-Rusizi rift, Africa: Tectonics, v. 8, p. 117–133.

Forsyth, D. W., 1985, Subsurface loading and estimates of the flexural rigidity of continental lithosphere: Journal of Geophysical Research, v. 90, p. 1263–12632.

Gardner, J. N., and Goff, F., 1984, Potassium-argon dates from the Jemez volcanic field: Implications for tectonic activity in the north-central Rio Grande rift: New Mexico Geological Society, 35th Field Conference, Guidebook, p. 75–81.

Gibbs, A. D., 1984, Structural evolution of extensional basin margins: Geological Society of London Journal, v. 141, p. 609–620.

Goff, F., and 6 others, 1983, Water geochemistry of the Lucero uplift, New Mexico: A geothermal investigation of low-temperature mineralized fluids: Los Alamos, New Mexico, Los Alamos National Laboratory Report, 26 p.

Hamilton, W., 1988, Detachment faulting in the Death Valley region, California and Nevada: U.S. Geological Survey Bulletin, v. 1790, p. 51–85.

Hammond, C. M., 1987, Geology of the Navajo Gap area between the Ladron Mountains and Mesa Sarca, Socorro County, New Mexico: A structural analysis [M.S. thesis]: Socorro, New Mexico Institute of Mining and Technology, 212 p.

Hayden, S. N., and Mawer, C. K., 1991, Dextral transtensional deformation along the western margin of the Rio Grande rift, New Mexico: Geological Society of America Abstracts with Programs, v. 23, p. 31.

Hintze, L. F., 1986, Stratigraphy and structure of the Beaver Dam Mountains, southwestern Utah, *in* Griffin, D., and Phillips, W. R., eds., Thrusting and extensional structures and mineralization in the Beaver Dam Mountains, southwestern Utah: Utah Geological Association Publication 15, p. 1–36.

Jicha, H. L., Jr., 1958, Geology and mineral resources of Mesa del Oro Quadrangle, Socorro and Valencia Counties, New Mexico: New Mexico Bureau of Mines and Mineral Resources Bulletin 56, p. 67.

Kamb, W. B., 1959, Ice petrofabric observations from the Blue Glacier, Washington, in relation to theory and experiment: Journal of Geophysical Research, v. 64, p. 1891–1909.

Kelley, S. A., 1990, Cooling history of the flanks of the northern Rio Grande rift, Colorado and New Mexico, based on fission-track analysis: Geological Society of America Abstracts with Programs, v. 22, p. A226.

Kelley, V. C., 1952, Tectonics of the Rio Grande depression of central New Mexico: New Mexico Geological Society, 3rd Field Conference, Guidebook, p. 93–105.

Kelley, V. C., 1977, Geology of Albuquerque Basin, New Mexico: New Mexico Bureau of Mines and Mineral Resources Memoir, 60 p.

Kelley, V. C., 1979, Tectonics, middle Rio Grande rift, New Mexico, *in* Riecker, R. E., ed., Rio Grande rift: Tectonics and magmatism: Washington, D.C., American Geophysical Union, p. 57–70.

Kelley, V. C., and Clinton, N. J., 1960, Fracture systems and tectonic elements of the Colorado Plateau: University of New Mexico Publications in Geology, 106 p.

Kelley, V. C., and Wood, G. H., 1946, Geologic map of the Lucero uplift: United States Geological Survey Oil and Gas Investigations, Preliminary Map, scale 1:48,000.

King, B. C., 1978, Structural and volcanic evolution of the Gregory rift valley, *in* Bishop, W. W., ed., Geological background to fossil man, Recent research in the Gregory rift valley, East Africa: Scottish Academic Press, p. 29–54.

Kruse, S., McNutt, M., Phipps-Morgan, J., Royden, L., and Wernicke, B., 1991, Lithospheric extension near Lake Mead, Nevada: A model for ductile flow in the lower crust: Journal of Geophysical Research, v. 96, p. 4435–4456.

Kusznir, N. J., and Egan, S. S., 1990, Simple-shear and pure-shear models of extensional sedimentary basin formation: Application to the Jeanne d'Arc Basin, Grand Banks of Newfoundland: American Association of Petroleum Geologists Memoir 46, p. 305–322.

Kusznir, N. J., Marsden, G., and Egan, S. S., 1991, A flexural-cantilever simple-shear/pure shear model of continental lithosphere extension: Applications to the Jeanne d'Arc Basin, Grand Banks and Viking graben, North Sea, *in* Roberts, A. M., Yielding, G., and Freeman, B., eds., The geometry of normal faults: Geological Society of London Special Publication No. 56, p. 41–60.

Lewis, C. J., 1989, Crustal extension and shoulder uplift in the Rio Grande rift, New Mexico: Half-grabens and accommodation zones in the Ladron Peak area [M.S. thesis]: Los Angeles, California State University, 78 p.

Lisenbee, A. L., Woodward, L. A., and Connolly, J. R., 1979, Tijeras-Cañoncito fault system—A major zone of recurrent movement in north-central New Mexico: New Mexico Geological Society, 30th Field Conference, Guidebook, p. 89–99.

Lozinsky, R. P., 1988, Stratigraphy, sedimentology, and sand petrology of the Santa Fe Group and pre–Santa Fe Tertiary deposits in the Albuquerque Basin, central New Mexico [Ph.D.thesis]: New Mexico Institute of Mining and Technology, 235 p.

Machette, M. N., 1982, Quaternary and Pliocene faults in the La Jencia and southern part of the Albuquerque-Belen Basins, New Mexico: Evidence of fault history from fault-scarp morphology and Quaternary geology: New Mexico Geological Society, 33rd Field Conference, Guidebook, p. 161–169.

Massengill, G. L., 1979, Geology of Riley-Puertecito area, southeastern margin of Colorado Plateau, Socorro County, New Mexico [Ph.D. thesis]: El Paso, University of Texas, 301 p.

McCarthy, J., 1986, Reflection profiles from the Snake Range metamorphic core complex: A window into the mid-crust, *in* Barazangi, M., and Brown, L., eds., Reflection seismology: The continental crust: American Geophysical Union Geodynamics Series, v. 14, p. 281–292.

Moore, R. T., 1972, Geology of the Virgin and Beaver Dam Mountains, Arizona: Arizona Bureau of Mines Bulletin 186, 65 p.

Morley, C. K., 1988, Variable extension in Lake Tanganyika: Tectonics, v. 7, p. 785–801.

Morley, C. K., 1989, Extension, detachments, and sedimentation in continental rifts (with particular reference to east Africa): Tectonics, v. 8, p. 1175–1192.

Morley, C. K., Nelson, R. A., Patton, T. L., and Munn, S. G., 1990, Transfer zones in the East Africa rift system and their relevance to hydrocarbon exploration in rifts: American Association of Petroleum Geologists Bulletin, v. 74, p. 1234–1253.

Nakiboglu, S. M., and Lambeck, K., 1983, A reevaluation of the isostatic rebound of Lake Bonneville: Journal of Geophysical Research, v. 88, p. 10439–10447.

Olsen, K. H., Baldridge, W. S., and Callendar, J. F., 1987, Rio Grande rift: An overview: Tectonophysics, v. 143, p. 119–139.

Osburn, G. R., 1984, Geology of Socorro County: New Mexico Bureau of Mines and Mineral Resources Open-File Map 238, scale 1:200,000.

Personius, S. F., and Machette, M. N., 1984, Quaternary and Pliocene faulting in the Taos Plateau region, northern New Mexico: New Mexico Geological Society, 35th Field Conference, Guidebook, p. 83–90.

Prescott, W. H., Savage, J. C., and Kinoshita, W. T., 1979, Strain accumulation rates in the western United States between 1970 and 1978: Journal of Geophysical Research, v. 84, p. 5423–5435.

Reiter, M., Mansure, A. J., and Shearer, C., 1979, Geothermal characteristics of the Rio Grande rift with the Southern Rocky Mountain complex, *in* Riecker, R. E., ed., Rio Grande rift: Tectonics and magmatism: Washington, D.C., American Geophysical Union, p. 253–267.

Ron, H., Aydin, A., and Nur, A., 1986, Strike-slip faulting and block rotation in the Lake Mead fault system: Geology, v. 14, p. 1020–1023.

Rosendahl, B. R., 1987, Architecture of continental rifts with special reference to east Africa: Annual Review of Earth and Planetary Sciences, v. 15, p. 445–503.

Rosendahl, B. R., and 7 others, 1986, Structural expressions of rifting: Lessons from Lake Tanganyika, Africa, *in* Frostick, L. E., Renaut, R. W., Reid, I., and Tiercelin, J. J., eds., Sedimentation in the African rifts: Geological Society of London Special Publication 25, p. 29–43.

Russell, L. R., and Snelson, S., 1990, Structural style and tectonic evolution of the Albuquerque Basin segment of the Rio Grande rift, *in* Pinet, B., and Bois, C., eds., The potential of deep seismic profiling for hydrocarbon exploration: Paris, Editions Technip, p. 175–207.

Sanford, A. R., and 6 others, 1977, Geophysical evidence for a magma body in the crust in the vicinity of Socorro, New Mexico, *in* Heacock, J. G., ed., The earth's crust: Its nature and physical properties: Washington, D.C., American Geophysical Union, Geophysical Monograph 20, p. 385–403.

Sanford, A. R., Olsen, K. H., and Jaksha, L. H., 1979, Seismicity of the Rio Grande rift, *in* Riecker, R. E., ed., Rio Grande rift: Tectonics and magmatism: Washington, D.C., American Geophysical Union, p. 145–168.

Seager, W. R., 1970, Low-angle gravity glide structures in the Northern Virgin Mountains, Nevada and Arizona: Geological Society of America Bulletin, v. 81, p. 1517–1538.

Seager, W. R., Shafiqullah, M., Hawley, J. W., and Marvin, R. F., 1984, New K-Ar dates from basalts and the evolution of the southern Rio Grande rift: Geological Society of America Bulletin, v. 95, p. 87–99.

Smith, R., and Bruhn, R. L., 1984, Intraplate extensional tectonics of the eastern Basin-Range: Inferences on structural style from seismic reflection data, regional tectonics, and thermal-mechanical models of brittle-ductile behavior: Journal of Geophysical Research, v. 89, p. 5733–5762.

Snyder, W. S., Dickinson, W. R., and Silberman, M. L., 1976, Tectonic implications of space-time patterns of Cenozoic magmatism in the western United States: Earth and Planetary Science Letters, v. 32, p. 91–106.

Spencer, J. E., 1984, Role of tectonic denudation in warping and uplift of low-angle normal faults: Geology, v. 12, p. 95–98.

Tonking, W. H., 1957, Geology of Puertecito Quadrangle, Socorro County, New Mexico: New Mexico Bureau of Mines and Mineral Resources Bulletin 41, 67 p.

Walcott, R. I., 1970, Flexural rigidity, thickness, and viscosity of the lithosphere: Journal of Geophysical Research, v. 75, p. 3941–3954.

Wdowinski, S., and Axen, G. J., 1992, Isostatic rebound due to tectonic denudation: A viscous flow model of a layered lithosphere: Tectonics, v. 11, p. 303–315.

Weissel, J. K., and Karner, G. D., 1989, Flexural uplift of rift flanks due to mechanical unloading of the lithosphere due to extension: Journal of Geophysical Research, v. 94, p. 13919–13950.

Wernicke, B., 1981, Low-angle normal faults in the Basin and Range province: Nappe tectonics in an extending orogen: Nature, v. 291, p. 645–648.

Wernicke, B., 1985, Uniform-sense normal simple shear of the continental lithosphere: Canadian Journal of Earth Sciences, v. 22, p. 108–125.

Wernicke, B. P., and Axen, G. J., 1988, On the role of isostasy in the evolution of normal fault systems: Geology, v. 16, p. 848–851.

Wernicke, B., and Burchfiel, B. C., 1982, Modes of extensional tectonics: Journal of Structural Geology, v. 4, p. 105–115.

Wernicke, B., Axen, G. J., and Snow, J. K., 1988, Basin and Range extensional tectonics at the latitude of Las Vegas, Nevada: Geological Society of America Bulletin, v. 100, p. 1738–1757.

Wright, H. E., 1946, Tertiary and Quaternary geology of the lower Rio Puerco area, New Mexico: Geological Society of America Bulletin, v. 57, p. 383–456.

Wu, Z., 1986, Shallow structure of the southern Albuquerque Basin (Rio Grande rift), New Mexico, from COCORP seismic reflection data, *in* Barazangi, M., and Brown, L., eds., Reflection seismology: A global perspective: Washington, D.C., American Geophysical Union, Geodynamics Series, v. 14, p. 293–304.

Zandt, G., and Owens, T. J., 1980, Crustal flexure associated with normal faulting, and implications for seismicity along the Wasatch front, Utah: Seismological Society of America Bulletin, v. 70, p. 1501–1520.

Zilinski, R. E., Jr., 1976, Geology of the central part of the Lucero uplift, Valencia County, New Mexico [M.S. thesis]: Albuquerque, University of New Mexico, 69 p.

Zoback, M. L., Anderson, R. E., and Thompson, G. A., 1981, Cainozoic evolution of the state of stress and style of tectonism of the Basin and Range province of the western United States: Philosophical Transactions of the Royal Society of London, v. 300, p. 401–434.

MANUSCRIPT ACCEPTED BY THE SOCIETY OCTOBER 13, 1993

Printed in U.S.A.

Geological Society of America
Special Paper 291
1994

Stratigraphic consequences of episodic extension in the Lemitar Mountains, central Rio Grande rift

S. M. Cather, R. M. Chamberlin, C. E. Chapin, and W. C. McIntosh
New Mexico Bureau of Mines and Mineral Resources, New Mexico Institute of Mining and Technology, Campus Station, Socorro, New Mexico 87801

ABSTRACT

Detailed stratal tilt information for radioisotopically dated Tertiary rocks in the Lemitar Mountains of the central Rio Grande rift allows discrimination of two episodes of extension in the late Oligocene (largely 28.6 to 27.4 Ma) and middle to late Miocene (16 to 10 Ma). Rapid deformation in the late Oligocene (at least 25% extension) produced narrow, wedge-shaped accumulations of ash-flow tuffs and mafic lavas within half grabens defined by closely spaced (2 to 5 km) planar-rotational normal faults. Volcanic accumulation rates during this episode were sufficient to bury the developing fault-block topography, resulting in preservation of only minor amounts of intercalated sedimentary rocks.

Following a period of weak extension in the early Miocene, rates of extensional strain again began to increase in the early middle Miocene (about 16 Ma) and continued to increase until about 10 Ma, when a more moderate rate ensued. During the time interval of 16 to 10 Ma, the Lemitar Mountains area was extended about 30% along widely spaced (5 to 15 km) planar-normal faults. Increased stratal-tilt rates in the middle Miocene caused the development of accommodation space to locally exceed sediment supply on the down-thrown side of a major intrabasinal fault (Silver Creek fault), resulting in topographic closure and the development of playa deposits within the lower Santa Fe Group (Popotosa Formation). As stratal-tilt rates further increased in the late Miocene, fluvial-lacustrine sedimentation offlapped toward narrow zones of maximum subsidence along active faults, causing the carving of widespread unconformities on older sediments on hanging-wall dip slopes. As tectonism waned in the latest Miocene–early Pliocene, sediment supply exceeded the development of accommodation space, resulting in onlap and aggradation of the upper Santa Fe Group (Sierra Ladrones Formation) that enabled the eventual integration of an axial-fluvial system (ancestral Rio Grande) between the longitudinally arrayed half grabens to the north and south of the Socorro Basin.

The late Oligocene episode of rapid extension appears to be related to strain focusing by local thermal weakening due to voluminous concurrent magmatism in the area. In contrast, the late Miocene extensional event may be of regional scale, as shown by evidence for contemporaneous tectonism in other areas and the development of widespread unconformities between Miocene and Pliocene deposits of the Santa Fe Group in many basins of the Rio Grande rift.

Cather, S. M., Chamberlin, R. M., Chapin, C. E., and McIntosh, W. C., 1994, Stratigraphic consequences of episodic extension in the Lemitar Mountains, central Rio Grande rift, *in* Keller, G. R., and Cather, S. M., eds., Basins of the Rio Grande Rift: Structure, Stratigraphy, and Tectonic Setting: Boulder, Colorado, Geological Society of America Special Paper 291.

INTRODUCTION

The Rio Grande rift is a northward-tapering area of late Cenozoic lithospheric extension in Colorado, New Mexico, west Texas, and northern Chihuahua (Seager and Morgan, 1979). The northern rift consists of a well-defined, narrow zone of extension that broadens near Socorro, New Mexico, and merges southward with the southern Basin and Range province. This study documents relationships between extensional deformation and stratigraphy in the Lemitar Mountains (Fig. 1), a desert range near Socorro that has been mapped in detail and within which the history of Tertiary volcanism, deformation, and sedimentation is relatively well studied and constrained by numerous radioisotopic ages.

The term "Lemitar Mountains" is herein used in the broad sense to designate the area between the La Jencia and Socorro Basins that is bounded to the north by the Rio Salado and to the south by the Socorro accommodation zone (Fig. 1). The Lemitar Mountains lie within the west-tilted array of half grabens and tilt blocks to the north of the horizontal to gently dipping strata of the Socorro accommodation zone (Chapin et al., 1978; Chapin, 1988). The Lemitar Mountains represent the structurally high, eastern part of the La Jencia Basin half-graben block, which is defined by major down-to-the-east faulting along the La Jencia–Cerro Colorado fault zone and the Socorro Canyon–Loma Peleda fault zone (Fig. 1). Down-to-the-east faulting along the Silver Creek fault repeats the section throughout much of the range. The central portion of the Lemitar Mountains is structurally higher than adjoining areas to the north and south, thus affording a perspective of differing stratigraphic and structural levels within the range.

CENOZOIC STRATIGRAPHY

Lithostratigraphic units

Cenozoic rocks of the Lemitar Mountains (Fig. 2) can be broadly divided into three stratigraphic sequences: (1) Eocene synorogenic deposits related to late Laramide crustal shortening (Baca Formation), (2) upper Eocene to upper Oligocene volcanic and volcaniclastic rocks that spanned the transition from waning Laramide shortening to mid-Tertiary extension, and (3) uppermost Oligocene(?) to lower Pleistocene syn-rift deposits and minor volcanics of the Santa Fe Group.

Conglomeratic beds of the Eocene Baca Formation fill broad paleocanyons incised into Pennsylvanian strata of the Laramide Sierra uplift (Chamberlin, 1980; Cather, 1983). Pebble imbrication shows evidence for westward paleotransport, toward the nearby Baca basin (Cather and Johnson, 1984, 1986). The Baca Formation is transitionally overlain by volcaniclastic deposits of the lower Datil Group (40 to 36 Ma) that record andesitic volcanism during the waning stage of Laramide crustal shortening (Cather, 1986, 1990a). The onset of bimodal (basaltic andesite-rhyolite) volcanism at about 36 Ma coincided with the beginning of weak crustal extension (Cather, 1990a). Major ignimbrite volcanism and extrusion of mafic to intermediate lavas continued for about 10 m.y. until largely supplanted by syn-rift sedimentation (Santa Fe Group) in the latest Oligocene–early Miocene.

The Santa Fe Group in the Socorro area consists of three distinct sedimentary facies associations with intercalated minor bimodal volcanic rocks (Osburn and Chapin, 1983; Cather et al., 1991), all of which were deposited during a semiarid paleoclimatic regime. In ascending stratigraphic order, these facies associations are:

1. Basal debris-flow deposits of the Popotosa Formation. These debris-flow and minor fluvial deposits consist of well-indurated volcaniclastic detritus derived in part from the Socorro accommodation zone, which was topographically positive throughout Popotosa time (Cather et al., 1991). Within the Lemitar Mountains, the basal debris-flow deposits of the Popotosa thicken to the south. These deposits have not been directly dated; their age is bracketed in the Lemitar Mountains between the 27.37 ± 0.07 Ma South Canyon Tuff (McIntosh et al., 1991) and the 16.2 ± 1.5 Ma Silver Creek andesite (Weber, 1971).

2. Fluviolacustrine deposits of the main body of the Popotosa Formation. These are the volumetrically dominant and the lithologically most diverse of the syn-rift deposits, ranging texturally from boulder conglomerate in proximal piedmont areas to thick playa mudstone near the basin center (Denny, 1940; Bruning, 1973; Chapin and Seager, 1975; Asher-Bolinder, 1988). Paleoflow indicators and clast-composition data (Chamberlin, 1980, 1982; Cather and Kugler, in preparation) show the main-body Popotosa deposits in the Lemitar Mountains were derived largely from the footwall block of the La Jencia–Cerro Colorado fault (i.e., the Magdalena Mountains, Bear Mountains, and Ladron Mountains areas) and areas farther to the west. The Lemitar Mountains were uplifted through the floor of the Popotosa basin in latest Miocene–early Pliocene time (Bruning, 1973; Chapin and Seager, 1975; Chamberlin, 1983), thus affording high-quality exposures of early rift deposits in the now-uplifted range. Numerous isotopic ages from intercalated ashes and lavas indicate deposition of the main body of the Popotosa occurred dominantly during middle and late Miocene time, although the basal part of the unit may be as old as latest Oligocene or early Miocene. A thick, unfaulted exposure of the Popotosa west of the Silver Creek fault near the Rio Salado shows a well-defined "fanning" of westerly dips, with the oldest beds inclined 30 to 35° and the youngest beds dipping only 15 to 20° (Cather and Kugler, in preparation).

3. Axial-fluvial and piedmont deposits of Sierra Ladrones Formation. Deposits of the ancestral Rio Grande in this unit mark the development of an integrated axial drainage within the Rio Grande rift beginning in the early Pliocene (Machette, 1978a, b). Associated piedmont deposits accumulated in a physiographic framework of basins and uplifts similar to that

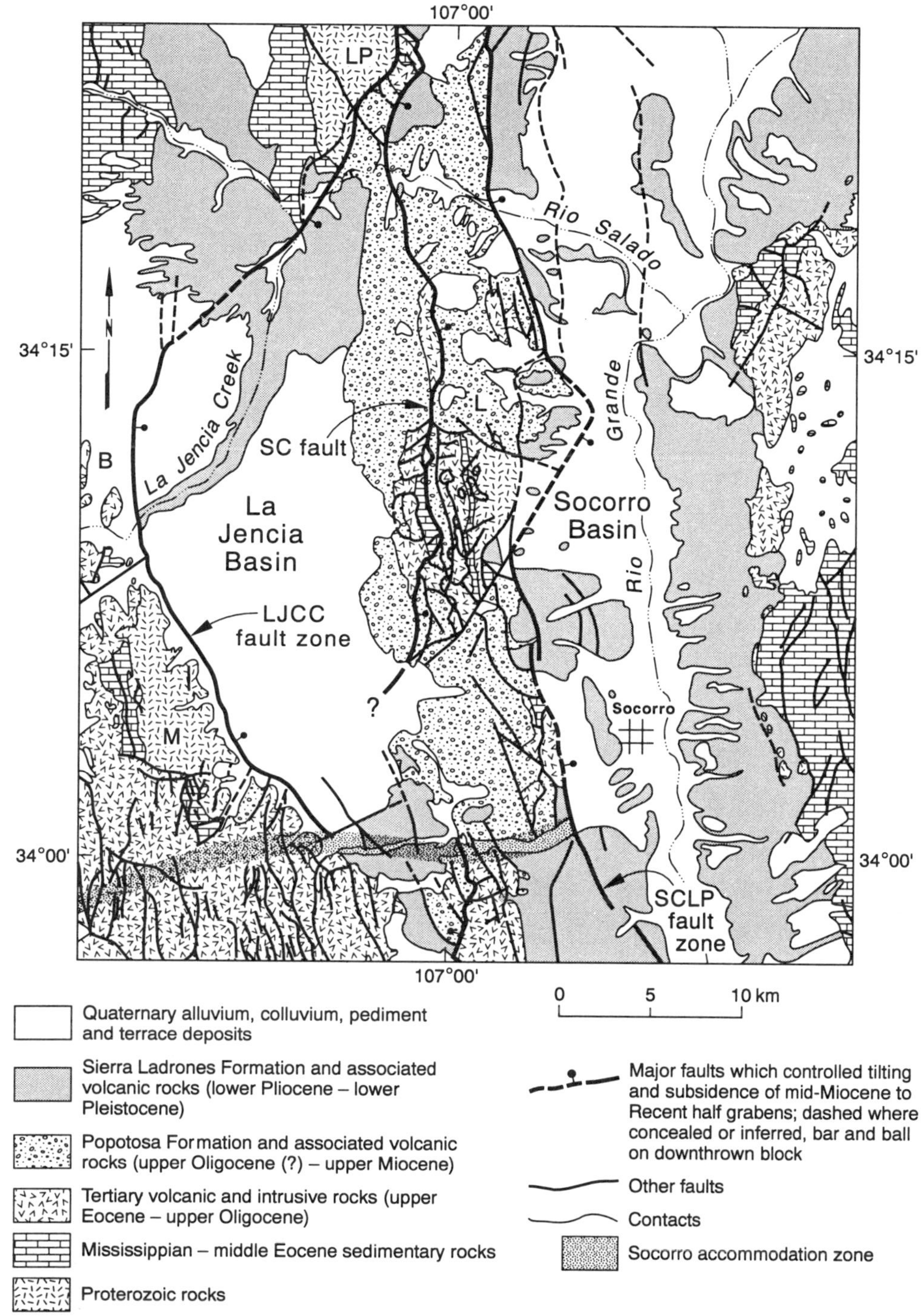

Figure 1. Simplified geologic map of the study area showing outcrop relations and selected structural features (modified from Osburn, 1984). Abbreviations are Ladron Peak (LP), Bear Mountains (B), Magdalena Mountains (M), Lemitar Mountains (L), La Jencia–Cerro Colorado (LJCC) fault zone, Silver Creek (SC) fault, Socorro Canyon–Loma Peleda (SCLP) fault zone.

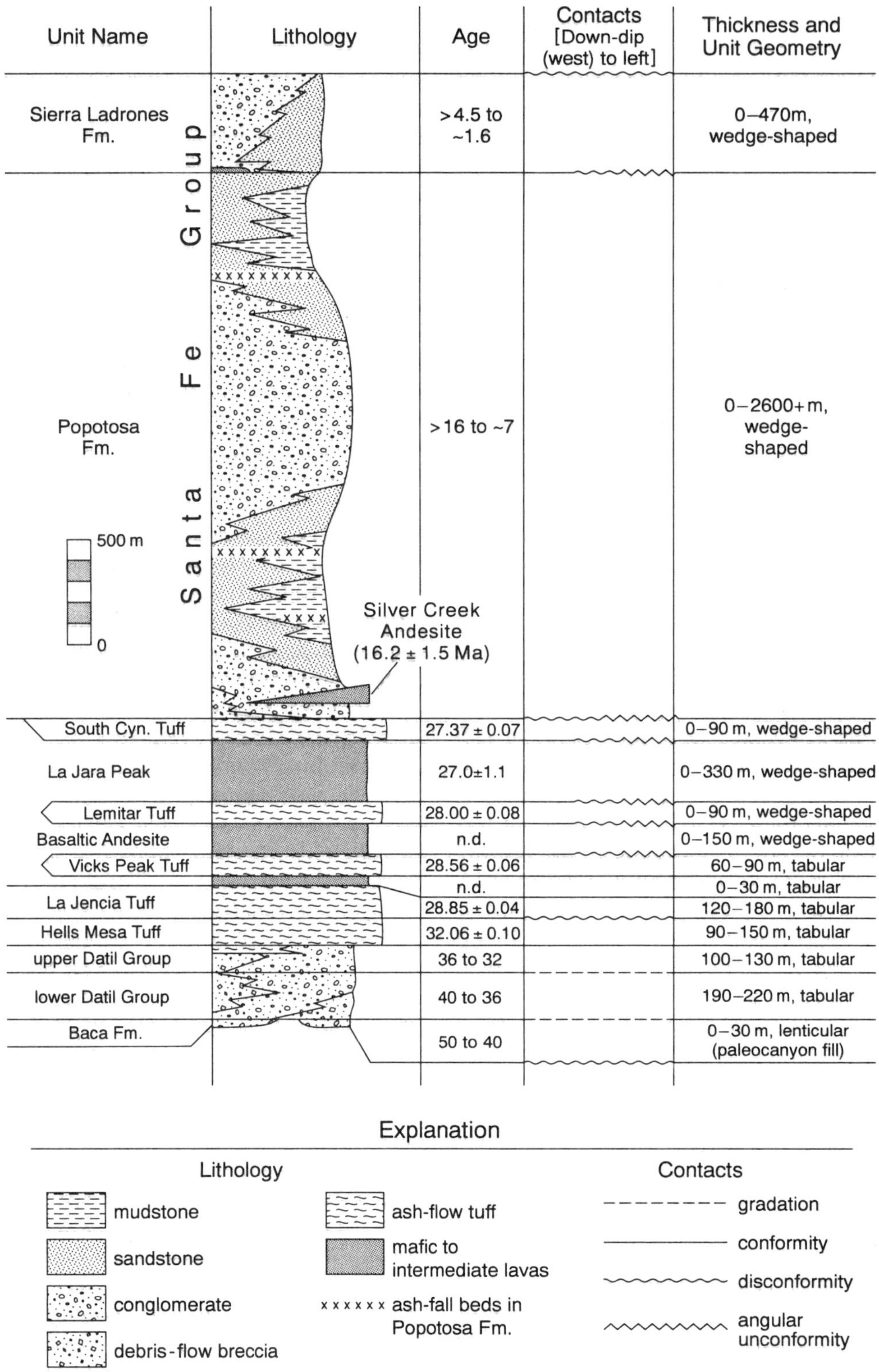

Figure 2. Stratigraphic column for Cenozoic rocks in Lemitar Mountains area showing stratigraphic nomenclature (Osburn and Chapin, 1983), lithology, age constraints (see text and Table 1), contact relations, and unit thickness and geometry (Chamberlin, 1983; Cather, 1986; Cather and Kugler, in preparation).

of today. The end of Santa Fe deposition in the Socorro Basin appears to be marked by the initial incision of the Rio Grande about 1.6 Ma, shortly prior to the eruption of the lower Bandelier Tuff (Cather and McIntosh, 1990).

Contact relations and unit geometries

A wide variety of contact relations exist among the Tertiary stratigraphic units of the Lemitar Mountains (Fig. 2). Most of the Tertiary units older than about 28.6 Ma exhibit tabular (sheetlike) geometries and are separated by conformable, commonly gradational, contacts. The principal exception is the disconformable contact between the Hells Mesa Tuff and the overlying La Jencia Tuff, which represents a widespread hiatus in ignimbrite volcanism in the Mogollon-Datil field (McIntosh et al., 1991, 1992). No evidence for angularity, however, has been noted between the Hells Mesa Tuff and the La Jencia Tuff.

Two principal episodes of development of major disconformable to angular unconformable contacts between Tertiary units occurred in the Lemitar Mountains. The first episode occurred in response to rapid rotation of narrow half grabens of late Oligocene–early Miocene(?) age in the Lemitar Mountains area, as evidenced by widespread development of narrow, wedge-shaped, unconformity-bounded accumulations of volcanic and sedimentary rocks (Chamberlin, 1983). Development of these wedge-shaped stratigraphic units and corresponding unconformities began with eruption of the Vicks Peak Tuff (28.56 ± 0.06 Ma; McIntosh et al., 1991) and ceased in most areas prior to, or during, deposition of basal debris flows of the Popotosa Formation (after deposition of the South Canyon Tuff [$^{40}Ar/^{39}Ar$ age 27.37 ± 0.07 Ma; McIntosh et al., 1991], but before eruption of the Silver Creek andesite [K/Ar age 16.2 ± 1.5 Ma; Weber, 1971; Fig. 2]). A distinct unconformity with as much as 20° of angularity is present at the base of the Popotosa Formation in the southern Lemitar Mountains. This unconformity grades to a disconformity in the northern Lemitar Mountains. The contact between the basal debris-flow deposits and the main body of the Popotosa is generally gradational in the northern part of the Lemitar Mountains (Cather and Kugler, in preparation), but is commonly disconformable to the south (Chamberlin, 1983).

The remaining major unconformity within the Tertiary strata of the Lemitar Mountains area is of late Miocene age and separates the Popotosa Formation from the overlying Sierra Ladrones Formation. This unconformity is typically angular, exhibiting as much as 10 to 20° of angular discordance over broad areas of the modern rift basins, but appears to grade to conformity in areas of maximum subsidence near major faults that bound the half grabens. Such a conformable sequence between the Popotosa and Sierra Ladrones is exposed east of the Silver Creek fault where it crosses the Rio Salado in the northern part of the study area (Fig. 1; Cather and Kugler, in preparation). The Popotosa and Sierra Ladrones Formations both show wedge-shaped geometries in the Lemitar Mountains area because of the effect of tilting and subsidence within the developing half grabens of the rift. These wedge-shaped units, however, are broader than those developed earlier during the late Oligocene episode of extension, due to an increase in the spacing of half-graben-bounding faults in the early Miocene (see below).

STRUCTURE

During the Eocene, stratigraphic and sedimentologic evidence indicate a large late Laramide uplift (Sierra uplift) encompassed much of the present areas of the Socorro Basin, Lemitar Mountains, and possibly the La Jencia Basin. Precambrian rocks were at least locally exposed on the Sierra uplift, which formed an orogenic highland between the late Laramide Baca and Carthage–La Joya basins to the west and east, respectively (Cather, 1983; Cather and Johnson, 1984, 1986). The Sierra uplift persisted during early Datil Group andesitic volcanism, but began to collapse during incipient extension about 36 Ma, coincident with the change to bimodal volcanism (Cather, 1990a). To the west of the study area in west-central New Mexico, this structural inversion was facilitated by displacement reversal along major fault zones (Cather, 1989, 1990a). The collapse of the Sierra uplift in the Lemitar Mountains vicinity may have also occurred by normal slip along former Laramide reverse faults. Evidence for Laramide precursors to the major fault zones depicted in Figure 1, however, is largely circumstantial due to the scarcity of exposure of Laramide deposits in the study area.

Rapid extension and pronounced stratal tilting began in the Lemitar Mountains about 28.6 Ma, although they were reported by Chamberlin (1983) to have begun about 31 Ma based on imprecise K/Ar ages for the Vicks Peak Tuff (Osburn and Chapin, 1983). Unconformity-bounded, wedge-shaped accumulations of volcanic and sedimentary rocks between closely spaced (2 to 5 km) normal faults provide strong evidence for development of narrow half grabens during the late Oligocene. Faulting during this early phase of rapid extension in the Lemitar Mountains was domino style, with older faults progressively flattening and being cut by younger faults as deformation proceeded (Chamberlin, 1983).

Half-graben width increased to 5 to 15 km by the early or middle Miocene, although age constraints are broad due to lack of dates from the basal Popotosa Formation. The increased spacing of half-graben-bounding faults through time may reflect the deepening of the brittle-ductile transition following the waning of major mid-Tertiary magmatism (Chamberlin and Osburn, 1986; Reiter et al., 1992). High-angle faults initiated during Miocene extension locally cut older faults that had rotated to moderate or low dips during the late Oligocene phase of extension (Chamberlin, 1983).

Major normal faults that controlled development of the broad, middle Miocene half grabens now constitute the modern range-bounding faults in the study area. These faults were initially high angle and most were not strongly listric as shown by lack of significant differences in stratal tilting across them. Except in areas where they cut unconsolidated sediments, most modern range-bounding faults have moderate dips (25 to 40°) due to the cumulative effects of block rotation since the middle Miocene. Geologic mapping in the northern part of the study area (Machette, 978a; Cather and Kugler, in preparation) indicates the internal deformation within these fault blocks is minor, with most faulting and folding confined to zones along the margins of the blocks.

Although sedimentologic data indicate that the footwall block of the La Jencia–Cerro Colorado fault (the Magdalena Mountains, Bear Mountains, and Ladron Peak) was a source area for Popotosa detritus throughout much of the Miocene, the Lemitar Mountains did not emerge through the floor of the greater Popotosa basin (e.g., Chapin and Seager, 1975) until after deposition of the Popotosa Formation ceased in the late Miocene. Facies relations and the restriction of playa deposits to the hanging-wall block of the Silver Creek fault (see below) indicates this fault was active as an intrabasinal growth fault during Popotosa deposition, and locally evolved into a mountain-front fault during late Miocene to early Pliocene time. A similar evolution is possible for the Socorro Canyon–Loma Peleda fault, but scarcity of exposure and lack of borehole data for Popotosa strata to the east of this fault preclude definitive analysis.

STRATAL TILTING THROUGH TIME

The geology of the Lemitar Mountains area has been mapped in detail, at scales ranging from 1:24,000 to 1:12,000 (Machette, 1978a; Chamberlin, 1980, 1982; Cather and Kugler, in preparation, Cather et al., in preparation). Tertiary strata exposed in the Lemitar Mountains area are well constrained by numerous isotopic ages, including many precise $^{40}Ar/^{39}Ar$ data (Table 1). Because numerous strike-and-dip data are available for radiometrically dated units, it is possible to quantitatively describe the stratal-tilt history (Fig. 3) of the Lemitar Mountains area. Figure 3 represents a new method by which to evaluate the strain history of part of the Rio Grande rift (see also the similar analysis by Ferguson, 1991). It is free from the vagaries of interpretation inherent in stratigraphic analysis and, because of the small relative error of the $^{40}Ar/^{39}Ar$ constraints, it is chronologically more precise than traditional structural methods or apatite fission-track cooling analysis.

Neogene extensional strain in the Socorro area of the Rio Grande rift averaged about 50% (Chamberlin and Osburn, 1984; Chapin, 1988). Extensional strain varied locally, however; in the Lemitar Mountains it was about 200% (Chamberlin, 1983; Chamberlin and Osburn, 1986). To diminish the effects of the regional variability of Neogene extensional strain and associated magnitude of half-graben tilting, stratal-tilt data in Figure 3 are derived from the Lemitar Mountains and adjacent areas within the La Jencia and Socorro Basins. The sole exception is for the youngest data point in Figure 3, which represents the San Antonio pumice in the southern part of the Socorro Basin (Table 1; Cather and McIntosh, 1990). The San Antonio pumice is included because it is the only radioisotopically dated deposit in the Socorro area that constrains the end of Santa Fe deposition. Stratal-tilt statistics (mean, standard deviation, and mode) were derived from geologic maps by compilation of strike-and-dip data from within dated volcanic units or ash beds, and from conformable strata directly above or below these ash beds. Because few dip data are available for upper Miocene and younger units in the study area, statistical treatment was not attempted and only the range of stratal tilts for these units are shown (dashed vertical lines in Fig. 3). Bedding-orientation data were compiled from within the major tilt blocks of the Lemitar Mountains area. Data were excluded from our compilation in areas where: (1) bedding has been disturbed by drag along faults or in local areas of anomalous eastward tilting within the study area, or (2) anomalously low stratal tilts are present on or near the Socorro accommodation zone.

The radioisotopic age constraints in Figure 3 consist of numerous $^{40}Ar/^{39}Ar$ and K/Ar ages, and one fission-track (zircon) age (Table 1). Except for the previously described entry for the San Antonio pumice, all of the post-Oligocene dates (i.e., those for the Popotosa and Sierra Ladrones Formations) are from the Lemitar Mountains area. Radioisotopic dates for the upper Eocene–Oligocene volcanic rocks depicted in Figure 3 are from units of regional extent that have been dated in nearby ranges and can be confidently correlated with units in the study area using criteria such as paleomagnetism (McIntosh and others, 1991), stratigraphic position, and petrologic characteristics (Osburn and Chapin, 1983). The horizontal brackets in Figure 3 indicate on standard deviation about the mean age of individual dated units. The only exception is for the oldest unit (the Datil Group) where the dashed horizontal line indicates the range of ages for numerous dated tuffs and conglomerate clasts from the unit in adjacent areas (see Cather et al., 1987). We have omitted the 27.0 ± 1.1 Ma La Jara Peak Basaltic Andesite of Machette (1978a) from our compilation because the dated sample is from a locality where strata exhibit anomalously low tilts due to their position on a large drag(?) fold in the footwall of the Loma Peleda fault. In other parts of the study area, however, the La Jara Peak Basaltic Andesite has stratal tilts similar to those of the dated ash-flow tuffs with which it interfingers (Chamberlin, 1980, 1982).

Not only do the data in Figure 3 depict an increase in stratal tilt with increasing stratal age that is an expectable consequence of accumulation of extensional strain by rotating half grabens, but a stepwise pattern is also apparent. The steep portions of this tilt-versus-time plot represent episodes of rapid extension and resultant rapid tilting of strata in the late Oligocene and middle to late Miocene. Both of these tectonic pulses correspond temporally with the development of major, commonly angular, unconformities in the Lemitar Mountains

TABLE 1. SELECTED RADIOISOTOPIC AGE CONSTRAINTS FOR TERTIARY UNITS IN LEMITAR MOUNTAINS AREA

Unit	Age (Ma)	Method	Reference	Comments
San Antonio pumice	1.64 ± 0.16	$^{40}Ar/^{39}Ar$ sanidine	This chapter	Pumiceous debris-flow deposit in southern Socorro Basin, correlated with eruption of lower Bandelier Tuff based on $^{40}Ar/^{39}Ar$ dating and paleomagnetism. Inset against upper part of Sierra Ladrones Formation.
Basalt of San Acacia	4.5 ± 0.1	K/Ar	Machette, 1978a Bachman and Mehnert, 1978	Interbedded with lower part of Sierra Ladrones Formation.
Basalt of Sedillo Hill	7.10 ± 0.18 7.15 ± 0.16	K/Ar, feldspar concentrate	This chapter	Interbedded between Popotosa and Sierra Ladrones Formations in sec. 14,T3S,R2W. Average of dates shown on Figure 3.
Socorro Peak Rhyolite	10.7 ± 0.8 11.5 ± 1.0	K/Ar (whole rock) K/Ar (whole rock)	Burke et al., 1963 Osburn and Chapin, 1983	Interbedded within main body of Popotosa Formation on Socorro Peak.
Ash-fall deposit	11.9 ± 0.8	Fission track (zircon)	K. Manley, unpublished data	Interbedded within main body of Popotosa Formation in SE/4, NW/4, sec.20,T2S,R1W.
Ash-fall deposit	14.55 ± 0.04	$^{40}Ar/^{39}Ar$ (biotite)	This chapter	Interbedded within main body of Popotosa Formation in SW/4,NE/4, SE/4,sec.21,T1N,R2W (unsurveyed; see Cather and Kugler, in preparation). Plateau age.
Ash-fall deposit	15.20 ± 0.05	$^{40}Ar/^{39}Ar$ (biotite)	This chapter	Interbedded within main body of Popotosa Formation about 2.3 km south-southwest of BM 5032 in southwest part of San Acacia 7.5' Quadrangle (see Machette, 1978a). Ash is within fluvial deposits which are approximately laterally equivalent to basal playa beds near Rio Salado. Plateau age.
Ash-fall deposit	15.54 ± 0.28	$^{40}Ar/^{39}Ar$ (biotite)	This chapter	Interbedded within main body of Popotosa Formation in SE/4,NE/4, NW/4,sec.22,T1N,R2W (unsurveyed, see Cather and Kugler, in preparation). $^{40}Ar/^{39}Ar$ spectrum failed to meet plateau criteria; reported age is average and standard deviation of heating steps containing bulk of ^{39}Ar.
Ash-fall deposit	15.62 ± 0.07	$^{40}Ar/^{39}Ar$ (biotite)	This chapter	Interbedded within main body of Popotosa Formation about 100 m above base of section in NW/4,NE/4,sec.23,T2S,R2W. Total gas age.
Silver Creek andesite	16.2 ± 1.5	K/Ar (whole rock)	Weber, 1971	Interbedded within lower part of main body of Popotosa Formation.
South Canyon Tuff	27.37 ± 0.07	$^{40}Ar/^{39}Ar$ (sanidine)	McIntosh et al., 1991	Average of three plateau ages.
Lemitar Tuff	28.00 ± 0.08	$^{40}Ar/^{39}Ar$ (sanidine)	McIntosh et al., 1991	Average of two plateau ages.
Vicks Peak Tuff	28.56 ± 0.06	$^{40}Ar/^{39}Ar$ (sanidine)	McIntosh et al., 1991	Average of three plateau ages. K/Ar date of 31.3 Ma was reported by Osburn and Chapin (1983) and cited by Chamberlin (1983).
La Jencia Tuff	28.85 ± 0.04	$^{40}Ar/^{39}Ar$ (sanidine)	McIntosh et al., 1991	Average of four plateau ages.
Hells Mesa Tuff	32.06 ± 0.10	$^{40}Ar/^{39}Ar$ (sanidine)	McIntosh et al., 1991	Average of four plateau ages.
Datil Group	40 to 32	Various techniques	See summary in Cather et al., 1987.	Age range for numerous dates from clasts and ash-fow tuffs.

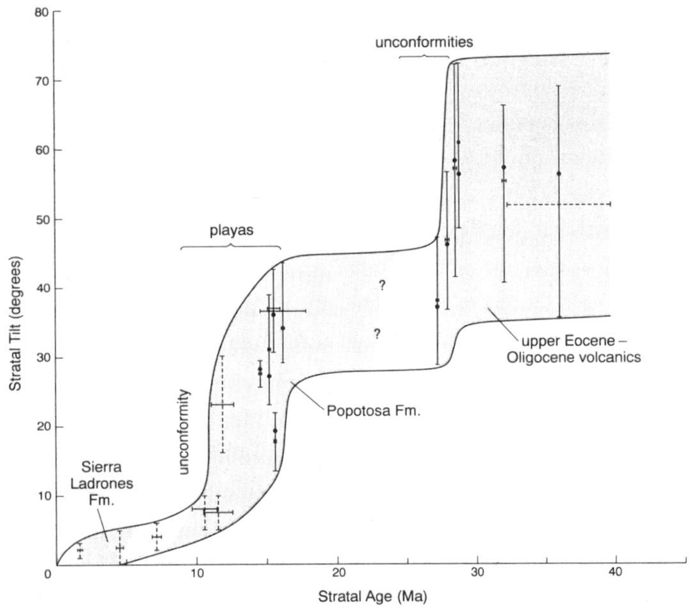

Figure 3. Plot of stratal tilt versus stratal age (Table 1) for post–40-Ma rocks in the Lemitar Mountains area. Brackets indicate one standard deviation about the mean age and mean tilt for dated units or beds. Dashed lines indicate ranges (see text for discussion). Dots are modes for tilt data. Note association between episodes of rapid tilting (steep portions of tilt-versus-age envelope) with development of interformational unconformities and playa-lake sedimentation in the Popotosa Formation.

area, and the younger event may also relate to the development of major playa-lake deposits in the Popotosa Formation during the middle and late Miocene.

STRATIGRAPHIC CONSEQUENCES OF EPISODIC EXTENSION

Two major episodes of rapid extension occurred in the Lemitar Mountains area during the Tertiary. Both of these episodes produced broadly similar amounts of stratal tilting (each caused about 20 to 30° of stratal rotation; Fig. 3). Half-graben width and nature of the basin fill, however, differed considerably between the two deformations.

Prior to the late Oligocene, extensional deformation was not sufficient to produce systematic differences in tilt among the volcanic units of the Lemitar Mountains (Fig. 3). The first episode of strong Tertiary extension and accompanying rapid stratal rotations began about 28.6 Ma in the Lemitar Mountains, and lasted for at least 1.2 m.y. (Fig. 3, Table 1). The early part of this deformation (28.6 to 27.4 Ma) is well constrained by differences in tilt between precisely dated ash-flow tuffs. During this time interval the Lemitar Mountains region was extended about 25% (calculated using method of Thompson, 1960; and Wernicke and Burchfiel, 1982), corresponding to an average strain rate of about 7×10^{-15}/sec. Total extension prior to basal Popotosa sedimentation (about 15.6 Ma, Table 1) in the central Lemitar Mountains, however, was as much as 90% (Chamberlin, 1983, fig. 5a–c), although some of this strain may be related to the early part of the middle to late Miocene deformation.

Topographic relief produced by rapid rotation of narrow tilt blocks during the late Oligocene was largely buried beneath wedge-shaped accumulations of upper Oligocene ash-flow tuffs and lavas. The accumulation rate of volcanic rocks was apparently sufficient to mask the developing half-graben topography, as is attested by the scarcity of interbedded sedimentary rocks. The early phase of rapid extension in the Lemitar Mountains ended prior to the beginning of Popotosa deposition in the northern part of the range, but basal Popotosa strata are locally influenced by early, closely spaced faults in the central part of the range. Structural and stratigraphic aspects of the late Oligocene extensional episode are thoroughly discussed by Chamberlin (1978, 1983) and will not be repeated here.

Despite the paucity of dated units of early Miocene age in the study area, similar amounts of tilting between the youngest dated ash-flow tuff below the Popotosa Formation and the oldest dated rocks within the Popotosa indicate little extensional deformation occurred in the Lemitar Mountains during the early Miocene (Fig. 3). The principal strata deposited during this episode of weak extension are the debris-flow breccias of the basal Popotosa Formation, which are bracketed within the broad interval of 27.4 to 16.2 Ma (see Cather et al., 1991, for discussion of age constraints in other areas).

Within the southern and central Lemitar Mountains, the debris-flow deposits of the basal Popotosa Formation are largely confined to north-trending strike valleys related to (or residual from) the narrow tilt blocks of the preceding deformation. These deposits are commonly bounded above and below by unconformities. Southward thickening and evidence for northerly paleoflow indicate the debris-flow and related fluvial deposits were derived largely from a highland along the Socorro accommodation zone. The accommodation zone continued to be topographically positive during middle Miocene time as shown by time-transgressive regional stratigraphic relations within the main body of the Popotosa (Cather et al., 1991).

In the northern Lemitar Mountains, debris-flow deposits of the basal Popotosa Formation fill *dip-oriented* paleocanyons incised into upper Oligocene volcanic rocks (Cather and Kugler, in preparation). In contrast to similar rocks to the south, these canyon-filling deposits typically grade upward into the main body of the Popotosa and show evidence of *westward* paleotransport. Such paleocanyon-fill deposits are present directly west of both the Silver Creek and Loma Peleda faults in the northern Lemitar Mountains. These deposits record aggradation within west-draining paleocanyons incised into the hanging-wall dip slopes of the developing broad, mid-Miocene half grabens, shortly prior to onlap of the basin-floor and east-facing piedmont deposits that characterize the bulk of the main body of the Popotosa. Given the broad constraints on timing, it is possible that the canyon-filling debris-flow deposits in the northern Lemitar Mountains may be genetically unrelated to the lithologically similar strike-valley-filling deposits to the south; the latter may be related to the final stages of rotation of narrow tilt blocks, whereas the former more likely were deposited in response to initial development of the broad, mid-Miocene half grabens. More work is needed to constrain the age and genesis of these deposits.

Rates of extensional strain again began to increase in the Lemitar Mountains during the early middle Miocene (about 16 Ma) and continued to increase until about 10 Ma, when a more moderate rate ensued. Apatite fission-track data (Kelley et al., 1992) for Proterozoic footwall rocks on Ladron Peak and in the Lemitar Mountains also indicate an episode of rapid uplift and cooling during the middle and late Miocene. Average extension in the Lemitar Mountains during the time interval 16 to 10 Ma was about 30%. This corresponds to a strain rate of about 1×10^{-15}/sec during the middle to late Miocene event, significantly slower than the preceding late Oligocene deformation. In contrast to the earlier extensional event, the middle to late Miocene deformation invoked a largely sedimentary response (i.e., deposition of fluvial-lacustrine sediments of the main body of the Popotosa Formation and development of its superjacent unconformity); volcanism was relatively minor.

Much has been written during recent years about the effects of tectonic tilting on the distribution of sedimentary facies within asymmetrical basins (e.g., Leeder and Gawthorpe, 1987; Alexander and Leeder, 1987; Blair and Bilodeau, 1988; Heller et al., 1988; Mack and Seager, 1990). A consensus seems to be developing that episodes of tectonic tilting within continental basins cause incursions of fine-grained, basin-axis sediments into the area of maximum subsidence near the faulted margin of the basin. Episodes of faulting and basin tilting are thus recorded by tongues of fine-grained, basin-axis deposits near the active margin of the basin, and not, as was traditionally supposed, by the coarse, scarp-derived deposits with which they interfinger. Basin-fill deposits of the Santa Fe Group in the northern Lemitar Mountains not only corroborate the importance of tectonic tilting in determining the distribution of sedimentary facies, but also illustrate the role of tilting and subsidence along *intrabasinal* faults. The Lemitar exposures also document that the "pivot line" (Leeder and Gawthorpe, 1987) between net erosion and net deposition on the hanging-wall dip slope can be highly migratory due to the effects of episodic tectonism, giving rise to widespread unconformities within the basin fill.

The majority of the main body of the Popotosa Formation in the Lemitar Mountains area comprises basin-axis (largely playa) and west-derived braid plain and alluvial-fan deposits. East-derived piedmont deposits are volumetrically minor in the Lemitar Mountains; most eastern piedmont deposits presumably have been buried by younger units within the Socorro Basin. Alluvial deposits of the western piedmont were derived from the footwall of the La Jencia–Cerro Colorado fault zone and from areas farther to the west that appear to have been drained by an ancestral La Jencia Creek and possibly an ancestral Rio Salado. During early Popotosa time such deposits locally extended far to the east, at least to the Socorro Canyon–Loma Peleda fault zone (Machette, 1978a). The widely dispersed nature of the western piedmont deposits was, in part, enabled by lack of emergence of the footwalls of the Silver Creek and Socorro Canyon–Loma Peleda faults; these faults were intrabasinal throughout Popotosa deposition. Additionally, because of the relatively slow strain rates during early Popotosa time, sediment supply from the west was able to keep pace with development of accommodation space, resulting in an extensive western piedmont.

Beginning about 16 to 15 Ma, playa-lake systems began to develop adjacent to and east of the Silver Creek fault, where they interfingered with alluvial deposits of the western piedmont. These playa deposits accumulated throughout the middle and late Miocene, a time of increasingly rapid strain rates (Fig. 3). The stratigraphically highest playa deposits locally appear to be transitional with the overlying Sierra Ladrones Formation. Major playa systems developed in what are now the structurally lowest parts of the Lemitar Mountains, near the Rio Salado and west of Socorro. Playa deposits consist primarily of variegated (commonly red) mudstone that is commonly calcareous and gypsiferous. Thickness of such deposits is as much as 500 m (Machette, 1978a).

The major playa deposits in the northern part of the study are closely bounded on the west by the Silver Creek fault, but nowhere have they been observed to cross the fault. This observation, in concert with evidence for intercalated, footwall-sourced alluvial fans and the presence of paleovalleys within Popotosa deposits of the footwall block (Cather and Kugler, in preparation), strongly suggest that the Silver Creek fault was an active intrabasinal element during Popotosa deposition (Fig. 4). Development of large playa systems adjacent to the Silver Creek fault indicates that generation of accommodation space by subsidence and tilting along this fault was able to overwhelm the rate of sediment supply to the distal western piedmont, which had dominated the area prior to playa development. Geologic mapping (Machette, 1978a; Cather and Kugler, in preparation) indicates pronounced stratigraphic variation in the along-strike width of the northern playa system near the Rio Salado. Our tilt-versus-time data in this area is not sufficiently detailed, however, to evaluate a possible tectonic origin for these second-order dimensional variations; climatic factors remain a viable alternative.

Playa systems extended eastward beyond the Socorro Canyon–Loma Peleda fault zone, as shown by the local truncation of playa deposits by the fault zone (Machette, 1978a; Chamberlin, 1980). Because of the scarcity of Popotosa exposure and lack of borehole data to the east of the fault zone, it is not known if this zone was active during Popotosa deposition as an intrabasin fault or if it entirely postdates the Popotosa Formation. Although sedimentologic data indicate the La

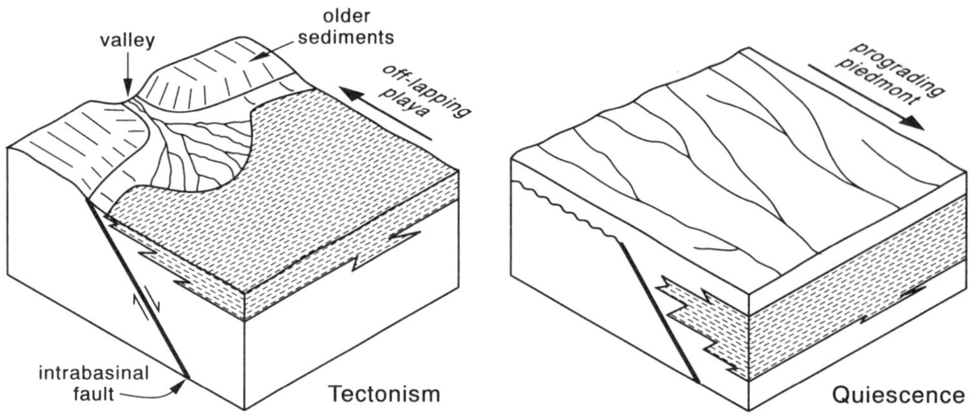

Figure 4. Schematic block diagram depicting facies control by episodic faulting and basin tilting along an intrabasinal fault. Diagram is derived from facies relations mapped along Silver Creek fault near the Rio Salado (Cather and Kugler, in preparation).

Jencia–Cerro Colorado fault zone was active during Popotosa deposition, tilting and subsidence along the fault in the northern La Jencia Basin was apparently insufficient (relative to sediment supply) to induce topographic closure and the development of lakes. Paleoflow during deposition of the main body of the Popotosa Formation in the northern La Jencia Basin (west of the Silver Creek fault) was uniformly eastward, with no evidence for drainage reversal or the development of major unconformities that would accompany the tectonic disruption of the prevailing western piedmont. It is possible that Popotosa-age tectonism produced lakes adjacent to the La Jencia–Cerro Colorado fault zone in the central or southern part of the La Jencia Basin, although no drilling data are available to test this hypothesis. In the northern part of the basin, however, such tectonic disruption did not occur until the late Miocene, at which time rapid tectonism caused the carving of a widespread unconformity that marked the end of Popotosa deposition.

In most areas, the Popotosa Formation is unconformably overlain by Plio-Pleistocene fluvial deposits of the Sierra Ladrones Formation. Angularity across the unconformity is commonly 10 to 20°, but passes westward into disconformable and conformable sequences adjacent to the major half-graben-bounding faults. The best example is exposed along the south bank of the Rio Salado just east of the Silver Creek fault (Fig. 1). Here conglomeratic beds identical to nearby piedmont deposits of the Sierra Ladrones Formation gradationally overlie fluvial and playa beds of the Popotosa Formation. Along the Rio Salado west of the Silver Creek fault, the contact passes westward from an unconformity with 10 to 15° of angularity to a disconformity with little or no apparent angularity near the Cerro Colorado fault (Cather and Kugler, in preparation). In the area of maximum subsidence in the subsurface to the southwest, this contact may become conformable.

In the northern La Jencia Basin between the Cerro Colorado and Silver Creek faults, beds of the Sierra Ladrones Formation lap eastward onto and successively pinch out against older and increasingly more tilted Popotosa strata below. Such onlapping relations are particularly well exposed in the lower box of the Rio Salado. Sierra Ladrones beds directly above the unconformity display evidence of temporary drainage reversal within the northern La Jencia Basin. Pebble imbrications within the basal beds locally show westward to southward paleotransport indicative of sediment derivation from the footwall of the Silver Creek fault and the rising Precambrian massif of Ladron Peak, respectively (Cather and Kugler, in preparation). These beds are, in turn, overlain by fine-grained, basin-axis deposits that lack a systematic paleodrainage pattern and contain a prominent travertine of spring, lacustrine, and shallow ground water origin (Barker, 1983). The basin-axis deposits are transitionally overlain to the west by scarp-derived piedmont deposits that show evidence for northeast to east paleoflow. Although no age constraints are present in the Sierra Ladrones Formation in the northern La Jencia Basin, a basalt within the formation to the east (Table 1) that dates 4.5 ± 0.1 Ma (Machette, 1978a; Bachman and Mehnert, 1978) indicates that aggradation of the Sierra Ladrones was underway during the early Pliocene in the Socorro Basin (Machette, 1978a).

Because of poor exposure, the basal-contact relations of the Sierra Ladrones Formation in the Socorro Basin are incompletely known. Piedmont deposits of the Sierra Ladrones Formation unconformably overlie rocks as old as Proterozoic along the eastern margin of the Socorro Basin (Fig. 1). The basal Sierra Ladrones contact is not exposed adjacent to the Socorro Canyon–Loma Peleda fault zone along the western flank of the basin, although this contact has been inferred by Hawley (1983) to be possibly transitional with the underlying Popotosa Formation. Integration of a through-going axial river system (ancestral Rio Grande) within the rift began during Sierra Ladrones deposition, about 4.1 to 4.5 Ma (Bachman and Mehnert, 1978). Early deposits of the ancestral Rio Grande dominantly occur along the west side of the Socorro Basin, directly east of the Socorro Canyon–Loma Peleda fault zone (Machette, 1978a; Chamberlin, 1980, 1982; Hawley, 1983). In contrast, the modern valley of the Rio Grande is farther to the east, near the geographic axis of the Socorro Basin. These relations suggest that a post-Miocene, eastward migration of component facies within the Sierra Ladrones Formation of the Socorro Basin (cf. DeBrine et al., 1963, fig. 2), similar to that of the northern La Jencia Basin, may have occurred prior to the incision of the modern Rio Grande Valley. In both of these basins, the eastward migration of facies within the Sierra Ladrones Formation appears to have resulted from displacement by prograding western piedmonts following the cessation of rapid extension in the late Miocene, in a manner analogous to that depicted by Mack and Seager (1990) for equivalent deposits in the Palomas Basin.

We interpret the development of the prominent unconformity between the Popotosa and Sierra Ladrones Formations to be a result of the sedimentary response to rapid tectonism and basin tilting during the late Miocene. This differs from the interpretations of some other workers, who apparently regard this and equivalent unconformities in the rift to represent a *lull* between episodes of active deformation and sedimentation. In our view, as stratal rotation rates increased during the late Miocene (Fig. 3), the rate of development of accommodation space exceeded the sediment-supply rate, causing offlap of sedimentation toward a narrow zone along the area of maximum subsidence and widespread erosion of older sediments on the hanging-wall dip slope (Fig. 5). Rapid tilting and subsidence enhanced topographic closure and caused incursion of the resulting playas toward the master faults (exposures of the uppermost playa deposits within the Popotosa Formation are less than a kilometer from the Silver Creek fault near the Rio Salado). As tectonism waned in the latest Miocene–early Pliocene, sediment supply exceeded the development of accommodation space, causing onlap and aggradation of the Sierra Ladrones Formation that enabled the eventual integration of an axial river system (ancestral Rio Grande) between the longitudinally arrayed half grabens to the north and south of the Socorro Basin. This integration also may have been abetted by increased precipitation resulting from enhanced late Miocene orographic effects (Chapin, 1979, 1988).

About 1.6 Ma, aggradation of the Santa Fe Group in the Socorro Basin appears to have ceased and incision of the Rio Grande began (Cather and McIntosh, 1990). Although lithospheric extension continues today, the incised nature of the Rio Grande and modest rates of extension have apparently masked any basin-scale effects of tilting on sedimentation since the early Pleistocene, with the possible exception of Recent uplift related to the Socorro magma body (Reilinger et al., 1979; Larsen and Reilinger, 1983; Ouchi, 1983).

REGIONAL RELATIONS

Normal faulting and incipient regional extension, at least in central and southern New Mexico, began in early Oligocene time (Cather, 1989, 1990a; Nightengale and Mack, 1992). Sedimentation within most early basins of the Rio Grande rift, however, did not begin until late Oligocene–early Miocene time (Chapin and Seager, 1975; Chapin, 1988). During the early stages of rifting, closely spaced, low-angle normal faults similar to those of late Oligocene age in the Lemitar Mountains developed in several local domains within the Rio Grande rift. However, geophysical surveys and well penetra-

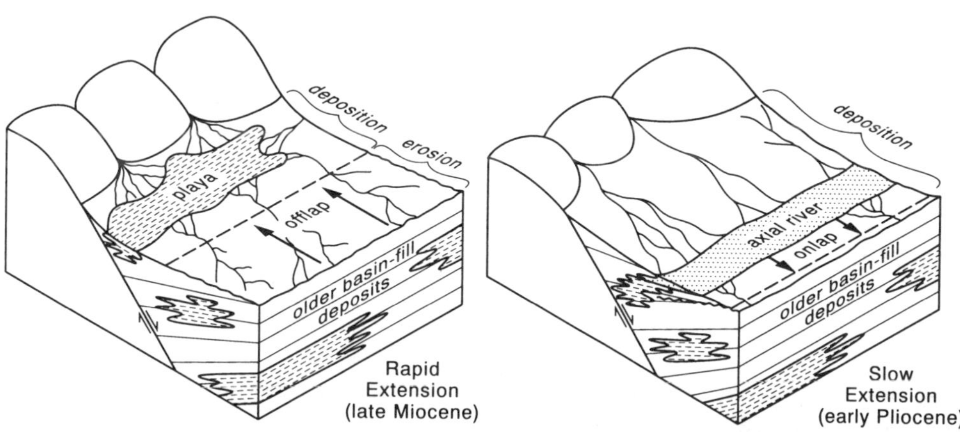

Figure 5. Model for development of Mio-Pliocene unconformity within Santa Fe Group by episodic basin tilting, based on composite data from the Lemitar Mountains area. Rapid extension in the late Miocene caused offlap of fluvial-lacustrine sedimentation toward area of maximum subsidence near active margin of basin and erosion of older basin-fill deposits on hanging-wall dip slope. Slow extension in early Pliocene caused sediment supply to exceed development of accommodation space, resulting in onlap and eventual integration of axial-river system (ancestral Rio Grande) between adjoining basins to north and south. Note that the unconformity within the Santa Fe Group gives way laterally to conformable and gradational beds in the area of maximum subsidence. See text for further discussion.

tions within the San Luis Basin (Brister and Gries, this volume; Kluth and Schaftenaar, this volume), Española Basin (Biehler et al., 1991; Cather, 1992), and Albuquerque Basin (Lozinsky, 1988; Russell and Snelson, 1990, and this volume) suggest that such faults did not play a major role in the early tectonic development of these basins. Instead, many of the presently active, moderate to steeply dipping faults appear to have also controlled early deposition within these basins.

Local domains of late Tertiary low-angle normal faults and associated steeply dipping strata are present in several areas of the Rio Grande rift (e.g., northern Franklin Mountains, Seager, 1981; Hagan embayment, Black, 1979; San Diego Mountain, Seager et al., 1971; East Potrillo Mountains, Seager and Mack, 1994). The chronologically best constrained examples of this structural style, however, are in the Lemitar Mountains (Chamberlin, 1983), near Questa in northern New Mexico (Meyer and Foland, 1991), and in the San Mateo Mountains about 60 km southwest of Socorro (Ferguson, 1991). Rapid extensional deformation in these three areas was not synchronous; early faulting in the Lemitars mostly occurred from 28.6 to 27.4 Ma, whereas similar deformation occurred at 25.7 to 24.6 Ma near Questa (Meyer and Foland, 1991) and 27.4 to 24.3 Ma in the San Mateos (Ferguson, 1991). The diachronous nature of early extension between these areas (particularly the lack of overlap in timing between the Lemitars and the other areas) argues against a causative, regional "pulse" of tectonism. Rather, it appears likely that ongoing, regional strain was focused by thermal weakening related to local, diachronous episodes of magmatism that occurred prior to and during tectonism in these areas.

The relatively slow rate of extension in the Lemitar Mountains area during the early Miocene approximately corresponds to a decrease in volcanism in the Socorro area (~24 to 18 Ma; Bobrow et al., 1983; Chapin, 1989), which, in turn, may explain the lack of datable rocks in the basal Popotosa Formation. The volcanic lull near Socorro is similar to, but not precisely synchronous with, similar early to middle Miocene volcanic lulls reported elsewhere in the Rio Grande rift (Chapin and Seager, 1975; Aldrich et al., 1986; Baldridge et al., 1980). The apparent correlation between tectonic and volcanic activity near Socorro is similar to that reported by workers in other areas (e.g., Anderson, 1971; Gans et al., 1989; Meyer and Foland, 1991).

In contrast to the preceding late Oligocene episode of tectonism, the culminatory late Miocene tectonism manifested in the Lemitar Mountains area may be of regional scope, as shown by widespread, commonly angular, unconformities that intervene between Miocene and Pliocene deposits in many basins of the Rio Grande rift (e.g., Hawley, 1978; Machette, 1978a; Manley, 1979; Lozinsky, 1988; Cather, 1990b). The regional synchronicity of these unconformities implies a common cause. Although data are insufficient to constrain the origin of the Mio-Pliocene unconformity in some areas, evidence for contemporaneous accelerated rifting in southern and central New Mexico (Seager, 1975; Chapin and Seager, 1975; Seager and others, 1984) and rapid basin tilting of late Miocene age in the Lemitar Mountains area suggests the unconformity is related to an episode of rapid regional extension.

Many workers in the Rio Grande rift have recognized evidence for increased strain rates during the late Miocene, but some have inferred that rapid extension continued into the early Pliocene. We are aware of no compelling stratigraphic, structural, or apatite-fission-track cooling data that indicate rapid extension continued after about 10 to 7 Ma anywhere in the Rio Grande rift. Indeed, the typical onlapping nature of the upper Santa Fe Group and the timing of initial integration of the ancestral Rio Grande in many basins of the rift attest to decreased rates of tectonic subsidence in the early Pliocene as compared to the late Miocene.

In many basins of the Rio Grande rift a lacuna of variable duration separates internally drained bolson deposits of upper Oligocene–Miocene age from axial river and associated piedmont deposits of Plio-Pleistocene age. It is this bipartite nature of the syn-rift deposits that gave rise to the common nomenclatural subdivision of the Santa Fe Group into upper and lower parts (each commonly consisting of one or more formation-rank units) by many workers in the Rio Grande rift (e.g., Kottlowski, 1958, 1960; Hawley et al., 1969; Hawley, 1978; Seager et al., 1984; Lozinsky and Hawley, 1986; Cather, 1990b). However, as noted by Chapin and Seager (1975, p. 313) ". . . the angular unconformities and compositional differences evident along the basin margins probably grade basinward to disconformities separating beds of similar composition." Indeed, the above-described exposures near the Rio Salado in the northern Lemitar Mountains indicate that unconformities may give way to entirely *transitional* sedimentary sequences in the area of maximum subsidence adjacent to the faulted margin of the basin. In these areas, the utility of the concept of upper versus lower Santa Fe Group is greatly compromised. J. W. Hawley (oral communication, 1990) has proposed using the term "middle Santa Fe Group" for beds transitional between the upper and lower units. Alternatively, in areas where data are insufficient to enable such a tripartite subdivision, we favor use of "Santa Fe Group, undivided."

In the broadest sense, the structural development of the Rio Grande rift may be viewed as the cumulative result of as much as 36 m.y. of weak to moderate rates of crustal extension punctuated by local, diachronous episodes of rapid magmatism-related extension in the late Oligocene–early Miocene and an apparently regional episode of strong deformation in the late Miocene. This latter deformation induced rapid tilting of half grabens within the rift, causing widespread erosion and development of unconformities that interrupted the otherwise relatively continuous late Oligocene–early Pleistocene aggradation of the Santa Fe Group and produced much of the present topography along the rift.

ACKNOWLEDGMENTS

We thank Ted Stans for access to the Sevilleta National Wildlife Refuge. The paper was improved by reviews from J. W. Hawley, G. H. Mack, and G. R. Osburn. Becky Titus drafted the figures and Lynne Hemenway typed the manuscript.

REFERENCES CITED

Aldrich, M. J., Jr., Chapin, C. E., and Laughlin, A. W., 1986, Stress history and tectonic development of the Rio Grande rift, New Mexico: Journal of Geophysical Research, v. 91, p. 6199–6211.

Alexander, J., and Leeder, M. R., 1987, Active control on alluvial architecture: Society of Economic Paleontologists and Mineralogists Special Publication 39, p. 243–252.

Anderson, R. E., 1971, Thin-skin distension in Tertiary rocks of southeastern Nevada: Geological Society of America Bulletin, v. 82, p. 43–58.

Asher-Bolinder, S., 1988, Stratigraphy of reference sections in the Popotosa Formation, Socorro County, New Mexico: U.S. Geological Survey Bulletin 1800, 22 p.

Bachman, G. O., and Mehnert, H. H., 1978, New K-Ar dates and late Pliocene to Holocene geomorphic history of the central Rio Grande region, New Mexico: Geological Society of America Bulletin, v. 89, p. 283–292.

Baldridge, W. S., Damon, P. E., Shafiqullah, M., and Bridwell, R. J., 1980, Evolution of the central Rio Grande rift, New Mexico, new potassium-argon ages: Earth and Planetary Science Letters, v. 51, p. 309–321.

Barker, J. M., 1983, Preliminary investigation of the origin of the Riley travertine, Socorro County, New Mexico: New Mexico Geological Society, Guidebook 34, p. 269–276.

Biehler, S., and 12 others, 1991, A geophysical model of the Española Basin, Rio Grande rift, New Mexico: Geophysics, v. 56, p. 340–353.

Black, B. A., 1979, Structure and stratigraphy of the Hagan embayment: A new look: New Mexico Geological Society, Guidebook 30, p. 101–105.

Blair, T. C., and Bilodeau, W. L., 1988, Development of tectonic cyclothems in rift, pull-apart, and foreland basins: Sedimentary response to episodic tectonism: Geology, v. 16, p. 517–520.

Bobrow, D. J., Kyle, P. R., and Osburn, G. R., 1983, Miocene rhyolitic volcanism in the Socorro area of New Mexico: New Mexico Geological Society, Guidebook 34, p. 211–217.

Bruning, J. E., 1973, Origin of the Popotosa Formation, north-central Socorro County, New Mexico [Ph.D. thesis]: Socorro, New Mexico Institute of Mining and Technology, 142 p.

Burke, W. H., Kenny, G. S., Otto, J. B., and Walker, R. D., 1963, Potassium-argon dates, Socorro and Sierra Counties, New Mexico: New Mexico Geological Society, Guidebook 14, p. 224.

Cather, S. M., 1983, Laramide Sierra uplift: Evidence for major pre-rift uplift in central and southern New Mexico: New Mexico Geological Society, Guidebook 34, p. 99–101.

Cather, S. M., 1986, Volcano-sedimentary evolution and tectonic implications of the Datil Group (latest Eocene–early Oligocene), west-central New Mexico [Ph.D. thesis]: Austin, University of Texas at Austin, 484 p.

Cather, S. M., 1989, Post-Laramide tectonic and volcanic transition in west-central New Mexico: New Mexico Geological Society, Guidebook 40, p. 91–97.

Cather, S. M., 1990a, Stress and volcanism in the northern Mogollon-Datil field, New Mexico: Effects of the post-Laramide tectonic transition: Geological Society of America Bulletin, v. 102, p. 1447–1458.

Cather, S. M., 1990b, Reconnaissance geology of the Santa Fe Group near Amalia, New Mexico: New Mexico Geological Society, Guidebook 41, p. 337–340.

Cather, S. M., 1992, Suggested revisions to the Tertiary tectonic history of north-central New Mexico: New Mexico Geological Society, Guidebook 43, p. 109–122.

Cather, S. M., and Johnson, B. D., 1984, Eocene tectonics and depositional setting of west-central New Mexico and eastern Arizona: New Mexico Bureau of Mines and Mineral Resources Circular 192, 33 p.

Cather, S. M., and Johnson, B. D., 1986, Eocene depositional systems and tectonics in west-central New Mexico and eastern Arizona: American Association of Petroleum Geologists Memoir 41, p. 623–652.

Cather, S. M., and McIntosh, W. C., 1990, Volcanogenic flood deposits near San Antonio, New Mexico: Depositional processes and implications: Sediments 1990; 13th International Sedimentological Congress, Abstracts, p. 80.

Cather, S. M., McIntosh, W. C., and Chapin, C. E., 1987, Stratigraphy, age, and rates of deposition of the Datil Group (upper Eocene–lower Oligocene), west-central New Mexico: New Mexico Geology, v. 9, p. 50–54.

Cather, S. M., Chapin, C. E., Chamberlin, R. M., McIntosh, W. C., and Kunk, M. J., 1991, Stratigraphy and age of the Santa Fe Group near Socorro, New Mexico: Geological Society of America Abstracts with Programs, v. 23, p. 10.

Chamberlin, R. M., 1978, Structural development of the Lemitar Mountains, an intrarift tilted fault block uplift, central New Mexico [abs.], *in* Proceedings, International Symposium on the Rio Grande Rift, Santa Fe, New Mexico, October 1978: Los Alamos Scientific Laboratory, Conference Proceedings, LA-7487-C, p. 22–24.

Chamberlin, R. M., 1980, Cenozoic stratigraphy and structure of the Socorro Peak volcanic center, central New Mexico [Ph.D. thesis]: Golden, Colorado School of Mines, 495 p.

Chamberlin, R. M., 1982, Geologic map, cross sections, and map units of the Lemitar Mountains, Socorro County, New Mexico: New Mexico Bureau of Mines and Mineral Resources Open-File Report 169, 3 plates, scale 1:12,000.

Chamberlin, R. M., 1983, Cenozoic domino-style crustal extension in the Lemitar Mountains, New Mexico: A summary: New Mexico Geological Society, Guidebook 34, p. 111–118.

Chamberlin, R. M., and Osburn, G. R., 1984, Character and evolution of extensional domains in the Socorro area of the Rio Grande rift, central New Mexico: Geological Society of America Abstracts with Programs, v. 16, p. 467.

Chamberlin, R. M., and Osburn, G. R., 1986, Tectonic framework, character, and evolution of upper crustal extensional domains in the Socorro area of the Rio Grande rift, New Mexico, *in* Beatty, B., and Wilkinson, P.A.K., eds., Frontiers in geology and ore deposits of Arizona and the southwest: Arizona Geological Society Digest, v. 16, p. 464.

Chapin, C. E., 1979, Evolution of the Rio Grande rift—A summary, *in* Riecker, R. E., ed., Rio Grande rift: Tectonics and magmatism: Washington, D.C., American Geophysical Union, p. 1–5.

Chapin, C. E., 1988, Axial basins of the northern and central Rio Grande rift, *in* Sloss, L. L., ed., Sedimentary cover—North American craton: U.S.: Geological Society of America, The Geology of North America, v. D-2, p. 165–170.

Chapin, C. E., 1989, Volcanism along the Socorro accommodation zone, Rio Grande rift, New Mexico: New Mexico Bureau of Mines and Mineral Resources, Memoir 46, p. 46–57.

Chapin, C. E., and Seager, W. R., 1975, Evolution of the Rio Grande rift in the Socorro and Las Cruces areas: New Mexico Geological Society, Guidebook 26, p. 297–321.

Chapin, C. E., Chamberlin, R. M., Osburn, G. R., Sanford, A. R., and White, D. L., 1978, Exploration framework of the Socorro geothermal area, New Mexico, in Chapin, C. E., and Elston, W. E., eds., Field guide to selected cauldrons and mining districts of the Datil-Mogollon volcanic field, New Mexico: New Mexico Geological Society Special Publication 7, p. 115–129.

DeBrine, B., Spiegel, Z., and William, D., 1963, Cenozoic sedimentary rocks in Socorro Valley, New Mexico: New Mexico Geological Society,

Guidebook 14, p. 123–131.

Denny, C. S., 1940, Tertiary geology of the San Acacia area, New Mexico: Journal of Geology, v. 48, p. 73–106.

Ferguson, C. A., 1991, Stratigraphic and structural studies in the Mt. Withington caldera, Grassy Lookout Quadrangle, Socorro County, New Mexico: New Mexico Geology, v. 13, p. 50–54.

Gans, P. B., Mahood, G. A., and Schermer, E., 1989, Synextensional magmatism in the Basin and Range province: A case study from the eastern Great Basin: Geological Society of America Special Paper 233, 53 p.

Hawley, J. W., 1978, Correlation chart 2—Middle to upper Cenozoic stratigraphic units in selected areas of the Rio Grande rift in New Mexico: New Mexico Bureau of Mines and Mineral Resources, Circular 163, p. 239.

Hawley, J. W., 1983, Geomorphic evolution of Socorro area of Rio Grande valley: New Mexico Geological Society, Guidebook 34, p. 13.

Hawley, J. W., Kottlowski, F. E., Seager, W. R., King, W. E., Strain, W. S., and LeMone, D. V., 1969, The Santa Fe Group in the south-central New Mexico border: New Mexico Bureau of Mines and Mineral Resources Circular 104, p. 52–76.

Heller, P. L., Angevine, C. L., Winslow, N. S., and Paola, C., 1988, Two-phase stratigraphic model of foreland-basin sequences: Geology, v. 16, p. 501–504.

Kelley, S. A., Chapin, C. E., and Corrigan, J., 1992, Late Mesozoic to Cenozoic cooling histories of the flanks of the northern and central Rio Grande rift, Colorado and New Mexico: New Mexico Bureau of Mines and Mineral Resources Bulletin 145, 39 p.

Kottlowski, F. E., 1958, Geological history of the Rio Grande near El Paso: West Texas Geological Society, Guidebook 1958, p. 46–54.

Kottlowski, F. E., 1960, Reconnaissance geologic map of Las Cruces 30-minute Quadrangle: New Mexico Bureau of Mines and Mineral Resources, Geologic Map 14, scale 1:126,720.

Larsen, S., and Reilinger, R., 1983, Recent measurements of crustal deformation related to the Socorro magma body, New Mexico: New Mexico Geological Society, Guidebook 34, p. 119–121.

Leeder, M. R., and Gawthorpe, R. L., 1987, Sedimentary models for extensional tilt-block/half graben basins: Geological Society of London Special Publication 28, p. 139–152.

Lozinsky, R. P., 1988, Stratigraphy, sedimentology, and sand petrology of the Santa Fe Group and pre–Santa Fe Tertiary deposits in the Albuquerque Basin, central New Mexico [Ph.D. thesis]: Socorro, New Mexico Institute of Mining and Technology, 298 p.

Lozinsky, R. P., and Hawley, J. W., 1986, The Palomas Formation of south-central New Mexico—A formal definition: New Mexico Geology, v. 8, no. 4, p. 73–82.

Machette, M. N., 1978a, Geologic map of the San Acacia Quadrangle, Socorro County, New Mexico: U.S. Geological Survey, Geological Quadrangle Map GQ-1415, scale 1:24,000.

Machette, M. N., 1978b, Late Cenozoic geology of the San Acacia–Bernardo area, *in* Hawley, J. W., ed., Guidebook to the Rio Grande rift in New Mexico and Colorado: New Mexico Bureau of Mines and Mineral Resources Circular 163, p. 135–137.

Mack, G. H., and Seager, W. R., 1990, Tectonic control on facies distribution of the Camp Rice and Palomas Formations (Pliocene-Pleistocene) in the southern Rio Grande rift: Geological Society of America Bulletin, v. 102, p. 45–53.

Manley, K., 1979, Stratigraphy and structure of the Española Basin, Rio Grande rift, New Mexico, *in* Reicker, R. E., ed., Rio Grande rift: Tectonics and magmatism: Washington, D.C., American Geophysical Union, p. 71–86.

McIntosh, W. C., Kedzie, L. L., and Sutter, J. F., 1991, Paleomagnetism and $^{40}Ar/^{39}Ar$ ages of ignimbrites, Mogollon-Datil volcanic field, southwestern New Mexico: New Mexico Bureau of Mines and Mineral Resources Bulletin 135, 79 p.

McIntosh, W. C., Chapin, C. E., Ratté, J. C., and Sutter, J. F., 1992, Time-stratigraphic framework for the Eocene-Oligocene Mogollon-Datil volcanic field, southwest New Mexico: Geological Society of America Bulletin, v. 104, p. 851–871.

Meyer, J., and Foland, K. A., 1991, Magmatic-tectonic interaction during early Rio Grande rift extension at Questa, New Mexico: Geological Society of America Bulletin, v. 103, p. 993–1006.

Nightengale, A. L., and Mack, G. H., 1992, Stratigraphy, sedimentology, and provenance of sedimentary rocks of the Bell Top Formation (Oligocene), Doña Ana County and Sierra County, New Mexico: New Mexico Geology, v. 14, p. 77–78.

Osburn, G. R., 1984, Geologic map of Socorro County, New Mexico: New Mexico Bureau of Mines and Mineral Resources Open-File Report 238, 13 p., scale 1:200,000.

Osburn, G. R., and Chapin, C. E., 1983, Nomenclature for Cenozoic rocks of northeast Mogollon-Datil volcanic field, New Mexico: New Mexico Bureau of Mines and Mineral Resources, Stratigraphic Chart 1.

Ouchi, S., 1983, Effects of uplift on the Rio Grande over the Socorro magma body, New Mexico: New Mexico Geological Society, Guidebook 34, p. 54–56.

Reilinger, R. E., Brown, L. D., Oliver, J. E.,and York, J. E., 1979, Recent vertical crustal movements from leveling observations in the vicinity of the Rio Grande rift, *in* Riecker, R. E., ed., Rio Grande rift: Tectonics and magmatism: Washington, D.C., American Geophysical Union, p. 223–236.

Reiter, M., Barroll, M. W., and Cather, S. M., 1992, Rotational buoyancy tectonics and models of simple half graben formation: Journal of Geophysical Research, v. 97, p. 8917–8926.

Russell, L. R., and Snelson, S., 1990, Structural style and tectonic evolution of the Albuquerque Basin segment of the Rio Grande rift, *in* Pinet, B., and Bois, C., eds., The potential for deep seismic profiling for hydrocarbon exploration: Paris, Editions Technip, p. 175–207.

Seager, W. R., 1975, Cenozoic tectonic evolution of the Las Cruces area, New Mexico: New Mexico Geological Society, Guidebook 26, p. 241–250.

Seager, W. R., 1981, Geology of Organ Mountains and southern San Andres Mountains, New Mexico: New Mexico Bureau of Mines and Mineral Resources Memoir 36, 97 p.

Seager, W. R., and Mack, G. H., 1994, Geology of the East Potrillo Mountains and vicinity, Doña Ana County, New Mexico: New Mexico Bureau of Mines and Mineral Resources Bulletin 113 (in press).

Seager, W. R., and Morgan, P., 1979, Rio Grande rift in southern New Mexico, west Texas, and northern Chihuahua, *in* Riecker, R. E., ed., Rio Grande rift: Tectonics and magmatism: Washington, D.C., American Geophysical Union, p. 87–106.

Seager, W. R., Hawley, J. W., and Clemons, R. E., 1971, Geology of the San Diego Mountain area, Doña Ana County, New Mexico: New Mexico Bureau of Mines and Mineral Resources Bulletin 97, 38 p.

Seager, W. R., Shafiqullah, M., Hawley, J. W., and Marvin, R., 1984, New K-Ar dates from basalts and the evolution of the southern Rio Grande rift: Geological Society of America Bulletin, v. 95, p. 87–99.

Thompson, G. A., 1960, Problem of late Cenozoic structure of the Basin Ranges: Proceedings of the 21st International Geological Congress, Copenhagen, v. 18, p. 62–68.

Weber, R. H., 1971, K-Ar ages of Tertiary igneous rocks in central and western New Mexico: Isochron/West, no. 71-1, p. 33–45.

Wernicke, B., and Burchfiel, B. C., 1982, Modes of extensional tectonics: Journal of Structural Geology, v. 4, p. 105–115.

Manuscript Accepted by the Society October 13, 1993

Printed in U.S.A.

Geological Society of America
Special Paper 291
1994

An integrated geophysical study of the Estancia Basin, central New Mexico

Ronald Barrow* and G. Randy Keller
Department of Geological Sciences, The University of Texas at El Paso, El Paso, Texas 79968

ABSTRACT

This study focuses on the interpretation of seismic reflection lines, gravity, and well data in the Estancia Basin, which is located along the eastern margin of the Rio Grande rift in central New Mexico. Major new data introduced are seven seismic reflection lines provided by the Mobil Exploration and Production Company. The basin began to form during Late Mississippian and Early Pennsylvanian uplift of the Ancestral Rocky Mountains. Laramide-age thrusting and folding occurred west of the basin and may have caused strike-slip movements in the basin. Limited packages of Tertiary rocks occur in local outcrops, and small normal faults associated with the Rio Grande rift are the latest structural features in the basin.

An integrated interpretation of the seismic reflection data and other geophysical and geological data make the following observations and interpretations possible: (1) the Pennsylvanian basin is a narrow, fault-bounded trough with a north-south trend; (2) the eastern boundary of this trough is a strike-slip fault that was probably active during formation of the basin; (3) this strike-slip fault was also probably active in the Laramide and possibly active in the Precambrian; (4) there are many prominent dipping reflectors in the Precambrian basement that may represent ductile shear zones, and (5) the effects of the Rio Grande rifting in the area were minor.

INTRODUCTION

The Estancia Basin is located approximately 80 km southeast of Albuquerque, New Mexico, between the Sandia-Manzano Mountains and the Pedernal Hills (Fig. 1). It is east of the physiographic expression of the Rio Grande rift, but may have been influenced by crustal extension during rifting (Chapin, 1971). The objectives of this study are to determine the subsurface structure of the Estancia Basin region and to formulate a model for its development through time. The approach taken consists of a careful integration of seismic reflection, gravity, aeromagnetic, well, and surface geologic data. Seven seismic reflection lines and about 160 gravity readings are the new data presented here.

*Present address: Marathon International, P.O. Box 3128, Houston, Texas 77253.

Several major tectonic events have affected the Estancia Basin. The best-documented early event was thrusting of Precambrian rocks coupled with metamorphism (Grambling et al., 1988, 1989). The seismic data show that many reflectors are present in the basement. Thus, the complex Precambrian history of the region was a more significant factor in our interpretations than was anticipated when the study began. Grambling et al. (1988) postulated that the Proterozoic history of the region was characterized by the assembly of as many as six tectonostratigraphic terranes. These authors suggested that at least some of the terrane boundaries were north-verging ductile thrusts. However, Grambling et al. (1989) demonstrated that extension also played a major role in the Proterozoic evolution of the region north of the Estancia Basin. Geochronologic studies (Dallmeyer et al., 1990) indicate that the extension occurred at about 1,440 Ma and that the thrusting occurred at about 1,350 Ma.

Thompson et al. (1991) discussed the Proterozoic tectonic

Barrow, R., and Keller, G. R., 1994, An integrated geophysical study of the Estancia Basin, central New Mexico, *in* Keller, G. R., and Cather, S. M., eds., Basins of the Rio Grande Rift: Structure, Stratigraphy, and Tectonic Setting: Boulder, Colorado, Geological Society of America Special Paper 291.

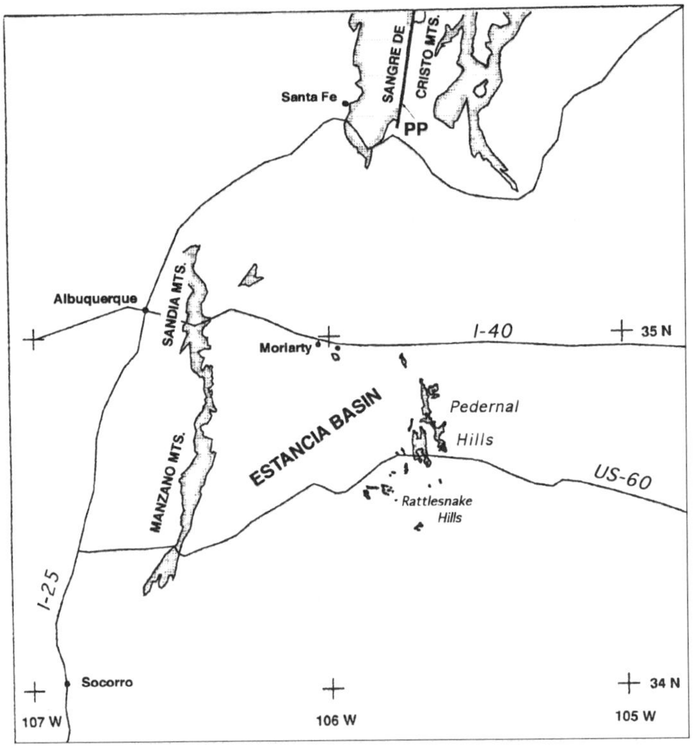

Figure 1. Regional setting of the Estancia Basin. Stipple, Precambrian. PP, Pecos-Picuris fault.

history of the Manzano Mountains (Fig. 2) where they mapped a major structure (the Monte Largo shear zone) in which amphibolite-grade rocks were thrust northward over greenschist-grade rocks. To the west of the Estancia Basin, a ductile shear zone separating a quartzite sequence from a metavolcanic sequence has been mapped (Fig. 2) in the Pedernal mountains (Bauer and Williams, 1985). Grambling et al. (1988) interpreted this feature as a terrane boundary.

A major strike-slip fault with a postulated 37 km of right-lateral displacement (the Picuris-Pecos fault) occurs north of the Estancia Basin (Fig. 1; Miller et al., 1963). This fault has been traced for 80 km and its trend is predominantly north-south. It juxtaposes Pennsylvanian sediments and Precambrian granites against Precambrian metamorphic rocks (Miller et al., 1963). This fault has experienced a complex history of movements with the most significant being Laramide in age (Chapin and Cather, 1981). However, rocks within the fault zone suggest the presence of ductile fabrics parallel to the fault. This observation indicates that some movements along this fault may have occurred under conditions favoring ductile shear (i.e., at considerable depth) and may be Precambrian in age.

After the Proterozoic, the next major tectonic event that can be documented in the Estancia Basin is the formation of the uplifts and basins of the Ancestral Rocky Mountains (e.g., Kluth and Coney, 1981; Dickinson, 1981). This middle Pennsylvanian

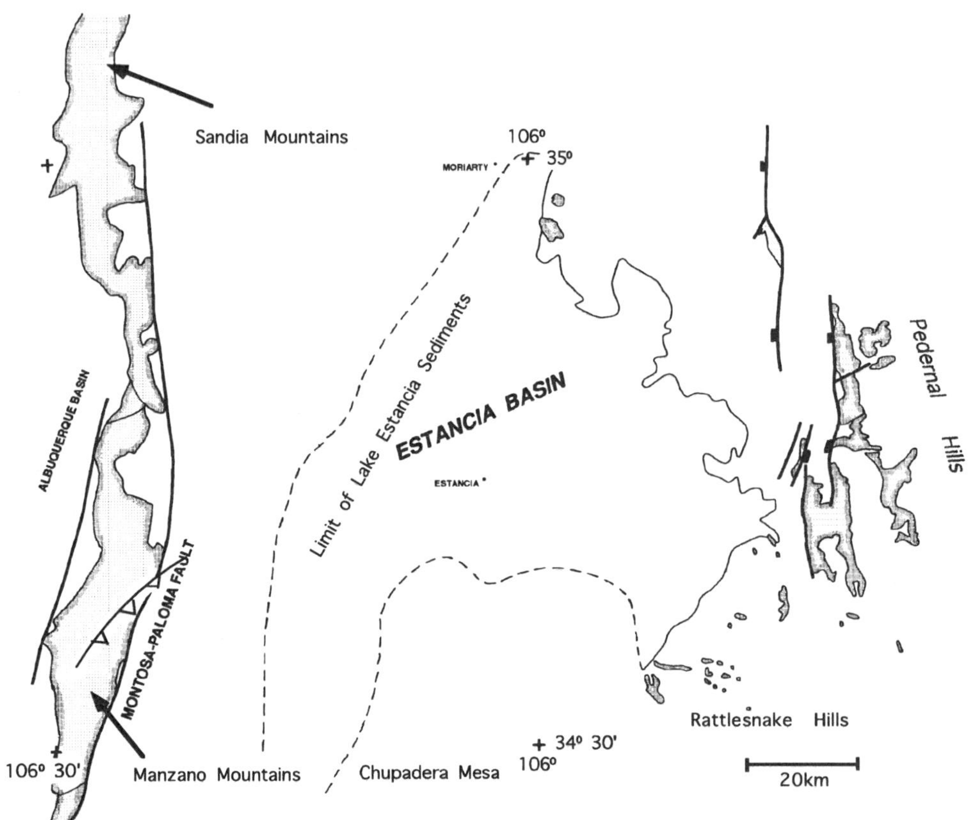

Figure 2. Local geologic setting of the Estancia Basin. Ductile thrust fault in the Manzano Mountains is shown as a line with teeth on the upper plate. Stipple, Precambrian.

activity is thought to be a result of the Ouachita-Marathon orogeny. Complex basement-involved structures formed and arkoses shed from highlands, such as the Pedernal Mountains, were deposited in developing basins. Tectonic activity associated with the formation of the Ancestral Rockies continued through the Late Pennsylvanian–Early Permian. Before the end of the Permian, however, the Pedernal highlands had been buried by the Permian Yeso Formation (Kelly, 1972).

The Late Cretaceous–Eocene Laramide orogeny was the next major tectonic event in the region. The Estancia Basin is located within the Rocky Mountain foreland (Woodward, 1976). Structures in the region are north-trending uplifts and intervening broad basins. Uplifts are usually fault-bounded with sedimentary strata draped over basement faults. Thrusting is thought to have extended as far east as the Manzano Mountains (Fig. 2) where the Paloma and Montosa faults formed. Another Laramide feature in the same area is the Oscura anticline (Fig. 2).

The Picuris-Pecos fault is believed to have been reactivated during the Laramide (Miller et al., 1963). A high-angle fault plane is thought to have been maintained with dip-slip movement causing the eastern block of the fault to be downdropped (Miller et al., 1963). Considerable strike-slip movement also occurred (Chapin and Cather, 1981).

The next major tectonism in the area was extension in the Rio Grande rift, which began at about 30 Ma. The Estancia Basin is east of the eastern flank of the Rio Grande rift. Chapin (1971) considered the Estancia Basin to be one of the largest of the inflection basins along the rift zone. Between 90 and 125 m (300 to 400 ft) of alluvial fill accumulated during the late Pliocene to Pleistocene (Smith, 1957) as a result of downwarping associated with the rift.

The Montosa fault, active in the Laramide (Fig. 2), is believed to have been reactivated with a normal sense of movement during the development of the rift (Bates et al., 1947). Where good exposures are found, a system of normal faults with the downthrown side toward the basin are observed. Throws on this fault system are generally less than 60 to 90 m (200 to 300 ft; Kelley, 1972).

GEOPHYSICAL DATA

Seismic reflection profiles

Seismic reflection profiles provided by Mobil Exploration and Production are the most important data set presented in this study (Fig. 3). These data were recorded at three different times with the oldest data being recorded in July 1980. Lines A and B belong to this generation of data, which employed a 67-channel recording system and a 24-fold common midpoint geometry. Geophones were arranged in 67-m (220-ft) groups with 48 geophones in each group. Vibrators were used as the

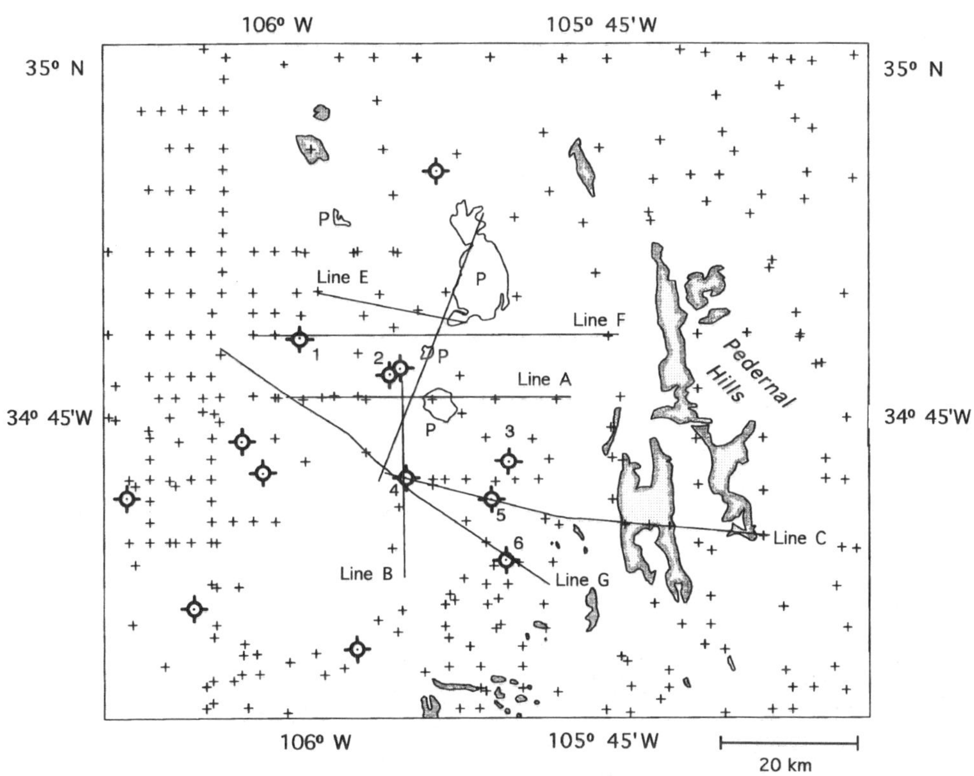

Figure 3. Index map of the Estancia Basin area showing locations of seismic reflection lines, wells, and gravity readings (small crosses). Numbered wells are keyed to Table 1. Stipple, Precambrian; P, Permian outcrops.

energy source with sweep frequencies of 56 to 14 Hz and a sweep length of 20 s.

The Line C was recorded in July of 1984. This line was recorded with a 48-fold common midpoint geometry and 24 geophones in each group at a spacing of 137 cm (4.5 ft). Vibrators were used as the energy source with a sweep length of 8 seconds and a sweep frequency of 72 to 8Hz.

Lines D, E, F, and G were recorded in October, 1988 using a 60-fold common midpoint geometry. Groups of geophones were positioned 23 m (75 ft) apart with 36 geophones in each group. Geophones within the groups were spaced 128 m (4.2 ft) apart. Vibrators were used as the energy source with a sweep length of 6 seconds, a 0.25 second taper, and a sweep frequency of 80 to 8 Hz.

A standard processing sequence was employed for each line followed by wave-equation migration. The first step in analyzing the migrated record sections involved examining data from wells in the area, finding formation tops from reports and scout tickets (Table 1), and converting the formation tops to time using velocities obtained from the sonic log of the Houston Oil and Minerals Federal #14-28 well (Fig. 4). This is the deepest and most recently drilled well in the area and is the only one with a full suite of well logs. Formation tops converted to time were located on the seismic sections and correlated with coherent reflections. Reflection boundaries were extrapolated throughout the basin with an emphasis on seismic character and known geology.

The data quality is generally good, but as is generally the case with land surveys, there is variability in the quality primarily due to surface conditions. All of these data are presented in Barrow (1992), but only the three lines that illustrate the major features of interest are shown here. These are presented in both interpreted and uninterpreted form so the reader can independently evaluate the proposed interpretations. Individual features are imaged with varying clarity and alternate interpretations are possible. However, the proposed interpretations tie between adjacent profiles and to the well data. Information from geologic mapping and gravity and magnetic data are also honored qualitatively. A detailed interpretation of these data is presented below but several key points are immediately obvious. First of all, as was expected, there is not a major rift basin present. Secondly, the thick Pennsylvanian strata that are known from drilling data are confined to a surprisingly narrow trough. Lastly, there are many interesting reflectors within the Precambrian basement.

TABLE 1. SIGNIFICANT OIL WELLS IN THE ESTANCIA BASIN*

Number and Name of Well	Formation	Depth to Top of Formation (ft)
1. Olson Drilling #1	Yeso	200
	Abo	780
	Madera	1,450
	Sandia	1,550
	Precambrian	3,170
2. California Exploration #1	San Andres	200
	Glorietta	500
	Yeso	1,000
	Abo	1,820
	TD	2,680
3. Gardner Petroleum #1	Abo	1,315
	Madera	1,865
	Sandia	5,370
	Precambrian	5,900
4. Houston Federal #14	Yeso	300
	Abo	1,100
	Madera	1,800
	Sandia	5,500
	Precambrian	8,700
5. Superior #28	Yeso	150
	Abo	940
	Sandia	1,410
	Precambrian	2,590
6. Powell-Walters #1	Yeso	150
	Abo	940
	Sandia	1,410
	Precambrian	2,590

*See Figure 3 for locations of wells.

Gravity and magnetic data

Gravity and magnetic data were employed to aid the seismic interpretations and to extrapolate below and beyond the extent of the seismic data. Gravity data used in this study were obtained from an extensive gravity data base maintained at the University of Texas at El Paso. Bouguer anomalies were calculated using the formulae of Cordell et al. (1982). Gravity values within the data base have all been tied to a common gravity datum. A reduction density of 2.67 g/cc and a sea-level datum were used for the Bouguer correction. Additional gravity measurements were taken in key areas where data points were sparse. The final distribution of gravity stations is shown in Figure 3.

Magnetic maps were produced from a merged data set from Cordell (1984). Four separate aeromagnetic surveys were involved. All of the flight lines trend east-west with 1-mi (1.6-km) spacing and were contracted by the U.S. Geological Survey. Flight-line elevations for these surveys ranged from 2,130 to 3,050 m (7,000 ft to 10,000 ft) above sea level (Cordell, 1984).

INTERPRETATION

The first step in the interpretation was to look at the regional setting via gravity and magnetic maps (Figs. 5 and 6). In Figure 5, a prominent gravity minimum associated with the Albuquerque Basin dominates the western portion of the map.

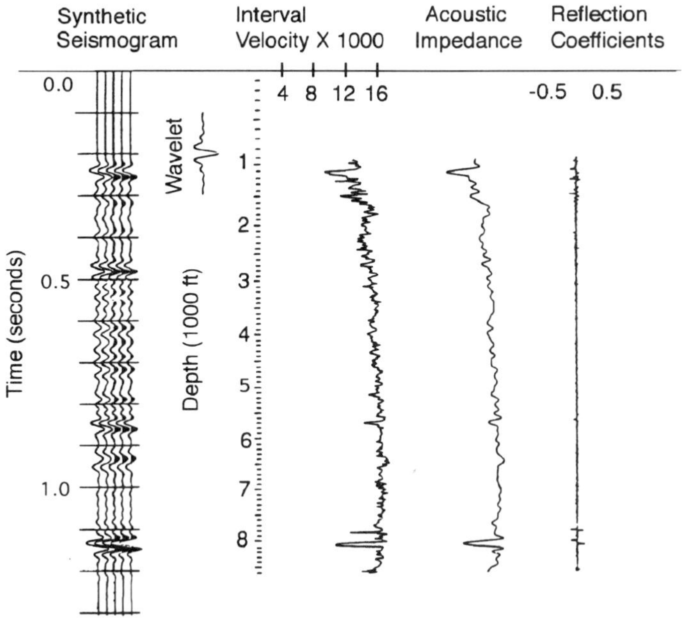

Figure 4. Synthetic seismogram from the Houston Oil and Minerals Federal #14-28 well.

The Estancia Basin is also associated with a prominent minimum, which in the context of the Rio Grande rift would indicate a substantial thickness of Cenozoic sediments. However, inspection of the drilling results in the area show that this is not the case. The seismic data were required to resolve this dilemma. In order to facilitate integration with seismic and drilling data, a gravity map of the immediate area around the Estancia Basin was constructed. This map (Fig. 7) shows that the thickest sedimentary section known from drilling is not located directly over the most negative of the gravity anomalies. In the western portion of the map (near 34°45′N, 106°W), there are two gravity lows within a larger gravity low. The gravity low to the southeast is associated with the area of thick Pennsylvanian sediments, but drilling results and seismic data show that the larger low to the northwest must be in part due to intrabasement density contrasts.

Although the drilling and seismic data document that the effects of the rift were minor in the Estancia Basin, there is some evidence for rift-related activity. The youngest feature that can be seen on the seismic data is a fault along the eastern side of line F (Figs. 8 and 9). This fault is exposed at the surface (Fig. 2). Line F shows that the fault is a low-angle normal fault with the downthrown side toward the deepest part of the Estancia Basin. This fault may be one of a series of small faults associated with the formation of the Rio Grande rift.

The thickest section of Pennsylvanian and Lower Permian sedimentary rocks in the Estancia Basin occurs in the Houston well (Fig. 3, Table 1) and can be seen most clearly on line F (Figs. 8 and 9) between 5 and 8 km from the western end of the line and on line A (Figs. 10 and 11) beginning at about 5 km from its western end. These sedimentary rocks appear as a distinct package of layered reflections that is bounded by high-angle faults and can also be seen on the record sections for lines B, D, G, and E. A westward dip is present in this package of reflectors, which probably occurred during late Paleozoic deformation.

West of the thick upper Paleozoic section, there are a series of faults. These faults occur on the western end of lines A, E, and F (Figs. 9 to 13) near an area where the reflections are

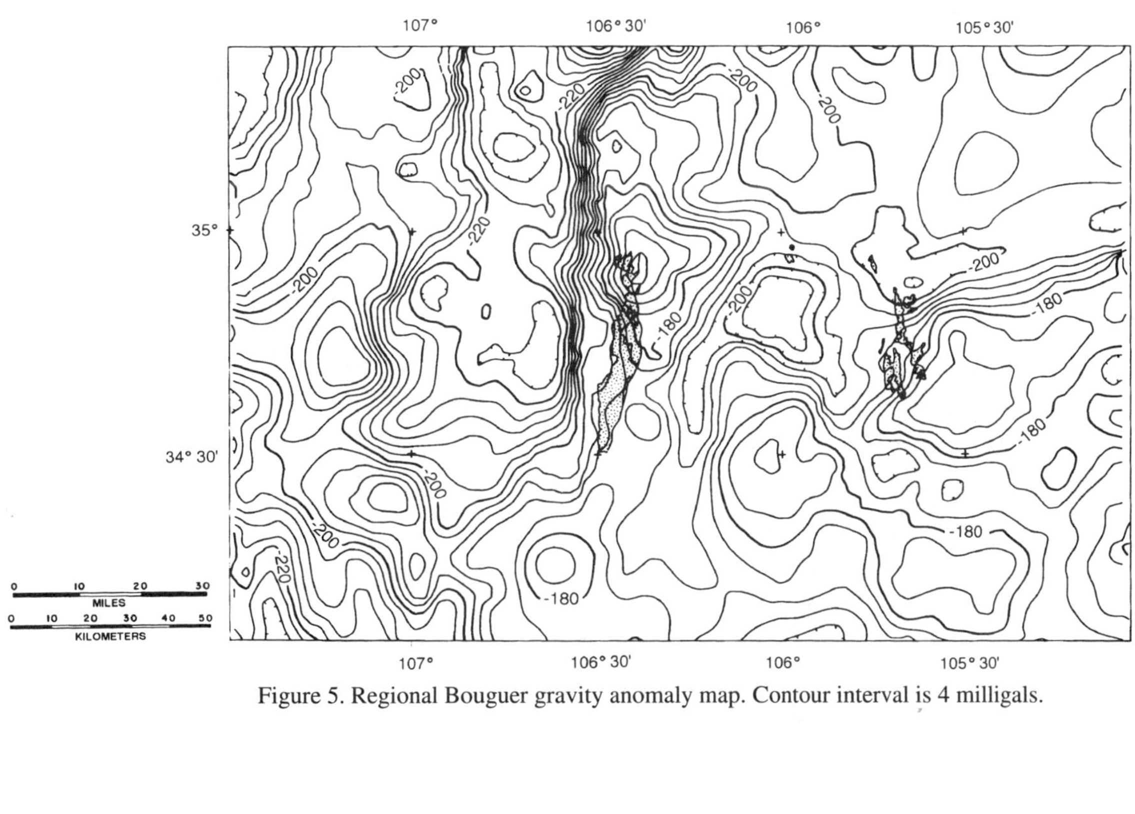

Figure 5. Regional Bouguer gravity anomaly map. Contour interval is 4 milligals.

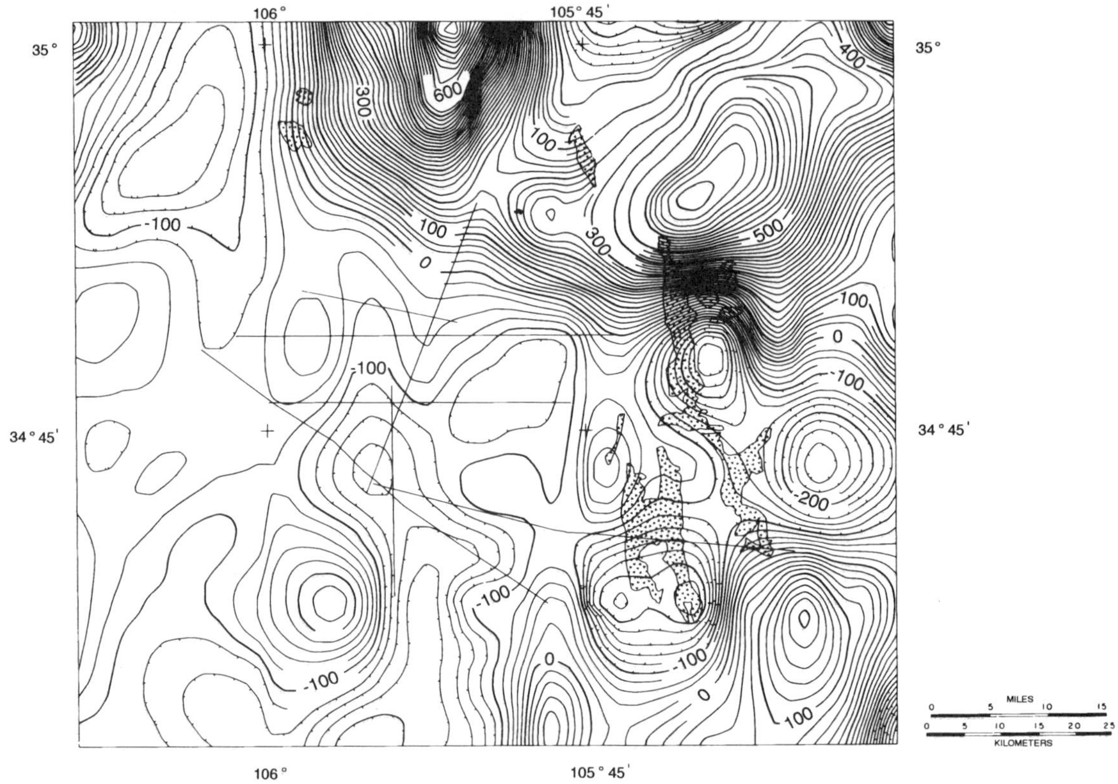

Figure 6. Magnetic anomaly map of the Estancia basin area. Contour interval is 20 gammas. Seismic reflection lines from Figure 3.

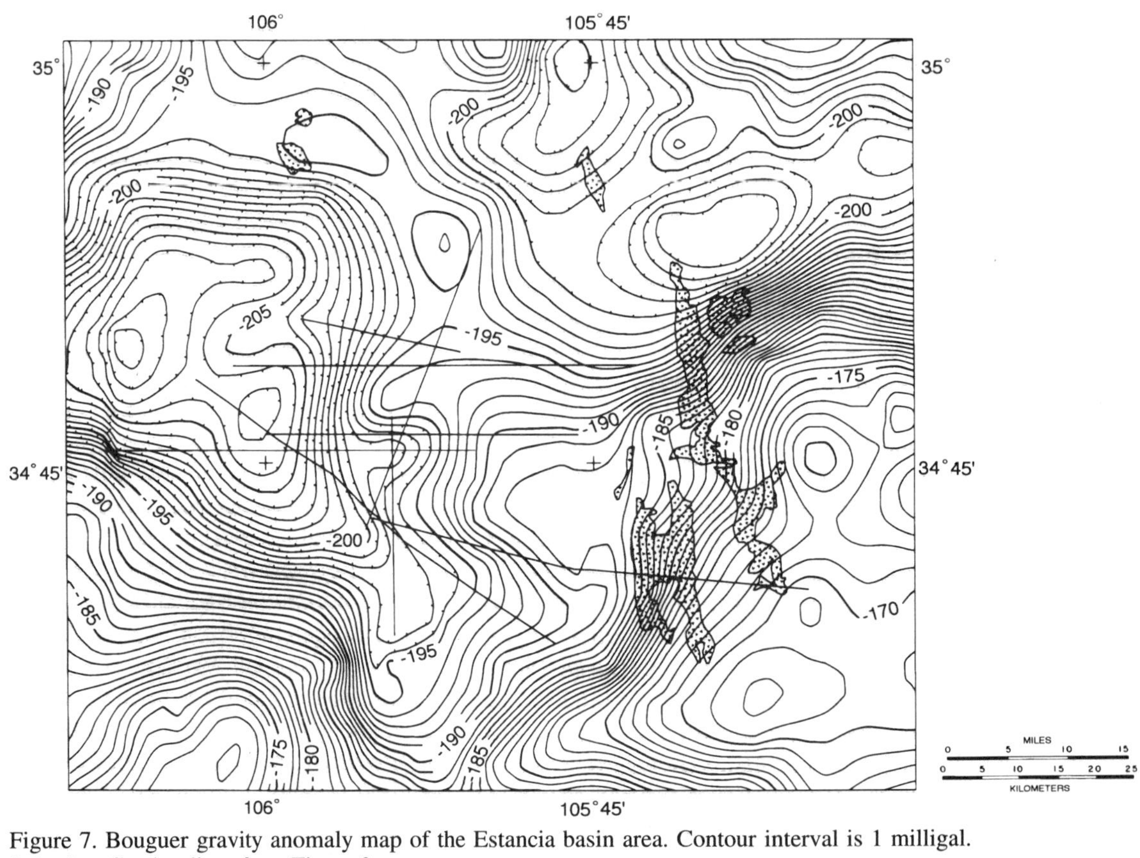

Figure 7. Bouguer gravity anomaly map of the Estancia basin area. Contour interval is 1 milligal. Seismic reflection lines from Figure 3.

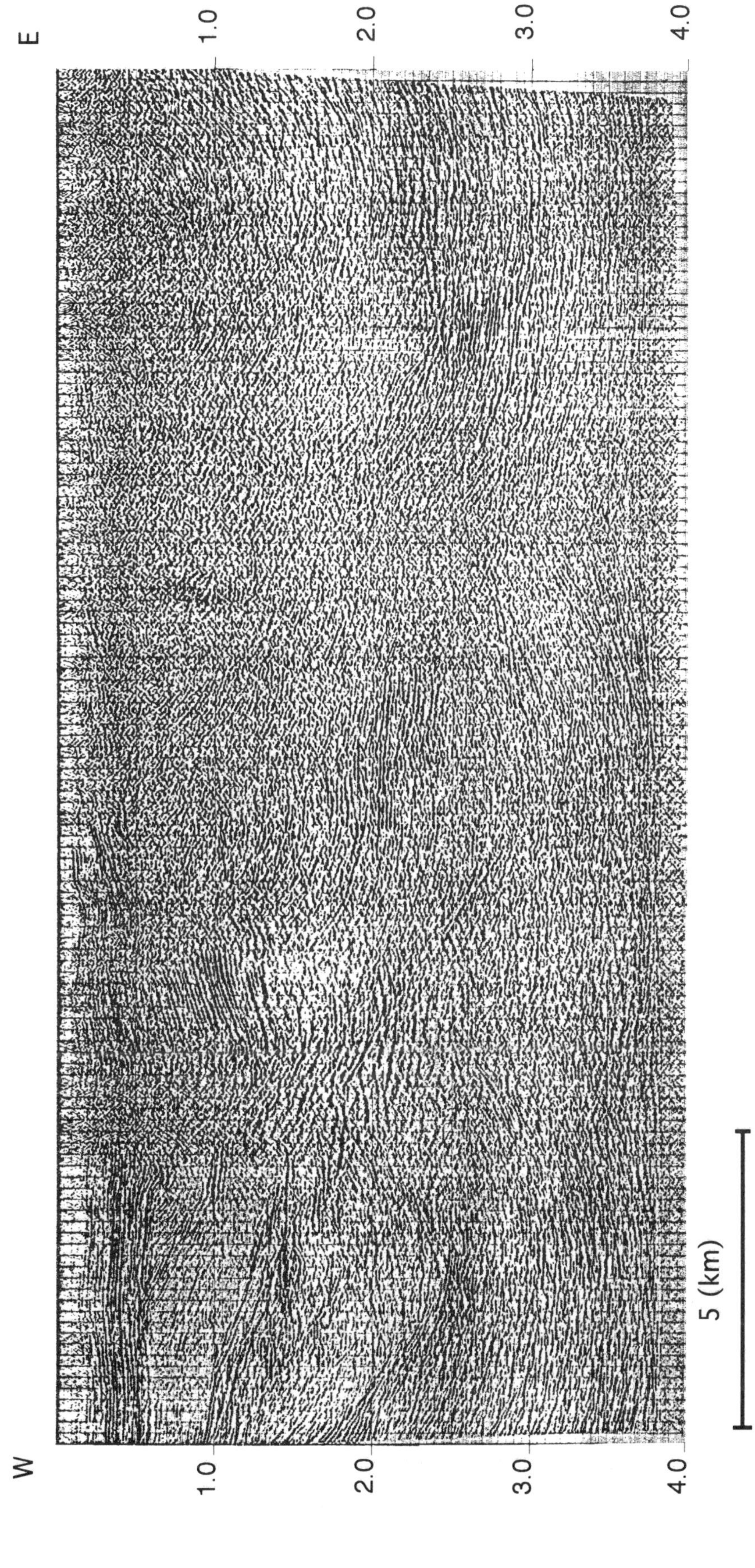

Figure 8. Uninterpreted record section for seismic Line F. Vertical scale is two-way travel time in seconds.

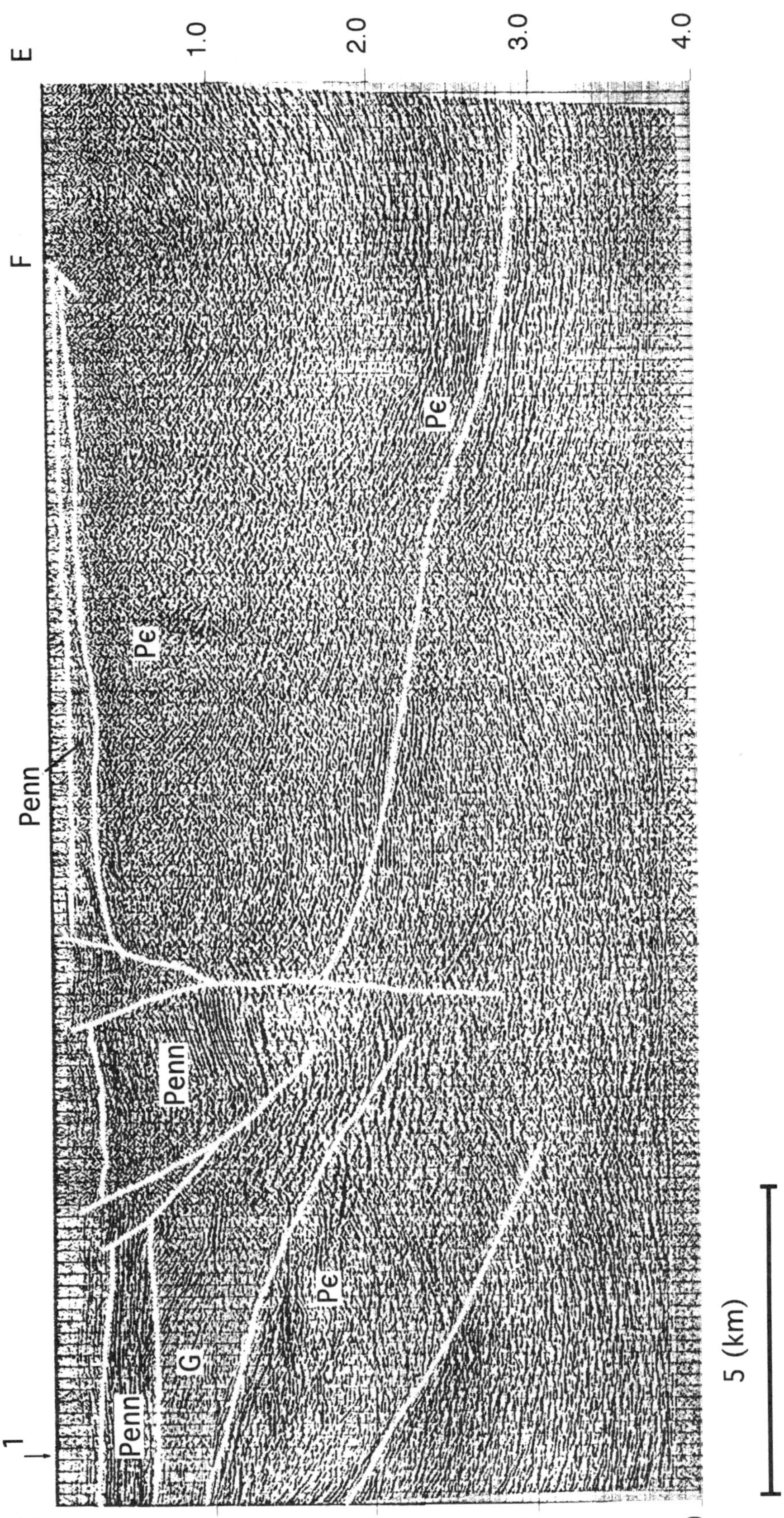

Figure 9. Interpreted record section for seismic Line F. Penn, Pennsylvanian sedimentary rocks that generally overlie the Precambrian basement. Interpreted ductile shear zones within the P€ are highlighted. G, granite body; 1, well (see Table 1); F, mapped fault. Vertical scale is two-way travel time in seconds.

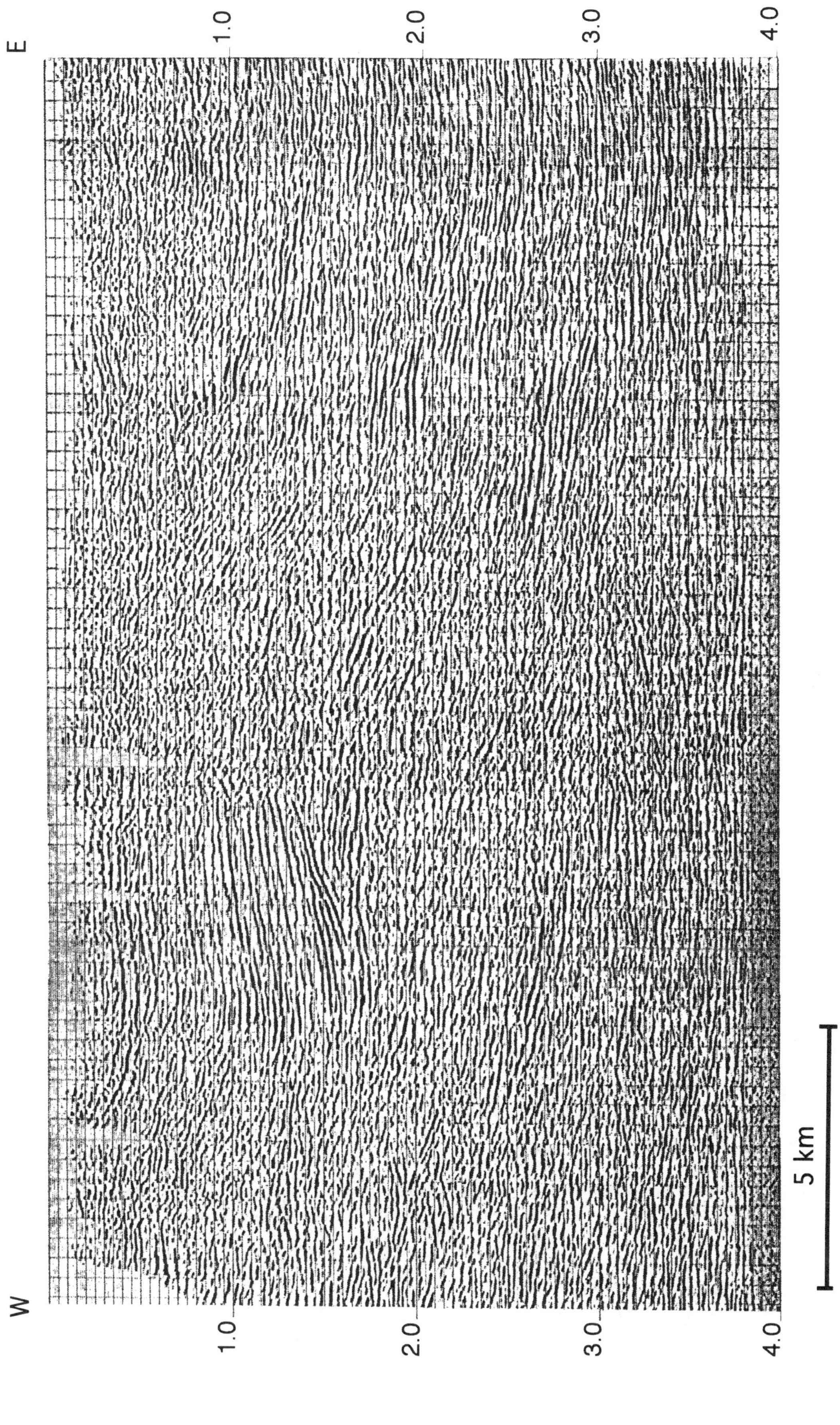

Figure 10. Uninterpreted record section for seismic line A. Vertical scale is two-way travel time in seconds.

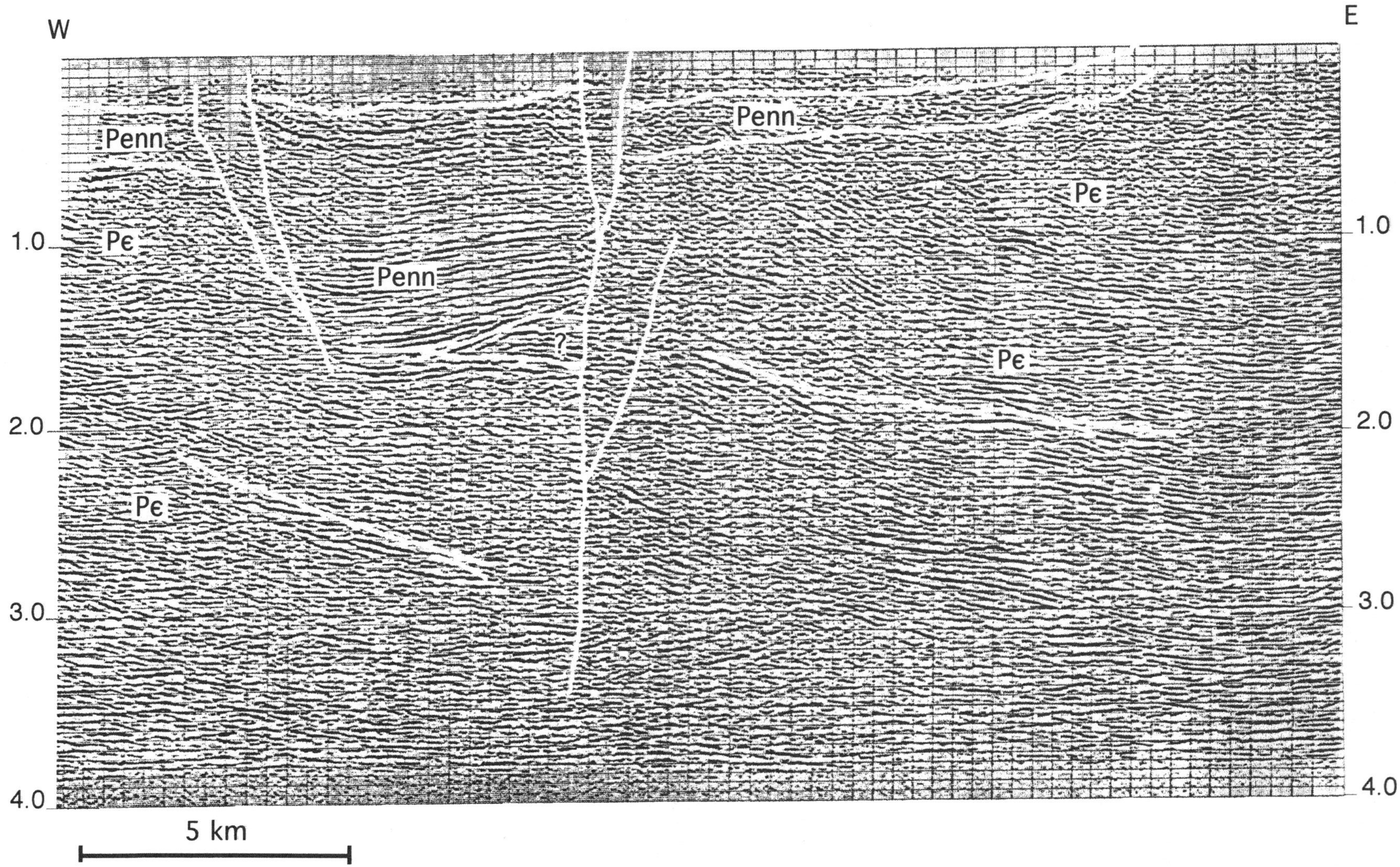

Figure 11. Interpreted record section for seismic line A. Penn, Pennsylvanian sedimentary rocks that overlie the Precambrian basement (P€). Interpreted ductile shear zones are highlighted. Vertical scale is two-way travel time in seconds.

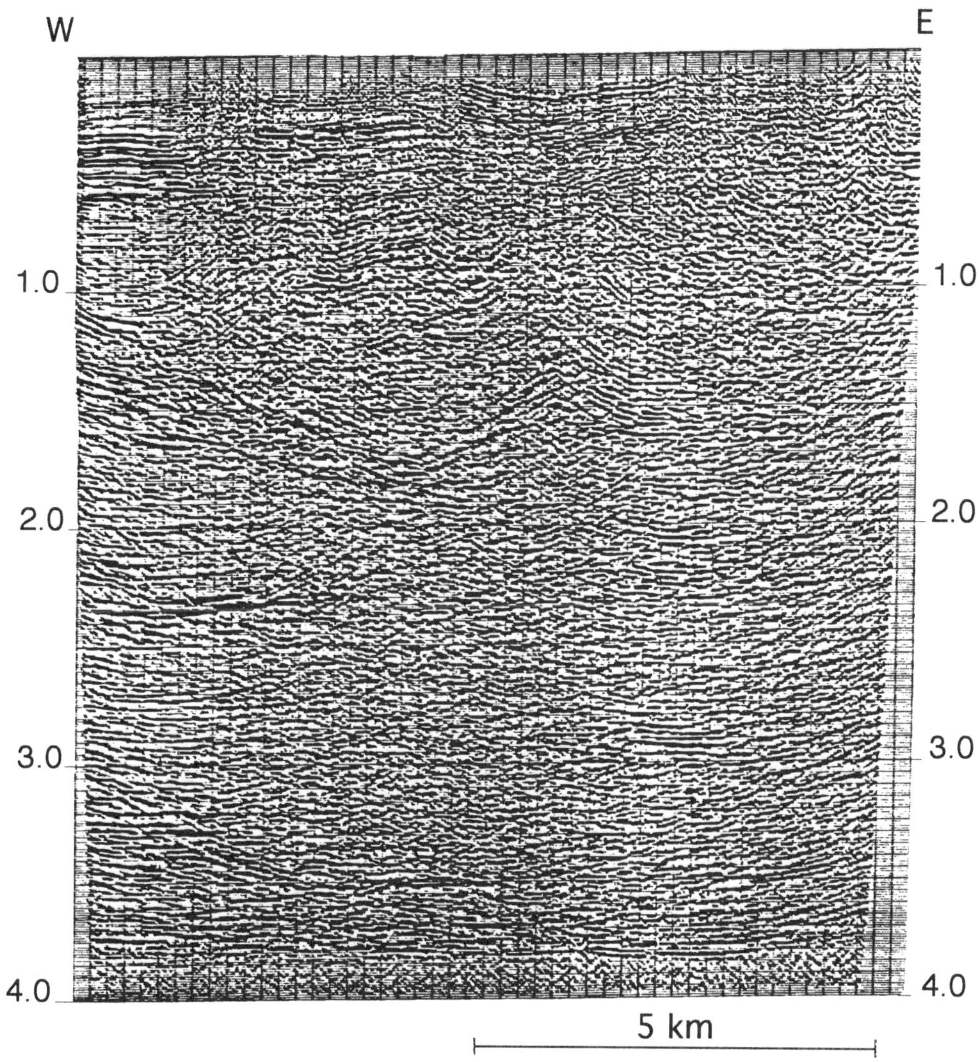

Figure 12. Uninterpreted record section for seismic line E. Vertical scale is two-way travel time in seconds.

weak. These faults are lower angle than those bounding the basin on the east and appear to have occurred contemporaneously with the deposition of the sediments.

The thick upper Paleozoic section is abruptly terminated on the east by a zone of disrupted reflectors. Farther east, much thinner sections of these rocks appear on the reflection data. This abrupt termination of the prominent reflectors marking the thick upper Paleozoic sequence, the change in depth to the basement shown by drilling results and seismic data, and the apparent flower structures seen on seismic data indicate that the eastern boundary of the thick late Paleozoic section is a strike-slip fault. These flower structures are important features and can be best seen on lines E, F, and G. They are negative features as defined by the downward merging normal faults (Harding, 1985).

The interpretation of the presence of flower structures is a key element of the results presented here. As discussed by Harding (1990), a number of criteria should be met before such an interpretation is proposed. We feel that the structures that we have identified meet most of Harding's (1990) criteria. Most importantly, there appears to be a steeply dipping master fault that offsets the Precambrian basement. This fault can be seen to some degree on all the record sections, and is best seen on line E (Fig. 12). Narrow fault slices with differing dips, which satisfy another of Harding's criteria, are also present in the upper part of the flower structures. Correlations of reflec-

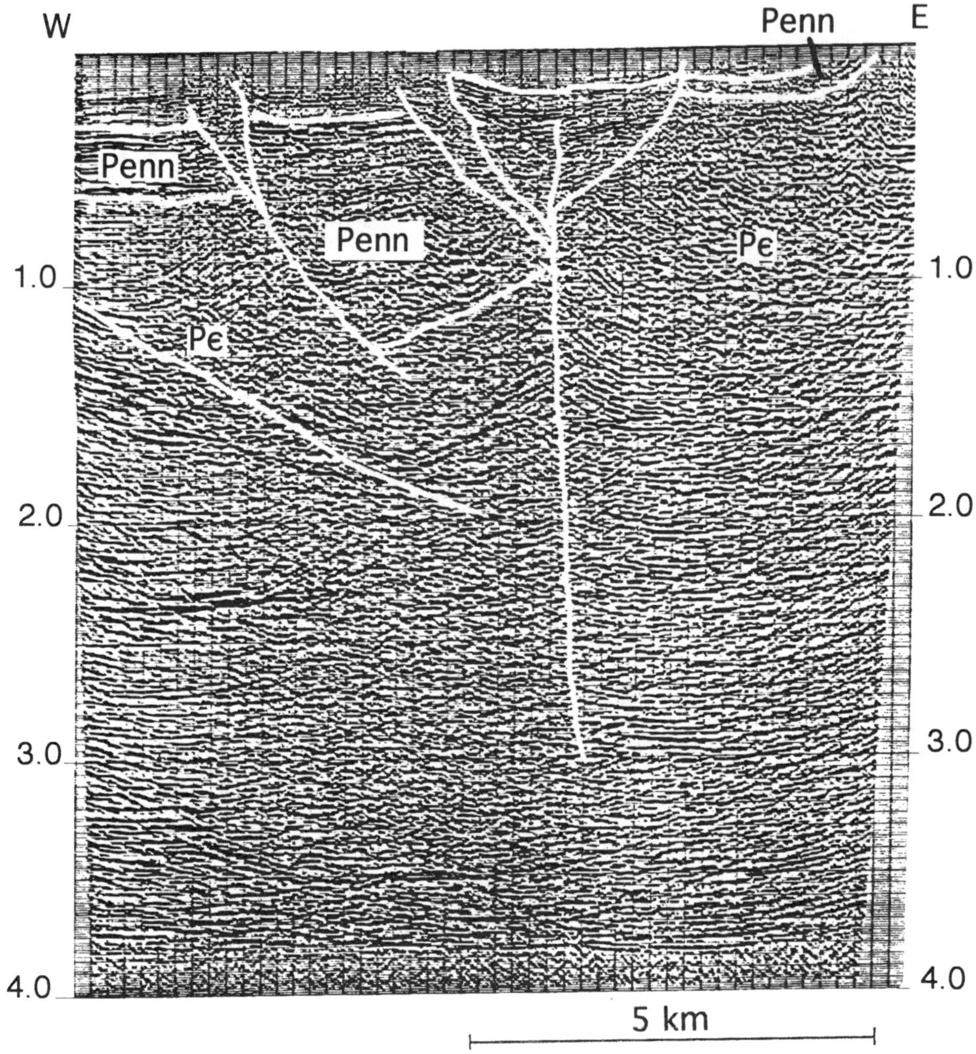

Figure 13. Interpreted record section for seismic line E. Penn, Pennsylvanian sedimentary rocks that generally overlie the Precambrian basement (P€). Interpreted ductile shear zones are highlighted. Vertical scale is two-way travel time in seconds.

tors across each structure are difficult, but there appear to be changes in the sense of movement along these slices. This criteria is difficult to apply with confidence because of the possibility of Laramide and rift-related reactivation of these faults.

A contour map of the two-way reflection time to the base of the Pennsylvanian section (top of the Precambrian basement) is shown in Figure 14. This picture of the basin's geometry indicates that it is a very steep sided trough. This geometry is consistent with what would be expected for a pull-apart basin (e.g., Aydin and Nur, 1982; Mann et al., 1983) along a left-lateral, strike-slip fault with a left stepover (Fig. 14). The left-lateral sense of movement is consistent with northwest-directed compression during Ancestral Rocky Mountains deformation (e.g., Kluth and Coney, 1981).

Based on surface mapping to the north, movement may have also occurred on this fault during the Laramide. Miller et al. (1963) indicated that in the region of the Picuris Mountains, a high-angle fault was downdropped to the east during the Laramide after strike-slip motion occurred along this fault during the Pennsylvanian and Precambrian. Chapin and Cather (1981) maintained that perhaps all of the movement along the Pecos-Picuris fault occurred during the Laramide. Seismic lines

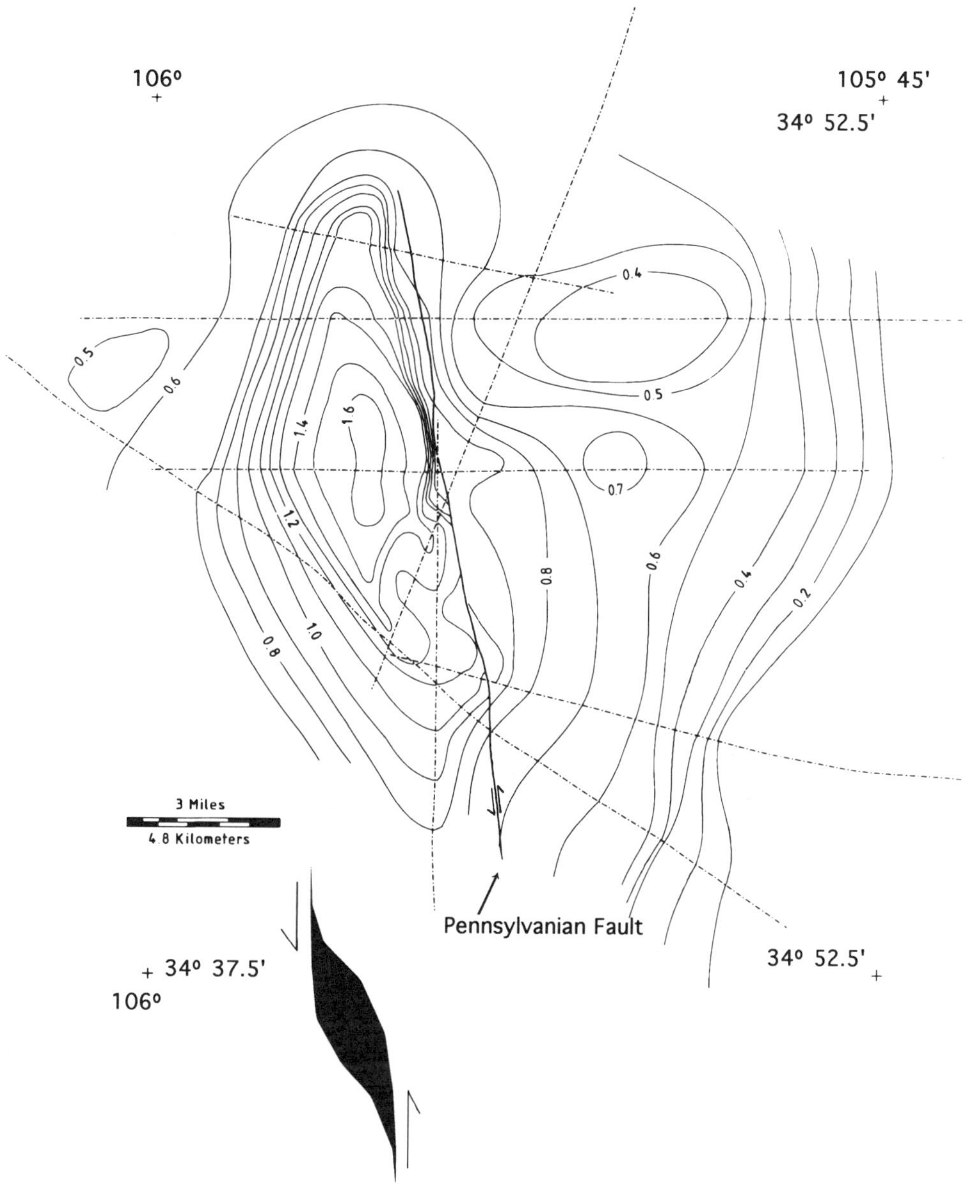

Figure 14. Time structure map of the interpreted Precambrian basement surface. Contour interval is 0.2 s. A time of 1.6 s converts to an approximate depth of 9,700 ft (3 km). Seismic lines from Figure 3 are dashed. A schematic diagram of the proposed pull-apart origin for the basin is shown as an inset at the bottom of the figure.

A, G, and F indicate that the eastern side of the Estancia Basin was downdropped after the deposition of the Permian sediments. The most likely time for this event was during the Laramide. The seismic data presented here imply older strike-slip movements but do not rule out Laramide strike-slip motion.

Reflectors associated with Permian rocks do not show much variation. The contact between the Permian and Pennsylvanian Systems is believed to be unconformable (Kelley, 1972). An unconformable relationship between reflectors can be seen on line F (Fig. 9) just east of the flower structure and at a time of 0.3 seconds. We interpret this relationship to correspond to the Pennsylvanian-Permian contact. The Permian sequence is nearly horizontal west of the series of faults that form the western boundary of the basin.

West of the strike-slip fault,the Pennsylvanian-Permian boundary is shallower than it is east of this structure. Data from the California Exploration #1 Well (Table 1) indicate a noticeable increase in the thickness of Yeso Formation (Permian) in this area. A pattern of reflectors indicating a thickening of the Yeso is present on all of the seismic sections east of the thick Pennsylvanian sequence. Since the Yeso contains very little carbonate material and large amounts of gypsum, this feature may be a collapse associated with dissolution of the evaporites within the Yeso Formation. The change in the depth of the Pennsylvanian-Permian boundary east of the thick Pennsylvanian sequence is probably due to movement along the strike-slip fault during the Laramide.

There is a remarkable amount of reflectivity within the Precambrian basement. The Proterozoic history of the New Mexico region is very complex and is the subject of much current research and debate (e.g., Bauer and Williams, 1985; Condie and Budding, 1979; Condie, 1986; Grambling et al., 1988,1989; Williams, 1990). There are a number of complex dipping reflectors in the basement. They are present to some extent on every seismic section. Since ductile shear zones are exposed in the nearby Pedernal Hills (Grambling et al., 1988) and in the Manzano Mountains (Thompson et al., 1991; Fig. 2), it seems plausible that these dipping reflectors (Figs. 8 to 13) are similar or even correlative features (Barrow, 1992). The basement structure indicated by such an interpretation is very similar to the interpreted cross section shown by Grambling et al. (1988) for northern New Mexico. It is beyond the scope of this paper to treat these features in detail, but they do attest to the complexity of the Precambrian basement in this area. This complexity explains some of the puzzling gravity anomalies in the area. As mentioned above, the gravity low in the Estancia Basin area only partly correlates with the thick Pennsylvanian section. The Olson Drilling #1 well encountered granite at a depth of only 1 km in an area where the gravity anomaly is still significantly negative (Figs. 3, 7). The seismic data show that this well penetrated a portion of the basement complex which lacks reflectivity (Fig. 15). Zones in the basement that are seismically transparent have often been found to be granitic bodies, and such bodies are usually associated with negative gravity anomalies. Barrow (1992) showed that the granitic body interpreted in Figure 15 is consistent with the gravity and seismic data and the fact that granitic intrusions are exposed in the Pedernal Hills (Bauer and Williams, 1985).

CONCLUSIONS

The main conclusion of this study is that strike-slip movement in the Pennsylvanian formed the Estancia Basin. The regional tectonic setting at this time would indicate left-lateral movement, and the geometry of the basin is consistent with this sense of movement on a fault with a left stepover (Fig. 14). However, we cannot presently identify any features that clearly establish this sense of movement. A similar sense of movement on late Paleozoic faults has been recognized in the nearby Joyita uplift (Beck and Chapin, this volume), and Hills (1970) has proposed that complex strike-slip movements occurred along faults in the Permian Basin in the late Paleozoic. Seismic reflection data indicate that the Estancia Basin is bounded by normal faults on the west, which suggests a pull-apart origin and the left stepover.

The possibility of Laramide movements on this fault is also of interest because of the strike-slip movements of this age proposed by Chapin and Cather (1981). The strike-slip fault(s) in the Estancia Basin and the Pecos-Picuris fault have a number of similar characteristics. The Pecos-Picuris fault has a northerly trend, a nearly vertical fault plane, right-lateral sense of displacement, and the eastern block was downdropped during the Laramide orogeny (Miller et al., 1963). Proximity and similarities in trend suggest that the two faults may be related. Miller et al. (1963) reported a minimum of 37 km of movement with episodes of movement during the Precambrian, Pennsylvanian, and the Laramide. Chapin and Cather (1981) suggested that all of the movement along the Pecos-Picuris fault may have occurred during the Laramide. It is unlikely that all of the movement along the strike-slip fault in the Estancia Basin occurred during the Laramide because structures within the Paleozoic section indicate movements that are contemporaneous with deposition. In addition, the normal faults forming the western boundary of the basin could not easily have occurred in a regional compressive regime without contemporaneous strike-slip movement to the east.

It is possible, however, that movement occurred along this fault during both the Pennsylvanian and the Laramide. The seismic data show that Paleozoic strata have been offset by vertical movements. Some faults in the interpreted flower structure demonstrate relative upward post-Paleozoic movement. This sense of movement is consistent with the Laramide regional stress regime and could be due in part to strike-slip motion. Evidence for strike-slip movements can be very elusive, but the post-Paleozoic structures present do not suggest that Laramide movements on this particular fault were large. Faults movements associated with the Rio Grande rift were also small.

Figure 15. Close-up interpreted record section for seismic line F, showing location of Olson Drilling #1 well (1) and the granitic body in Proterozoic basement that it penetrated. Dipping reflectors beneath the granitic body are ductile shear zones. Vertical scale is two-way travel time in seconds.

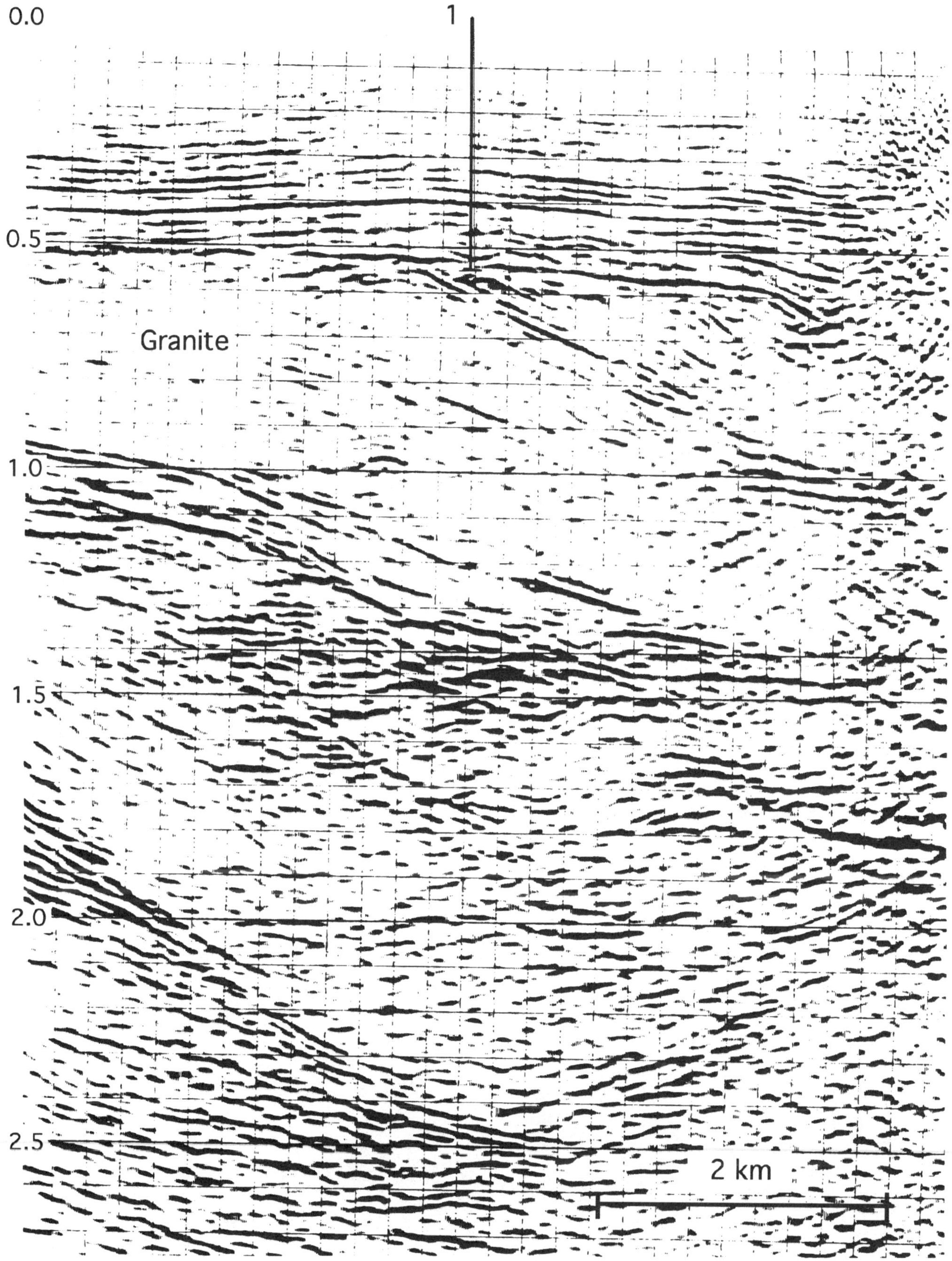
0.0
1
0.5
Granite
1.0
1.5
2.0
2.5
2 km

ACKNOWLEDGMENTS

We want to thank Ben Donegan for encouraging this study and helping us obtain the data we employed. Mobil Exploration and Production and Grant-Norpac (now Grant-Tensor) graciously made their data available for the study. Ron Broadhead of the New Mexico Bureau of Mines and Mineral Resources was very helpful with the well data. Jeff Grambling, Charles Chapin, and Steve Cather provided valuable geologic insight and the reviews of Jeff Grambling, Steve Cather, and Judson May were very helpful. Betsy Julian was a great help with early drafts of the manuscript.

REFERENCES CITED

Aydin, A., and Nur, A., 1982, Evolution of pull-apart basins and their scale independence: Tectonics, v. 1, p. 91–105.

Barrow, R. K., 1992, An integrated geophysical study of the Estancia Basin, central New Mexico [M.S. thesis]: University of Texas at El Paso, 83 p.

Bates, R. L., Wilpolt, R. H., MacAlpin, A. J., and Vorbe, G., 1947, Geology of the Gran Quiviva Quadrangle: New Mexico Bureau of Mines and Mineral Resources Bulletin 26, 57 p.

Bauer, P. W., and Williams, M. L., 1985, Structural relationships and mylonites in Proterozoic rocks of the Northern Pedernal Hills, central New Mexico: New Mexico Geological Society, 36th Field Conference, Guidebook, p. 141–145.

Chapin, C. E., 1971, The Rio Grande rift, Part 1: Modifications and additions: New Mexico Geological Society, 22nd Field Conference, Guidebook, p. 191–202.

Chapin, C. E., and Cather, S. M.,1981, Eocene tectonics and sedimentation in the Colorado Plateau–Rocky Mountain area: Arizona Geological Society Digest, v. 14, p. 13–198.

Condie, C., 1986, Geochemistry and tectonic setting of Early Proterozoic supracrustal rocks in the southwestern United States: Journal of Geology, v. 94, p. 845–854.

Condie, C., and Budding, J., 1979, Geology and geochemistry of Precambrian rocks, central and south-central New Mexico: New Mexico Bureau of Mines and Mineral Resources Memoir 35, 58 p.

Cordell, L., 1984, Composite residual total intensity aeromagnetic map of New Mexico: National Oceanographic and Atmospheric Administration, National Geophysical Data Center, scale 1:500,000.

Cordell, L., Keller, G. R., and Hildenbrand, T. G., 1982, Bouguer gravity map of the Rio Grande rift: U.S. Geological Survey, Map GP-949, scale 1:1,000,000.

Dallmeyer, R. D., Grambling, J. A., and Thompson, A. G., 1990, Age and character of Proterozoic polymetamorphism in New Mexico: Geological Society of America Abstracts with Programs, v. 22, no. 7, p. 113.

Dickinson, W. R., 1981, Plate tectonic evolution of the Southern Cordillera: Arizona Geological Society Digest, v. 14, p. 113–135.

Grambling, J. A., Williams, M. L., and Mawer, C. K., 1988, The Proterozoic tectonic assembly of New Mexico: Geology, v. 16, p. 724–727.

Grambling, J. A., Williams, M. L., Smith, R. F., and Mawer, C. K., 1989, The role of crustal extension in the metamorphism of Proterozoic rocks in New Mexico: Geological Society of America Special Paper 235, p. 87–110.

Harding, T. P., 1985, Seismic characteristics and identification of negative structures, positive flower structures, and positive structural inversion: American Association of Petroleum Geologists Bulletin, v. 69, p. 582–600.

Harding, T. P., 1990, Identification of wrench faults using subsurface structural data: Criteria and pitfalls: American Association of Petroleum Geologists Bulletin, v. 74, p. 1590–1609.

Hills, J. M., 1970, Late Paleozoic structural directions in southern Permian Basin, west Texas and southeastern New Mexico: American Association of Petroleum Geologists Bulletin, v. 54, p. 809–827.

Kelley, V. C., 1972, Geology of the Fort Summer Sheet, New Mexico: New Mexico Bureau of Mines and Mineral Resources Bulletin 98, 55 p.

Kluth, C. F., and Coney, P. J., 1981, Plate tectonics of the Ancestral Rocky Mountains: Geology, v. 9, p. 10–15.

Mann, P., Hempton, M. R., Bradley, D. C., and Burke, K., 1983, Development of pull-apart basins: Journal of Geology, v. 91, p. 529–554.

Miller, J. P., Montgomery, A., and Sutherland, P. K., 1963, Geology of part of the southern Sangre de Cristo Mountains, New Mexico: New Mexico Bureau of Mines and Mineral Resources Memoir 11, 106 p.

Smith, R. E., 1957, Geology and ground-water resources of Torrance County, New Mexico: New Mexico Bureau of Mines and Mineral Resources Bulletin 97, 186 p.

Thompson, A. G., Grambling, J. A., and Dallmeyer, R. D., 1991, Proterozoic tectonic history of the Manzano Mountains, central New Mexico: New Mexico Bureau of Mines and Mineral Resources Bulletin 137, p. 71–77.

Williams, M. L., 1990, Proterozoic geology of northern New Mexico: Recent advances and ongoing questions: New Mexico Geological Society, 41st Field Conference, Guidebook, p. 151–159.

Woodward, L. A., 1976, Laramide deformation of Rocky Mountain foreland: Geometry and mechanics: New Mexico Geological Society Special Publication 6, p. 11–17.

MANUSCRIPT ACCEPTED BY THE SOCIETY OCTOBER 13, 1993

Printed in U.S.A.

Geological Society of America
Special Paper 291
1994

Structural and tectonic evolution of the Joyita Hills, central New Mexico: Implications of basement control on Rio Grande rift

William C. Beck* and Charles E. Chapin
New Mexico Bureau of Mines and Mineral Resources, Campus Station, Socorro, New Mexico 87801

ABSTRACT

Phanerozoic sedimentary and volcanic rocks exposed in the Joyita Hills area (Socorro County, New Mexico) record structures that developed in response to three deformational episodes: ancestral Rocky Mountain, Laramide, and Rio Grande rift. Additional exposures of Proterozoic gneiss in the core of the Joyita Hills reveal three preferred orientations of high-angle to vertical, gneissic to mylonitic foliations. These strike north, northwest, and east-northeast. Proterozoic amphibolite dikes parallel the east-northeast foliations. Comparison of Phanerozoic fault and dike orientations with attitudes of Proterozoic structures indicates that reactivation of basement flaws was a factor common to each Phanerozoic orogenic event.

Reactivation of Proterozoic foliations during ancestral Rocky Mountain tectonism resulted in basins and uplifts of north and northwest trend. Normal, left-oblique slip on north-striking faults and normal slip on northwest-striking faults, when combined with the orientation and spatial distribution of basins and uplifts, indicates that late Paleozoic deformation in central New Mexico was the result of north-trending, divergent, sinistral wrench faulting.

Laramide tectonism reactivated Proterozoic structures in a strike-slip sense. North-striking (synthetic) and east-northeast–striking (antithetic) faults define a north-trending, dextral wrench fault system. Other dextral wrench faults strike northeast, and parallel the Montosa and Del Curto fault zones. Thrust faulting and associated folds were a lesser component of Laramide deformation, and were apparently related to Laramide wrench fault systems.

Extensional reactivation of Proterozoic structures influenced the development of basins and uplifts of the Rio Grande rift. Northwest-striking and east-northeast–striking normal fault and dike systems paralleled Proterozoic foliations and amphibolite dikes. North-striking foliations and ancestral Rocky Mountain faults observed along the East Joyita fault influenced the development of this major, down-to-the-east normal fault (approximately 3 km stratigraphic throw). Moreover, north-striking foliations and ancestral Rocky Mountain structures (the north-trending Lucero basin and adjacent uplift) influenced the location and orientation of the Albuquerque Basin segment of the Rio Grande rift.

*Present address: Belfort Engineering & Environmental Services, Inc., One Bank Street, Suite 150, Gaithersburg, Maryland 20878.

Beck, W. C., and Chapin, C. E., 1994, Structural and tectonic evolution of the Joyita Hills, central New Mexico: Implications of basement control on Rio Grande rift, *in* Keller, G. R., and Cather, S. M., eds., Basins of the Rio Grande Rift: Structure, Stratigraphy, and Tectonic Setting: Boulder, Colorado, Geological Society of America Special Paper 291.

INTRODUCTION

The Joyita Hills (Los Cañoncitos) area is a north-north-east–trending, west-tilted, normal-fault block at the southern end of the Albuquerque Basin (Fig. 1). Approximately 4 km of inclined Phanerozoic sedimentary and volcanic strata are well exposed in the Joyita Hills and adjacent areas to the east and south (El Valle de La Joya; Figs. 1 and 2). Proterozoic basement rocks (augen gneiss) are also exposed in the core of the Joyita Hills. These rock sequences represent much of the known section in central New Mexico, and record the faults, folds, and other structures that formed in response to a minimum of three recognized Phanerozoic orogenic events. Previous stratigraphic and biostratigraphic investigations (Read and Wood, 1947; Kottlowski, 1960; Kottlowski and Stewart, 1970; Baars, 1982; Siemers, 1978, 1983) have identified the Joyita Hills area as the location of an ancestral Rocky Mountain uplift (the late Paleozoic Joyita uplift). Structural and tectonic interpretations of the Joyita uplift have been previously discussed by Beck and Chapin (1991) and Beck and Johnson (1992). In addition to ancestral Rocky Mountain deformation, partial omission of late Paleozoic and Mesozoic strata west of the East Joyita fault (Figs. 1 and 2) is indicative of Laramide uplift and erosion, and west-tilted Oligocene volcanic strata attest to deformation associated with Rio Grande rift extension.

Resultant overprinting of diverse structural styles has resulted in a complex array of fault types and orientations. Observed structures include those associated with crustal extension, wrench tectonics, and to a lesser extent, thrust faulting. Fortuitously, the stratigraphic units preserved in the Joyita Hills and El Valle de La Joya (Fig. 2) represent pre-, syn-, and post-tectonic sequences that roughly bracket each recognized tectonic episode. For example, middle Permian (Leonardian and Guadalupian) and Mesozoic sedimentary units postdate ancestral Rocky Mountain deformation and predate Laramide tectonism; Oligocene volcanic units postdate Laramide deformation and represent pre- and syntectonic units with respect to Rio Grande rift extension. These stratigraphic sequences record the differing structures that developed during each suc-

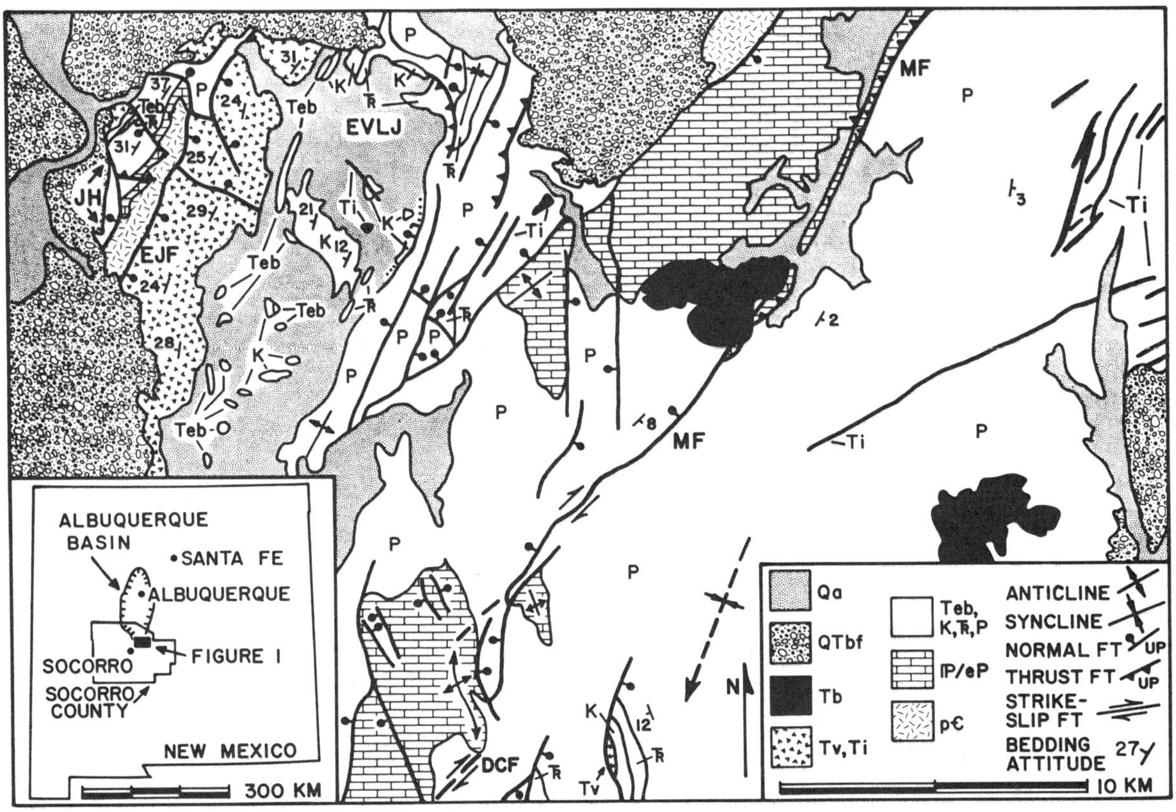

Figure 1. Generalized geologic map of northeast Socorro County, New Mexico. Data presented herein were collected from Proterozoic and Phanerozoic exposures in the Joyita Hills (JH) and El Valle de La Joya (EVLJ). Regional faults include the East Joyita fault (EJF), Del Curto fault (DCF), and Montosa fault (MF). pЄ, Proterozoic; IP/eP, Pennsylvanian and earliest Permian strata; P, Permian strata; TR, Triassic strata; K, Cretaceous strata; Teb, Eocene basin deposits; Tv, latest Eocene–Oligocene volcanics; Ti, Oligocene intrusives (dikes and plugs); Tb, Pliocene basalt flows; QTbf, Quaternary and Tertiary basin fill; Qa, Quaternary alluvium. Modified from Osburn (1984) and Myers et al. (1986).

cessive orogeny, and the stratigraphic interval that a given structural style overprints has been used to correlate structures with specific tectonic events.

The purpose of this paper is to present structural data associated with each of the foregoing deformational episodes and to formulate structural and tectonic interpretations within a temporal perspective. Although both ductile (Proterozoic) and brittle (Phanerozoic) structures are well exposed, emphasis has been placed on structures associated with Phanerozoic deformation. Fault, fault-striae, dike, and fold-axis orientations have been grouped into compatible systems in order to evaluate structural styles, associated kinematics, and, where possible, to infer applied stress orientations. Reactivation of existing structures was a fundamental aspect of these deformational episodes. Orientations of Proterozoic foliations and amphibolite dikes have been plotted in order to evaluate the influence of basement structures during Phanerozoic deformation and the extent to which reactivation has occurred.

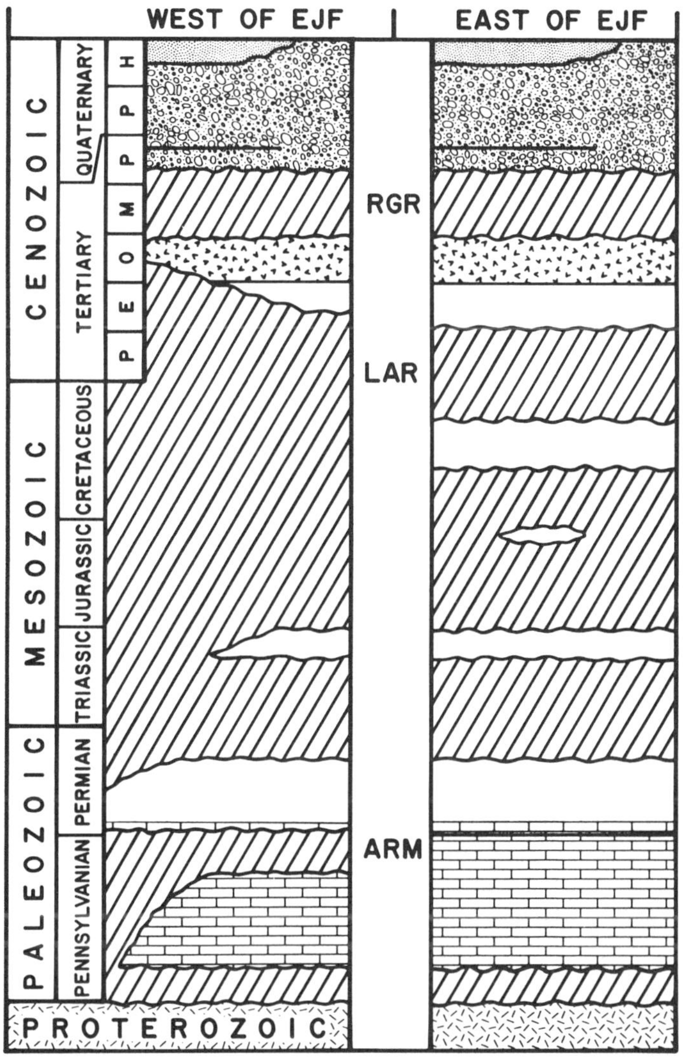

Figure 2. Generalized stratigraphy exposed to the west (Joyita Hills) and east (El Valle de La Joya) of the East Joyita fault (EJF). Initially horizontal sedimentary and/or volcanic sequences were deposited prior to and after each recognized tectonic episode, allowing a direct correlation between observed structural styles and specific orogenic events. RGR, Rio Grande rift; LAR, Laramide; ARM, ancestral Rocky Mountain. Continuous deposition of Pennsylvanian and Permian strata to the east of the East Joyita fault (not observed in field area) inferred from Read and Wood (1947), Kottlowski (1960), and Peterson (1980). Continuous deposition during this interval also occurred to the west and south of the Joyita Hills (Kottlowski, 1960). Lithologic patterns correlated with those of Figure 1; diagonal rule indicates absence (either nondeposition or erosion) of stratigraphic record. Modified from Osburn and Lochman-Balk (1983).

METHOD OF ANALYSIS

Evaluation of Phanerozoic structures and kinematics has been accomplished largely through fracture analyses. Data collection has been coordinated with detailed mapping (minimum scale of 1:12,000; 1:6,000 where structurally complex), and stratigraphic relationships have been used to supplement structural and kinematic interpretations. Structural attitudes, including bedding, fault, fault-striae, and dike orientations, were described and recorded in the field. Displacement (net slips) along individual faults range from less than 1 cm to several kilometers (as along the East Joyita fault, which displays approximately 3 km of down-to-the-east displacement; Fig. 1). Most faults observed and used in the following analyses, however, exhibit net slips of a few meters to 100 m.

Fault and dike orientations have been used to infer applied stress orientations in a manner similar to that proposed by Anderson (1951) for shallow-crustal, brittle deformation. Anderson (1951) proposed that in near-surface settings, one of the principal stress axes should be vertical, and accordingly, conjugate fault orientations would be equally disposed to horizontal. However, field observation in the Joyita Hills area indicated that what appeared to be conjugate fault orientations were not equally disposed to horizontal, but rather were equally inclined with respect to bedding. For example, faults rather consistently cut across bedding at approximately 30° (thrust faults), 60° (normal faults), or 90° (strike-slip faults), regardless of bedding orientation. Similarly, extensional fractures (mid-Tertiary mafic dikes) were consistently oriented at roughly 90° to bedding. These relationships are interpreted to indicate that these structures were superimposed on bedding prior to appreciable rotation associated with a given tectonic event, and that rotation has removed early formed fractures from their original orientation.

The consistency between fracture orientation and bedding has been demonstrated by plotting structural elements with respect to bedding. This was accomplished by restoring inclined bedding to horizontal, and rotating contained structural elements by a comparable amount. The analytical method, therefore, treats the foregoing structural features as early formed elements within a deformational episode, and produces orientations of faults, fault striae, and inferred stress axes that are compatible with observed structures, brittle deformation, and Andersonian theory.

The concept of brittle failure prior to rotation of strata (initial fracturing) is not new. Previous investigations (e.g., Melton, 1929; Parker, 1942; Reches, 1976; Beck, 1984; Beck and Burford, 1985) have also treated tectonically induced fractures in this manner. Neither is it the only means by which the consistency between fracture orientation and bedding may be explained. Alternatively, the consistency may be explained by assuming that frictionless bedding-plane slip occurred during deformation. In this latter approach, the bedding plane would be by definition a principal plane of stress, and one of the principal stress axes would be consistently oriented normal to the bedding plane, regardless of bedding orientation. Analyses in which fracture patterns have been evaluated in light of this hypothesis have been provided by Jackson and Pollard (1990). Similar analyses, in which the symmetry of fractures about bedding was attributed to bedding-parallel compression and/or extension, have been provided by Reches (1978).

The analyses that follow take the initial fracturing approach, and brittle structures are treated as early formed structural elements within a given deformational episode. This analytical method is preferred for its inherent simplicity: Since tectonically induced fractures should be preferentially oriented with respect to the applied stress field, the consistency between fracture orientation and bedding would be most readily explained if fractures were superimposed on bedding while bedding was consistently oriented, rather than calling for multiple orientations of the applied stress field. A more thorough discussion of the reasoning behind this approach, and a more detailed analysis of brittle structures in the Joyita Hills area, has been provided by Beck (1993). The analyses of Reches (1978) and Jackson and Pollard (1990) are cited in the interest of providing alternative interpretations to the approach adopted herein.

This analytical technique has been applied to fractures associated with Rio Grande rift and Laramide tectonisms, and applied stress orientations have been inferred for both of these events. In contrast, ancestral Rocky Mountain structures have been overprinted and reactivated by these younger deformational events, which has made application of the foregoing analytical technique to late Paleozoic structures difficult. Fortunately, ancestral Rocky Mountain structures have been determined through field observation and recognition of buried structure. Late Paleozoic structures (faults and striae), as well as Proterozoic structures (foliations and amphibolite dikes), have been restored to their approximate original orientations by using regional bedding.

PROTEROZOIC BASEMENT STRUCTURES

Proterozoic rocks (augen gneiss) in the core of the Joyita Hills (Fig. 1) host a number of planar structures. These include joints, faults, foliations, and amphibolite, pegmatite, and aplite dikes. Previous petrologic, structural, and descriptive interpretations of these structures have been provided by Herber (1963a, b).

Foliations are gneissic to mylonitic, strongly lineated, and occur along three common trends: north, northwest, and east-northeast. Stretching lineations are well developed, and locally, as in the southern Joyita Hills, the stretching lineation is the dominant structural feature of the augen gneiss complex (an L-tectonite or L>S tectonite). Amphibolite dikes and some aplite dikes (deformed aplites) are also foliated and lineated. Where observed, foliations and lineations within these dikes parallel those in the surrounding gneiss, and accordingly they are interpreted as pre- or synmetamorphic structures. In contrast, pegmatite and other aplite dikes (undeformed aplites) are unfoliated structures that both parallel and crosscut the older foliations.

Restoration of foliation and amphibolite dike orientations has been accomplished by removing the combined effects of Phanerozoic rotations (29° of down-to-the-west rotation about a N32°E, horizontal axis; Fig. 3A). Restored foliations (Fig. 3B) are high-angle to vertical structures that strike north, northwest, and east-northeast. North- and east-northeast–striking foliations are most common. Restored amphibolites (Fig. 3C) are also high-angle structures that parallel the east-northeast–striking foliations. Given the limited rotations associated with ancestral Rocky Mountain and Laramide tectonisms (Fig. 3A), the orientations of restored foliations and amphibolite dikes (Fig. 3B, C) approximate the attitudes of basement structures in the underlying basement rocks of the Joyita Hills prior to mid-Tertiary rotation.

ANCESTRAL ROCKY MOUNTAIN DEFORMATION

General stratigraphic relationships

Regional aspects of late Paleozoic sedimentation and tectonism have been previously discussed by Peterson (1980), Kluth and Coney (1981), and Kluth (1986). In New Mexico, deformation resulted in basins and uplifts of dominantly north and northwest trend. These trends began to develop as early as latest Mississippian (Chesterian) time (Armstrong, 1962; Armstrong and Mamet, 1988; Kluth, 1986), persisted throughout the Pennsylvanian (Kottlowski, 1960), and into the early Permian (Peterson, 1980). On a more local scale, stratigraphic investigations (Read and Wood, 1947; Kottlowski, 1960; Kottlowski and Stewart, 1970; Baars, 1982; Siemers, 1978, 1983) have identified the Joyita Hills area as the location of an ancestral Rocky Mountain uplift (the Joyita uplift).

Accumulations of Pennsylvanian sediments in depositional basins to the west, south, and east of the Joyita Hills attain thicknesses of up to 820 m (Fig. 4; Read and Wood, 1947; Kottlowski, 1960; Baars, 1982; Siemers, 1983). These strata are typically shallow water (continental shelf) carbonates and subordinate clastics (Kottlowski, 1960). The sequences represent nearly continuous sedimentation that spans Atokan through Virgilian time (Kottlowski, 1960; Siemers, 1983), and therefore subsidence must have kept pace with sedimentation. In contrast, Pennsylvanian sequences in the Joyita Hills are anomalously thin (0 to 127 m, Kottlowski and Stewart, 1970; 0 to 112 m, Siemers, 1983). These sequences also consist of shallow-water (restricted lagoon, intertidal, and inner to outer shelf) carbonates and subordinate clastics that were deposited during Atokan through Missourian time. Kottlowski and Stewart (1970) inferred that Virgilian strata were also deposited in the Joyita Hills, but subsequently were removed by Late Pennsylvanian or Early Permian uplift and erosion. Several stratigraphic discontinuities have been observed in this section (Kottlowski and Stewart, 1970). Diastems occur in early and late Atokan sequences, along the Atokan/Desmoinesian boundary, and in early Desmoinesian strata. The latter two are indicative of localized uplift and erosion, as evidenced by clasts of augen gneiss, feldspar, quartz, and reworked limestone. A disconformity occurs between early Desmoinesian and Missourian strata, which has also been attributed to uplift

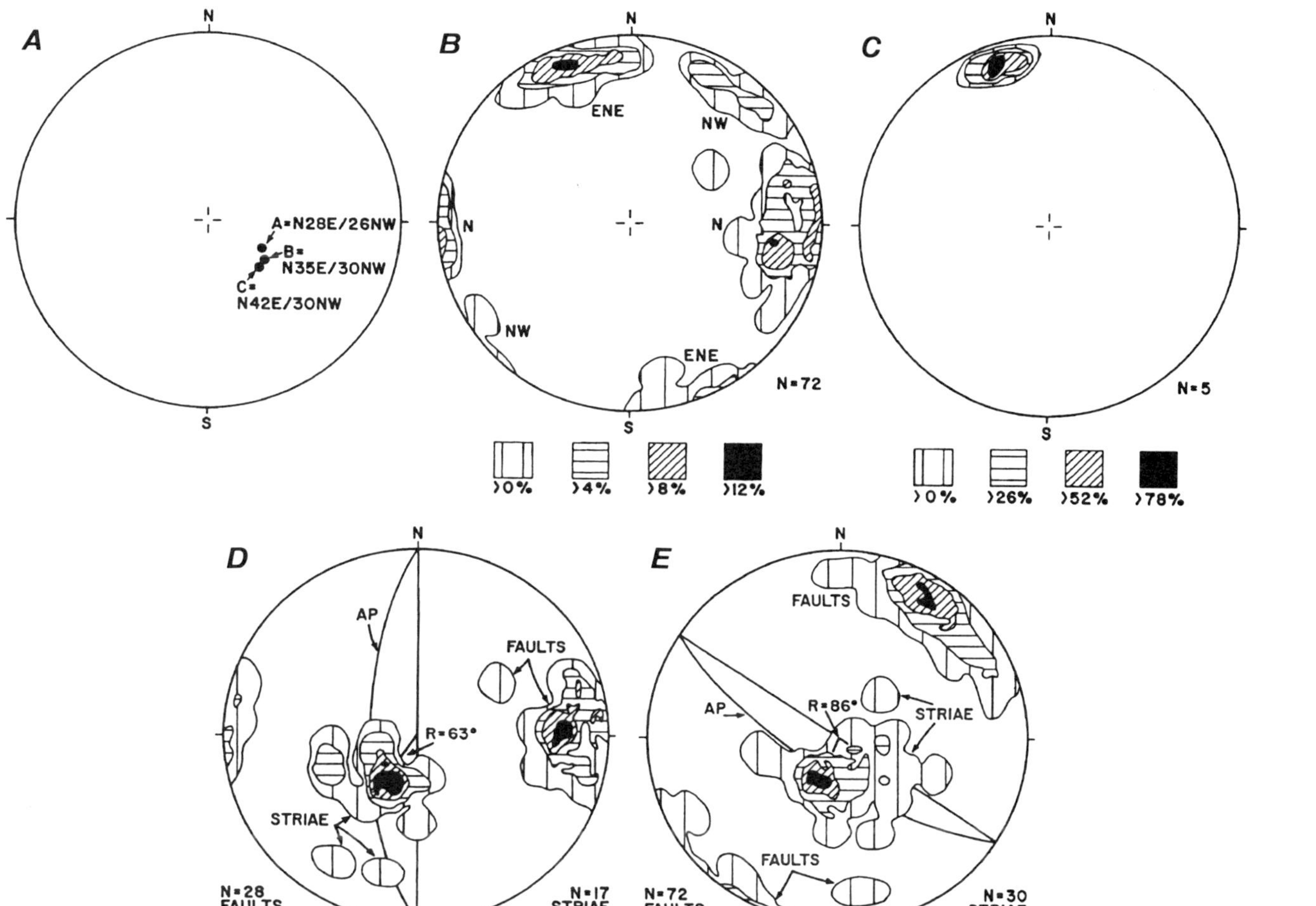

Figure 3. Lower-hemisphere, equal-area projections of regional bedding, Proterozoic and late Paleozoic structural orientations. A, Regional bedding, where A represents average orientation of Tertiary volcanic rocks (post-Laramide, pre– and syn–Rio Grande rift sequences; refer to Fig. 2); B represents average orientation of middle-Permian through Eocene sedimentary rocks (post–ancestral Rocky Mountain, pre- and syn-Laramide sequences); and C represents average orientation of Pennsylvanian and earliest Permian strata (pre- and syn–ancestral Rocky Mountain sequences). B, Restored orientations of Proterozoic foliations, which strike north (N), northwest (NW), and east-northeast (ENE); C, restored Proterozoic amphibolite dikes, which also strike east-northeast; D,E, restored north-striking and northwest-striking ancestral Rocky Mountain faults and contained striae, respectively. AP denotes average fault plane; R denotes approximate rake of striae. Poles to planes (B, C) and composite poles to planes and lineations (D,E) contoured at specific intervals per 1% area.

and erosion during middle Desmoinesian time (Kottlowski and Stewart, 1970; see also Kluth, 1986). In addition, evidence for an early (Atokan) episode of faulting has been observed in the Joyita Hills (Beck and Johnson, 1992). Down-to-the-east displacements along subvertical, north-striking faults offset early Atokan strata (Sandia Formation), but are overlain by unfaulted, late Atokan strata (basal Madera Formation).

Accordingly, the anomalously thin Pennsylvanian section in the Joyita Hills can be attributed to tectonic adjustments during sedimentation, rather than uplift and erosion of an initially thick Pennsylvanian section. The foregoing stratigraphic discontinuities and evidence of Atokan deformation are thought to reflect these adjustments, which served to elevate the Joyita uplift as a structurally positive block amid subsiding basins throughout the Pennsylvanian.

Ancestral Rocky Mountain structures

Ancestral Rocky Mountain tectonism in the Joyita Hills reached its culmination during early Wolfcampian (earliest Permian) time (Kottlowski and Stewart, 1970; Beck and Johnson, 1992). Deformation in the Joyita Hills occurred along two fault sets: north-striking normal faults and northwest-striking normal faults (Beck and Chapin, 1991). North-striking faults have been found within or along the fault-bounded margins of Proterozoic through early Wolfcampian rocks, and typically occur as wide (3 to 7 m) gouge zones that are conspicuously marked by south-plunging fault striae and mullion structures. Most faults of this orientation have been reactivated and display a down-to-the-east sense of displacement. These reactivated faults are found in proximity

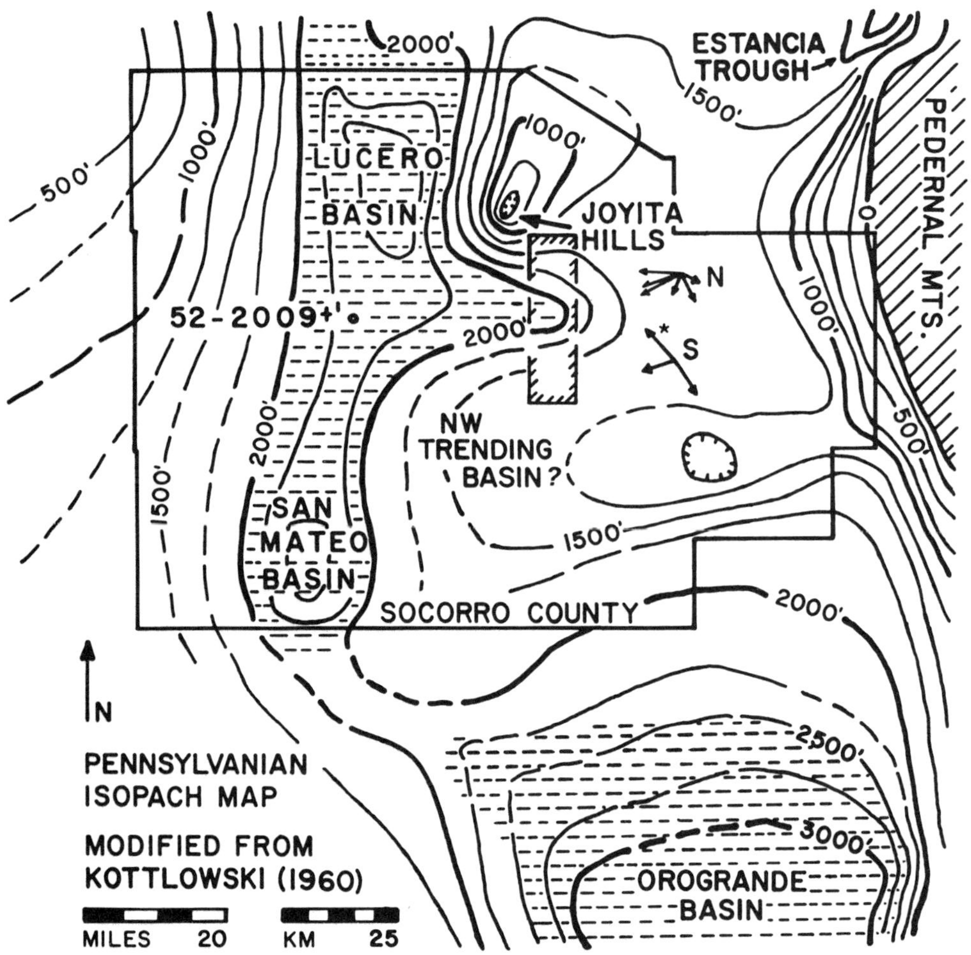

Figure 4. Modified Pennsylvanian isopach map superimposed with early Wolfcampian paleocurrent data (Altares, 1990). Hachured box outlines location of Altares' (1990) field area. Paleocurrent directions (vector means) labelled N and S are from the northern and southern parts of Altares' field area; length of each vector is proportional to vector strength. Drill-core data (Krewedl, 1974; C. E. Chapin, unpublished core log; control point 52, Kottlowski, 1960) documents stratigraphic thickness to be in excess of 610 m (2,000 ft), and redefines the Lucero and San Mateo basins as one continuous, north-trending basin. Hachured contours denote relative uplifts: respectively, these are the Joyita uplift and the Oscura Mountains in north-central and southeastern Socorro County. Contour interval = 250 ft.

to the East Joyita fault (which exhibits approximately 3 km of mid-Tertiary, down-to-the-east displacement), and reactivation is thought to have occurred as a result of mid-Tertiary extension (discussed below). Other faults of this orientation display a down-to-the-west displacement, and do not appear to have been reactivated. One of these faults (the fault is unnamed; Fig. 5) is comparatively well exposed. Along this fault, Proterozoic through early Wolfcampian rocks display a stratigraphic offset of approximately 100 m. Syntectonic strata (Bursum Formation) are anomalously thick and coarse grained in the hanging wall (west block, downthrown) and apparently were deposited along the exposed fault scarp. In the footwall (east block, upthrown), syntectonic clastics overlie erosionally thinned Pennsylvanian (Madera Formation) limestones. Both the fault and offset strata are overlain by depositionally thinned, unfaulted mid-Wolfcampanian strata (Abo Formation), and the fault is therefore interpreted as a buried ancestral Rocky Mountain fault that has not been reactivated.

Northwest-striking faults are more problematic, in that all faults of this orientation appear to have been reactivated during mid-Tertiary extension (discussed below). However, subtle differences in fault character and orientation indicate that these faults were also active during late Paleozoic tectonism. Similar to the north-striking faults, northwest-striking normal faults superimposed on early Wolfcampian (Bursum Formation) and older rocks also occur as wide (3 to 7 m) gouge zones, are marked by well-developed fault striae and mullion structures, and dip consistently to the southwest. In contrast, northwest-striking normal faults superimposed on mid-Wolfcampian (Abo Formation) and younger strata occur as discrete fault planes that exhibit limited brecciation, poorly developed fault striations, and occur as conjugate fault orientations. These changes in fault character occur at the boundary between early and mid-Wolfcampian strata. Although interpretive, it is thought that the disparity in observed strain over such a short stratigraphic interval cannot readily be attributed to a single deformational episode. Rather, it is thought that the greater strain observed in early Wolfcampian and older rocks is attributable to structural deformation during two different orogenic episodes (ancestral Rocky Mountain and Rio Grande rift).

Restoration of both fault sets has been accomplished by using regional bedding (removing 29° of down-to-the-west rotation about a N32°E, horizontal axis; Fig. 3A). Restoration defines the fault sets as north-striking, west-dipping to subvertical normal faults Fig. 3D) and northwest-striking, southwest-dipping normal faults (Fig. 3E). Restored fault striae define

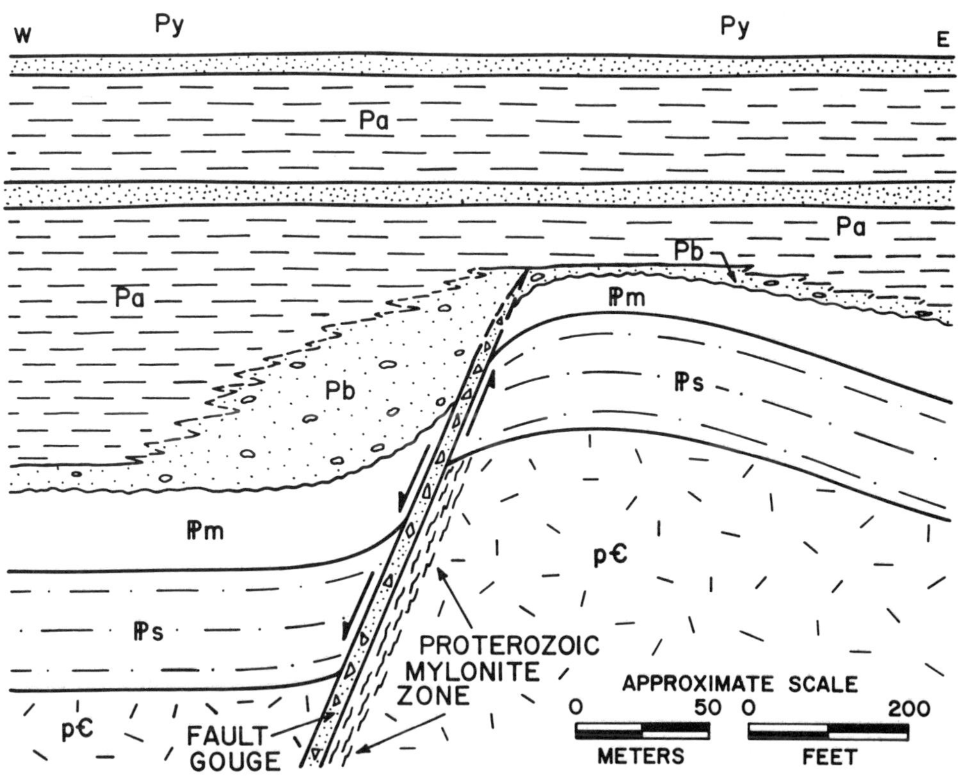

Figure 5. Generalized cross section normal to one of the north-striking, ancestral Rocky Mountain normal faults that has not been reactivated. Cross section has been restored to its post–ancestral Rocky Mountain orientation by removing 29° of down-to-the-west rotation about a N32°E, horizontal axis (Fig. 3A). Py, Yeso; Pa, Abo; Pb, Bursum; IPm, Madera; IPs, Sandia; p€, Proterozoic basement. Fault and contacts dashed where obscured. The location of this fault has been provided by Beck and Chapin (1991) and Beck and Johnson (1992).

fault displacements as normal, left-oblique slip along north-striking faults (striae rake approximately 63° southward), and as dip slip along northwest-striking normal faults.

Comparison of ancestral Rocky Mountain fault orientations (Figs. 3D, E) with those of north- and northwest-striking Proterozoic foliations (Fig. 3B) indicates that orientations are similar. Reactivation of north-striking foliations has been verified in the field along the north-striking ancestral Rocky Mountain fault depicted in Figure 5. Reactivation of northwest-striking foliations is also thought to have occurred, given the comparable orientations, although this correlation has not been verified by field observation.

Interpretation of ancestral Rocky Mountain deformation

Modified Pennsylvanian isopach maps (Fig. 4) support the foregoing fault orientations and define the Joyita uplift as a north-trending structural element. The uplift was bordered to the west and southwest by depositional basins (the Lucero and San Mateo basins). Modifications to the isopach map (originally drawn by Kottlowski, 1960) consist of redefining contour patterns between the Lucero and San Mateo basins. These changes are the result of more recent drill-core data from a mineral exploration hole on the west flank of the Magdalena Mountains (Krewedl, 1974; C. E. Chapin, unpublished core log), which encountered 612+ m (2009+ ft) of Pennsylvanian strata. This drill-core data indicates that the San Mateo and Lucero basins were one continuous, north-trending basin. Both the Joyita uplift and modified San Mateo/Lucero basin probably extended northward as continuous structures to the Sierra Nacimiento area and the adjacent Cabezon sag (to the west of Sierra Nacimiento; the Peñasco uplift of Kottlowski, 1960, Fig. 30; see also Baars, 1982; Read and Wood, 1947).

Although they do not bear directly on Pennsylvanian rocks, paleocurrent data from early Wolfcampian strata (Altares, 1990) have also been superimposed on the Pennsylvanian isopach map (Fig. 4). Altares' (1990) field area spans the north-south dimension of a poorly delineated prong extending from the San Mateo/Lucero basin. Northwest-trending isopachs along the northern margin of the prong are reasonably well defined by control points (Kottlowski, 1960) and are compatible with the northwest-striking normal faults observed in the Joyita Hills (Fig. 3E). In contrast, isopach trends along the southern margin of the prong, as well as between the San Mateo/Lucero and Orogrande basins in general, are poorly constrained due to a paucity of control points (Kottlowski, 1960). Altares' (1990) paleocurrent data to both north and south of the prong indicate general paleoflow directions to the south and southwest, indicating that the depositional axis of the prong lay farther to the south and/or southwest than originally drawn by Kottlowski (1960). Although the northwest-trending paleocurrent vector (marked with an asterisk in Fig. 4) may not appear compatible with this interpretation, it should be noted that this vector occurs near the top of the Bursum section (Altares, 1990). The northwesterly directed paleoflow indicated by this vector may be indicative of changing tectonic and/or paleoflow patterns during the latter stages of Bursum deposition. All other paleocurrent vectors occur comparatively lower in the Bursum section, and are compatible with the foregoing interpretation.

The limited amount of data between the San Mateo/Lucero and Orogrande basins, in terms of both paleocurrent vectors and control points for delineating isopach trends, makes interpretation difficult. However, the combination of southeast-trending isopachs at the southern end of the Joyita Hills (Fig. 4), northwest-striking normal faults in the Joyita Hills (Fig. 3E), paleoflow directions to the south and southwest, and northwest-trending isopachs between the southern end of the San Mateo/Lucero basin and the northern end of the Orogrande basin indicate that the two basins were connected by a northwest-trending basin (Figs. 4 and 6) during late Paleozoic deformation.

North-striking faults in the Joyita Hills are comparable, in terms of both orientation and sense of offset, to other late Paleozoic structures in the Sierra Nacimiento area to the north (Baars, 1982), and in the Sacramento Mountains to the south (Pray, 1961). Similarly, the timing of deformation in these three areas was approximately synchronous. The dominant phase of uplift in the Joyita Hills was early Wolfcampian (Kottlowski and Stewart, 1970; Beck and Johnson, 1992). Several episodes of late Paleozoic deformation in the Sierra Nacimiento area were reported by Baars (1982), the youngest in post-Madera, pre-Abo time (compare with Fig. 5). Pray (1961) also concluded that the dominant phase of deformation in the Sacramento Mountains occurred during Late Pennsylvanian to early Wolfcampian time. The compatibility of observed structural styles and comparable timing of deformation in the three areas indicate that a large part of the crust in New Mexico was subjected to the same style of deformation during ancestral Rocky Mountain tectonism.

Integration of the foregoing fault orientations, associated kinematics, and the spatial distribution and orientation of known and inferred basins suggests that late Paleozoic deformation in central New Mexico was the result of sinistral wrench faulting (Fig. 6). The north-trending San Mateo/Lucero and Orogrande basins are arranged in a left-stepping, en echelon pattern and represent the principal zones of sinistral shear. The two basins were probably connected by a northwest-trending basin (Figs. 4 and 6), which is interpreted as either a rhomb-graben or a releasing bend (Ramsay and Huber, 1987, Figs. 23.37 and 23.38) between the two shear zones. The former alternative is presented in Figure 6. Bounding structures between basins and adjacent uplifts were high-angle to vertical normal faults (Fig. 3D, E). The predominance of normal displacement along both fault sets indicates that the inferred wrench fault system must have been largely divergent, with sinistral offset along north-striking faults being a lesser component of deformation. The impetus for sinistral displacement is not clear, but most likely was the result of continental suturing between North America and South America–Africa (Kluth, 1986). Resultant, northerly directed compression along the Marathon-Ouachita belt of west Texas, Oklahoma, and Arkansas (to the east and southeast of New Mexico) reactivated existing basement flaws (Proterozoic foliations; Fig. 3B) of both north and northwest trend, and resulted in a north-trending, divergent, sinistral wrench fault system in central and southern New Mexico during the late Paleozoic.

LARAMIDE DEFORMATION

General stratigraphic relationships

Late Paleozoic (middle Permian) and Mesozoic sedimentary rocks represent post-ancestral Rocky Mountain, pre-Laramide strata (Fig. 2). Middle to late Wolfcampian rocks of the Abo Formation are composed of red, arkosic mudstones, shales, and lesser channel sandstone beds. These strata represent subaerial sedimentation during the waning stages of late Paleozoic tectonism (Beck and Johnson, 1992). Overlying

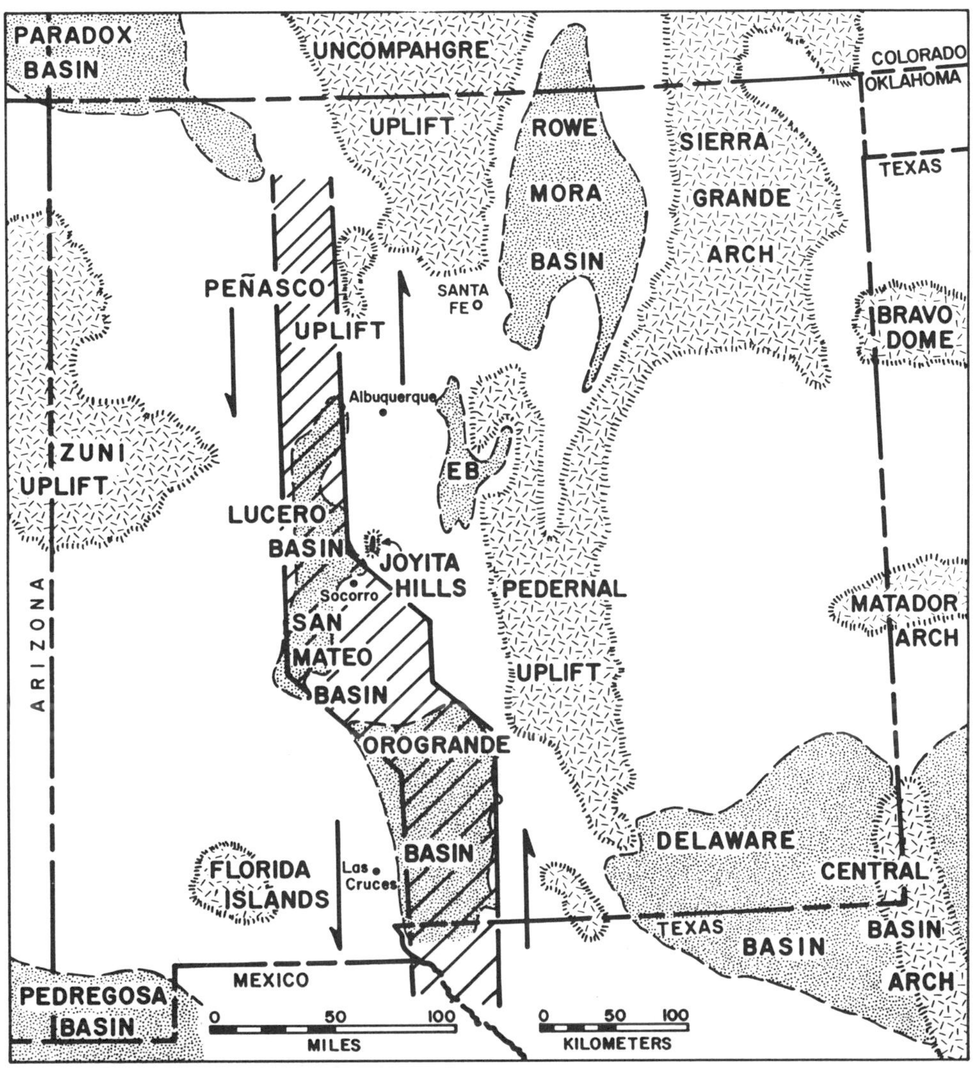

Figure 6. Inferred divergent, sinistral wrench fault system superimposed on Pennsylvanian paleogeographic map. The north-trending San Mateo/Lucero and Orogrande basins developed within principal zones of sinistral shear. The inferred northwest-trending basin is envisioned as a rhomb-graben between two left-stepping, sinistral shear zones. Paleogeographic base map modified from Kottlowski and Stewart (1970). Note that the Sierra Nacimiento area (used in text) is coincident with the Peñasco uplift of Kottlowski and Stewart (1970).

Leonardian and Guadalupian strata of the Yeso, Glorieta, and San Andres Formations formed extensive, sheetlike deposits of clean carbonates and mature clastics that eventually buried the late Paleozoic uplifts (Peterson, 1980), and signalled an end to ancestral Rocky Mountain tectonism in central New Mexico.

Triassic strata consist of texturally and compositionally immature conglomerates, sandstones (micaceous litharenites and graywackes), and mudstones of the Moenkopi and San Pedro Arroyo Formations (Lucas, 1991). Jurassic strata have been observed only locally. Mature sediments (claystones, siltstones, and micrites) of the Brushy Basin Member of the Morrison Formation (Hayden et al., 1990) occur as erosional remnants between Triassic and Cretaceous strata along the eastern margin of El Valle de La Joya. Cretaceous strata (396+ m; 1,300+ ft) are composed of interstratified, well-sorted sandstones, siltstones, shales, and thin coal beds of the Dakota Sandstone, Mancos Shale, Tres Hermanos Formation, Gallup Sandstone, and Crevasse Canyon Formation (Baker, 1981).

Where observed, strata deposited during each of the foregoing periods exhibit disconformable contact relationships, and with the exception of Triassic strata, all sequences are composed of well-sorted clastics and clean carbonates indicative of relative tectonic stability. Although Smith (1983) has inferred a post-middle Permian, pre–Late Triassic deformational event in the Socorro Region, no field evidence has been recognized to support this interpretation in the Joyita Hills area. Rather, the typically mature sediments and disconformable contact relationships between the foregoing stratigraphic sequences indicate that the interval between ancestral Rocky Mountain and Laramide tectonisms was one of relative tectonic quiescence in the Joyita Hills area.

In contrast, syntectonic strata of the Eocene Baca Formation are indicative of tectonic instability and late Laramide uplift and erosion (Cather, 1980; Cather and Johnson, 1986). These sequences consist of compositionally and texturally immature conglomerates, sandstones (lithic arkoses, feldspathic litharenites, and sublitharenites), and mudstones that unconformably overlie Cretaceous, Triassic, and Permian rocks. More specifically, fundamentally different stratigraphic columns in the western and eastern portions of the study area at-

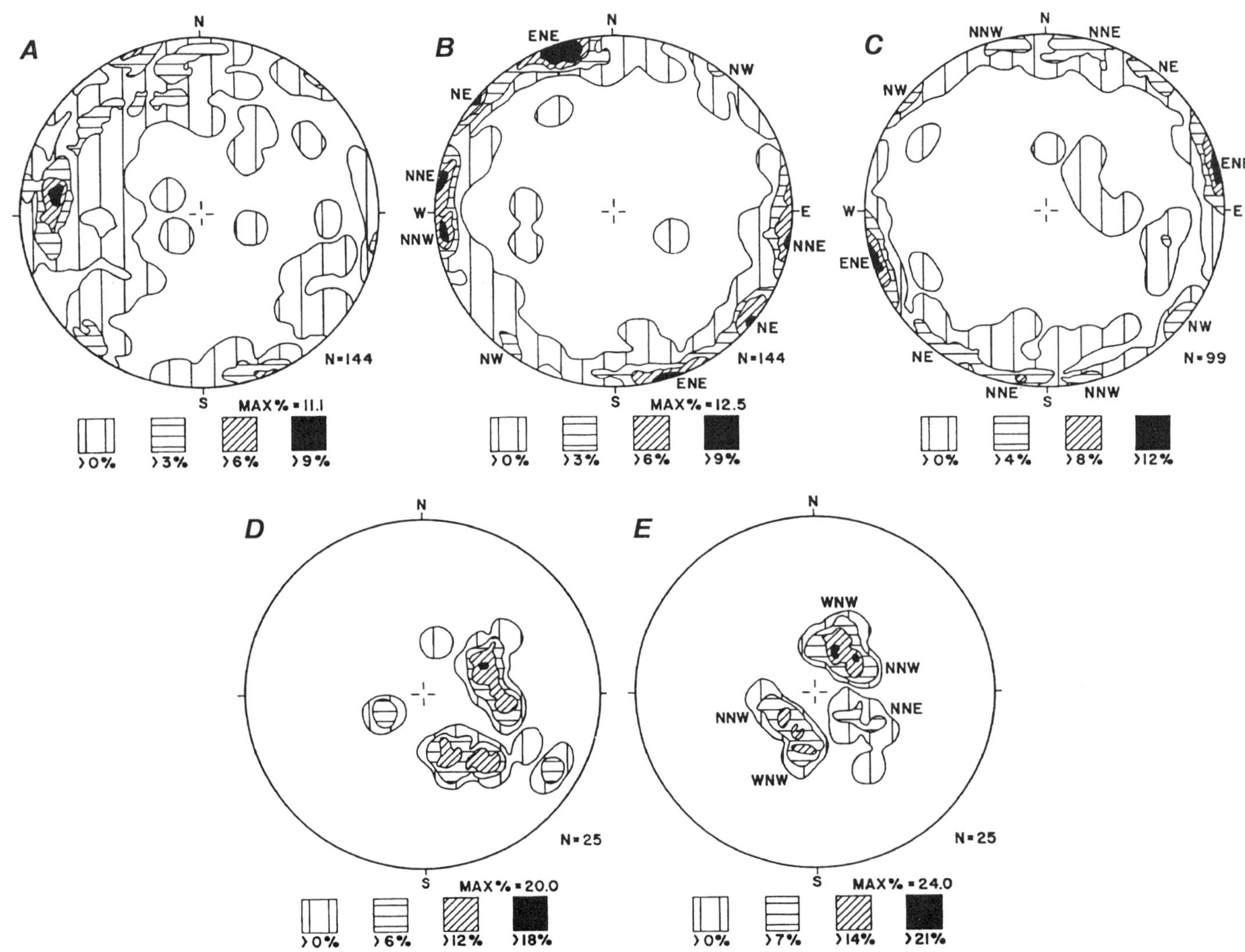

test to proximal uplift (Figs. 1 and 2). To the east (El Valle de La Joya), Baca sediments overlie Cretaceous rocks through disconformable to mildly angular contact relationships. To the west (Joyita Hills), however, Cretaceous strata are absent, and conglomerates of the Baca Formation overlie erosionally thinned Triassic and Permian (Leonardian and Guadalupian) rocks along an angular unconformity. The north-northeast– to northeast-striking East Joyita fault appears to be the boundary between these two distinct stratigraphic columns, and the partial omission of strata is indicative of Laramide uplift and erosion west of the East Joyita fault.

Laramide structures

Laramide deformation in the Joyita Hills area resulted in both wrench faulting and subordinate thrust faulting (Beck, 1993). Strike-slip faults have been observed predominantly in Proterozoic through Cretaceous rocks, and only a limited number (<2%) of strike-slip faults have been observed in post-Laramide strata (Tertiary volcanic rocks). Thrust faults and associated folds have been found exclusively in Cretaceous and older strata, and no folds (other than fault drag) have been observed in Tertiary volcanic rocks. Accordingly, both strike-slip faults and thrust faults are attributed to Laramide deformation.

Field observations indicate that strike-slip faults most commonly occur as high-angle to vertical structures (Fig. 7A) marked by low-rake (typically 0 to 30°) fault striae. Observation also indicates that these strike-slip faults are consistently oriented at roughly 90° to bedding, regardless of present bedding orientation. The constancy between fault orientation and bedding has been demonstrated by plotting the strike-slip faults with respect to bedding (herein referred to as "restored" orientations). This was accomplished by restoring inclined bedding to horizontal (using the specific bedding orientation at the locality where the faults were observed) and rotating (restoring) the faults by a comparable amount. The analyses, therefore, treat the strike-slip faults as early formed structural elements that predate appreciable rotation of strata.

Comparison of Figures 7A and B indicates that when restored (Fig. 7B), strike-slip faults plot as subvertical to vertical planes in accordance with Andersonian theory for strike-slip deformation. Restored strike-slip faults form five concentrations (preferred orientations or maxima) that are both enhanced as distinct concentrations and increased in maxima percent value. These preferred orientations strike north-northwest, north-northeast, northeast, east-northeast, and less commonly northwest. Comparably restored fault striae (Fig. 7C) plot as horizontal to subhorizontal lineations that correspond to each observed concentration of strike-slip fault orientations (Fig. 7B). Interpretation of fault steps (Durney and Ramsay, 1973) on the fault surface indicate dextral displacements on faults that strike north-northwest, north-northeast, and northeast, and sinistral displacements on east-northeast–striking faults. The sense of displacement on northwest-striking faults has not been determined.

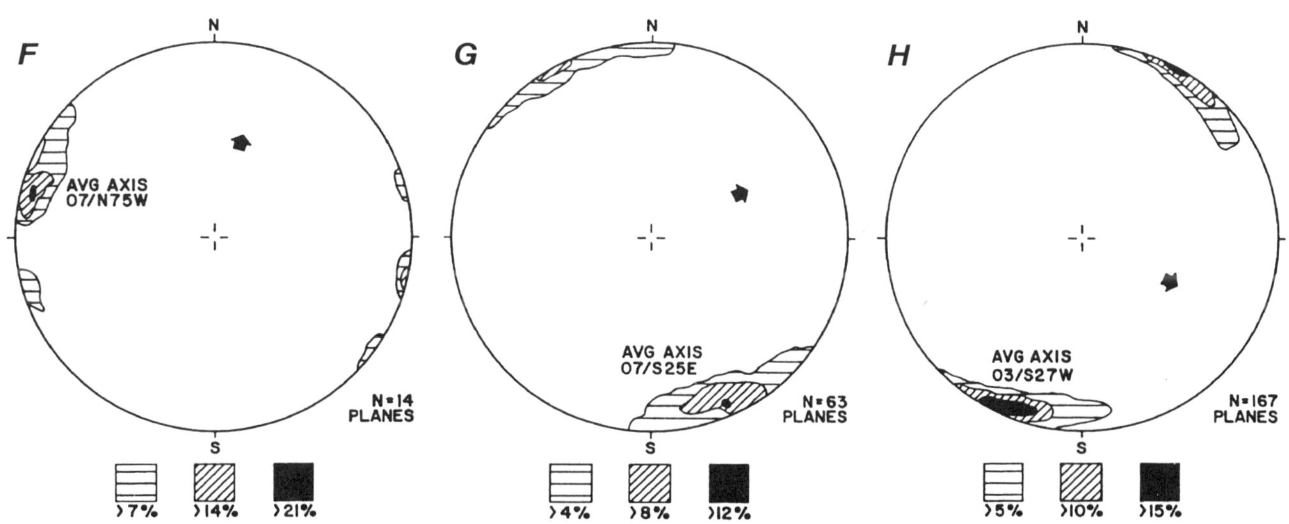

Figure 7 (on this and facing page). Lower-hemisphere, equal-area projections of Laramide strike-slip fault, thrust fault, and fold-axis orientations. A, Field-recorded strike-slip fault orientations; B, restored strike-slip faults; preferred orientations strike north-northwest (NNW), north-northeast (NNE), northeast (NE), east-northeast (ENE), and northwest (NW); C, restored fault striae; preferred orientations correspond to each observed concentration of restored strike-slip faults in B; D, field-recorded thrust fault orientations; E, restored thrust faults; preferred orientations strike to west-northwest (WNW), north-northwest (NNW), and north-northeast (NNE; singular maxima); F, G, and H, restored fold axes trend WNW, SSE (NNW), and SSW (NNE), respectively. Black arrows denote observed vergence direction. Poles to planes (A, B, D, and E), lineations (C), and contoured beta diagrams (F, G, and H) contoured at specified intervals per 1% area.

In addition to strike-slip faults, Laramide deformation resulted in reverse faulting and associated folding of strata. Observed reverse faults occurred as moderate- to high-angle structures of diverse strike (Fig. 7D). However, observation also indicated that these reverse faults were consistently oriented at 25 to 30° to bedding. Accordingly, reverse faults have been evaluated in a manner similar to that used for strike-slip faults. Comparison of field-recorded (Fig. 7D) and restored (Fig. 7E) reverse faults indicates that when restored, reverse faults plot as shallow-dipping structures, or true thrust faults in accordance with Andersonian theory. When restored, observed maxima are enhanced as distinct concentrations and/or increased in maxima percent value, and form three distinct fault systems. Two of these systems are represented by diametrically opposed concentrations that define conjugate thrust fault orientations. These partially overlapping concentrations strike west-northwest and north-northwest. The third fault system, striking north-northeast, consists of a singular fault set.

Folds are relatively uncommon in the Joyita Hills area. Where observed, folds occur as broad, open structures or asymmetric anticlinal/synclinal pairs or fold trains that occupy a small percentage of the field area. Folds have been restored by using the regional bedding orientation determined from Tertiary volcanic strata (26° of down-to-the-west rotation about a horizontal, N28°E-trending axis of rotation to compensate for rotation associated with Rio Grande rift deformation; Fig. 3A). When restored, Laramide fold axes plot as subhorizontal fold axes that trend west-northwest, north-northwest, and north-northeast (Figs. 7F, G, H) and correspond to the strike direction of each inferred thrust fault system (Fig. 7E).

Interpretation of Laramide deformation

Strike-slip faulting was the dominant style of Laramide deformation in the Joyita Hills area. Although thrust faults and folds are present, thrust faults are comparatively less numerous and folds are either poorly developed or of limited extent. In some cases, however, the orientations of fold and thrust fault systems can be directly related to strike-slip fault systems.

Preferred orientations of restored strike-slip faults are interpreted as respective parts of two separate dextral wrench fault systems. Dextral faults that strike north-northeast and north-northwest, and sinistral faults that strike east-northeast are interpreted as respective elements of a north-trending, dextral wrench fault system. Respectively, these faults are interpreted as synthetic, secondary synthetic and antithetic shears (or, using Riedel shear terminology, R, P,and R′ shears; Christie-Blick and Biddle, 1985, Fig. 5). Here, allowance is made for the differences between pure shear (Andersonian) and simple shear (wrench) models (e.g., Sylvester, 1988, p. 1673). Inferred regional average stress axes were oriented such that σ_1 was horizontal, trending northeast; σ_2 was vertical; and σ_3 was horizontal, trending northwest. As defined, the inferred dextral wrench fault system is compatible with late Paleocene to early Eocene, dextral wrench fault interpretations of Chapin and Cather (1981) and Chapin (1983) for Laramide deformation in New Mexico.

The second wrench fault system is based on the combination of dextral, northeast-striking faults and the north-northwest–trending fold and thrust fault system. Dextral displacement has been documented along the Montosa and Del Curto fault zones (to the east and southeast of the study area, Fig. 1; Brown, 1987; Hayden, 1991). Spatial relationships between these two fault zones and a north-northwest–trending anticlinal flexure (Fig. 1) may be interpreted as a convergent overstep structure associated with dextral wrenching along the left-stepping, northeast-striking faults. This interpretation is also compatible with late Laramide dextral wrench fault interpretations of Chapin and Cather (1981) and Chapin (1983). The dextral, northeast-striking faults and north-northwest–trending folds observed in the Joyita Hills and El Valle de La Joya are thought to represent similar (but smaller scale) structures as those observed along the Montosa and Del Curto fault systems.

Comparison of restored strike-slip fault orientations (Fig. 7B) with Proterozoic foliations and amphibolite dikes (Fig. 3B, C) indicates that orientations are similar, and fault striae indicative of strike-slip displacements have occasionally been observed along foliation planes. Accordingly, it is inferred that the orientations of Laramide strike-slip faults were controlled by reactivation of existing planar weaknesses in the basement rocks of the Joyita Hills area. These consist of north-northwest- and north-northeast–striking dextral faults (collectively influenced by the generally north-striking foliations) and the sinistral east-northeast–striking faults (controlled by both foliations and amphibolite dikes). Reactivation of northwest-striking foliations during Laramide tectonism is also inferred. However, strike-slip faults of this orientation are comparatively uncommon, and their tectonic significance is not fully understood at this time.

Observed thrust faults (Fig. 7E) are interpreted to be the result of three separate fault systems. These systems strike west-northwest, north-northwest, and north-northeast. Each inferred fault system corresponds to the restored orientations of Laramie fold axes (Fig. 7F, G, H), and the combination of fault sets and fold axes is thought to represent realistic orientations of applied stress fields during Laramide deformation. The inferred stress fields were oriented such that in each system σ_1 was horizontal and normal to fault strike and fold axis trend, σ_2 was horizontal and parallel to fault strike and fold axis trend, and σ_3 was vertical.

Laramide uplift to the west of the East Joyita fault is due, at least in part, to deformation associated with the west-northwest fold and thrust fault system. Uplift along a west-northwest–striking thrust fault (Fig. 8) resulted in erosional stripping of Cretaceous and older rocks. Restoration of bedding orientations (using the average orientation of Oligocene

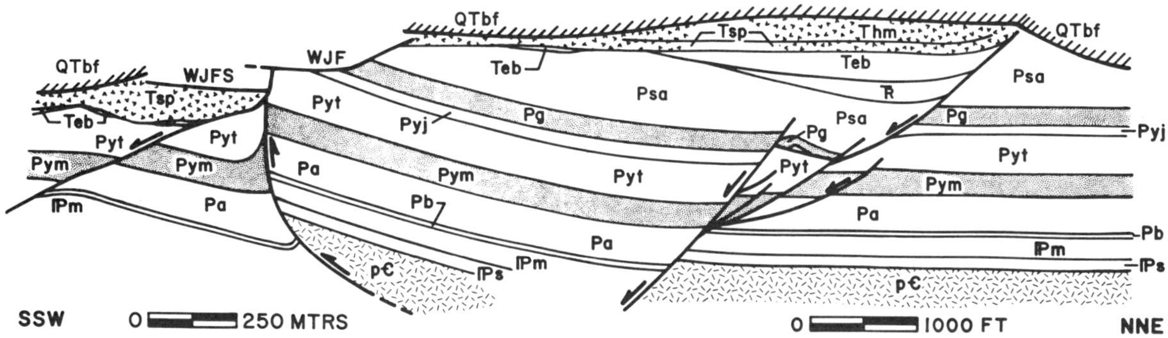

Figure 8. Generalized structural profile of Laramide uplift west of the East Joyita fault (compare with Figs. 2 and 7E, F). Profile oriented N25°E, 60°SE to compensate for down-to-the-west rotation associated with mid-Tertiary extension. Normal faults are of mid-Tertiary (Rio Grande rift) origin, thrust fault is of Laramide origin. Northeastward tilting, absence of Cretaceous strata and southwestward truncation of Triassic (₮) and middle Permian rocks by the overlying, syntectonic Baca Formation (Teb) attest to Laramide uplift and erosion prior to deposition of mid-Tertiary volcanic sequences (Tsp, Spears Formation; Thm, Hells Mesa Tuff). Uppermost extent of section obscured by down-to-the-west displacement along the West Joyita fault (WJF), a subparallel fault splay (WJFS), and basinfill sediments (QTbf). pЄ, Proterozoic basement; IPs, Sandia; IPm, Madera; Pb, Bursum; Pa, Abo; Pym, Pyt, and Pyj, Meseta Blanca, Torres, and Joyita Members of the Yeso Formation; Pg, Glorieta; Psa, San Andres.

volcanic rocks, Fig. 3A) indicates that Paleozoic and remnant Triassic strata in proximity to the fault were inclined an average of 8 to 10° to the north-northeast. These strata were bevelled and partially buried by syntectonic deposits of the Eocene Baca Formation, which overlie and progressively truncate older sequences from north-northeast to south-southwest. This uplift (Fig. 8) has been observed only to the west of the East Joyita fault, and only a few small-displacement (<1 m) thrust faults of comparable orientation have been observed in El Valle de La Joya. The East Joyita fault appears to bound the uplift to the east, and although prominently marked by dip-slip fault striae (the East Joyita fault was also active during Rio Grande rift tectonism, discussed below), overprinted fault striae indicative of strike-slip displacement have been observed along various segments of this fault. The East Joyita fault is subparallel to the dextral Montosa fault, and it is inferred that dextral displacement also occurred along the East Joyita fault. It is thought that the observed west-northwest–striking thrust fault (Fig. 8) originated as a result of compression associated with dextral displacement along the East Joyita fault (Fig. 9).

However, this deformation need not be the only cause of Laramide uplift to the west of the East Joyita fault. Folds asso-

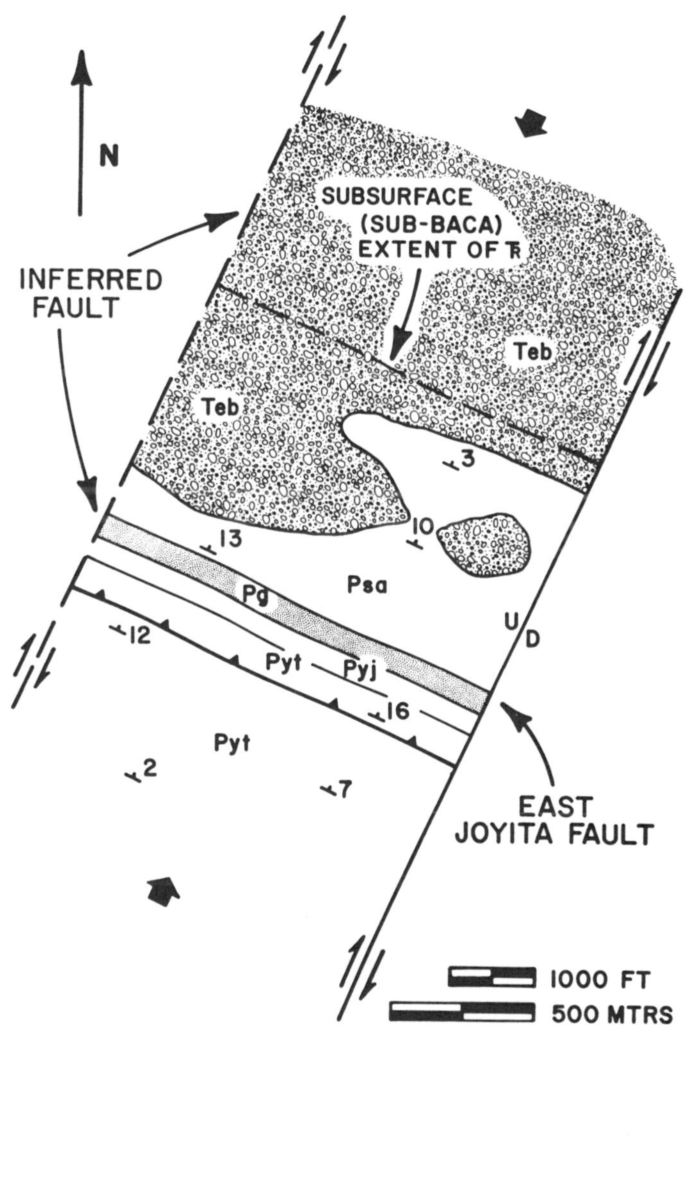

Figure 9. Plan view of structural relationships west of the East Joyita fault after Laramide uplift and prior to deposition of the Spears Formation (compare with Fig. 8). West-northwest–striking bedding and thrust fault are interpreted to be the result of compression directed north-northeast/south-southwest as a result of dextral displacement along the East Joyita fault. Both the existence and the location of the dextral fault to the west-northwest of the East Joyita fault are inferred. Formation symbols keyed to those of Figure 8.

ciated with the north-northeast fold and thrust fault system have been observed in proximity to the East Joyita fault and in El Valle de La Joya. Folds of this orientation become more pronounced to the east and northeast of the study area (Fig. 1). Myers et al. (1986) mapped comparably oriented structures between El Valle de La Joya and the Montosa fault, and documented an early (Laramide?) episode of up-to-the-west reverse faulting along the Montosa fault zone (interpreted as transpressional, dextral reverse slip by Hayden, 1991). The East Joyita fault is a north-northeast– to northeast-striking structure, and conceivably, the up-to-the-west displacement may be the result of compressional deformation associated with the north-northeast–striking fold and thrust fault system. Here, analogy is made between observed displacements in the Joyita Hills area and seemingly similar deformational styles along the Montosa fault zone (Myers et al., 1986; Hayden, 1991). Conceivably, Laramide uplift of the Joyita Hills block may be the result of both west-northwest and north-northeast fold and thrust fault systems.

RIO GRANDE RIFT DEFORMATION

General stratigraphic relationships

Syntectonic strata of the Baca Formation are depositionally overlain by approximately 2 km of Tertiary volcanic sequences. The volcanic section of the Joyita Hills has previously been mapped, measured, and described by Spradlin (1976). However, the stratigraphic nomenclature has been largely redefined (Osburn and Chapin, 1983). Interbedded volcaniclastics, lava flows, and ash-flow tuffs of the Spears Formation (latest Eocene–early Oligocene; Cather, 1986) form the basal unit of the volcanic section. The Spears Formation is overlain by a series of interstratified, Oligocene rhyolitic ash-flow tuffs (Hells Mesa Tuff, Tuff of South Crosby Peak, and La Jencia, Vicks Peak, Lemitar, and South Canyon tuffs) and basaltic andesite flows (collectively, the La Jara Peak Basaltic Andesite). Detailed descriptions, paleomagnetic analyses, and time-stratigraphic ($^{40}Ar/^{39}Ar$) assignments of these rocks have been provided by Spradlin (1976), Osburn and Chapin (1983), McIntosh (1983), and McIntosh et al. (1992).

Stratigraphic differences in the volcanic section also occur to the west and east of the East Joyita fault. To the east, thick accumulations of the Spears Formation overlie the Baca Formation. Cather (1986) reported 740 to 1,480 m (2,428 to 4,856 ft) of the Spears Formation, with abrupt thickening across an inferred northwest-striking reverse fault. Spradlin (1976, Plate 2) reported approximately 885 m (2,900 ft) of Spears. In contrast, accumulations of the Spears are comparatively thin to the west of the East Joyita fault, and rest in depositional contact on rocks of Eocene or middle Permian age. Relationships between the Spears and underlying Eocene and/or Permian strata and overlying Hells Mesa Tuff are commonly obscured by mid-Tertiary normal faults, making an accurate determination of the thickness of the Spears Formation difficult. However, two exposures have been observed in which it is thought that reliable thickness determinations can be made. These yield stratigraphic thicknesses of 38 m (125 ft) and approximately 125 m (410 ft) for the Spears (to the north and south of the thrust fault depicted in Fig. 8, respectively; the latter aided by a marker horizon in proximity to the West Joyita fault). The anomalously thin Spears section west of the East Joyita fault is thought to reflect the persistence of the uplifted Joyita fault block (Laramide deformation) as a topographic high through most of Spears time (latest Eocene–early Oligocene). However, it is also possible that additional, down-to-the-east displacements along the East Joyita fault occurred during Spears deposition. If so, indicators of these additional displacements have gone unrecognized. In contrast, overlying Oligocene ash-flow tuffs and basaltic andesites apparently blanketed terrane to both east and west of the East Joyita fault.

Rio Grande rift structures

Mid-Tertiary extension in the Joyita Hills area resulted in both high- and low-angle normal faults. These faults offset Tertiary volcanic and older rocks, and exhibit a variety of orientations. Field observations indicate that most of these faults are consistently oriented at roughly 60° to bedding, regardless of present bedding orientation. Accordingly, normal faults have been evaluated in a manner identical to that used for strike-slip and thrust faults of Laramide origin; that is, normal faults were evaluated as both pre- and post-rotational structures. Comparison of field-recorded (Fig. 10A) and restored (Fig. 10B) normal fault orientations indicates that when restored, most normal faults plot as diametrically opposed (conjugate) orientations that are both enhanced as distinct concentrations and/or increased in maxima percent value. These faults define three distinct fault systems that respectively strike north-northeast, northwest, and less commonly, east-northeast. Restored fault striae (Fig. 10C) indicate that fault displacements were predominantly dip slip.

Other structures of mid-Tertiary origin include mafic (basaltic andesite) dikes. Restoration of these extensional structures (Fig. 10D) indicates that when restored, mid-Tertiary dikes plot as subvertical structures that preferentially strike north-northeast, northwest, and east-northeast, bisecting the acute angle formed by each inferred fault pair.

An additional set of normal faults have also been observed in the Joyita Hills. In contrast to the foregoing faults, these faults occurred as shallow (0 to 30°), east-dipping planes of north and north-northeast strike (Fig. 10E), and typically showed no consistent orientation with respect to bedding. Geologic mapping and field observations indicate that these low-angle normal faults are commonly associated with allochthonous terrane. Allochthonous blocks composed of Oligocene volcanic and older rocks form the upper plates of these faults. Additionally, these relatively flat faults display dif-

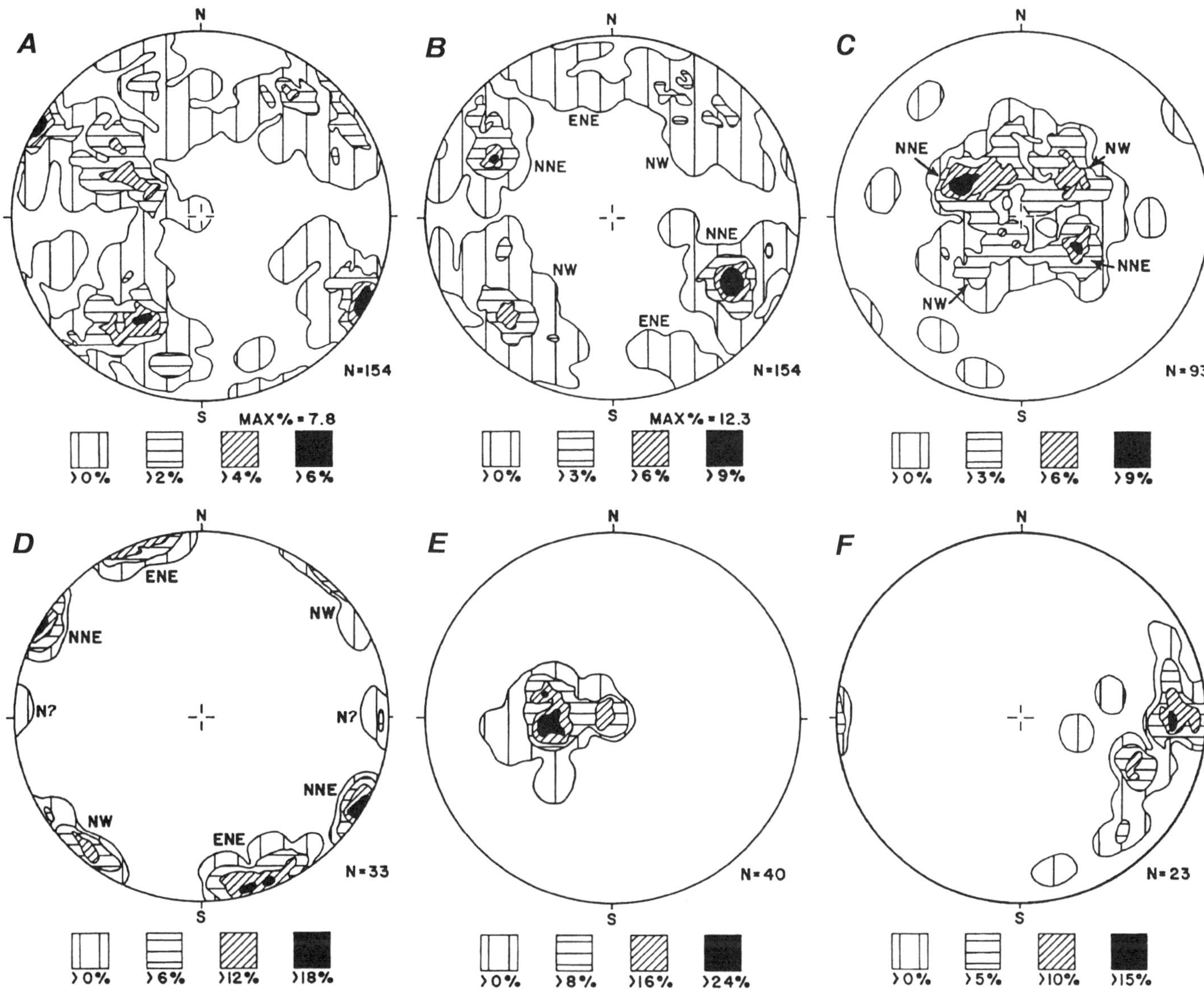

Figure 10. Lower-hemisphere, equal-area projections of Rio Grande rift normal fault, dike, and detachment fault orientations. A, Field-recorded normal fault orientations; B, restored normal faults; conjugate pairs strike north-northeast (NNE), northwest (NW), and less commonly east-northeast (ENE); C, restored normal fault striae (labelled with respect to the fault system, rather than trend of striae); D, restored mid-Tertiary mafic dikes; E, field-recorded detachment faults; F, field-recorded detachment fault striae. Poles to planes (A, B, D, and E) and lineations (C and F) contoured at specified intervals per 1% area.

ferences in fault character that have not been observed on the originally high angle faults. Some of the shallow-dipping faults exhibit microbrecciated fault surfaces (2 to 5 cm; 1 to 2 in) and locally, as along the southernmost exposure of the East Joyita fault, are overlain by 0 to 30 m (0 to 100 ft) of tectonic mélange.

Interpretation of Rio Grande rift deformation

Comparative analysis of mid-Tertiary normal fault orientations as both post-tilt (Fig. 10A) and pretilt (Fig. 10B) structures indicates that most faults were originally pretilt, high-angle (60 to 65°) normal faults that experienced dip-slip displacements. Preferred orientations of mid-Tertiary mafic dikes respectively bisect the acute angle formed by each inferred fault system. The combination of restored shear and extensional orientations yields three distinct fracture systems compatible with Andersonian (1951) theory for shallow crustal deformation. These fracture systems are therefore interpreted as early formed structural elements during Rio Grande rift tectonism. Inferred stress axes were oriented such that in each system σ_1 was vertical, σ_2 was horizontal and parallel to strike of dikes and faults, and σ_3 was horizontal and normal to strike of dikes and faults. Each stress field orientation and resultant fracture pattern can be correlated to reactivation of existing structures (discussed below). Therefore, these three apparently distinct stress field orientations may be indicative of one overall stress field where $\sigma_1 > \sigma_2 \sim \sigma_3$. In this interpretation, small changes in relative magnitudes of σ_2 and σ_3 would allow significant changes in the direction of extension, with fracture patterns preferentially aligned parallel to existing structural grains.

Mid-Tertiary normal fault and dike orientations (Fig. 10B, D) are similar to the orientations of Proterozoic foliations and amphibolite dikes (Fig. 3B, C). It is interpreted that northwest- and east-northeast–striking normal fault and dike systems were controlled by, and represent reactivation of, northwest- and east-northeast–striking anisotropies in the Proterozoic basement. Here, extensional orientations are subparallel to, and conjugate shears symmetrically disposed about, the preferred orientations of Proterozoic structures. To date, no comparable correlation can be made for the north-northeast–striking normal fault and dike system. Rather, it is inferred that this normal fault and dike system represents reactivation of the north-northeast–striking thrust fault and fold system of Laramide origin. Tertiary volcanic rocks strike north-northeast and are inclined 25 to 30° to the west-northwest. Accordingly, the shallow-dipping thrust faults of north-northeast strike may have influenced or enhanced rotation about a north-northeast–trending, horizontal axis during Rio Grande rifting.

Reactivation of the East Joyita fault also occurred during Rio Grande rift deformation. Down-to-the-west rotation (25 to 30°, Fig. 3A) reoriented the initially subvertical East Joyita fault to a steeply east dipping orientation. Renewed down-to-the-east displacements then juxtaposed Oligocene ash-flow units against the Proterozoic core of the Joyita Hills (Fig. 1). Combined Rio Grande rift and Laramide down-to-the-east offsets approximate 3 km of stratigraphic throw. The modern-day East Joyita fault exhibits a very complex geometry. Although a generally north-northeast–striking structure, the fault is actually composed of a series of right-stepping, north-striking, steeply (55 to 85°) east dipping planes. Field observations indicate that these fault segments are reactivated ancestral Rocky Mountain faults and Proterozoic foliation planes. In addition, a mid-Tertiary rhyolite dike (31.3 ± 1.2 Ma, Aldrich et al., 1986, their stress indicator #324) was injected along both the East Joyita fault and the foregoing, late Paleozoic faults and Proterozoic foliation planes. Elsewhere, as in the southernmost exposure of the Proterozoic core, the East Joyita fault displays fairly shallow (30°) east-dipping fault surfaces. Fault striae observed on both shallow and steeply dipping fault segments indicate that mid-Tertiary displacements along the East Joyita fault were predominantly dip slip.

Other normal faults were not originally high angle (Fig. 10E). These faults typically occur as shallow (0 to 30°), east-dipping faults. Consideration of these faults as pretilt structures indicates that they could not have been steeper than 25 to 55°. In contrast to originally high angle normal faults, low-angle faults are associated with and underlie allochthonous terrane. These low-angle normal faults are interpreted as extensional detachment faults, and are thought to represent the near-surface segments of the detachment faults detected in seismic sections across the Albuquerque Basin (e.g., Cape et al., 1983; De Voogd et al., 1988; Russell and Snelson, 1990). Field observation indicates that originally high angle normal faults are abruptly truncated by the detachment faults. The detachment faults are therefore interpreted as comparatively young structures that cut the older, originally high angle faults in a manner analogous to excising detachment faults (Lister and Davis, 1989). It is inferred that normal displacement along the detachment faults accommodated rotation and developed contemporaneously with rotation. Rotation of bedding and originally high angle faults resulted in domino-style faulting in extensionally detached terrane (see also Chamberlin, 1983). Westward tilting of the Phanerozoic section (Fig. 3A), when combined with the orientation of faults and fault striae (Fig. 10E, F), indicate a hanging-wall displacement direction to the east and east-southeast.

Unfaulted, subhorizontal basin-fill deposits of the Sierra Ladrones Formation of the Santa Fe Group (Fig. 2; Plio-Pleistocene, Osburn and Chapin, 1983) overlie allochthonous terrane, detachment faults, and rotated and eroded volcanic strata as young as the South Canyon Tuff (27.4 Ma, McIntosh et al., 1992). Therefore, mid-Tertiary deformation in the Joyita Hills can be broadly constrained as an Oligocene-Miocene event.

SUMMARY

Structures associated with three Phanerozoic orogenic events have been recorded and preserved in the sedimentary and volcanic strata of the Joyita Hills area. Reactivation of existing structures, most notably the foliations and amphibolite dikes in the Proterozoic basement, was the rule rather than the exception. Resultantly, fault and dike orientations preferentially developed along north, northwest, and east-northeast trends.

Ancestral Rocky Mountain deformation is thought to be the result of divergent, sinistral displacement along a north–striking wrench fault system (Fig. 6). Late Paleozoic fault orientations represent reactivation of Proterozoic foliations of both north and northwest strike. Northward thrusting in the Marathon-Ouachita orogenic belt (Kluth, 1986) resulted in sinistral displacements along north-striking zones of weakness, resulting in basins and uplifts of dominantly north and subordinate northwest trend. The north-trending Joyita uplift and adjacent San Mateo/Lucero basin probably extended northward through the Albuquerque Basin to the Sierra Nacimiento area, as earlier proposed by Baars (1982) and Read and Wood (1947).

The main obstacle to conclusively demonstrating a late Paleozoic sinistral wrench fault system is the result of overprinting by younger deformation. North-striking dextral wrench faulting of Laramide origin has been documented in New Mexico by Chapin and Cather (1981) and Chapin (1983), along the Nacimiento fault by Baltz (1967) and in the Joyita Hills (Fig. 7B). Laramide dextral wrenching reactivated Proterozoic basement structures of north and east-northeast strike, the former in a dextral (synthetic) sense, the latter in a sinistral (antithetic) sense. Northwest-striking foliations were also reactivated in a strike-slip sense at this time, although the causal mechanisms for faults of this orientation have not been recognized.

Additional Laramide structures include northeast-striking dextral faults, thrust faults, and folds. Both northeast-striking dextral faults and north-northwest–oriented thrusts and folds are interpreted to be related to dextral wrench faulting along the Montosa and Del Curto fault zones (Fig. 1; Brown, 1987; Hayden, 1991).

Laramide uplift in the Joyita Hills area occurred to the west of the East Joyita fault. Uplift resulted in partial stripping of Mesozoic and middle Permian strata prior to deposition of the syntectonic Eocene Baca sediments (Fig. 8). The East Joyita fault during Laramide time was a north-northeast–striking, subvertical fault, and was subparallel to the dextral Montosa fault system. It is inferred that the East Joyita fault originated during Laramide deformation as a dextral strike-slip fault. Remnant strike-slip fault striae support this interpretation, although younger, normal displacements have overprinted and removed all but a few strike-slip fault striae.

Anomalously thin (38 to 125 m; 125 to 410 ft) strata of the Spears Formation occur to the west of the East Joyita fault, indicating that Laramide uplift persisted as a topographic high through much of Spears time (latest Eocene–early Oligocene). The East Joyita fault was reactivated as a down-to-the-east normal fault during Rio Grande rift tectonism. Although a generally north-northeast–striking structure, down-to-the-east displacement reactivated numerous north-striking ancestral Rocky Mountain fault planes and Proterozoic foliation planes. Combined Laramide and mid-Tertiary, down-to-the-east displacements along the East Joyita fault approximate 3 km of stratigraphic throw.

Deformation associated with Rio Grande rifting resulted in both high-angle normal faults and low-angle, extensional detachment faults. Originally high-angle faults developed along three directions, two of which represent reactivation of Proterozoic structures of northwest and east-northeast strike. The third normal fault system is thought to have reactivated the north-northeast–striking fold and thrust fault system of Laramide origin. Extensional detachment faults occur as low-angle structures (Fig. 10E) that are commonly associated with microbrecciated fault surfaces, tectonic mélange, and allochthonous terrane. These structures are interpreted as the shallow crustal analogs to the deeper detachment faults in the Albuquerque Basin detected by seismic investigations (e.g., Russell and Snelson, 1990; De Voogd et al., 1988). An extensional model compatible with Lister and Davis (1989) is favored, in which comparatively younger, excising detachment faults cut older, originally high angle normal faults. Associated rotation of both bedding and high-angle normal faults resulted in domino-style faulting within allochthonous terrane. Both high- and low-angle normal faults offset Oligocene and older rocks, but are overlain by unfaulted strata of the Sierra Ladrones Formation of the Santa Fe Group. Therefore, mid-Tertiary extensional deformation in the Joyita Hills area can be constrained as an Oligocene-Miocene event.

ACKNOWLEDGMENTS

The authors thank S. M. Cather, W. C. Haneberg, S. N. Hayden, and S. Ralser for editorial comments and critical reviews of the original manuscript. Their efforts improved the original draft and were much appreciated. Additional thanks are due to R. M. Chamberlin, S. G. Lucas, S. N. Hayden, and O.J. Anderson for their help with the stratigraphy; to the Sevilleta National Wildlife Refuge (T. M. Stans, Manager) for granting access to the study area; to M. W. Knoper for his aid in adapting computer programs; and to many others too numerous to name for their help with various aspects of this endeavor. This paper constitutes part of the senior author's dissertation, and he would like to thank the New Mexico Bureau of Mines and Mineral Resources (C. E. Chapin, Director) and the New Mexico Geological Society for their financial support throughout this project.

REFERENCES CITED

Aldrich, M. J., Jr., Chapin, C. E., and Laughlin, A. W., 1986, Stress history and tectonic development of the Rio Grande rift, New Mexico: Journal of Geophysical Research, v. 91, no. B6, p. 6199–6211.

Altares, T., III, 1990, Stratigraphic description and paleoenvironments of the Bursum Formation, Socorro County, New Mexico [M.S. thesis]: Socorro, New Mexico Institute of Mining and Technology, 189 p.

Anderson, E. M., 1951, The dynamics of faulting and dyke formation with applications to Britain: Edinburgh, Oliver and Boyd, 206 p.

Armstrong, A. K., 1962, Stratigraphy and paleontology of the Mississippian System in southwestern New Mexico and adjacent southeastern Arizona: New Mexico Bureau of Mines and Mineral Resources Memoir 8, 99 p.

Armstrong, A. K., and Mamet, B. L., 1988, Mississippian (lower Carboniferous) biostratigraphy, facies, and microfossils, Pedregosa basin, southeastern Arizona and southwestern New Mexico: U.S. Geological Survey Bulletin 1826, 40 p.

Baars, D. L., 1982, Paleozoic history of the Albuquerque trough: Implications of basement control on Rio Grande rift: New Mexico Geological Society, 33rd Field Conference, Guidebook, p. 153–157.

Baker, B. W., 1981, Geology and depositional environments of Upper Cretaceous rocks, Sevilleta Grant, Socorro County, New Mexico [M.S. thesis]: Socorro, New Mexico Institute of Mining and Technology, 159 p.

Baltz, E. H., 1967, Stratigraphy and regional tectonic implications of part of Upper Cretaceous and Tertiary rocks, east-central San Juan Basin, New Mexico: U.S. Geological Survey Professional Paper 552, 101 p.

Beck, W. C., 1984, Joint analyses of the Casper Mountain/Emigrant Gap anticline juncture, Natrona County, Wyoming [M.S. thesis]: Akron, The University of Akron, 101 p.

Beck, W. C., 1993, Structural evolution of the Joyita Hills, Socorro County,

New Mexico [Ph.D. thesis]: Socorro, New Mexico Institute of Mining and Technology, 187 p.

Beck, W. C., and Burford, A. E., 1985, Stress analysis of the Casper Mountain—Emigrant Gap anticline juncture, Natrona County, Wyoming: Wyoming Geological Association, 36th Field Conference, Guidebook, p. 59–65.

Beck, W. C., and Chapin, C. E., 1991, Structural data from the Joyita uplift: Implications for ancestral Rocky Mountain deformation within central and southern New Mexico: New Mexico Geological Society, 42nd Field Conference, Guidebook, p. 183–190.

Beck, W. C., and Johnson, D. B., 1992, New fusulinid data and multiple episodes of ancestral Rocky Mountain deformation in the Joyita Hills, Socorro County, New Mexico: New Mexico Geology, v. 14, no. 3, p. 53–59.

Brown, K. B., 1987, Geology of the southern Canoncito de la Uva area, Socorro County, New Mexico [M.S. thesis]: Socorro, New Mexico Institute of Mining and Technology, 89 p.

Cape, C. D., McGeary, S., and Thompson, G. A., 1983, Cenozoic normal faulting and the shallow structure of the Rio Grande rift near Socorro, New Mexico: Geological Society of America Bulletin, v. 94, p. 3–14.

Cather, S. M., 1980, Petrology, diagenesis, and genetic stratigraphy of the Eocene Baca Formation, Alamo Navajo Reservation and vicinity, Socorro County, New Mexico [M.A. thesis]: Austin, The University of Texas at Austin, 244 p.

Cather, S. M., 1986, Volcano-sedimentary evolution and tectonic implications of the Datil Group (latest Eocene, early Oligocene), west-central New Mexico [Ph.D. thesis]: Austin, The University of Texas at Austin, 484 p.

Cather, S. M., and Johnson, B. D., 1986, Eocene depositional systems and tectonic framework of west-central New Mexico and eastern Arizona, *in* Peterson, J. A., ed., Paleotectonics and sedimentation: American Association of Petroleum Geologists Memoir 41, p. 623–652.

Chamberlin, R. M., 1983, Cenozoic domino-style crustal extension in the Lemitar Mountains, New Mexico: A summary: New Mexico Geological Society, 34th Field Conference, Guidebook, p. 111–118.

Chapin, C. E., 1983, An overview of Laramide wrench faulting in the southern Rocky Mountains with emphasis on petroleum exploration, *in* Lowell, J. D., ed., Rocky Mountain foreland basins and uplifts: Rocky Mountain Association of Geologists, p. 169–179.

Chapin, C. E., and Cather, S. M., 1981, Eocene tectonics and sedimentation in the Colorado Plateau–Rocky Mountain area, *in* Dickinson, W. R., and Payne, M. D., eds., Relations of tectonics to ore deposits in the Southern Cordillera: Arizona Geological Society Digest, v. 14, p. 173–198.

Christie-Blick, N., and Biddle, K. T., 1985, Deformation and basin formation along strike-slip faults, *in* Biddle, K. T., and Christie-Blick, N., eds., Strike-slip deformation, basin formation, and sedimentation: Society of Economic Paleontologists and Mineralogists Special Publication 37, p. 1–34.

De Voogd, B., Serpa, L., and Brown, L., 1988, Crustal extension and magmatic processes: COCORP profiles from Death Valley and the Rio Grande rift: Geological Society of America Bulletin, v. 100, p. 1550–1567.

Durney, D. W., and Ramsay, J. G., 1973, Incremental strains measured by syntectonic crystal growths, *in* De Jong, K. A., and Scholten, R., eds., Gravity and tectonics: New York, John Wiley & Sons, p. 67–96.

Hayden, S. N., 1991, Dextral oblique-slip deformation along the Montosa fault zone at Abo Pass, Valencia and Socorro Counties, New Mexico [abs.]: New Mexico Geology, v. 13, no. 3, p. 64.

Hayden, S. N., Lucas, S. G., Hunt, A. P., and Beck, W. C., 1990, Triassic-Jurassic stratigraphy, Palo Duro Canyon, Sevilleta Grant, Socorro County, New Mexico [abs.]: New Mexico Geology, v. 12, no. 3, p. 65.

Herber, L. J., 1963a, Structural petrology and economic features of the Precambrian rocks of La Joyita Hills [M.S. thesis]: Socorro, New Mexico Institute of Mining and Technology, 36 p.

Herber, L. J., 1963b, Precambrian rocks of La Joyita Hills: New Mexico Geological Society, 14th Field Conference, Guidebook, p. 180–184.

Jackson, M. D., and Pollard, D. D., 1990, Flexure and faulting of sedimentary host rocks during growth of igneous domes, Henry Mountains, Utah: Journal of Structural Geology, v. 12, p. 185–206.

Kluth, C. F., 1986, Plate tectonics of the ancestral Rocky Mountains, *in* Peterson, J. A., ed., Paleotectonics and sedimentation: American Association of Petroleum Geologists Memoir 41, p. 353–369.

Kluth, C. F., and Coney, P. J., 1981, Plate tectonics of the ancestral Rocky Mountains: Geology, v. 9, p. 10–15.

Kottlowski, F. E., 1960, Summary of Pennsylvanian sections in southwestern New Mexico and southeastern Arizona: New Mexico Bureau of Mines and Mineral Resources Bulletin 66, 187 p.

Kottlowski, F. E., and Stewart, W. J., 1970, The Wolfcampian Joyita uplift in central New Mexico: New Mexico Bureau of Mines and Mineral Resources Memoir 23, Part I, 31 p.

Krewedl, D. A., 1974, Geology of the central Magdalena Mountains, Socorro County, New Mexico [Ph.D. thesis]: Tucson, University of Arizona, 128 p.

Lister, G. S., and Davis, G. A., 1989, The origin of metamorphic core complexes and detachment faults formed during Tertiary continental extension in the northern Colorado River region, U.S.A.: Journal of Structural Geology, v. 11, no. 1/2, p. 65–94.

Lucas, S. G., 1991, Triassic stratigraphy, paleontology and correlation, south-central New Mexico: New Mexico Geological Society, 42nd Field Conference, Guidebook, p. 243–259.

McIntosh, W. C., 1983, Preliminary results from a paleo- and rock-magnetic study of Oligocene ash-flow tuffs in Socorro County, New Mexico: New Mexico Geological Society, 34th Field Conference, Guidebook, p. 205–210.

McIntosh, W. C., Chapin, C. E., Ratté, J. C., and Sutter, J. F., 1992, Time-stratigraphic framework for the Eocene-Oligocene Mogollon-Datil volcanic field, southwest New Mexico: Geological Society of America Bulletin, v. 104, p. 851–871.

Melton, F. A., 1929, A reconnaissance of the joint-systems in the Ouachita Mountains and central plains of Oklahoma: Journal of Geology, v. 37, p. 729–746.

Myers, D. A., Sharps, J. A., and McKay, E. J., 1986, Geologic map of the Becker SW and Cerro Montoso Quadrangles, Socorro County, New

Mexico: U.S. Geological Survey, Map I-1567, scale 1:24,000.

Osburn, G. R., 1984, Socorro County geologic map: New Mexico Bureau of Mines and Mineral Resources Open-File Report #238, scale 1:200,000.

Osburn, G. R., and Chapin, C. E., 1983, Nomenclature for Cenozoic rocks of northeast Mogollon-Datil volcanic field, New Mexico: New Mexico Bureau of Mines and Mineral Resources, Stratigraphic Chart 1.

Osburn, G. R., and Lochman-Balk, C., 1983, Stratigraphic nomenclature chart: New Mexico Geological Society, 34th Field Conference, Guidebook, p. 98.

Parker, J. M., III, 1942, Regional systematic jointing in slightly deformed sedimentary rocks: Geological Society of America Bulletin, v. 53, p. 381–408.

Peterson, J. A., 1980, Permian paleogeography and sedimentary provinces, west-central United States, *in* Fouch, T. D., and Magathan, E. R., eds., Paleozoic paleogeography of west-central United States: Society of Economic Paleontologists and Mineralogists, Rocky Mountain Section, Symposium 1, p. 271–292.

Pray, L. C., 1961, Geology of the Sacramento Mountains escarpment, Otero County, New Mexico: New Mexico Bureau of Mines and Mineral Resources Bulletin 35, 144 p.

Ramsay, J. G., and Huber, M. I., 1987, The techniques of modern structural geology, Volume 2: Folds and fractures: New York, Academic Press, 391 p.

Read, C. B., and Wood, G. H., 1947, Distribution and correlation of Pennsylvanian rocks in late Paleozoic sedimentary basins of northern New Mexico: Journal of Geology, v. 55, p. 220–236.

Reches, Z., 1976, Analysis of joints in two monoclines in Israel: Geological Society of America Bulletin, v. 87, p. 1654–1662.

Reches, Z., 1978, Development of monoclines: Part I, Structure of the Palisades Creek branch of the East Kaibab monocline, Grand Canyon, Arizona, *in* Matthews, V., III, ed., Laramide folding associated with basement block faulting in the western United States: Geological Society of America Memoir 151, p. 235–271.

Russell, L. R., and Snelson, S., 1990, Structural style and tectonic evolution of the Albuquerque Basin segment of the Rio Grande rift, *in* Pinet, B., and Bois, C., eds., The potential of deep seismic profiling for hydrocarbon exploration: Paris, Editions Technip, p. 175–207.

Siemers, W. T., 1978, The stratigraphy, petrology, and paleoenvironments of Pennsylvanian rocks in west-central New Mexico [Ph.D. thesis]: Socorro, New Mexico Institute of Mining and Technology, 275 p.

Siemers, W. T., 1983, The Pennsylvanian system, Socorro region, New Mexico: Stratigraphy, petrology, depositional environments: New Mexico Geological Society, 34th Field Conference, Guidebook, p. 147–155.

Smith, C. T., 1983, Structural problems along the east side of the Socorro constriction, Rio Grande rift: New Mexico Geological Society, 34th Field Conference, Guidebook, p. 103–109.

Spradlin, E. J., 1976, Stratigraphy of tertiary volcanic rocks, Joyita Hills area, Socorro County, New Mexico [M.S. thesis]: Albuquerque, The University of New Mexico, 73 p.

Sylvester, A. G., 1988, Strike-slip faults: Geological Society of America Bulletin, v. 100, p. 1666–1703.

Manuscript Accepted by the Society October 13, 1993

Printed in U.S.A.

Geological Society of America
Special Paper 291
1994

Crustal structure of the western margin of the Rio Grande rift and Mogollon-Datil volcanic field, southwestern New Mexico and southeastern Arizona

Robert V. Schneider* and G. Randy Keller
Department of Geological Sciences, The University of Texas at El Paso, El Paso, Texas 79968

ABSTRACT

The southeastern margin of the Colorado Plateau (CP) lies in southwestern New Mexico and southeastern Arizona. It is defined as the boundary between the CP, the Rio Grande rift, and the Basin and Range (BRP) provinces. Along its western and southern margins, the CP is physically distinguishable from the BRP by major escarpments. Because of a thick cover of volcanic material associated with the Mogollon-Datil volcanic field (MDVF), no such demarcation exists in the study area. To accurately determine lithospheric structure in this area, two seismic refraction lines that intersect within the MDVF were studied. In addition, gravity data along these lines were simultaneously analyzed. Based on these data, velocity and density profiles were created via an iterative forward modeling processes. Major crustal boundaries for each paired gravity/refraction profile were constrained to agree at the point of intersection and with other geological and geophysical control. Structurally, the CP margin appears as an abrupt northward deepening of the Moho just south of Datil, New Mexico. This lies beneath a major low density–low velocity upper crustal body, interpreted to be a plutonic complex approximately 230 by 150 by 6 km. It is very likely that this complex provided the source material for the MDVF. Stoping appears to be the vehicle for emplacement of this body, implying a major crustal reorganization in middle Tertiary time.

INTRODUCTION

The transition zone between the Rio Grande rift (RGR), Basin and Range province (BRP), and the Colorado Plateau (CP) is located in southwestern New Mexico (Fig. 1). The neighboring BRP/CP crustal boundary in central Arizona appears to be gradual beneath a major surface escarpment known as the Mogollon Rim (Brumbaugh, 1986; Hendricks and Plescia, 1991; McCarthy and Kohler, 1990). We know very little about the deep structure of this transition in New Mexico, but there is no major physiographic demarcation (Aldrich and Laughlin, 1984).

The lack of a physiographic boundary in southeastern Arizona and southwestern New Mexico is due in part to intense mid-Tertiary volcanic activity. However, large late Cenozoic faults should be identifiable if present. Extrusive material associated with the Mogollon-Datil volcanic field (MDVF) is at least 3 km thick (Elston and others, 1976), and may increase to more than 5 km (Gish and others, 1981). Prevolcanic geology is almost totally obscured by these deposits. To the north and northwest of the MDVF, geologic features associated with the CP predominate, while BRP structures are present to the southwest. The MDVF is truncated along its eastern and southeastern margin by the RGR.

A major objective of this study was to employ a combination of geophysical techniques and regional geologic data to study the nature and history of the transition zone between the

*Present address: AMOCO Production Company, 501 WestLake Park Blvd., P.O. Box 3902, Houston, Texas 77253.

Schneider, R. V., and Keller, G. R., 1994, Crustal structure of the western margin of the Rio Grande rift and Mogollon-Datil volcanic field, southwestern New Mexico and southeastern Arizona, *in* Keller, G. R., and Cather, S. M., eds., Basins of the Rio Grande Rift: Structure, Stratigraphy, and Tectonic Setting: Boulder, Colorado, Geological Society of America Special Paper 291.

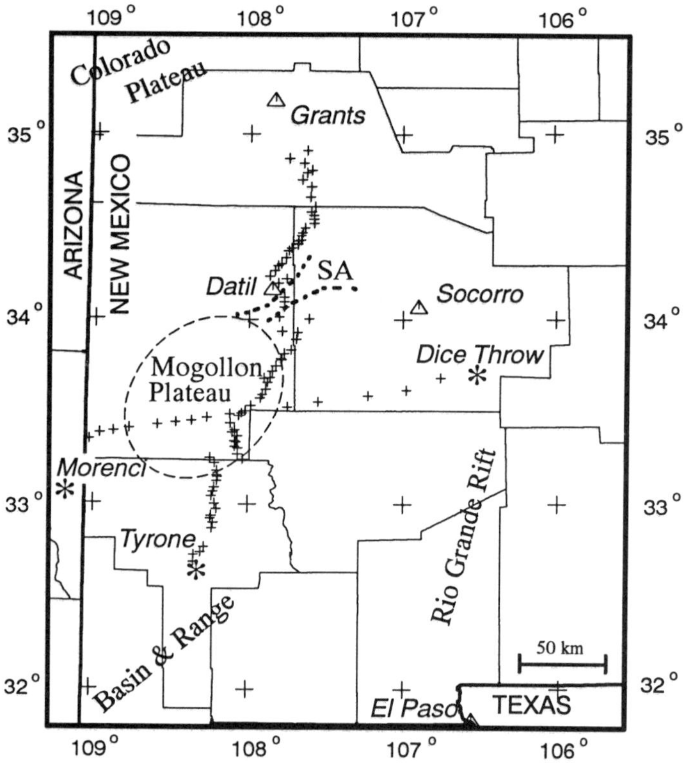

Figure 1. Location of seismic refraction lines used in the present study. The east-west–trending line is from Jaksha (1982). Locations of seismic recording stations are indicated by a (+). Seismic sources are shown by stars. SA, San Agustin plains (outlined by dark dotted lines).

southeastern Colorado Plateau and the Rio Grande rift/Basin and Range province. The new data presented here consist of a north-northeast trending unreversed seismic refraction line occupied in western New Mexico. This line crossed a previously recorded reversed refraction line (Jaksha, 1982) that was reinterpreted as part of this study. Gravity data were also used in the integrated analysis that produced the final earth models.

GEOLOGIC BACKGROUND

Pre-mid-Tertiary

As in central and western Arizona (Hendricks and Plescia, 1991), the New Mexico transition zone mainly reflects post-Laramide events. Whereas the Colorado Plateau has remained as a relatively coherent, undeformed block, the Basin and Range province area in Arizona has experienced intense tectonic activity intermittently since at least the Paleozoic (Coney, 1978a). Many of these features developed along previously existing structures as did many pre–middle Tertiary structures in southern New Mexico (e.g., Aldrich and Laughlin, 1984; Bilodeau, 1984; Seager and Mack, 1986). Because the region has been so dominated by mid-Tertiary and younger events, our discussion will be restricted to this time frame. For a more complete guide to the pre–mid-Tertiary tectonic history see Coney (1978b) and Seager and Mack (1986).

Mid-Tertiary

The end of the Laramide orogeny is generally regarded to be nearly synchronous with initiation of mid-Tertiary volcanism. This extensive magmatic event involved the ascent of melts into the upper crust, creating large volcano-plutonic complexes throughout the western United States and Mexico (Elston and Bornhorst, 1979). Silicic volcanism occurred as far east as the Trans-Pecos volcanic field (Barker, 1979), and coeval metamorphic core complexes formed in southern Arizona (Spencer and Reynolds, 1989). The earliest mid-Tertiary volcanic flows and ash falls were interbedded with the uppermost sediments in the Eocene Baca Formation (Seager and Mack, 1986; Cather and Chapin, 1989), and grade transitionally into the Spears Formation, which is composed of latest Eocene andesitic volcaniclastic material.

Mogollon-Datil volcanic field

The Mogollon-Datil volcanic field (MDVF), whose central region is called the Mogollon Plateau, overlies the transitions between the CP, BRP, and RGR in southwestern New Mexico (Fig. 1) and was of major interest in this study. Some associated volcanic centers appear to the south and southeast in the extensional provinces, but none are present to the north and northwest on the CP other than the San Juan volcanic field. Elston (1989) in fact considers the MDVF to be an outlier of the larger Sierra Madre Occidental to the south in Mexico. Volcanism in the MDVF produced large volumes of intermediate lavas and volcaniclastic rocks overlain by a bimodal sequence of silicic ash-flow tuffs and basaltic andesite lava flows. The ash flows were vented by caldera-type volcanism as part of the ignimbrite flareup in western North America. McIntosh and others (1992) have established a detailed time-stratigraphic framework for the MDVF ignimbrites, and recent research results in the area summarized in Ratté and others (1989). Ten cauldrons are well documented and up to 35 may exist (Elston, 1989). These volcanic rocks are found in and around a structural feature known as the Mogollon Plateau (MP; Fig. 1).

The bulk of silicic magmatic activity occurred in late Eocene–early Miocene time (36.2 to 24.3 Ma). McIntosh and others (1992) used $^{40}Ar/^{39}Ar$ chronology to document the existence of four major volcanic episodes, each separated by a hiatus. In general, volcanic activity progressed from east to west and south to north. Three of the four episodes correlate in time with volcanism in the San Juan volcanic field in Colorado, implying consistent tectonic control on the east side of the mid-Tertiary Colorado Plateau.

Volcanism was initiated with the extrusion of the calc-alkalic andesite to rhyolite group (Elston and others, 1976).

Minor amounts of this suite began to vent at 43 Ma, with a significant increase in volume at 36 Ma (McIntosh, 1989). Source magmas for these rocks may have been derived from the remnants of a subducted slab, with probable upper mantle and/or lower crustal contamination.

Overlapping this intermediate suite in time (36.2 to 24.3 Ma) were the rocks of the high silica alkalic rhyolite suites (Elston, 1976; Elston and others, 1976; McIntosh, 1989; McIntosh and others, 1992). These form a distinctive group of rocks with respect to chemistry, location of volcanism, and possibly source location and composition. With minor exceptions, they occur entirely within the Mogollon Plateau.

With the exception of about 2.5 km^2 of exposed underlying material, the MP is covered by volcanic rocks of Oligocene-Pliocene age (Coney, 1976). The eruptive style for these rocks was that of a series of overlapping flows, very minor ash-falls, and pyroclastic and epiclastic debris (Elston and others, 1976; Rhodes, 1976). The total thickness of the volcanic pile averages 3.1 km (Coney, 1976; Rhodes, 1976). These units originate from overlapping resurgent cauldrons, ring-dike complexes, and rhyolite domes.

There are several noteworthy geological and geophysical aspects of the Mogollon Plateau area:

1. With the exception of two minor centers of high-silica extrusives (Red Mountain, Arizona, and the Mule Creek caldera, New Mexico), all of the high silica rocks are clustered in a region to the south of the Plains of San Agustin (Fig. 1) and to the north of Silver City (Elston and others, 1976).

2. Elston and others (1976) state that the high-silica volcanics were emplaced in a neutral stress field.

3. Gravity studies (Coney, 1976; Krohn, 1976) show low values over the MP that cannot be completely explained by surface geology. They interpret the anomaly to represent a mass deficiency in the upper crust, but did not determine its boundaries and thickness.

4. The exterior grabens surrounding the MP are extensional features. However, the grabens do not strike north-northwest as do other Basin and Range/Rio Grande rifts faults in the region but appear to wrap around the MP. Elston and others (1976) interpret this to be an indication of some type of barrier to through-going faulting. Taking these observations together, Elston and others (1976) deduced that a shallow granitic plutonic complex exists under the MP.

Post mid-Tertiary

Block faulting, bimodal (mostly basaltic) volcanism, and extension has been dominant in the southeastern BRP and RGR from the late Tertiary to present time (Coney, 1978a, b; Henry and Price, 1986; Morgan and others, 1986; Seager and others, 1984). Crustal thinning has also occurred (Warren, 1969; Cook and others, 1979; Olsen and others, 1979; Sinno and others, 1981, 1986; Daggett and others, 1986; Perry and others, 1988). Volcanic fields are associated with the southern margin of the Colorado Plateau in Arizona (Luedke and Smith, 1978) and the Jemez lineament in New Mexico (Aldrich and Laughlin, 1984). The tectonic style associated with this latest stage of deformation appears to be distinct from the previous ignimbrite episode, and the chemistry of the volcanics indicates a different source as well.

Seager and others (1984) have determined that two periods of extension have occurred in the BRP and RGR immediately to the southeast of the Mogollon-Datil volcanic field. The first phase (29 to 18 Ma) produced northwest-trending moderately deep basins bounded by low-angle faults. The later phase of extension (9 to 3 Ma), was associated with north-south–trending block faulting. This latter period was also associated with 1 to 2 km of regional uplift (Morgan and Swanberg, 1985), lithospheric thinning, increased heat flow, and the appearance of minor true basaltic volcanism (Baldridge and others, 1989). BRP style faulting is not present in either the CP or the MP (Elston and others, 1976; Coney, 1978a). This may indicate that the Mogollon Plateau rests on the Colorado Plateau, but normal faulting occurs between the CP and MP (Dane and Bachman, 1965; Wilson and others, 1969).

GEOPHYSICAL FRAMEWORK

Previous seismic (Fig. 2) and gravity studies in the region provide a good framework on which to base this study. The Rio Grande rift is the most intensely studied feature in the area, and Keller and others (1990) provided a synthesis of geophysical and geologic data in the southern rift. High heat flow (about 80 mWm^{-2}), thin crust (about 30 km), low upper mantle seismic velocities (Pn about 7.7 km/s), deep basins, and Quaternary faulting and volcanism distinguish this feature from adjacent provinces including to some extent the Basin and Range province. However, our knowledge of the deep structure of the boundaries of the rift is limited.

The BRP immediately west of the RGR appears to be less tectonically active and to have slightly thicker crust than the RGR (Sinno and others, 1986; Daggett and others, 1986). Farther west, the BRP in southern Arizona was uplifted in mid-Tertiary time along with the Colorado Plateau (Peirce and others, 1979). As with the RGR, low seismic velocities in the BRP indicate elevated upper mantle temperatures at the present time. Some evidence exists for modification of the BRP crust in recent time. Sinno and others (1986) suggest that crustal thinning in southern Arizona has occurred over the past 5 Ma, roughly equivalent to the time that north-south structures were fully developed in the southern Rio Grande rift (Seager and others, 1984) and regional uplift occurred over a broad area of southwestern North America (Morgan and Swanberg, 1985).

The base of the crust (Moho) under the CP has been estimated by seismic refraction experiments and surface wave studies to be 40 to 45 km below sea level (Roller, 1965; Toppozada and Sanford, 1976; Keller and others, 1979). How-

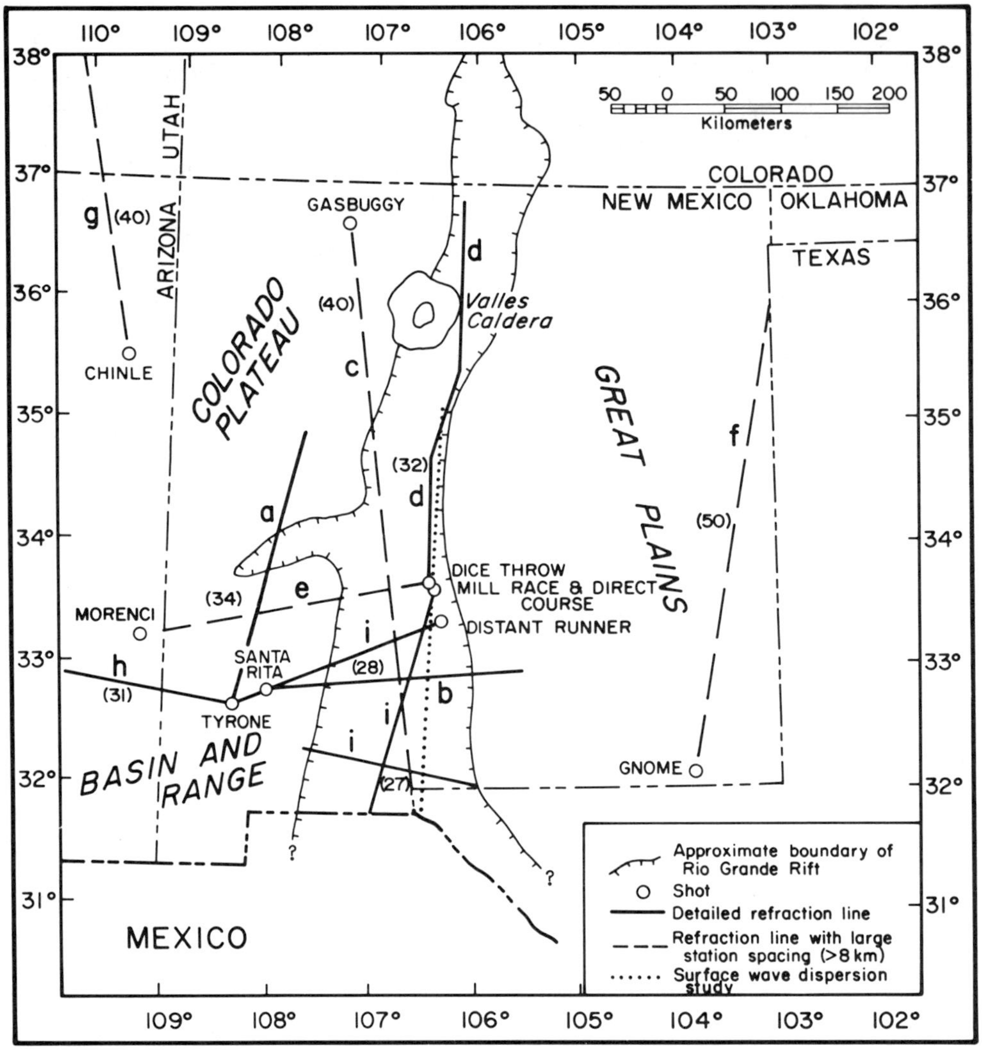

Figure 2. Index map of previous seismic studies in the region. Sources for refraction data: a, this study; b, Cook and others (1979); c, Toppozada and Sanford (1976); d, Olsen and others (1979); e, Jaksha (1982); f, Stewart and Pakiser (1962); g, Roller (1965); h, Gish and others (1981); i, Sinno and others (1986). Numbers in parentheses are crustal thicknesses in kilometers.

ever, the crustal thickness was estimated to be 50 km by deep reflection profiling (Hauser and Lundy, 1989). Considering the 2-km average elevation of the plateau that should be added to the older values to obtain crustal thickness and the differences in the techniques, this discrepancy is not major. The uppermost mantle P-wave velocity (Pn) determined by Roller (1965) was slow (7.8 km/sec) but earthquake data (Beghoul and Barazangi, 1989) indicate that average velocities in the mantle lithosphere are more typical of the stable North American craton (about 8.1 km/s). The CP has near-zero mean free-air gravity anomalies indicating isostatic equilibrium (Keller and others, 1979; Morgan and Swanberg, 1985). Heat-flow data (Reiter and others, 1975, 1979; Blackwell, 1978; Thompson and Zoback, 1979; Shearer and Reiter, 1981; Reiter and Clarkson, 1983; Swanberg and Morgan, 1985) show the interior of the Colorado Plateau to have moderate heat flow, about 63 to 75 mWm^{-2}. This is bounded by a 100-km-wide zone of increasing heat flow towards the BRP and RGR (Lachenbruch and Sass, 1978; Swanberg and Morgan, 1985). The moderate surface heat flow is not consistent with high-temperature conditions in the upper mantle implied by low seismic velocities and electrical conductivity (Morgan and Swanberg, 1985). It is possible that the heat-flow values on the plateau are artificially low due to ground-water circulation (Shearer and Reiter, 1981; Sass and others, 1982). It is also possible that the base of the crust is presently the site of a thermal anomaly, but that the thermal pulse has not yet reached the surface. Recent seismic results, however, suggest that uppermost mantle velocities may not be low and thus may not indicate high temperatures in the mantle (Beghoul and Barazangi, 1989; Hauser and Lundy, 1989; McCarthy and others, 1990; Wolf and Cipar, 1991).

The present-day stress field of the Basin and Range province and Rio Grande rift appears to encroach on the margins of the Colorado Plateau (Humphrey and Wong, 1983; Aldrich and Laughlin, 1984). Geophysical studies along the Great Basin–Colorado Plateau boundary in Utah (Keller and others, 1975) indicate lithospheric thinning and Basin and Range structures inside the plateau's physiographic margin. These studies suggest that the BRP is presently growing at the expense of the CP.

The transition zone between the CP and RGR/BRP in southwestern New Mexico and southeastern Arizona is enigmatic in that the surface geology that might provide clues to its nature is buried. However, the structural boundary must occur under the Mogollon-Datil volcanic field, since identifiable tectonic features of each province surround it.

The physiographic transition between the Basin and Range and Colorado Plateau provinces in east-central Arizona has been placed at the Mogollon Rim (e.g., Mayer, 1979; Peirce and others, 1979; Hendricks and Plescia, 1991). Based on several reversed seismic refraction profiles, Warren (1969) concluded that the Mogollon Rim lis above an abrupt change in crustal thickness of approximately 4 km in central Arizona. Additional thickening occurs from southwest to northeast as a result of a 0.5° dip on the Moho. Farther west, the recent Pacific to Arizona Crustal Experiment (PACE) indicates that the boundary is associated with a gradual thickening of the crust (McCarthy and others, 1990).

Aldrich and Laughlin (1984) propose that the current plateau margin in New Mexico is located along the Jemez lineament, an alignment of young volcanic features. Southeast of this lineament, northwest-trending dikes dated at 29 to 27 Ma (Laughlin and others, 1983) indicate extension beginning in the mid-Tertiary, and currently this region displays a transitional stress regime (Aldrich and Laughlin, 1984).

DATA COLLECTION AND COMPILATION

Seismic refraction

Two seismic refraction profiles have been recorded across the Mogollon-Datil volcanic field (Fig. 1) and form the basis of this study. Jaksha (1982) recorded a partially reversed east-west profile between Morenci, Arizona, and the White Sands Missile Range, New Mexico. Data from this profile were reanalyzed as part of this study. The new data presented here consist of additional seismograms gathered along the profile of Harden (1982). The total number of stations deployed along this profile constituted a twofold increase in data over this previous study, and an increase of approximately 125% over the MDVF as a whole. Seismic sources for this survey were quarry blasts as the Phelps dodge open-pit copper mine at Tyrone, New Mexico. A base station was installed on the mine property near the pit to determine origin times. Blast locations were provided within ±30 m by the company. Distances from the blasts to this base station never exceeded 3 km. The velocity of P-waves within the pit was measured at 5.5 km/sec (S. H. Harder, personal communication, 1987). Origin times were thus calculated using this velocity and the distance from the blast to the base station. Based on calibrations in the pit, origin-time errors are estimated to less than 0.1 sec.

Recording stations were occupied by Sprengnether MEQ-800 portable seismographs. Portable FM tape recorders were connected to the MEQs to record the signal. Sprengnether S-7000 and L4 seismometers, both with 1 Hz natural frequency, were used in this study. Origin times and station locations are presented in Schneider (1990).

Fifty-nine distinct, good-quality seismograms were recorded. Several sites were reoccupied to check origin time accuracy and to insure good signal-to-noise ratios. Where possible, the FM tapes were digitized by an analog to digital converter system. The remaining seismograms were hand digitized and resampled using a cubic spline interpolation routine. The digitized sample interval was either 0.01 or 0.02 sec. Each record was normalized to its greatest amplitude for display. These records were then merged with the data from Harden (1982) to produce a composite record section (Fig. 3). Arrivals of significant, recognizable phases were picked for each site and input to a ray-tracing program for analysis.

The new seismic data and gravity results justified a reanalysis of the data from the Jaksha (1982) cross line (Fig. 4). Jaksha (1982) employed a simple layered earth to interpret his data, but we used ray tracing so that more complex structures could be modeled. Arrivals were repicked from his record sections, but our interpreted phase arrivals differ from Jaksha (1982) by less than 0.2 s.

Gravity

The gravity data base at the University of Texas at El Paso was utilized to make gravity maps of the study area and to extract gravity values along each of the seismic refraction profiles described above. Complete Bouguer gravity anomaly values were retrieved from a region enclosed by 31.0 to 36.5° N latitudes and 105.5 to 112.5° W longitudes. Over 40,000 readings were available in this area. A sea-level datum and density of 2.67 gm cm^{-3} were used in the Bouguer and terrain corrections for this study. Interpolated gravity values were derived at evenly spaced (4 km) locations in the region using the technique of minimum curvature (Briggs, 1974; Swain, 1976). To insure stability in frequency-domain filtering, a third-order polynomial surface was fitted to, and subsequently subtracted from the data. These data were then filtered using the technique of Peeples and others (1986).

Daggett and others (1986) found that broad trends reflecting Moho relief and large crustal features were best enhanced by a 125-km low-pass filter in an area overlapping that of the

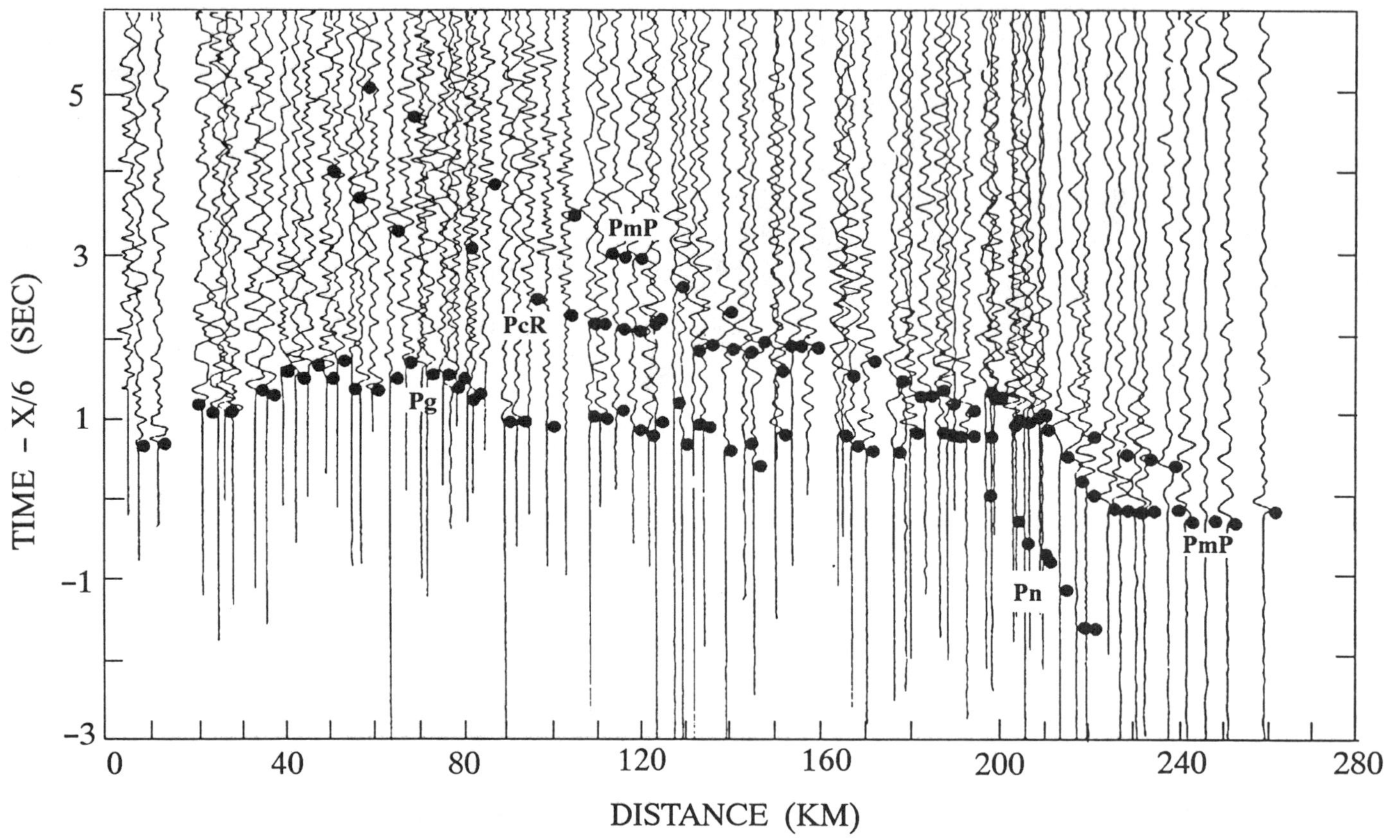

Figure 3. Seismic record section for the Tyrone-Acoma line. A 1- to 15-Hz, band-pass filter was applied to the data. Reduction velocity is 6 km/s. Arrival time picks of phases are shown by dots. These picks are shown on ray-trace models in Figures 11 and 13. The nomenclature for the phases is discussed in the text.

present study. Thus this same wavelength filter was applied in this study and the results are shown in Figure 5. Other wavelength cutoffs were also tested but produced less useful results. This map delineates a large gravity minima that is associated with the MDVF. A north-south–trending maxima is associated with the Rio Grande rift.

DATA ANALYSES, MODELING, AND DISCUSSION

Seismic refraction and gravity profiling

The seismic modeling program used to determine earth structure was based on the algorithm described by Cerveny and Ravindra (1971). This technique, as implemented by Leutgert (personal communication, 1989; Hill and others, 1985), generates a model consisting of two-dimensional varying earth structures. Many workers have used this ray tracing technique, called asymptotic ray theory (ART), and it is well represented in the literature.

Gravity readings within 1 km of each seismic profile were chosen by projection along a line perpendicular to the profile so that gravity data could be used as an independent constraint in the integrated analysis. The objective of this analysis was to create an earth model satisfying both of these data sets, the desired result being an image of present-day crustal structure. The computer program used to interpret the gravity profiles across the area was based on a 2.5-dimensional algorithm, based on the technique of Cady (1980). It was written by S. F. Lai at the University of Texas at Dallas, and modified extensively by D. Roberts (personal communication, 1989).

Depth to major velocity/density interfaces were kept the same for the seismic and gravity models at the point of intersection of the Tyrone-Acoma and Morenci–Dice Throw profiles, except for the near-surface, upper-crustal boundary. This exception was necessary because: (1) the seismic ray-tracing program is capable of including topography, while gravity data processing assumes reduction to a datum plane effectively at the lowest station elevation; and (2) gravity and seismic stations do not coincide. Thus, gravity readings may reflect local structure that does not influence the seismic ray path. In general, the near-surface structure is smoother in the seismic model than in the gravity models because of limitations in the ART approach. In order to honor geologic constraints, the near-surface structure in the gravity models is

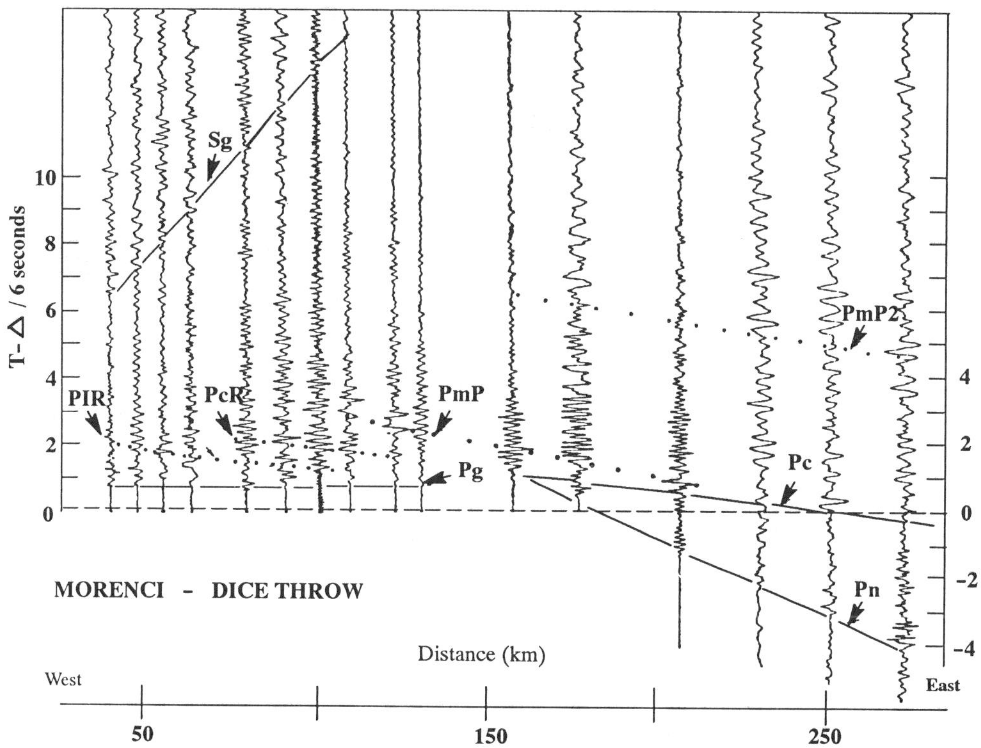

Figure 4. Seismic record section for the Morenci–Dice Throw refraction line (Jaksha, 1982) with our picks for the major phases. These picks are shown on the ray trace model in Figure 15. Modified from Jaksha (1982).

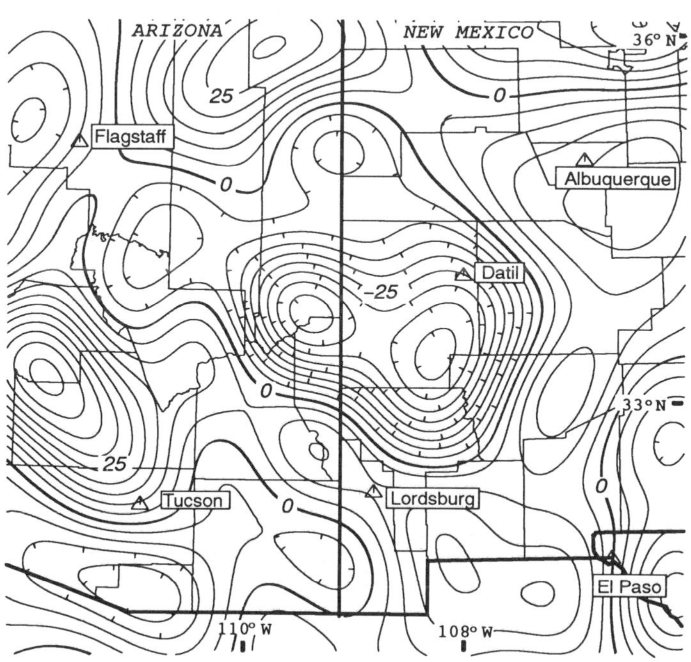

Figure 5. Regional Bouguer gravity anomaly map of eastern Arizona and western New Mexico produced by a 125-km, low-pass filter in the wave-number domain. Contour interval is 5 mgal.

more complex than required by either the gravity or seismic data alone. Thus, the gravity models should be considered to represent large-scale geologic cross sections.

When alterations to structure were indicated for a particular profile (gravity or seismic, Tyrone-Acoma or Morenci–Dice Throw) during modeling, the effects on the other profiles were examined. Thus every modification required running two ART and two gravity-modeling iterations to observe its effect. More than 250 repetitions of these computer programs were run before a satisfactory earth structure was derived. The final models satisfy both refraction and gravity observations, tie to nearby results (Olsen and others, 1979; Gish and others, 1981; Sinno and others, 1986), and are compatible with surface geologic features. They are not unique, but 250 iterations gave us a chance to thoroughly review the ranges of possible earth models. Thus, we are confident that the final models presented are valid in terms of their major implications. The worst resolution is that of the geometry and physical properties of the lower crust.

Refraction observations

Arrivals for major crustal velocity interfaces are shown for the Tyrone-Acoma and Jaksha (1982) data in Figures 3 and 4, respectively. For this discussion, the following notation is used to identify different phases: Pg and Pn are critically refracted P waves from the top of the upper crust (crystalline basement) and upper mantle, respectively; and PIR and PcR indicate reflections from the top of the intermediate and lower crust, respectively. PmP corresponds to the P-wave reflection from the M-discontinuity.

The Jaksha profile (Fig. 4) shows Pg to be a first arrival to a distance of 155 km, after which the Pn branch becomes the first arrival. PIR, PcR, and PmP never become first arrivals on this line, yet their arrival times combined with gravity data and the intersection with the Tyrone-Acoma line are important contributors to determining crustal structure. Upper-mantle velocity structure and placement of the Moho is supported by the use of the reversed Pn branch for Jaksha's (1982) profile.

The greater density of seismic readings along the Tyrone-Acoma profile provided added confidence in the identification and timing of arrivals. Pg arrivals are delayed in the interval from 30 km to 85 km, indicating decreased upper crustal velocities. Local variations in the Pg arrival curve are interpreted to be caused by topography and near-surface inhomogeneities in the volcanic pile within the southern Mogollon Plateau. PcR never occurs as a first arrival, but it provides a valuable constraint for the model derived for this profile. Pn becomes the first arrival at approximately 170 km, and ceases to be observed beyond 220 km.

The most striking feature in the Tyrone-Acoma profile is the time shift in PmP arrivals at distances between 210 to 225 km. Across this interval, PmP arrivals step almost 1.5 sec forward on the reduced (T–X/6 sec) travel time plot. A step of this magnitude requires a change of fundamental importance within the crust, and must be accounted for in the final integrated model.

Gravity observations

Gravity profiles for both refraction lines are shown in Figure 6. A gravity low is apparent in the center of Jaksha's (1982) line (Fig. 6A). It is approximately 80 mgal lower than the west end and 70 mgal lower than the east end (disregarding the Rio Grande rift basin anomaly). By itself, this low does not constrain the density difference or depth of the causitive anomalous body. However, the half wavelength (about 250 km) suggests a deep, possibly Moho-related feature. A depression or sag is observed within this low at distances between 50 km to 165 km. This feature, about 115 km in half wavelength, may be caused by a low-density, upper-crustal body. The Tyrone-Acoma gravity field also displays a low in the center of the profile (Fig. 6B). Values in the center of this feature are nearly 100 mgal lower than those at the south end, and nearly 50 mgal lower than the north end. Thus, it is superimposed on a slope in the regional gravity field that becomes more negative to the north.

Model generation and evolution

Jaksha (1982) used a flat-lying, homogeneous-layered model to interpret his results for the Morenci–Dice Throw profile. His model was approximately duplicated using ART modeling, and although this structure is adequate in reproducing his observed arrivals, it yields a Bouguer anomaly field that fails to match observed values by about 50 mgal (Fig. 7). Since the gravity field indicates greater complexity within the crust and upper mantle than proposed by Jaksha (1982), alternative models were generated.

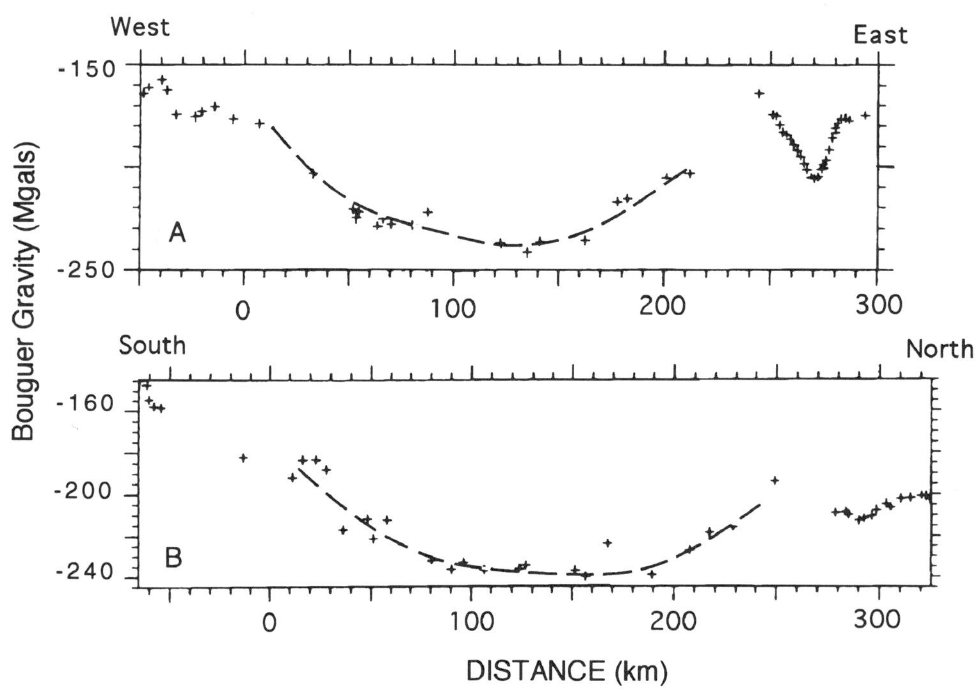

Figure 6. A, Gravity profile along Jaksha's (1982) refraction line. B, Gravity profile along the Tyrone-Acoma refraction line. +, observed gravity values.

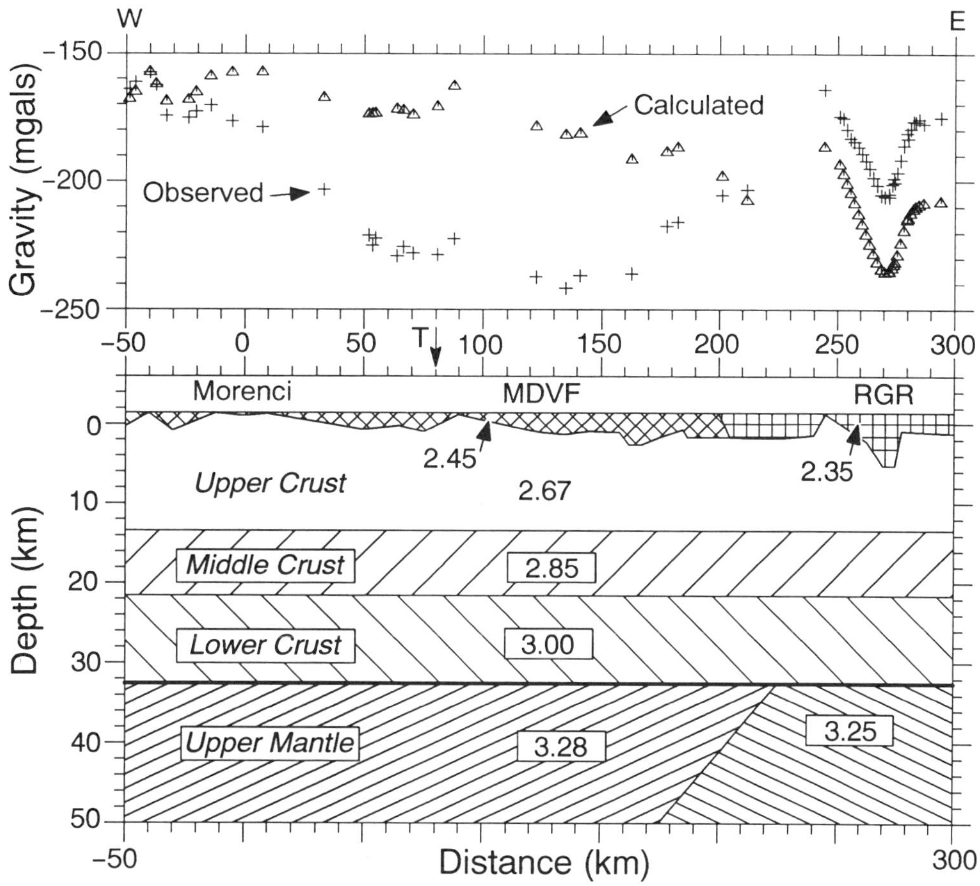

Figure 7. Gravity model based on Jaksha's (1982) earth model. The flat-lying structure obviously does not fit the observed data. Values for density are in gm/cm^3. T, tie to Tyrone-Acoma profile. MDVF, Mogollon-Datil volcanic field; RGR, Rio Grande rift.

The first alternative was one in which the gravity anomaly is derived entirely by crustal thickening (Fig. 8). This model shows that a near-surface pluton is not necessary to satisfy gravity observations. This model was also used in the ART seismic ray-tracing program and fitted the observations reasonably well. Thus, for the Morenci–Dice Throw profile, seismic and gravity data indicate that crustal thickening is a viable explanation of geophysical anomalies in the region. As modeled in this approach, the maximum thickness of the crust does not exceed 37 km along this line, which is intermediate between the Colorado Plateau and Basin and Range/Rio Grande rift provinces. The margins of crustal thickening correspond approximately to the –20 mgal contour line of the low-pass gravity map in Figure 5. This contour extends nearly as far south as Silver City, New Mexico.

The second alternative investigated was whether a near-surface pluton could account for the gravity anomaly instead of crustal thickening. Of primary concern in this model was the tradeoff between density contrast and volume. The density of a granitic pluton is almost always lower than that of the displaced country rock (Bott and Smithson, 1967), usually within the range of –0.05 to –0.18 gm/cm^3 and rarely more than –0.2 gm/cm^3. For this study, the density contrast of the pluton was set to –0.1 gm/cm^3 with respect to country rock (in a lateral sense). By picking a fairly large contrast, the volumes derived are conservative estimates (i.e., a smaller density contrast would require a larger volume to produce the required mass anomaly).

The volume expected for a buried plutonic complex is difficult to estimate because venting of a pluton into a caldera volcanic complex may remove from 10 to 90% of the original volume of the magma (E. A. Anthony, personal communication, 1989). If the plutonic complex of Rhodes (1976) and Elston and others (1976) covers the same lateral extent as the caldera complex, the thickness of the underlying magma chamber can be calculated using the formula for the gravitational attraction of a cylinder. Given an average thickness of 3.1 km for the volcanic pile, 90% removal of liquid from the underlying chamber requires an approximately 0.33-km-thick pluton. A pluton of this size is too thin to have significant effect on the gravity field or the wide-angle seismic measurements. Venting of only 10% of the magma chamber yields a residual pluton about 27.9 km thick. Since this represents almost the entire thickness of the crust, the 10% scenario was also discarded for this study. Two models were therefore de-

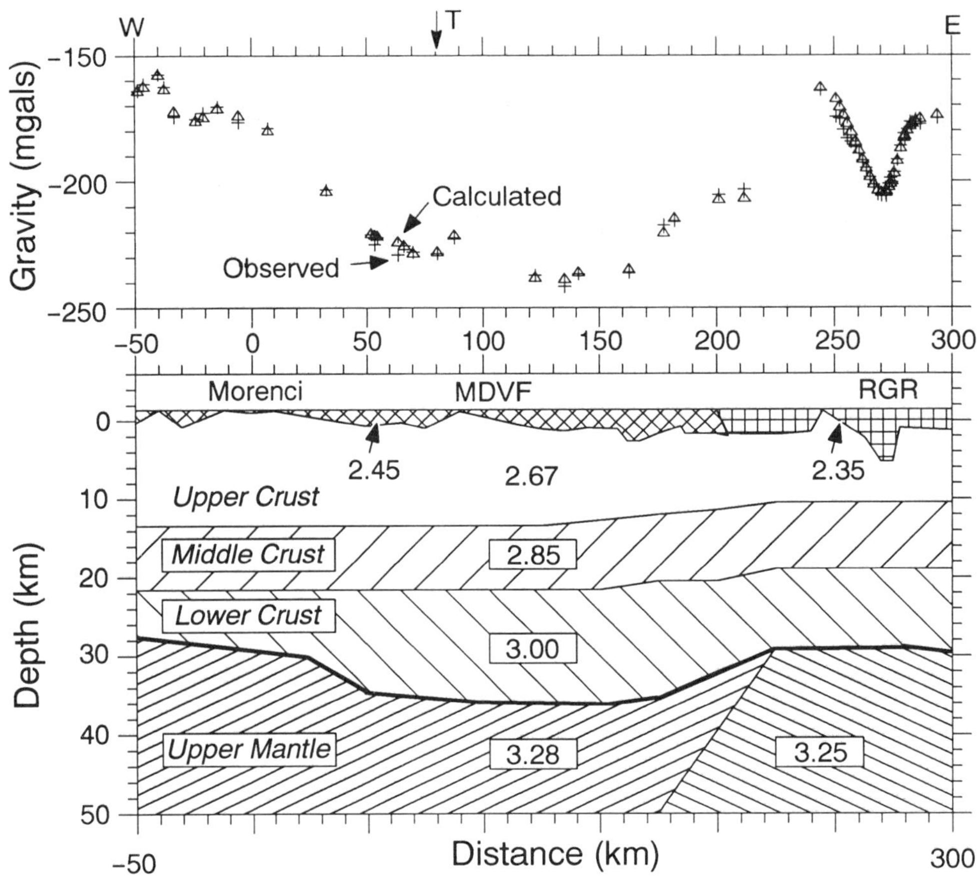

Figure 8. Gravity model for thickened crust showing the fit of predicted values to observed data along the Jaksha (1982) refraction line. Crustal thickening alone is sufficient to account for the observed anomaly. Values for density are in gm/cm^3. T, tie to Tyrone-Acoma profile. MDVF, Mogollon-Datil volcanic field; RGR, Rio Grande rift.

veloped with plutons of intermediate thicknesses: (1) 50% of the magma was vented to the surface, leaving a 3.1-km-thick pluton in the subsurface; and (2) 20% of the magma was removed, yielding a 12.5-km-thick pluton. The top of each pluton was set at 4.5 km below the present surface, primarily to leave room for the volcanic pile that now exists in the area.

When a 3.1-km pluton is placed into the upper crust, the gravity anomaly produced is insufficient to explain the observed readings by about 50 mgal. Thus, this model would also require significant crustal thickening to supply the mass deficiency needed to produce the gravity anomaly. However, the mass deficiency of a pluton about 12.5 km thick is sufficient to reproduce the gravity low between Morenci and Dice Throw (Fig. 9) and produces a match to the observations similar to crustal thickening. However, one would expect evidence of significant doming if such a large body were emplaced into the crust. No significant radial dikes have been observed, and there is no evidence for large differential uplift across the region. Thus, a smaller pluton seems more likely.

The Morenci–Dice Throw line alone was thus not sufficient to resolve the relative importance of crustal thickening and a batholith. However, the results from the analysis of these data acted as a guide for analysis of the Tyrone-Acoma line. Thus, three models were tested for the Tyrone-Acoma line: a thickened crust, a 3.1-km-thick pluton, and a 12.-5-km-thick pluton. The depth to major interfaces as well as velocity and density were matched at the intersection point of the two profiles. That point is in Corduroy Canyon, approximately 80 km from Tyrone and 125 km from Morenci. Crustal structure and velocity were also constrained near Tyrone by the previous work of Sinno and others (1986).

The calculated gravity results along the Tyrone-Acoma line for the two plutonic models produced results similar to those for the Morenci–Dice Throw profile. The north end of the pluton was placed 220 km from Tyrone, the edge of the sag in gravity values. Initially, the northern extent of the pluton was placed 150 km north of Tyrone, as mapped by Elston and others (1976) and suggested by Krohn (1976). Again, the 3.1-km-thick pluton was insufficient to account for the entire gravity anomaly. The 12.-5-km pluton created a gravity field that is relatively close to the observed readings. However, it is clear that the gravity low extends farther north along the pro-

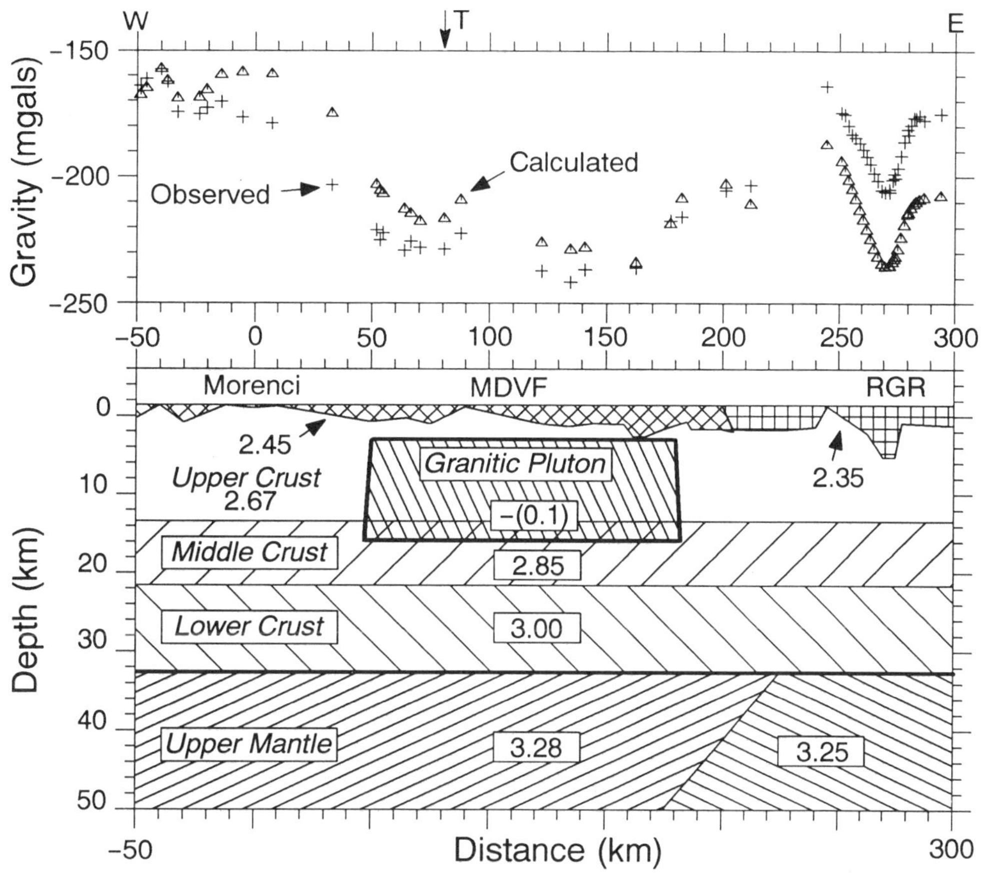

Figure 9. Gravity for the Jaksha (1982) refraction line in which a 12.5 km-thick pluton is placed in the upper crust. Values for density are in gm/cm^3. T, tie to Tyrone-Acoma profile. MDVF, Mogollon-Datil volcanic field; RGR, Rio Grande rift.

file along the profile than predicted by Elston and others (1976). Reconciling this problem requires extending the low-density body or increasing the thickness of the crust to the north. Based on gravity analysis alone, the crustal thickening and 12.5-km-thick pluton models provide equally acceptable solutions, and the model explaining the major gravity anomaly by variations in crustal thickness alone is shown in Figure 10.

The validity of each of these models was tested by the use of the ART ray-tracing program to see if they could match the observed arrival times of major seismic phases. A major aspect of the observed data was the shift in observed PmP arrivals between 200 and 220 km (Fig. 3). This advance of the arrivals was more than 1.0 sec and indicates a major structure at or above the Moho. As shown in Figure 11, the rays arriving just before the shift in the PmP phase reflect off the Moho about 100 km from Tyrone. If the travel-time anomaly is caused by thinning of the crust, it must begin just beyond 100 km from Tyrone. The model in Figure 11 shows that extreme thinning to 23 km is required to recreate this step in the PmP phase. However, such thinning produces a discrepancy between the predicted and observed gravity fields of almost 50 mgal. Clearly, the gravity field does not support localized crustal thinning of such magnitude along this profile.

The time step in the PmP arrivals must therefore be produced by lateral velocity variations within the crust. With an average velocity of 4 km/sec in the near-surface layer and a thickness of 3 km, the travel time for a vertical seismic ray through this layer is 0.75 sec. If this layer is removed and replaced by upper crustal material with a velocity of 6 km/sec, a difference in travel time of only 0.25 sec is expected. Thus, the PmP arrival time anomaly cannot be simply caused by variations in the near surface.

The travel-time anomaly could be produced by thickening the lower crustal layers in the north at the expense of the upper crustal layer but would violate the gravity constraints and is not supported by other seismic results in the region.

The presence of a near-surface plutonic complex under the Mogollon Plateau could also create the time shift in PmP. Granitic batholiths are usually less dense than the surrounding upper crustal rocks implying that the seismic velocity is also lower. The travel-time step could then indicate the point where the pluton no longer affects the ray paths, suggesting that the

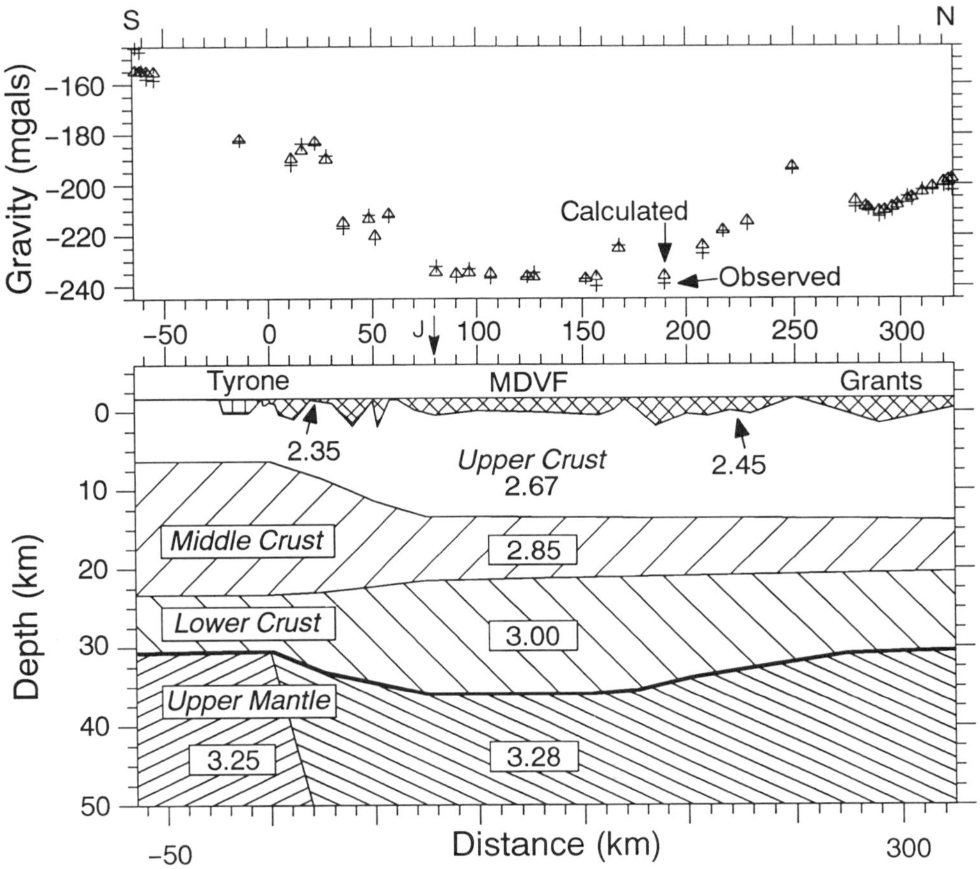

Figure 10. Gravity model along the Tyrone-Acoma refraction line involving crustal thickening alone. Values for density are in gm/cm^3. J, tie to Jaksha (1982) profile. MDVF, Mogollon-Datil volcanic field; RGR, Rio Grande rift.

plutonic complex extends approximately 50 km north of Elston and others (1976) boundary. This position also corresponds to the north end of the sag superimposed on the regional Bouguer gravity low at that point.

Structures north of the PmP travel-time shift were another concern. This region is not constrained directly by the cross profile or any other seismic results, and the lack of reversed coverage produces ambiguities in the explanation for the arrivals of the Pn phase. A number of combinations of crustal thickness and upper-mantle velocities could satisfy the observations. An attempt was made to model the Pn arrivals by upper-mantle velocity changes without crustal thickening, but no reasonable model could be found that was consistent with the constraints provided by nearby seismic profiles. Seismic results in Arizona and northern New Mexico and the gravity data also suggested continued thickening to the north under the Colorado Plateau.

The emplacement of a large plutonic complex has profound significance on the overall structure of the crust. Of critical importance is the question of the fate of the displaced upper crustal material. Bott and Smithson (1967) observed that when a pluton is inserted into the upper crust, a like volume of material is displaced regardless of where the magma is derived. Forceful or active injection of fluids implies that country rock is uplifted, pushed aside, or is displaced downward by stoping. If the primary displacement is upward, then the 20% model should create a structural high of 12.5 km under the Mogollon Plateau relative to its surroundings. No uplift of this magnitude has been reported in the region. If displacement is primarily outward, significant compressive features should be seen in the area surrounding the Mogollon Plateau. However, southwestern New Mexico switched from regional compression to tension during the time of emplacement, and there are no outward-radiating indications of stress from the Mogollon Plateau in the mid-Tertiary. Instead, radially directed tension cracks occur in a wide region around the Mogollon Plateau, with the center of the pattern somewhere to the southwest of Magdalena, New Mexico (Elston and others, 1976). Furthermore, they report that the calderas formed in a neutral stress field, and presumably the pluton emplacement occurred in the

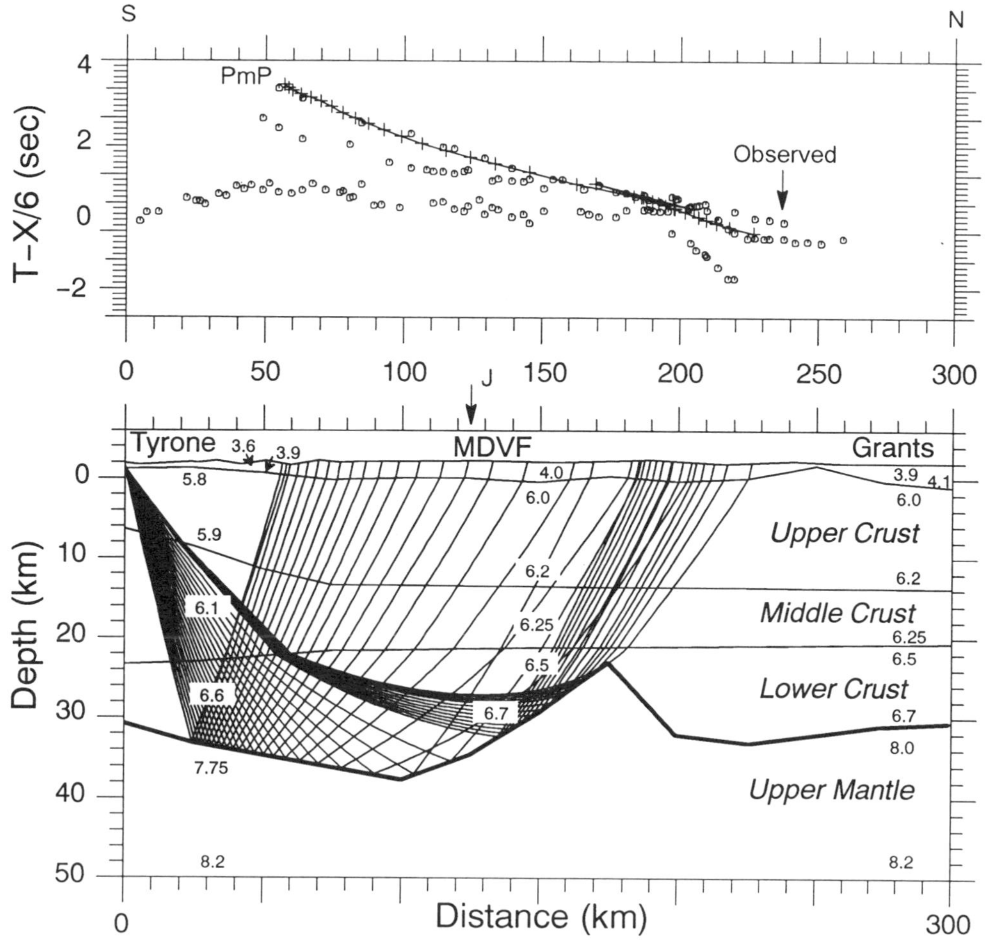

Figure 11. Ray-tracing model of the Tyrone-Acoma line, attempting to model the advance in PmP arrivals by crustal thinning. Even by thinning to 23 km, the step is not reproduced. Velocities are in km/sec. Symbols: o, observed arrivals; +, calculated reflected arrivals; J, tie to Jaksha (1982) profile. MDVF, Mogollon-Datil volcanic field.

same field. Therefore, stoping must be the dominant factor in movement of country rock during emplacement of the proposed mid-Tertiary pluton in southwest New Mexico. Elston and others (1976) propose that the source region for these fluids was in the lower crust. Melting was due to heat from asthenospheric basalts that pooled at the base of the crust.

With these geological constraints in mind, the key feature of the final models became apparent: a near-surface pluton, extending through the Plains of San Agustin must be present in the crust. It must be sufficiently thick to produce a fundamental velocity change throughout the crust via stoping, and account for (at least) the sag within the regional Bouguer gravity low. It must also satisfy both the Tyrone-Acoma and Jaksha (1982) refraction and gravity data. The final integration of these data sets proved to be difficult and time consuming (i.e., 250 iterations were required) since perturbing one model had a cascading effect on the other three.

The bottom of the pluton in the final gravity model for the Tyrone-Acoma profile is estimated to be at a depth of 10 km, as shown in Figures 12 and 13. The stoping effect is shown in this model as a depression recorded by the intermediate crustal boundaries. The northern boundary of the depression is more abrupt than that to the south, and may imply a major fault. A major crustal-scale fault and an abrupt sag under the pluton both provide the same result; they bring faster velocity material nearer to the surface north of the boundary. In the final seismic model, both features are necessary to create the change in arrival time of the PmP phase (Fig. 13).

To insure the viability of the model developed for the Tyrone-Acoma data, it was tested against the Morenci–Dice Throw seismic and gravity profiles. The integrated results are shown in Figures 14 and 15. These models are compatible with Jaksha's (1982) data, including the reversed Pn branch for the Morenci–Dice Throw study (Fig. 16). They explain the

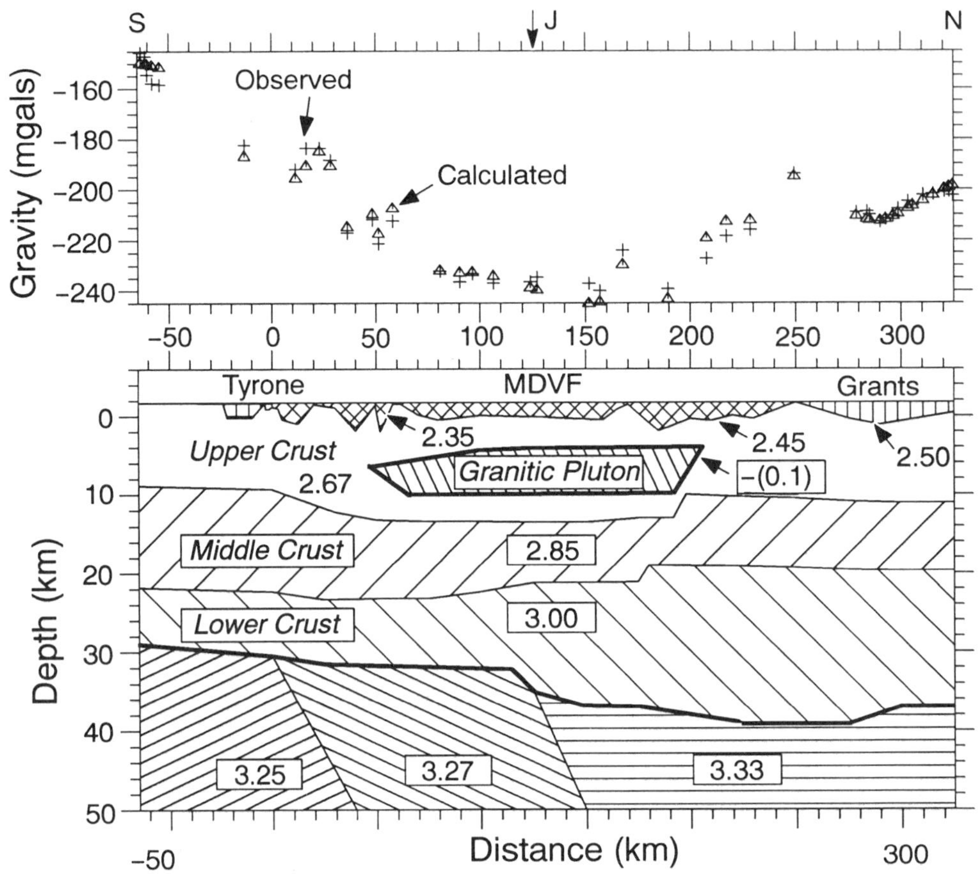

Figure 12. Final model for gravity profile along Tyrone-Acoma refraction line. Values for density are in gm/cm^3, except for the granitic pluton that displays a density contrast with its country rock. J, tie to Jaksha (1982) profile. MDVF, Mogollon-Datil volcanic field.

gravity anomaly and tie to the Tyrone-Acoma models. Thus, a model with a combination of a pluton and crustal thickening is the preferred way to explain Jaksha's (1982) results.

The crustal structure models that result from the combined analysis of seismic and gravity data, along with our knowledge of the geology in the region, indicate a complex Tertiary history of crustal evolution. Any one of the seismic or gravity profiles alone cannot resolve this complex crustal structure, and minor details are not uniquely determined by this study. However, we estimate that the use of both data sets constrains placement of major crustal boundaries with an error estimated at ±2 km.

CONCLUSIONS

This study attempts to answer two fundamental problems in southeastern Arizona and southwestern New Mexico: (1) the present-day crustal geometry, and (2) the nature, extent, and post-Laramide tectonic history of the transition zone between the Rio Grande rift, the Basin and Range province, and the Colorado Plateau province. A complicating factor in this study was the presence of the mid-Tertiary Mogollon-Datil volcanic field (Elston and others, 1968, 1970, 1976). The volcanics hide previous features and have influenced the structural response to subsequent tectonic activity. Thus, solutions to these problems could only come from the combined use of seismic refraction and gravity data along with the use of available geologic data. The integrated use of seismic refraction and gravity information resolves crustal structure to an extent not possible by employing either type of data alone. The most significant features of the earth models derived are the following:

1. A major low-velocity, low-density body is present in the upper crust of southwestern New Mexico and southeastern Arizona. The spatial correlation of this body with the Mogollon-Datil volcanic field supports the interpretation that it represents a pluton which was the source for the volcanics. However, the geophysical data indicate that the pluton extends about 50 km north of the boundaries suggested by Elston and others (1976), Krohn (1976), and McIntosh (1989). The ap-

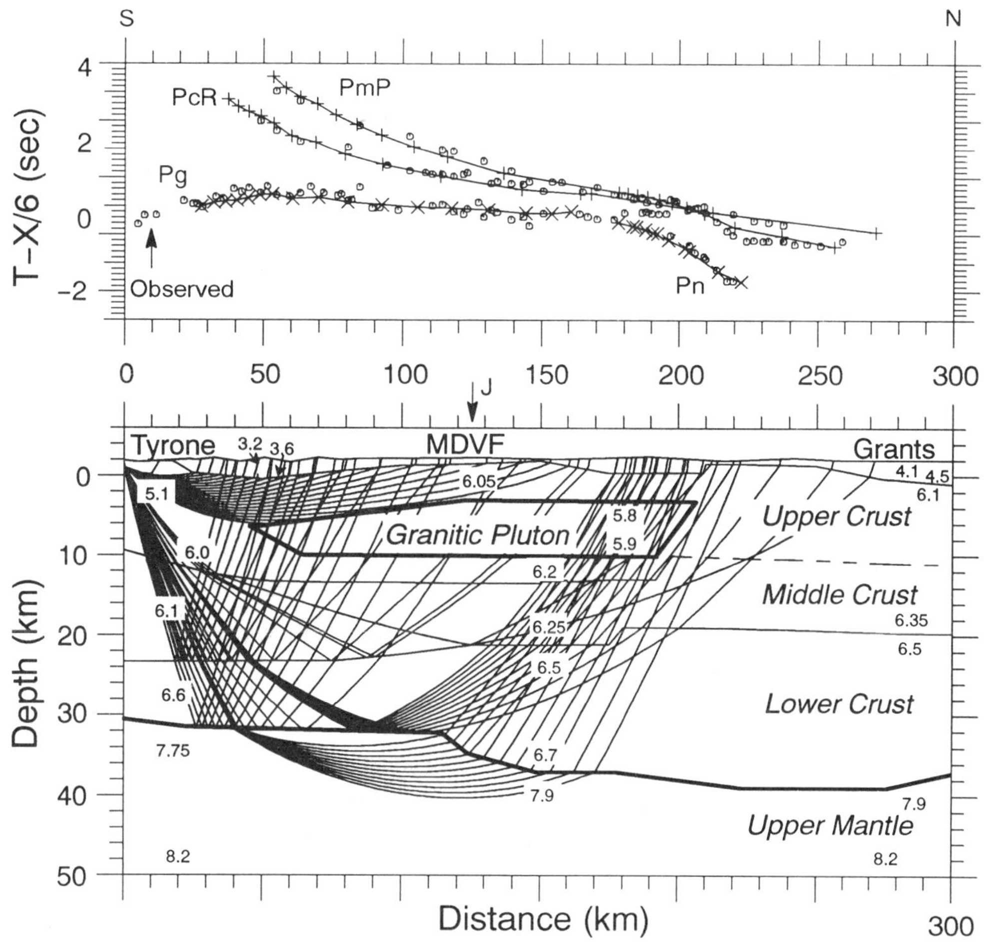

Figure 13. Ray-tracing model for Tyrone-Acoma line, final configuration. Velocities are in km/sec. Symbols: o, observed arrivals; +, calculated reflected arrivals; x, calculated refracted arrivals; J, tie to Jaksha (1982) profile. MDVF, Mogollon-Datil volcanic field.

proximate extent of this pluton is shown in Figure 17. The volume occupied by this complex combined with the volcanic material vented during the mid-Tertiary implies a major crustal and probably lithospheric reorganization during that time.

2. The Moho deepens fairly abruptly approximately 90 km north of Silver City, from 32 to 37 km over a distance of about 35 km. This region is located about midway between the north and south ends of the plutonic complex. This deepening does not correlate with any known late Tertiary features suggesting that it predates the MDVF. The best-available seismic image of the transition zone in Arizona (the PACE experiment results, McCarthy and Kohler, 1990) indicates a gradual thickening of the crust from the Basin and Range into the Colorado Plateau. However, Warren (1969) found evidence for an abrupt thickening of the crust in central Arizona near the Mogollon Rim. His result is similar to that observed in this study. These steps in the Moho are shown in Figure 17. If one accepts the Colorado Plateau boundary of Aldrich and Laughlin (1984) based on stress analysis, then the results of this study would simply suggest a broader transition zone in southwestern New Mexico as shown in Figure 17 with the step in the Moho being in the central portion of this zone.

3. In spite of these steps in the Moho, the overall impression is that there is a gradual transition in crustal structure that correlates with the broad transition zone observed at the surface. The MDVF basically covers this transition zone in southwestern New Mexico. The transition zone in crustal structure along the western margin of the CP in Utah is much more abrupt (Keller and others, 1975).

Clearly mid-Tertiary time was one of large-scale reorganization of the crust in our study area and the following evolutionary sequence is proposed:

1. Magmas formed from excess heat in the lower crust worked their way to the near-surface portion of the upper

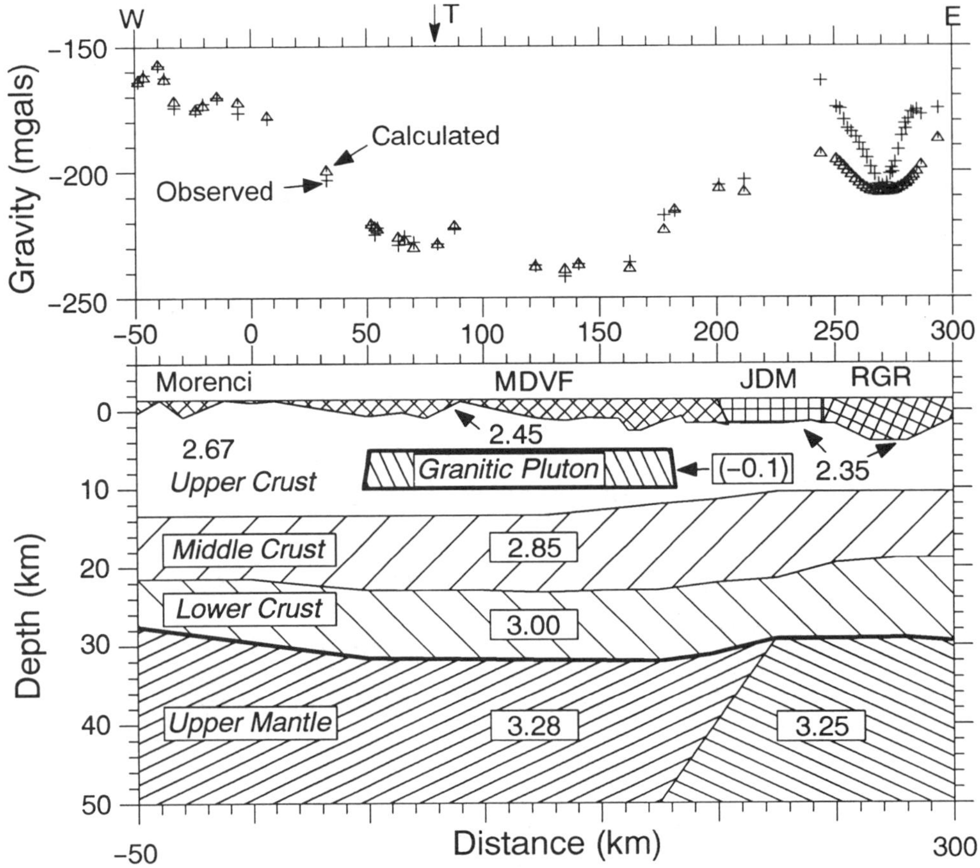

Figure 14. Final model for gravity profile along Jaksha (1982) cross line. Values for density are in gm/cm^3, except for the granitic pluton that displays a density contrast with its country rock. T, tie to Tyrone - Acoma profile. MDVF, Mogollon-Datil volcanic field; JDM, Jornada del Muerte; RGR, Rio Grande rift.

crust. Movement must have been rapid, as Elston and others (1976) find little evidence of crustal contamination.

2. It is commonly held that a basaltic source rock yields 10% of its volume to acidic fractionation (Cox and others, 1979). To account for this volume reduction during the evolution of the MDVF, the lower crust was therefore either reduced in total volume, or the loss of magma to the upper crust was replaced by basalt derived from the mantle (underplated). If the lower crust has been reduced in total volume, it is possible that this region was part of the Colorado Plateau which has thicker crust prior to the heating event. If the fluid loss was replaced by new material, the Mogollon-Datil volcanic field may have formed off the Colorado Plateau. The solution to this dilemma is beyond this study.

3. The crustal column between the near-surface pluton and the base of the crust was displaced downward by stoping. This implies that a negative density anomaly was introduced at all levels of the crust in this column, and slower-velocity material was pushed downward throughout this region.

4. Venting of the near-surface pluton, especially with caldera collapse, provided a near-surface stoping effect, completing a total reorganization of the crust.

5. If the initial heating event was caused by the introduction of asthenospheric fluids at the base of the crust, it follows that the lithosphere has been disturbed in addition to the crust. Thus, the mid-Tertiary events in southeastern Arizona and southwestern New Mexico actually represent reorganization of the entire lithosphere.

Rhyolites are mapped in the Datil Mountains, continuing north to near the outcrops of Mesozoic sediments (New Mexico Geological Society, 1982). However, these are not of the high-silica type found on the Mogollon Plateau (Elston and others, 1976). If these rocks were also vented from the pluton, a chemical differentiation existed and/or developed during the time of extrusion. The presence of a pluton with greater extent than the region directly under the Mogollon Plateau has some significant implications. For example, Elston and others (1976) suggest that the outer grabens that wrap around the Mogollon

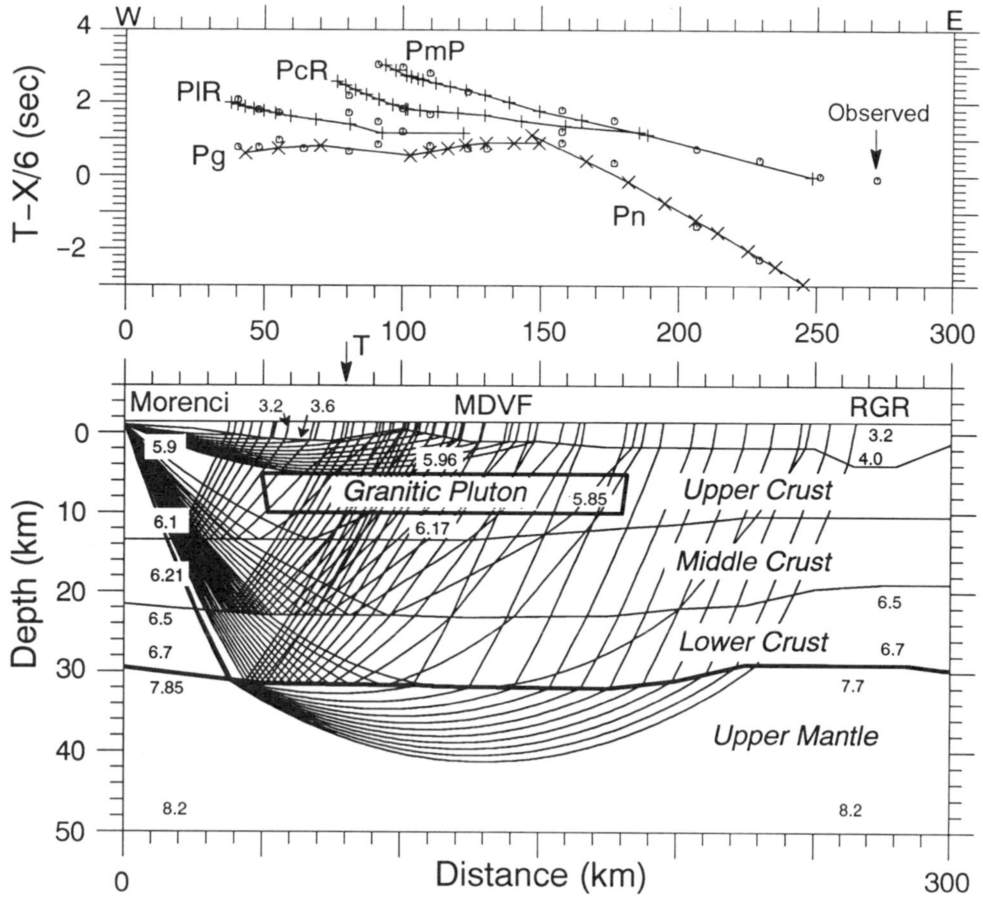

Figure 15. Ray-tracing model for the Jaksha (1982) cross line, final configuration. Velocities are in km/sec. Symbols: o, observed arrivals; +, calculated reflected arrivals; x, calculated refracted arrivals; T, tie to Tyrone - Acoma profile. MDVF, Mogollon-Datil volcanic field; RGR, Rio Grande rift.

Plateau are caused by Basin and Range type extension, and that the wrapping is caused by resistance of the pluton to throughgoing faulting. Since the pluton clearly extends north of the Plains of San Agustin, this interpretation may be doubtful.

The spatial relationship between the high-silica rhyolite source and the rest of the complex cannot be uniquely determined in this study. An alternative interpretation to a unique high-silica magma chamber is one in which the high-silica fluids are derived from a larger magma chamber. From sedimentary evidence, one may conclude that the crust under the Colorado Plateau in New Mexico was lower in elevation relative to the south immediately prior to and during initial stages of mid-Tertiary volcanism (Cather and Johnson, 1986). This structural trend has subsequently reversed (Pearce and others, 1979; Morgan and Swanberg, 1985). It is possible that the most felsic components, including volatiles, worked their way to the top of the chamber as it developed. These pooled under the Mogollon Plateau for the simple reason that it was the highest part of the pluton at the time. Subsequent cauldron-producing volcanic events were above the lightest, hottest material. This pooling effect may have caused a minor local uplift over the Mogollon Plateau, creating the radial tension cracks reported by Elston and others (1976).

The two high-silica cauldron events outside the Mogollon Plateau (Ratté, 1989; Rhodes and Smith, 1972) may reflect local maxima in the roof of the magma chamber. The density/chemistry stratification would satisfy the requirement that only high-silica rhyolites were vented within the Mogollon Plateau during that time.

The Colorado Plateau appears to be resistant to thinning with respect to the Basin and Range Province. This may imply a fundamental chemical difference in the lower crust between the Colorado Plateau, Rio Grande rift, and Basin-Range Provinces. If present-day crustal geology reflects that of mid-Tertiary time, then perhaps the high-silica rhyolite suite is a product of this difference in crustal chemistry.

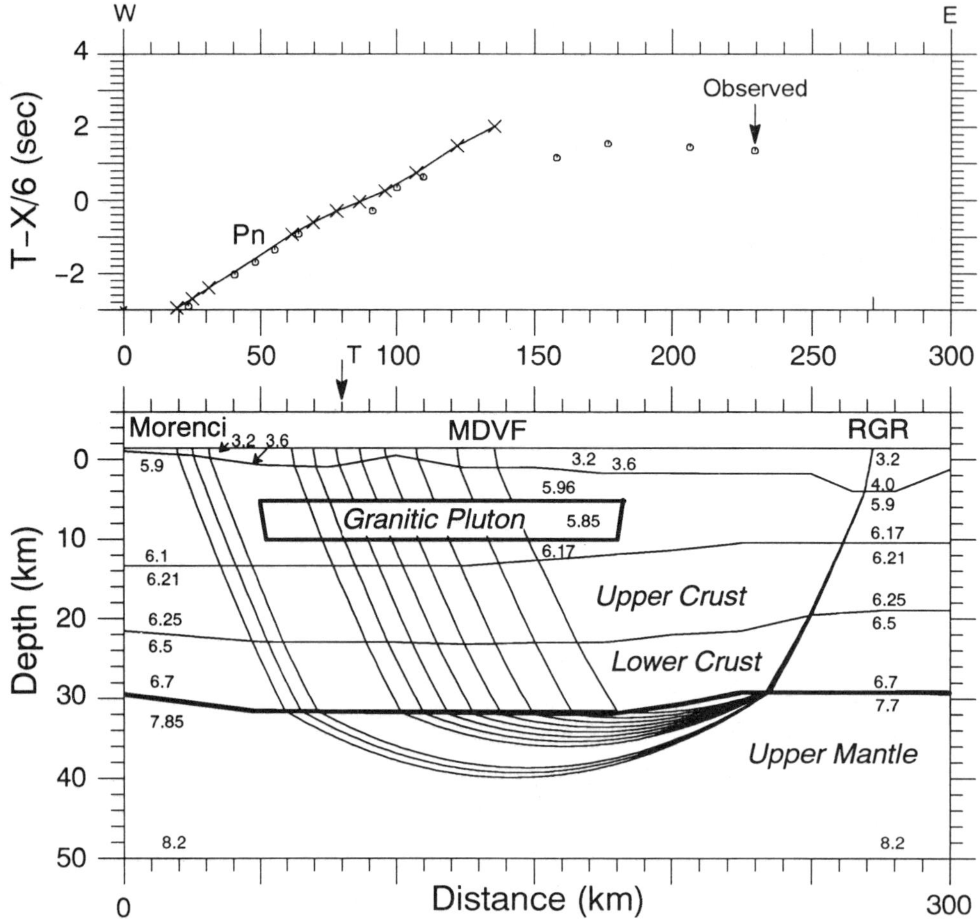

Figure 16. Ray-tracing model for the Jaksha (1982) cross line, reversed Pn branch, final configuration. Velocities are in km/sec. Symbols: o, observed arrivals; x, calculated refracted arrivals; T, tie to Tyrone - Acoma profile. MDVF, Mogollon-Datil volcanic field; RGR, Rio Grande rift.

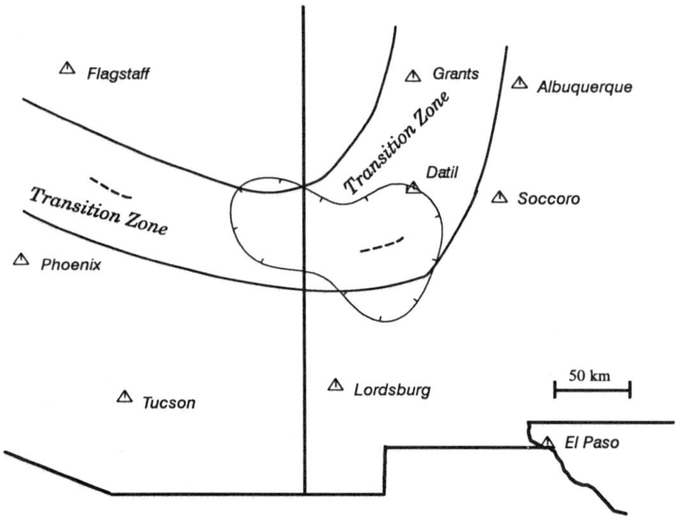

Figure 17. Summary diagram showing the extent (hachured line) of the proposed upper crustal pluton as the –25 mgal contour of the 125-km low-pass gravity anomaly. The steps in the Moho detected by Warren (1969) and this study are shown by dashed lines. The proposed boundaries of the Transition Zone are shown by solid lines. The Arizona boundaries are generalized from Hendricks and Plescia (1991) and the northern boundary in New Mexico is generalized from Aldrich and Laughlin (1984).

ACKNOWLEDGMENTS

This study was supported by Air Force Office of Scientific Research Grant Number F49620-92-J-0438. Elizabeth Anthony provided valuable advice for the integration of petrologic and geophysical data. We gratefully acknowledge the cooperation of the Phelps-Dodge staff at the Tyrone mine and the landowners in the region. John Schlue, Allan Sanford, and Charles Chapin provided helpful reviews.

REFERENCES CITED

Aldrich, M. J., Jr., and Laughlin, A. W., 1984, A model for the tectonic development of the southeastern Colorado Plateau boundary: Journal of Geophysical Research, v. 89, p. 10207–10218.

Baldridge, W. S., and 9 others, 1989, Magmatism associated with lithospheric extension: Middle to late Cenozoic magmatism of the southeastern Colorado Plateau and central Rio Grande rift, New Mexico and Arizona: New Mexico Bureau of Mines and Mineral Resources Memoir 46, p. 187–202.

Barker, D. S., 1979, Magmatic evolution in the Trans-Pecos province: University of Texas at Austin, Bureau of Economic Geology, Guidebook 19, p. 4–9.

Beghoul, N., and Barazangi, M., 1989, Mapping high Pn velocity beneath the Colorado Plateau constrains uplift models: Journal of Geophysical Research, v. 94, p. 7083–7104.

Bilodeau, W. L., 1984, Laramide sedimentation and tectonics in southeastern Arizona and southwestern New Mexico: Reactivation of pre-Laramide basement structures: Geological Society of America Abstracts with Programs, v. 16, p. 445.

Blackwell, D. D., 1978, Heat flow and energy loss in the western United States, *in* Smith, R. B., and Eaton, G. P., eds., Cenozoic tectonics and regional geophysics of the Western Cordillera: Geological Society of America Memoir 152, p. 175–208.

Bott, M.H.P.,and Smithson, S. B., 1967, Gravity investigations of subsurface shape and mass distributions of granite batholiths: Geological Society of America Bulletin, v. 78, p. 859–878.

Briggs, I. C., 1974, Machine contouring using minimum curvature: Geophysics, v. 39, p. 39–48.

Brumbaugh, D. S., 1987, A tectonic boundary for the southern Colorado Plateau: Tectonophysics, v. 136, p. 125–136.

Cady, J. W., 1980, Calculation of gravity and magnetic anomalies of finite-length right polygonal prisms: Geophysics, v. 45, p. 1507–1512.

Cather, S. M., and Chapin, C. E., 1989, Field guide to upper Eocene and lower Oligocene volcaniclastic rocks of the northern Mogollon-Datil volcanic field: New Mexico Bureau of Mines and Mineral Resources Memoir 46, p. 60–68.

Cather, S. M., and Johnson, B. D., 1986,Eocene depositional systems and tectonic framework of west-central New Mexico and eastern Arizona: American Association of Petroleum Geologists Memoir 41, p. 623–652.

Cerveny, V., and Ravindra, R., 1971, Theory of seismic head waves: Toronto, Canada, University of Toronto Press, p. 312.

Coney, P. J., 1976, Structure, volcanic stratigraphy, and gravity across the Mogollon Plateau, New Mexico: New Mexico Geological Society Special Publication 5, p. 29–41.

Coney, P. J., 1978a, Mesozoic-Cenozoic Cordilleran plate tectonics, *in* Smith, R. B., and Eaton, G. B., eds., Cenozoic tectonics and regional geophysics of Western Cordillera: Geological Society of America Memoir 152, p. 33–50.

Coney, P. J., 1978b, The plate tectonic setting of southeastern Arizona: New Mexico Geological Society 29th Field Conference, Guidebook, p. 285–290.

Cook, F. A., McCullar, D. B., Decker, E. R., and Smithson, S. B., 1979, Crustal structure and evolution of the southern Rio Grande rift, *in* Riecker, R. E., ed., Rio Grande rift: Tectonics and magmatism: Washington, D.C., EOS Transactions of the American Geophysical Union, p. 195–208.

Cox, K. G., Bell, J. D., and Pankhurst, R. J., 1979, The Interpretation of igneous rocks: London, George Allen & Unwin, 450 p.

Daggett, P. H., Keller, G. R., Morgan, P., and Wen, C. L., 1986, Structure of the southern Rio Grande rift from gravity interpretation: Journal of Geophysical Research, v. 91, p. 6157–6167.

Dane, C. H., and Bachman, G. O., 1965, Geologic map of New Mexico: U.S. Geological Survey, 2 sheets, scale 1:500,000.

Elston, W. E., 1976, Tectonic significance of mid-Tertiary volcanism in the Basin and Range province: A critical review with special reference to New Mexico: New Mexico Geological Society Special Publication 5, p. 93–102.

Elston, W. E., 1989, Overview of the Mogollon-Datil volcanic field: New Mexico Bureau of Mines and Mineral Resources Memoir 46, p. 43–44.

Elston, W. E., and Bornhorst, T. I., 1979, The Rio Grande rift in context of regional post–40 m.y. volcanic and tectonic events, *in* Riecker, R. E., ed., Rio Grande rift: Tectonics and magmatism: Washington, D.C., EOS Transactions of the American Geophysical Union, p. 416–438.

Elston, W. E., Coney, P. J., and Rhodes, R. C., 1968, A progress report on the Mogollon Plateau volcanic province, southwestern New Mexico: Colorado School Mines Quarterly, v. 63, p. 261–287.

Elston, W. E., Coney, P. J., and Rhodes, R. C., 1970, Progress report on the Mogollon Plateau volcanic province, southwestern New Mexico, No. 2: New Mexico Geological Society, 21st Field Conference, Tyron–Big Hatchet Mountains–Florida Mountains Region, Guidebook, p. 75–86.

Elston, W. E., Rhodes, R. C., Coney, P. J., and Deal, E. G., 1976, Progress report on the Mogollon Plateau volcanic field, southwestern New Mexico, No. 3—Surface expression of a pluton: New Mexico Geological Society Special Publication 5, p. 3–28.

Gish, D. M., Keller, G. R., and Sbar, M. L., 1981, A refraction study of deep crustal structure in the Basin and Range: Colorado Plateau of eastern Arizona: Journal of Geophysical Research, v. 86, p. 6029–6038.

Harden, S., 1982, A seismic refraction study of west-central New Mexico [M.S. thesis]: El Paso, Texas, University of Texas at El Paso, 58 p.

Hauser, E. C., and Lundy, J., 1989, COCORP deep reflections: Moho at 50 km (16 s) beneath the Colorado Plateau: Journal of Geophysical Research, v. 94, p. 7071–7081.

Hendricks, J. D., and Plescia, J. B., 1991, A review of the regional geophysics of the Arizona transition zone: Journal of Geophysical Research, v. 96, p. 12351–12373.

Henry, C. D., and Price, J. G., 1986, Early Basin and Range development in Trans-Pecos Texas and adjacent Chihuahua: Magmatism and orientation, timing, and style of extension: Journal of Geophysical Research, v. 91, p. 6213–6224.

Hill, D. P., Kissling, E., Luetgert, J. H., and Kradolfer, U., 1985, Constraints on the upper crustal structure of the Long Valley–Mono Craters volcanic complex, eastern California, from seismic refraction measurements: Journal of Geophysical Research, v. 90, p. 11135–11150.

Humphrey, J. R., and Wong, I. G., 1983, Recent seismicity near Capitol Reef National Park, Utah, and its tectonic implications: Geology, v. 11, p. 447–451.

Jaksha, L. H., 1982, Reconnaissance seismic refraction-reflection surveys in southwestern New Mexico: Geological Society of America Bulletin, v. 93, p. 1030–1037.

Keller, G. R., Smith, R. B., and Braile, L. W., 1975, Crustal structure along the Great Basin–Colorado Plateau transition from seismic refraction studies: Journal of Geophysical Research, v. 80, p. 1093–1098.

Keller, G. R., Braile, L. W., and Morgan, P., 1979, Crustal structures, geophysical models and contemporary tectonism of the Colorado Plateau: Tectonophysics, v. 61, p. 131–147.

Keller, G. R., Morgan, P., and Seager, W. R., 1990, Crustal structure, gravity anomalies and heat flow in the southern Rio Grande rift and their rela-

tionship to extensional tectonics: Tectonophysics, v. 61, p. 131–147.
Krohn, D. H., 1976, Gravity survey of the Mogollon Plateau volcanic province, southwestern New Mexico: New Mexico Geological Society Special Publication 5, p. 113–116.
Lachenbruch, A. H., and Sass, J. H., 1978, Models of an extending lithosphere and heat flow in the Basin and Range province, *in* Smith, R. B., and Eaton, G. P., eds., Cenozoic tectonics and regional geophysics of the Western Cordillera: Geological Society of America Memoir 152, p. 209–250.
Laughlin, A. W., Aldrich, M. J., and Vaniman, D. T., 1983, Tectonic implications of mid-Tertiary dikes in west-central New Mexico: Geology, v. 11, p. 45–48.
Luedke, R. G., and Smith, R. L., 1978, Map showing distribution, composition, and age of late Cenozoic volcanic centers in Arizona and New Mexico: United States Geological Survey Miscellaneous Investigations Series, Map I-1091-A, scale 1:1,000,000.
Mayer, L., 1979, Evolution of the Mogollon Rim in central Arizona: Tectonophysics, v. 61, p. 49–62.
McCarthy, J., and Kohler, W., 1990, Preliminary two-dimensional modeling of crustal structure across the Colorado Plateau–Basin and Range transition, Arizona: EOS Transactions of the American Geophysical Union, v. 71, p. 1564.
McCarthy, J., Hajnal, Z., Howie, J., Johnson, R. A., Keller, G. R., and Wolf, L. W., 1990, PACE lithospheric investigations of the Colorado Plateau: EOS Transactions of the American Geophysical Union, v. 71, p. 1564.
McIntosh, W. C., 1989, Timing and distribution of ignimbrite volcanism in the Eocene-Miocene Mogollon-Datil volcanic field: New Mexico Bureau of Mines and Mineral Resources Memoir 46, p. 58–59.
McIntosh, W. C., Chapin, C. E., Ratté, J. C., and Sutter, J. F., 1992, Time-stratigraphic framework for the Eocene-Oligocene Mogollon-Datil volcanic field, southwest New Mexico: Geological Society of America Bulletin, v. 104, p. 851–871.
Morgan, P., and Swanberg, C. A., 1985, On the Cenozoic uplift and tectonic stability of the Colorado Plateau: Journal of Geodynamics, v. 3, p. 39–63.
Morgan, P., Seager, W. R., and Golombek, M. P., 1986, Cenozoic thermal, mechanical and tectonic evolution of the Rio Grande rift: Journal of Geophysical Research, v. 91, p. 6262–6276.
New Mexico Geological Society, 1982, New Mexico highway geologic map: Scale 1:1,000,000.
Olsen, K. H., Keller, G. R., and Stewart, J. N., 1979, Crustal structure along the Rio Grande rift from seismic refraction profiles, *in* Riecker, R. E., ed., Rio Grande rift: Tectonics and magmatism: Washington, D.C., EOS Transactions of the American Geophysical Union, p. 127–143.
Peeples, W. J., Coultrip, R. L., and Keller, G.R., 1986, Quasi-ideal spatial filters for large maps: Annales Geophysicae, v. 4, p. 547–554.
Peirce, H. W., Damon, P. E., and Shafiqullah, M., 1979, An Oligocene (?) Colorado Plateau edge in Arizona: Tectonophysics, v. 61, p. 1–24.
Perry, F. V., Baldridge, W. S., and DePaolo, D. J., 1988, Chemical and isotopic evidence for lithospheric thinning beneath the Rio Grande rift: Nature, v. 332, p. 432–434.
Ratté, J. C., 1989, Selected volcanic features of the western Mogollon-Datil volcanic field: New Mexico Bureau of Mines and Mineral Resources Memoir 46, p. 68–91.
Ratté, J. C., Cather, S. M., Chapin, C. E., Duffield, W. A., Elston, W. E., and McIntosh, W. C., 1989, Eocene-Miocene Mogollon-Datil volcanic field, New Mexico: New Mexico Bureau of Mines and Mineral Resources Memoir 46, p. 43–119.
Reiter, M., and Clarkson, G., 1983, A note on terrestrial heat flow in the Colorado Plateau: Geophysical Research Letters, v. 10, p. 929–932.
Reiter, M., Edwards, C. L., Hartman, H., and Weidman, C., 1975, Terrestrial heat flow along the Rio Grande rift, New Mexico and southern Colorado: Geological Society of America Bulletin, v. 86, p. 811–818.
Reiter, M., Mansure, A. J., and Shearer, C., 1979, Geothermal characteristics of the Colorado Plateau: Tectonophysics, v. 61, p. 183–195.
Rhodes, R. C., 1976, Petrologic framework of the Mogollon Plateau volcanic ring complex, New Mexico—Surface expression of a major batholith: New Mexico Geological Society Special Publication 5, p. 103–112.
Rhodes, R. C., and Smith, E. I., 1972, Geology and tectonic setting of the Mule Creek caldera, New Mexico, U.S.A.: Bulletin of Volcanology, v. 36, p. 401–411.
Roller, J. C., 1965, Crustal structure in the eastern Colorado Plateaus province from seismic refraction measurements: Seismological Society of America Bulletin, v. 55, p. 107–119.
Sass, J. H., Stone, C., and Bills, D. J., 1982, Shallow subsurface temperatures and some estimates of heat flow from the Colorado Plateau of northeastern Arizona: United States Geological Survey Open-File Report 82–994, 112 p.
Schneider, R. V., 1990, An integrated geophysical and geological investigation of the transition zone between the Colorado Plateau, Rio Grande rift and Basin and Range provinces: Arizona and New Mexico [Ph.D. thesis]: University of Texas at El Paso, 204 p.
Seager, W. R., and Mack, G. H., 1986, Laramide paleotectonics of southern New Mexico: American Association of Petroleum Geologists Memoir 41, p. 669–685.
Seager, W. R., Shafiqullah, M., Hawley, J. W., and Marvin, R. F., 1984, New K-Ar dates from basalts and the evolution of the southern Rio Grande rift: Geological Society of America Bulletin, v. 95, p. 87–99.
Shearer, C., and Reiter, M., 1981, Terrestrial heat flow in Arizona: Journal of Geophysical Research, v. 86, p. 6249–6260.
Sinno, Y. A., Keller, G. R., and Sbar, M. L., 1981, A crustal seismic refraction study in west-central Arizona: Journal of Geophysical Research, v. 86, p. 5023–5038.
Sinno, Y. A., Daggett, P. H., Keller, G. R., Morgan, P., and Harder, S. H., 1986, Crustal structure of the southern Rio Grande rift determined from seismic refraction profiling: Journal of Geophysical Research, v. 91, p. 6143–6156.
Spencer, J. E., and Reynolds, S. J., 1989, Middle Tertiary tectonics of Arizona Geological Society Digest 17, p. 539–574.
Stewart, S. W., and Pakiser, L. C., 1962, Crustal structure in eastern New Mexico interpreted from the Gnome explosion: Bulletin of the Seismological Society of America, v. 55, p. 877–886.
Swain, C. J., 1976, A FORTRAN IV program for interpolating irregularly spaced data using the difference equations for minimum curvature: Computers in the Geosciences, v. 1, p. 231–240.
Swanberg, C. A., and Morgan, P., 1985, Silica heat flow estimates and heat flow in the Colorado Plateau and adjacent areas: Journal of Geodynamics, v. 3, p. 65–85.
Thompson, G. A., and Zoback, M. L., 1979, Regional geophysics of the Colorado Plateau: Tectonophysics, v. 61, p. 149–181.
Toppozada, T. R., and Sanford, A. R., 1976, Crustal structure in central New Mexico interpreted from the Gasbuggy explosion: Seismological Society of America Bulletin, v. 66, p. 877–886.
Warren, D. H., 1969, A seismic-refraction survey of crustal structure in central Arizona: Geological Society of America Bulletin, v. 80, p. 257–282.
Wilson, E. D., Moore, E. T., and Cooper, J. R., 1969, Geologic map of Arizona: Arizona Bureau of Mines and United States Geological Survey, scale 1:500,000.
Wolf, L. W., and Cipar, J. J., 1991, A new crustal model for the Colorado Plateau: Results from the 1989 Pacific to Arizona Experiment: EOS Transactions of the American Geophysical Union, v. 72, p. 327.

MANUSCRIPT ACCEPTED BY THE SOCIETY OCTOBER 13, 1993

Printed in U.S.A.

Geological Society of America
Special Paper 291
1994

Winston graben: Stratigraphy, structure, and tectonic setting

Richard W. Harrison*
New Mexico Institute of Mining and Technology, Socorro, New Mexico 87801

ABSTRACT

The Winston graben is located on the western margin of the Rio Grande rift in south-central New Mexico in an area of intense Neogene faulting. It is a symmetrical graben, 5 to 10 km wide, approximately 56 km long, and trends north-northwest to north-northeast. The structure is bounded by high-angle normal faults with about 2 km of stratigraphic separation across both margins. Most of its eastern boundary fault is believed to be a reactivated Eocene strike-slip fault. Northeast-trending accommodation zones (structural highs) terminate the graben at both north and south ends. Rocks exposed within the graben include Pennsylvanian and Permian sedimentary formations, several Eocene-Oligocene volcanic and volcaniclastic units, the upper Oligocene–Quaternary Santa Fe Group, a Miocene andesite flow that is intercalated with the Santa Fe Group, and a Pliocene basalt flow. Some of the volcanic units occur only within the Winston graben. Initial development of the graben began in the late Oligocene, but most of its growth was in the Neogene. Boundary faults of the Winston graben have been inactive for at least the past 4.8 m.y. When placed in a tectonic setting with surrounding structures of the Rio Grande rift, the Winston graben is interpreted as having formed over a deep detachment surface.

INTRODUCTION

In south-central New Mexico, the Rio Grande rift consists of north-south–oriented basins separated by uplifts (Fig. 1). The Winston graben developed along the western margin of the southern rift in an area of intense Neogene faulting. It lies between the Black Range to the west and the Sierra Cuchillo Range and San Mateo Mountains to the east, and is entirely within the early Tertiary Mogollon-Datil volcanic field. Its name comes from the small community of Winston, New Mexico, where most of the people in this very sparsely populated area live.

The Winston graben is long and narrow, approximately 5 to 10 km wide and 56 km in length. There is approximately 2 km of stratigraphic separation along the western margin and 1.8 km along the eastern margin. It has a slight dogleg shape; its southern half trends north-northwest and its northern half trends north-northeast. The northern terminus occurs at a structural high along the Morenci lineament of Chapin et al. (1978a). Its southern terminus occurs along another structural high, the Chise lineament of Harrison (1990), which separates the Winston graben from the Las Animas graben.

STRATIGRAPHY

Rocks within the Winston graben are dominantly those of the lower Oligocene–Quaternary Santa Fe Group. However, at the southern end of the graben, Oligocene volcanic units crop out, and along the structurally complex southeastern margin, rocks of the Eocene Rubio Peak Formation, Permian Abo Formation, and Pennsylvanian Madera Formation are exposed. Figure 2 is a composite stratigraphic section for the Winston graben. The following descriptions are of units found in the Winston graben and depicted in Figures 2 and 3. The regional cross section presented in Figure 4 contains units found outside the Winston graben that are not described in this text. Also, some of the units described in the section on Tertiary volcanic units (Tu) are differentiated in Figure 4.

*Present address: 555 Parnassus Road, Linden, Virginia 22642.

Harrison, R. W., 1994, Winston graben: Stratigraphy, structure, and tectonic setting, *in* Keller, G. R., and Cather, S. M., eds., Basins of the Rio Grande Rift: Structure, Stratigraphy, and Tectonic Setting: Boulder, Colorado, Geological Society of America Special Paper 291.

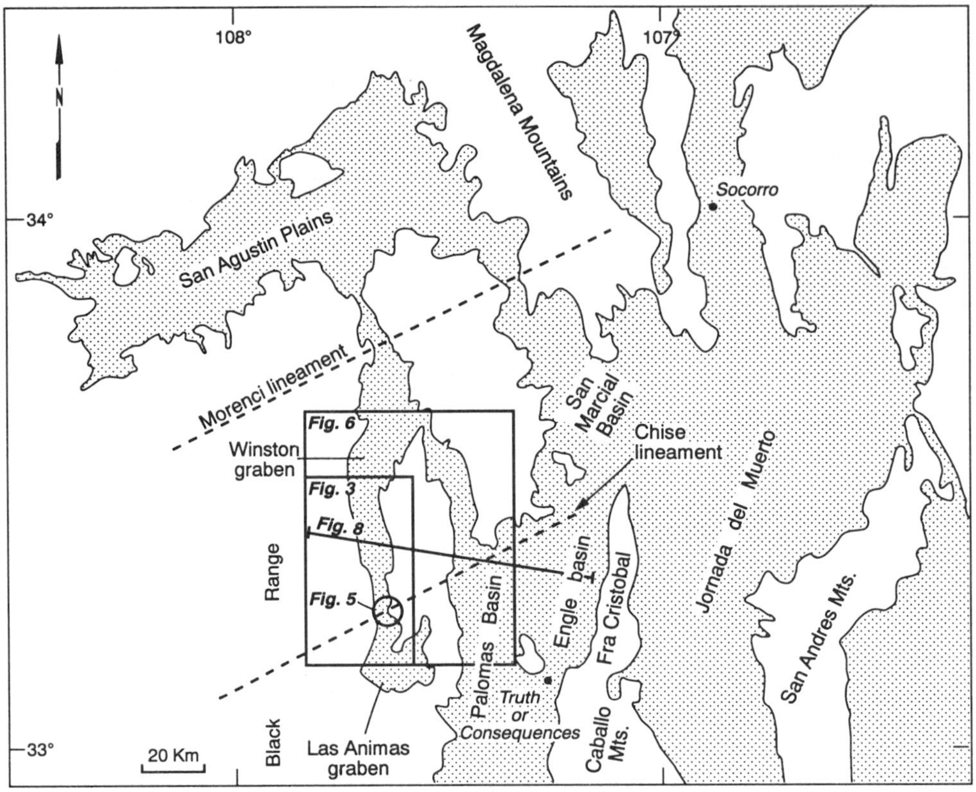

Figure 1. Location map for the Winston graben, southwestern New Mexico.

Paleozoic rocks

Outcrops of the Pennsylvanian Madera Formation (ℙ) and Permian Abo Formation (Pa) occur in thin, fault-bounded outcrops within the southeastern portion of the Winston graben (Pa-ℙm on Fig. 2). The Madera Formation is a thin- to medium-bedded, fossiliferous limestone. The Abo Formation consists of red, cross-bedded siltstone and fine-grained red sandstone.

Eocene Rubio Formation (Trp)

In the Winston graben, the Eocene Rubio Peak Formation (Trp on Fig. 2) consists dominantly of volcaniclastic debris flows and sandstones. Locally, masses of intrusive hornblende monzonite occur that are age equivalent to the Rubio Peak Formation. All of these lithologies are widespread in the Rubio Peak Formation of south-central and southwestern New Mexico (Harrison, 1990). Radiometric age dates and geologic constraints for the Rubio Peak Formation indicate deposition from approximately 45 to 36 Ma (Harrison, 1990; McIntosh, 1989).

Debris-flow deposits within the Rubio Peak Formation are massive, poorly to nonbedded, matrix supported, and contain intermediate volcanic clasts. Clasts are usually heterolithic, but locally monolithic intervals of hornblende andesite are found. The size of clasts in these deposits ranges from lapilli to large boulders a meter or more in diameter. Matrix consists of fragmental detritus, generally 0.5 to 0.2 mm in size; there is paucity of clay-sized material. Debris-flow deposits in the Winston graben show all of the characteristics ascribed to this rock type under the classification of Smith (1986). These characteristics include en masse mode of deposition, lack of stratification, variable grading, matrix support, and a lack of imbrication.

Rubio Peak sandstones within the Winston graben are massive, medium to fine grained, poorly to moderately bedded, poorly to nonsorted, and lack cross-stratification as well as other typical fluvial sedimentary features. These sandstones closely fit descriptions of hyperconcentrated-flow deposits by Smith (1986, 1987, 1988) and Scott (1988). They consist of detrital grains of predominantly plagioclase, with abundant lithic fragments, less abundant altered ferromagnesium minerals, minor sphene, and rare quartz.

Tertiary volcanic units (Tv)

Several Tertiary volcanic units crop out at the southern end of the Winston graben. Because of structural complexities and the thin nature of these units, they are grouped together on Figure 3. However, on the cross section shown in Figure 4,

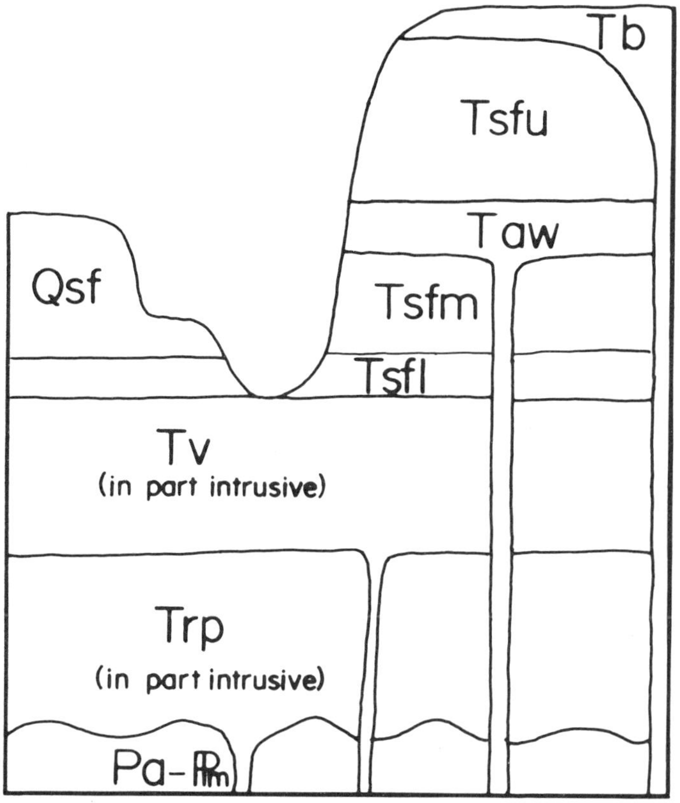

Figure 2. Generalized stratigraphic column for the Winston graben: IPm, Pennsylvanian Madera Formation; Pa, Permian Abo Formation; Trp, Eocene Rubio Peak Formation; Tv, Eocene-Oligocene volcanic units; Tsfl, lower Santa Fe Group; Tsfm, middle Santa Fe Group; Taw, andesite of the Winston graben; Tsfu, upper Santa Fe Group; Tb, Pliocene basalt of Tabletop Mountain; Qsf, Quaternary Santa Fe Group.

they are partially differentiated. In ascending order, these units include: tuff of Rocque Ramos Canyon, tuff of Stone House Ranch, Kneeling Nun Tuff, Cuchillo Negro volcanic complex, basaltic andesite of Poverty Creek, tuff of Little Mineral Creek, Vicks Peak Tuff, and unnamed water-laid and zeolite-altered tuffs. These later tuffs are intercalated with basal Santa Fe Group sediments. Thin volcaniclastic beds typically occur between the volcanic rocks, particularly the lower units.

The tuff of Rocque Ramos Canyon is a regional unit named by McIntosh (1989) for exposures along that canyon in the eastern Sierra Cuchillo Range. It is a moderately to poorly welded, moderately crystal-rich ash-flow tuff. Phenocrysts constitute 10 to 20% of the rock with subequal amounts of plagioclase and sanidine, and 1 to 2% biotite. Small (1.5 to 2.0 cm) pumice is common and lithic fragments of andesite are present, but rare. Chemical analyses by Eggleston (1987) and Harrison (1990) indicate that the tuff of Rocque Ramos Canyon is trachytic to rhyolitic in composition (classification scheme by Le Maitre, 1984). McIntosh (1989) reported a ^{40}Ar/^{39}Ar age of 34.93 ± 0.12 Ma for this unit and correlated it with the Bell Top 4 tuff in the Goodsight–Cedar Hills area.

The tuff of Stone House Ranch is a minor unit, whose only known occurrence is in the southern Winston graben. It was named by Harrison (1990) for exposures in the center of Sec. 11, T.12S., R.8W. in the Winston 7½′ Quadrangle. There it is approximately 30 m thick and is separated from the underlying tuff of Rocque Ramos Canyon by about 10 m of fine-grained volcaniclastic sandstone, and is separated from the overlying Kneeling Nun Tuff by about 20 m of fine-grained volcaniclastic sandstone. Tuff of Stone House Ranch consists of two members. The upper member is buff-white in color, poorly welded, and contains approximately 5% phenocrysts of altered feldspar and traces of biotite. Small, dark (mafic?) pumice fragments are conspicuous in this member. The lower member is reddish in color, moderately welded, pumice rich, and contains approximately 15% phenocrysts of plagioclase, sanidine, and minor biotite. The two members are welded together with a gradational contact. Paleomagnetic data of McIntosh (1989) indicates that the tuff of Stone House Ranch is distinct from other known ignimbrites in the Mogollon-Datil volcanic field. Based on the close stratigraphic position of the unit to the Kneeling Nun Tuff, similar major-element chemistry, and its close proximity to the Emory caldera (source of the Kneeling Nun Tuff), Harrison (1990) concluded that the tuff of Stone House Ranch is probably a precursor unit related to the Kneeling Nun Tuff.

The Kneeling Nun Tuff (Tkn on Fig. 4) is one of the most widespread units in southwestern and south-central New Mexico. A radiometric K-Ar age of 34.2 ± 1.0 Ma (recalculated using new IUGS constants) was reported by McDowell (1971) for the Kneeling Nun Tuff. Using high-precision ^{40}Ar-^{39}Ar techniques, McIntosh (1989) determined an age of 34.9 Ma. In the Winston graben, Kneeling Nun Tuff is a multiple-flow, compound cooling unit of poorly to moderately welded, crystal-rich ash-flow tuff. It contains from 20 to 45% phenocrysts of quartz, sanidine, plagioclase, and biotite. It has normal zonation, with a very crystal-rich, quartz-rich and biotite-poor base and a quartz-poor, more-biotite-rich top. Outcrops of Kneeling Nun Tuff in the Winston graben are less than 20 km from the northern margin of the Emory caldera.

The Cuchillo Negro complex (included with the Tkn unit on Fig. 4) is an intrusive-extrusive complex that is found only in the southern part of the Winston graben and the northern part of the Las Animas graben (see location on Figs. 1 and 3). It was named by Harrison (1990) for exposures along Cuchillo Negro Creek in Secs. 13, 14, 22, and 23, of T.12S., R.8W. in the Winston and Chise 7½′ Quadrangles. Outcrops in this area provide excellent exposures of vent and near-vent deposits. Figure 5 depicts a generalized cross section through this area.

The exact stratigraphic position of the Cuchillo Negro complex is not clear from field relationships. An intrusive member of this unit distinctly crosscuts rocks of the Rubio

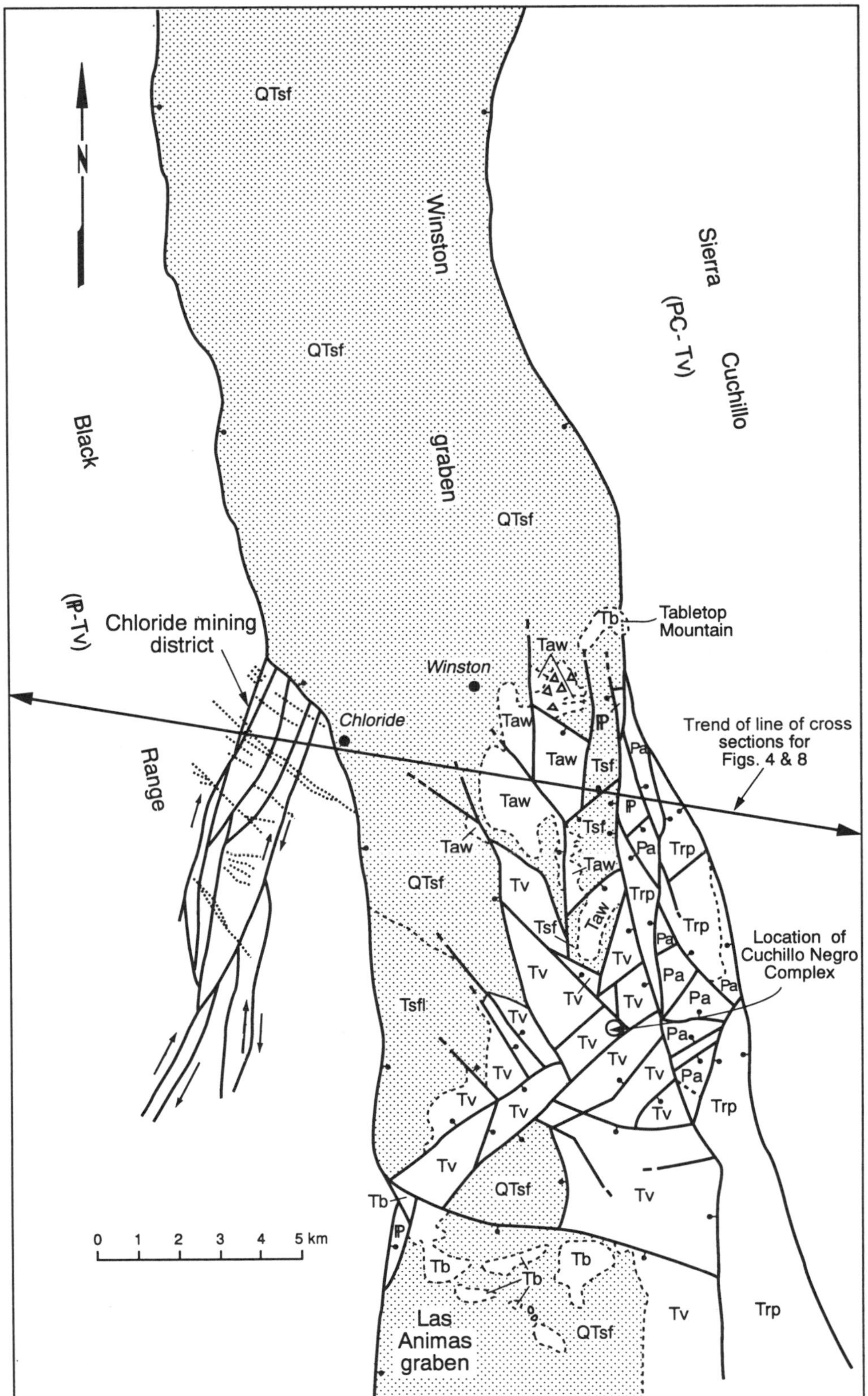
N
QTsf
Winston
graben
Sierra
Cuchillo
(PC- Tv)
Black
Range
(ℙ-Tv)
Chloride mining
district
Chloride
Winston
Tabletop
Mountain
Trend of line of cross
sections for
Figs. 4 & 8
Location of
Cuchillo Negro
Complex
Taw
Tsf
Tsfl
Tv
Trp
Pa
Tb
ℙ
0 1 2 3 4 5 km
Las
Animas
graben

Figure 3. Generalized geologic map of the southern half of the Winston graben showing boundary faults and some of the larger faults in the graben. Extensive faulting in the Black Range and Sierra Cuchillo is not shown; some of the strike-slip faults (lines with arrows indicating motion) and mineralized faults (dotted lines) in the Black Range are shown only to illustrate the cross-cutting relationship of the Winston graben to them. IP, Pennsylvanian rocks, undivided; Pa, Permian Abo Formation; P€, Precambrian rocks; Trp, Rubio Peak Formation; Tv, undifferentiated Tertiary volcanic rocks; Tsfl, Tertiary lower Santa Fe Group; Tsf, undifferentiated Tertiary Santa Fe Group; Taw, andesite of the Winston graben; Tb, Pliocene basalt (includes basalt of Tabletop Mountain); QTsf, undifferentiated Quaternary-Tertiary Santa Fe Group; triangle pattern, gravity slide block of Vicks Peak Tuff.

Peak Formation and tuff of Rocque Ramos Canyon, but relationships to younger rocks are not straightforward. A whole-rock K-Ar age of 34.7 ± 0.8 Ma was determined for this unit (C. E. Chapin, personal communication, 1988); however, the error range is such that Cuchillo Negro complex could be older or younger than the Kneeling Nun Tuff. The Cuchillo Negro complex is probably slightly younger than the Kneeling Nun Tuff,a relationship reflected in Figure 5. If so, it occupies the same stratigraphic position as the Mimbres Peak Formation described by Jicha (1954), Elston (1957), Elston et al. (1975), and Seager et al. (1982), which is interpreted as moat and ring-fracture deposits of the Emory caldera. Rhyolite members of the Cuchillo Negro complex strongly resemble rhyolite lavas in the Mimbres Peak Formation.

The Cuchillo Negro complex consists of three intrusive rock types, three ash-flow tuffs intercalated with volcaniclastic sediments, and one rhyolitic lava flow. Major-element analyses for the Cuchillo Negro complex are given by Harrison (1990).

Intrusive rock types in the Cuchillo Negro complex are crystal-poor rhyolite, moderately crystal-rich basaltic andesite, and amygdule-rich trachyandesite. The intrusive rhyolite member is flow-banded and contains 5 to 10% phenocrysts of plagioclase and sanidine. The intrusive basaltic andesite member contains 20 to 25% large (up to 5 mm) phenocrysts of plagioclase and apple-green pyroxene. This member also contains 2 to 3% magnetite and is characteristically magnetic. The trachyandesite member crosscuts all other members and contains numerous, irregular-shaped amygdules as much as 10 mm in diameter that are commonly lined with calcite, quartz, and unidentified zeolites. The trachyandesite member also contains 10 to 15% phenocrysts of plagioclase.

Extrusive members of the Cuchillo Negro complex consist of three dacite ash-flow tuffs that are intercalated with volcaniclastic sediments and a flow-banded rhyolite lava which is identical to the intrusive rhyolite. The lower ash-flow tuff is tan, nonwelded lithic rich, pumice rich, and crystal poor. It contains about 2% biotite and less than 5% plagioclase phenocrysts. Lithic fragments are red-purple aphanitic volcanic rocks that range from 5 mm to 3 cm in diameter. The middle ash-flow tuff is white, nonwelded, crystal poor, and contains few lithic fragments. It contains approximately 3% conspicuous green biotite and 2 to 3% plagioclase. The upper ash-flow tuff is flesh colored, slightly to moderately welded, and lithic rich. It contains 3 to 4% biotite, and 5 to 10% plagioclase phenocrysts. Lithic fragments are similar to those found in the lower tuff.

Volcaniclastic sediments occur intercalated between all three ash-flow tuffs. A thin (6 to 10 cm), horizontally bedded sandstone occurs between the lower and middle tuffs, and contains abundant lithic fragments and biotite. Between the middle and upper tuffs is a paleochannel filled with a near-vent, lag-breccia deposit that consists of chaotic, nonsorted, crudely stratified, heterolithic pebble and boulder detritus. Clasts are dominantly Rubio Peak–like volcanic rocks with sparse Paleozoic and Precambrian rock types. This lag breccia is clast supported and has a pumiceous matrix. In the classification of Walker (1985), lag breccias form as co-ignimbrite deposits within a proximal deflation zone around a collapsing eruption column. Overlying the upper ash-flow tuff is a relatively thick (18 m) sequence of fine-grained sandstone, mudstone, and conglomerate beds. This sequence probably represents moat-fill deposits, related to near-vent subsidence after eruption of the ash-flow tuffs.

The uppermost extrusive member is a flow-banded rhyolite lava similar to the intrusive, vent-filling rhyolite. It is the thickest extrusive member of the Cuchillo Negro complex (30 to 50 m) and contains approximately 5% phenocrysts of plagioclase and sanidine.

Elston et al. (197) and Coney (1976) assigned the name basaltic andesite of Poverty Creek (Tpc on Fig. 4) to a sequence of dark, aphanitic lava flows exposed in the north-central Black Range. Harrison (1990) correlated this unit to other units described throughout south-central and southwestern New Mexico and recognized that together they formed an important time-stratigraphic interval which blanketed much of the Mogollon-Datil volcanic field. Flow domes, dikes, and outflow lavas of basaltic andesite of Poverty Creek are found within the Winston graben. A K-Ar age of 28.8 ± 0.6 Ma was determined for a vent within the graben located in the center of Sec. 2, T.12S., R.8W. in the Winston 7½′ Quadrangle (C. E. Chapin, personal communication, 1988). This age is very similar to other units that are correlated with the basaltic andesite of Poverty Creek elsewhere in southwestern New Mexico (Harrison, 1990). Chemical analyses by Harrison (1990) indicate that within the Winston graben this unit is an andesite to trachyandesite (under the classification scheme of Le Maitre, 1984).

Eruption of the basaltic andesite of Poverty Creek and correlative rocks marks the end of a 5 to 6 m.y. hiatus in volcanic activity in south-central New Mexico, and the beginning of a relatively brief, but very extensive and voluminous volcanic episode. Intermediate rocks of this period have recently been referred to as the Southern Cordilleran basaltic andesite suite (SCORBA) by Cameron et al. (1989). In south-central

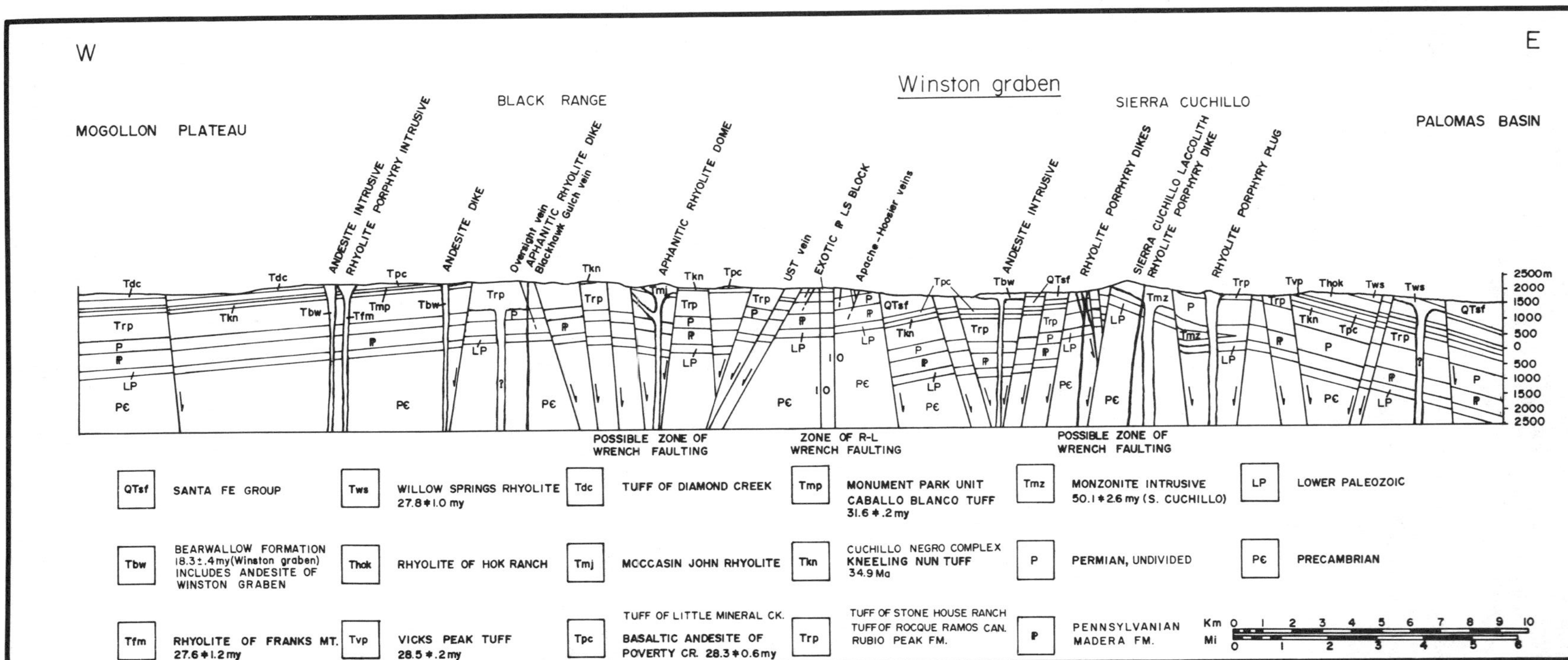

Figure 4. East-west cross section through a part of south-central New Mexico (see Fig. 6). Descriptions of units in the Winston graben are provided in this paper; for descriptions of units outside the Winston graben see Harrison (1990), Abitz (1989), and Eggleston (1987).

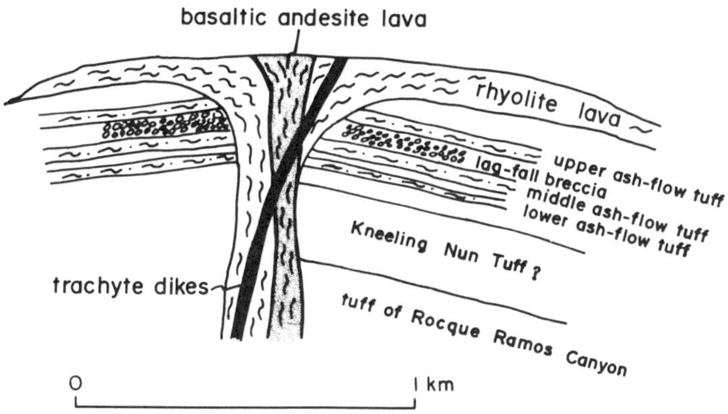

Figure 5. Schematic cross section through the Cuchillo Negro complex showing the relationships between various members. Intrusive phases are known to cross cut the tuff of Rocque Ramos Canyon; however, relationships with the Kneeling Nun Tuff are not clear from field exposures and are inferred from age dates. See Figure 1 and 3 for location of complex.

New Mexico, these rocks have generally been regarded as the products of the initial phase of extension related to the Rio Grande rift (Chapin and Seager, 1975; Seager and Morgan, 1979; Elston and Bornhorst, 1979; Seager et al., 1984). The fact that the basaltic andesite of Poverty Creek and correlative units are blanketlike in occurrence and are not interfingered with basin-related sediments of the Santa Fe Group or Gila Conglomerate indicates that major faulting with significant uplifts and basins postdates the Poverty Creek interval.

The tuff of Little Mineral Creek (part of TpC on Fig. 4) is a relatively minor unit that crops out only in the north-central Black Range and the southern portion of the Winston graben. It has a source area in the area of Moccasin John and Cobb Mountains (Moccasin John flow dome complex). McIntosh (1989) obtained a $^{40}Ar/^{39}Ar$ age of 29.10 ± 0.10 Ma for this unit. Tuff of Little Mineral Creek is a poorly welded, lithic-rich, crystal-poor, ash-flow tuff. Red and gray, angular, aphanitic flow-banded lithic fragments of Moccasin John Rhyolite make up 10 to 50% of the rock. Sparse (1 to 4%) broken crystals of quartz, sanidine, and rare biotite occur in a highly pumiceous matrix. Interbedded intervals of air-fall deposits and stream-reworked volcanics are common. Zeolitic alteration of volcanic glass to clinoptilolite in the tuff of Little Mineral Creek is locally pervasive in the southernmost portion of the Winston graben (Bowie and Barker, 1986).

Vicks Peak Tuff (Tvp on Fig. 4) is a regionally extensive ignimbrite that erupted from the Nogal Canyon cauldron in the southern San Mateo Mountains (Deal and Rhodes, 1976). A mean $^{40}Ar/^{39}Ar$ age of 28.56 ± .06 Ma was obtained for the Vicks Peak Tuff by McIntosh (1989). Vicks Peak Tuff is a massive, moderately to densely welded, very crystal-poor, ash-flow tuff. Phenocrysts consist of 1 to 3% sanidine and rare quartz in a vitroclastic groundmass. Large, elongated pumice containing vapor-phase mineralization with a granular texture are characteristic of this unit. Occurrences of Vicks Peak Tuff in the Winston graben are limited, but important because this unit is totally absent from the adjacent Black Range. A small, fault-bounded outcrop of Vicks Peak Tuff occurs in SE¼ SE¼ Sec. 32, T.9S., R.9W. in the Iron Mountain 7½′ Quadrangle (Harrison, 1990). Approximately 5 km east of the fault-bounded outcrop, unpublished Gulf Minerals Exploration well logs indicate as much as 100 m of Vicks Peak Tuff exists directly below the Santa Fe Group (P. Willard, personal communication, 1987).

In the southern Winston graben, isolated, thin (15 m or less) outcrops of Vicks Peak Tuff rest with angular unconformity upon the basaltic andesite of Poverty Creek and tuff of Little Mineral Creek. This angular unconformity, coupled with nondeposition of the Vicks Peak Tuff in the bordering Black Range, are the first indications in the stratigraphic record of subsidence in the Winston graben.

An additional occurrence of the Vicks Peak Tuff is as a large gravity-slide block that is surrounded by the Santa Fe Group, out of stratigraphic position. This occurrence is described in the section on the Santa Fe Group.

Multiple intervals of water-laid tuffs and poorly to nonwelded, crystal-poor, ash-flow tuffs occur along the South Fork of Cuchillo Negro Creek in the southern Winston graben. These pyroclastic rocks interfinger with the basal Santa Fe Group sediments. The thickest and most continuous bed in this interval is an ash-flow tuff that occurs at or near the top of the sequence and bears a strong resemblance to exposures of the South Canyon Tuff found interbedded with the sandstone of Inman Ranch along Scales and Stiver Canyons in the Taylor Creek tin district (see Eggleston, 1987). McIntosh (1989) reported a $^{40}Ar/^{39}Ar$ age of 27.28 ± 0.08 Ma for the South Canyon Tuff in the Taylor Creek tin district. If Harrison's (1990) tentative correlation of these two tuffs is correct, it indicates a late Oligocene age for the beginning of deposition of the Santa Fe Group in the Winston graben.

Santa Fe Group (Tsfl, Tsfm, Tsfu, Qsf), andesite of the Winston graben (Taw)

The Santa Fe Group (QTsf on Fig. 4) within the Winston graben was informally referred to as the "Winston beds" by Jahns (1955) and Chapin et al. (1978a). Jahns (1955) incorrectly stated that the Winston beds could be traced into the type section of the "Palomas gravel" of Gordon and Graton (1907), Harley (1934),and Kottlowski (1953, 1955), which was later redefined as the Palomas Formation by Lozinsky and Hawley (1986). Along both margins of the Winston graben, the Santa Fe Group is in fault contact with older Tertiary volcanic units or Paleozoic rocks (Fig. 4). The same is true for the Las Animas graben to the south. In fact, the Santa Fe Group within the Winston and Las Animas grabens is physically separated from outcrops of the Palomas Formation to the east by older rocks in the Sierra Cuchillo and Animas uplifts, which are 6 to 20 km wide. To the west, they are separated from outcrops of the Gila Conglomerate of Gilbert (1875) and Elston (1976) by an approximately 25-km-wide exposure of Tertiary volcanic rocks in the Black Range. Thus, the Santa Fe Group within the Winston and Las Animas grabens are geographically isolated.

Within the Winston graben, the Santa Fe Group can be divided into four units: Tertiary lower (Tsfl on Fig. 2), middle (Tsfm on Fig. 2), and upper units (Tsfu on Fig. 2), and a Quaternary unit (Qsf on Fig. 2). All of these units are somewhat similar in appearance and at many localities are impossible to separate; yet at other localities, distinctive physical and compositional characteristics allow subdivision of the Santa Fe Group. Therefore, these units are represented together on the maps and cross sections, but are described individually. The break between the Tertiary middle and upper units is marked by a rather thick lava flow informally named the andesite of the Winston graben (Taw on Fig. 2 and Tbw on Fig. 6) by Harrison (1990). The group is capped by a Pliocene basalt flow on Tabletop Mountain (Tb on Fig. 2). Total thickness of Santa Fe Group varies within the Winston graben. At its southern and northern ends, it pinches out and underlying volcanic units are exposed. Maximum thickness occurs at the center of graben, where it is at least 400 m thick.

Tertiary lower unit of the Santa Fe Group (Tsfl). The Tertiary lower unit of the Santa Fe Group (Tsfl on Fig. 2) in the Winston graben is intercalated with Oligocene volcanic rocks and is well exposed along the banks of South Fork of Cuchillo Negro Creek. These lower sediments are well-indurated, boulder-to-pebble conglomerates and coarse-grained sandstones with a large ash content in their matrix. They are clast supported and poorly sorted. Locally, large-scale cross-beds and scour-and-fill features are present. Clast lithologies are heterolithic and consist dominantly of rhyolitic lavas and ash-flow tuffs. Both clasts and matrix commonly are altered to zeolites (Bowie and Barker, 1986). Transport directions as indicated from pebble imbrications are rather chaotic, yielding northerly, westerly, and southwesterly directions. Characteristically, the lower Santa Fe Group dips 25 to 30° and is overlain in angular unconformity by subhorizontal beds of the Tertiary middle unit and the Quaternary unit.

Tertiary middle unit of the Santa Fe Group (Tsfm). Volcaniclastic sediments of the Tertiary middle unit of the Santa Fe Group (Tsfm on Fig. 2) are poorly consolidated, poorly sorted pebble conglomerates, coarse-grained sandstones, and minor mudstones. Clast lithologies are similar to those in the lower Santa Fe unit, yet with a distinctively greater content of intermediate rock types. As with the lower unit, there is a conspicuous absence of pre-Cenozoic rock types in the middle unit.

Surrounded by beds of the middle unit is an important gravity-slide block of the Vicks Peak Tuff. Exposures of this block occur near the center of the Winston graben approximately 1 km east of the town of Winston in Secs. 13, 14, 23, and 24, T.11S., R.8W. There, a blocklike mass of Vicks Peak Tuff crops out over an area of about 2 km^2. Well-developed pumice foliations in this tuff are vertical to near vertical and strike north-south. Intense, chaotic brecciation occurs around the margins of this block and the base is not exposed. The upper contact is a pronounced angular unconformity with overlying near-horizontal beds of the Tertiary middle unit of the Santa Fe Group and the andesite of the Winston graben. This intriguing block is thought to have slipped gravitationally from the eastern margin of the Winston graben and undergone rotation during transport (Harrison, 1990).

Andesite of the Winston graben (Taw, Fig. 2; included with Tbw on Fig. 4). The andesite of the Winston graben was informally named by Harrison (1990) for the large flow-dome complex and related outflow lava that occurs in the south-central portion of the graben. This andesite is loosely correlated with the Bearwallow Mountain Formation (Tbw on Fig. 4) of Elston (1968, 1976). Andesite of the Winston graben separates the Tertiary middle and upper units of the Santa Fe Group. It has been dated by the K-Ar method at 18.3 ± 0.4 Ma by Seager et al. (1984). This is the only known occurrence of the Bearwallow Mountain Formation interbedded with either Santa Fe Group or Gila Conglomerate in south-central New Mexico. It is also the youngest and easternmost occurrence in south-central New Mexico and is restricted to the Winston graben.

Andesite of the Winston graben is dark gray, fine grained to aphanitic, and vesicular. Flows typically display autobrecciated tops and bottoms. Phenocrysts from 1 to 3 mm in diameter of plagioclase, olivine, pyroxene, and biotite are rare. Microscopically, these phenocrysts are set in a pilotaxitic groundmass of plagioclase (An35-40) microlites.

A well-exposed breccia pipe that cross-cuts the basaltic andesite of Poverty Creek at the extreme northeast corner of Sec. 34, T.11S., R.8W. is believed to be one vent area related to eruption of andesite of the Winston graben. Extrusive breccias, similar to those in the breccia pipe, crop out at the base of outflow lava adjacent of the pipe. Other vents for this unit are inferred to exist beneath thick occurrences of lava within the graben.

Tertiary upper unit of the Santa Fe Group (Tsfu). The upper unit of the Santa Fe Group in the Winston graben (Tsfu on Fig. 2) is similar to the middle unit in that it consists of poorly consolidated, poorly sorted, pebble conglomerates and coarse-grained sandstones with minor mudstone intervals. It differs from the middle unit in that it contains occasional clasts of Pennsylvanian, Permian, and Cretaceous rock types (Abo, Madera, and Dakota ? Formations) in its upper beds along the eastern side of the graben. These occurrences clearly indicate the initial exposure and erosion of pre-Cenozoic strata in the adjacent Sierra Cuchillo uplift.

Quaternary unit of the Santa Fe Group (Qsf). A problem is presented by the presence of early Pleistocene vertebrate and invertebrate fossils reported by Jahns (1955) at two localities in the center of the Winston graben, and by Quaternary fresh-water mollusks also reported in that area by Chapin et al. (1978b). The beds that contain these fossils lie a few hundred meters below the elevation of the previously described Pliocene basalt flow and pediment surface. The most likely explanation is that down-cutting and reworking of older Santa Fe beds must have been well underway in the Pleistocene. Unfortunately, contacts between the poorly consolidated Quaternary and Tertiary deposits are virtually impossible to detect.

Other Quaternary deposits within the Winston graben consist of alluvium in active stream channels. Prominent terraces from 10 to 20 m above present valley floors exist in all larger drainages.

Pliocene basalt of Tabletop Mountain (Tb)

A Pliocene basalt flow caps Tabletop Mountain (Tb on Fig. 2) in Sec. 12, T.11S., R.8W. in the Chise 7½′ Quadrangle (shown in Fig. 3). This flow unconformably overlies the upper part unit of the Santa Fe Group, crosses the boundary fault between the Winston graben and Sierra Cuchillo Range without offset, and is dated by the K-Ar method at 4.8 ± 0.1 Ma (Seager et al., 1984).Lava is domed over the boundary fault, which presumedly served as a conduit for the flow. Major-element chemical analysis of this flow by Fodor (1975) indicates that it is an alkaline basalt. Phenocrysts of plagioclase and olivine occur in the basalt, with the olivine typically altered to iddingsite. Xenolithic inclusions of spinel lherzolite, clinopyroxene, and gabbro found in the basalt were analyzed by Fodor (1978) and were interpreted as representative of upper-mantle material from which the basalt was probably derived at a minimum formation depth of 33 to 45 km.

Basalt flows similar to that at Tabletop Mountain occur along the juncture of the Winston and Las Animas grabens (Fig. 3). The map of late Tertiary–Quaternary tectonic and volcanic activity in New Mexico by Callender et al. (1983) shows that these lie in a zone of widespread volcanic activity along the southern Rio Grande rift during this time period.

The basalt flow on Tabletop Mountain was deposited upon an extensive pediment surface (now deeply eroded) that developed on top of the Tertiary upper unit of the Santa Fe Group within the Winston graben and on parts of the Eocene-Oligocene volcanic pile in the adjacent Black Range. This surface crosses the western boundary fault of the Winston graben and is not offset by this fault (Jahns, 1955; Chapin et al., 1978b; Seager et al., 1984). Thus, the 4.8-Ma age of the Tabletop Mountain basalt gives a minimum age for the latest faulting along both margins of the Winston graben.

STRUCTURE

The Winston graben is one of the major structures of the Rio Grande rift in south-central New Mexico (Fig. 1). It is approximately 56 km long and from 5 to 10 km wide. It trends north-northeast to north-northwest, with a dogleg near its center. Geophysical data of Ericksen et al. (1970) suggest that it is deepest near the dogleg. The northern end of the Winston graben occurs at a structural high, the Morenci lineament of Chapin et al. (1978a) or the Socorro accommodation zone of Chapin (1989; Fig. 1). The southern end occurs at a similar structural high or accommodation zone, the Chise lineament of Harrison (1990). The Chise lineament separates the Winston and the Las Animas grabens, which share common boundary faults. There has been abundant Pliocene basaltic volcanism along the Chise lineament between the Winston and Las Animas grabens.

In contrast to most basins of the Rio Grande rift, the Winston graben is nearly symmetrical, with approximately 2 km of stratigraphic separation along the western fault margin and 1.8 km along the eastern fault margin. Displacement along the western margin occurs along a single fault, while the eastern margin consists of several braided fault strands, particularly near the southern end of the graben. Numerous faults within the graben displace all units except the Pliocene basalt and Quaternary Santa Fe Group. Figure 3 is a geologic map of the southern half of the Winston graben where faults are most prominently displayed through offset of the Tertiary volcanic units. Similar faulting extends to the north in the graben, but is difficult to follow in the poorly consolidated, undifferentiated Santa Fe Group.

Both boundary faults and the majority of faults in the graben are high-angle normal and oblique-slip structures. In a few locations, minor east-west–striking, horizontally striated tear faults with small displacement are found. The attitude of beds in the graben is subhorizontal everywhere, except at the extreme southern end, where the Tertiary lower unit of the Santa Fe Group and Oligocene volcanic units dip 25 to 37° to the west.

Eocene and Oligocene faults in the vicinity of the Winston graben

The vicinity of the Winston graben has experienced pre-rift Cenozoic faulting that is important to the full understanding of the development of the graben. During the Eocene, there was an episode of dextral, strike-slip faulting in south-central New Mexico (Harrison, 1988, 1989a, 1989b, 1990; Harrison and Chapin, 1990), which was part of an extensive northerly trending fault zone recognized by Chapin and Cather (1981) and Chapin (1983) along the eastern margin of the Colorado Plateau during a late stage of the Laramide orogeny. This Eocene faulting is important to the development of Winston graben in that it imparted a pervasive north-northeast and north-northwest structural fabric that in part controlled the location and style of subsequent faulting. Indeed, the eastern margin of the Winston graben appears to be a reactivated strike-slip fault. Evidence for this interpretation lies in scattered relict features in the Sierra Cuchillo uplift (and its southern extension, the Animas uplift) indicative of strike-slip faulting: (1) antithetic R′ shears striking east-northeast; (2) folds with northwest-striking axes; (3) horizontally striated slickenside fault surfaces striking north-northwest; and (4) high-angle reverse faults striking north-northwest (relict positive flower structures) still present in the fault zone (Fig. 4). The western boundary fault of the graben, on the other hand, truncates north-northeast–striking segments of strike-slip faults in the Chloride mining district (Fig. 3).

In southern New Mexico, the initial phases of the Rio Grande rift began in the late Oligocene (Chapin and Seager, 1975). In the area of the Winston graben, this activity was expressed as northwest-striking epithermal veins in the southern Chloride mining district (Harrison, 1986). (In the northern Chloride mining district not shown on Figure 3, vein trends are dominantly north-south.) All veins occupy high-angle normal faults from which vein adularia have yielded K-Ar ages of 26.2 to 28.9 Ma (Harrison, 1986). Along the western boundary fault of the graben, several northwest-striking jogs occur along reactivated segments of epithermal vein structures as evidenced by brecciated vein material. One of the more prominent jogs occurs northwest of the town of Chloride (Fig. 3). Elsewhere along the western boundary fault, north-northeast-striking fault segments cross cut the vein structures (Fig. 3). Although documentation is lacking, the strong tilting of lower Santa Fe Group sediments and Tertiary volcanics within the southern Winston graben is probably related to late Oligocene faulting.

Neogene faults

The Winston graben is largely a product of Neogene faulting, which is the dominant stage of structural deformation in south-central New Mexico. Figure 6 is a tectonic sketch map of major Neogene structures in this part of south-central New Mexico. An east-west cross section, at the approximate latitude of Winston, is shown in Figure 4. As indicated by these two figures, a structural boundary between strongly faulted and virtually nonfaulted terranes closely follows an extension of the Santa Rita–Hanover axis (postulated position shown on Fig. 4) of Elston et al. (1968, 1970) and Aldrich (1972, 1974, 1976) northward through the north-central Black Range. This boundary is the western margin of the Rio Grande rift. West of this boundary, relatively unfaulted Cenozoic rocks dip gently westward into the Mogollon Plateau. East of this boundary, Neogene faulting was intense, forming numerous north-south–elongated fault blocks, with grabens filled by the Santa Fe Group. The Winston and Las Animas grabens are the westernmost rift basins that contain Santa Fe Group sediments.

Figure 6 shows that the Neogene structural grain in the vicinity of the Winston graben is dominantly north-northeast to north-south and lesser northwest fabric. Figure 7A is a compass-rose diagram for more than 400 km of high-angle normal and oblique-slip faults in this area. Oblique-slip and dip-slip slickenside striations and mullions (Fig. 7B) on these structures indicate an east-west to east-southeast–west-northwest, horizontal least principal stress (σ_3). Figure 7C is a compass-rose diagram for orientations of the direction of horizontal least principal stress (σ_3) based on the assumption that the intermediate principal stress (σ_2) is normal to displacement in the fault plane and that the maximum principal stress (σ_1) is approximately 30° to the fault plane and in the plane containing the slip vector.

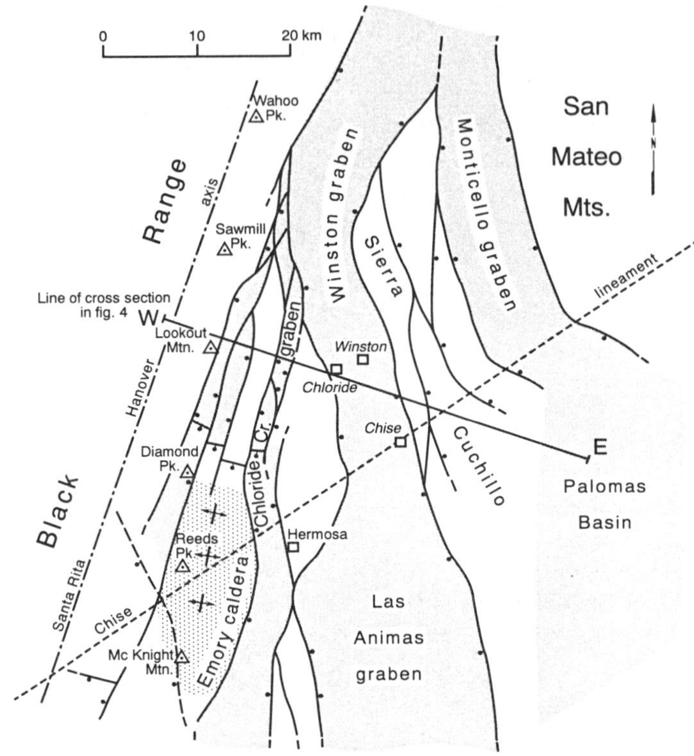

Figure 6. Sketch map of major Neogene structures in a portion of south-central New Mexico. Shaded areas represent grabens of the Rio Grande rift; nonshaded areas are uplifts or horst blocks. Figure 4 is an east-west cross section for this area at the approximate latitude of Winston, New Mexico. Virtually all structural deformation stops at the Santa Rita–Hanover axis as indicated on this map; Cenozoic rocks west of this line dip gently westward with very minor faulting.

TECTONIC IMPLICATIONS

The Winston graben is a major structure in the tectonic framework of the Rio Grande rift in south-central New Mexico. Individual fault blocks (Black Range uplift, Winston graben, Sierra Cuchillo uplift, Palomas basin, and Engle basin) can be united into a broad extensional model in which the blocks overlie a major detachment fault in the basement (Fig. 8).

The Palomas and Engle basins are asymmetrical, east-tilted half grabens (Lozinsky, 1987) derived from large-scale hanging-wall rotation above the detachment fault. Such asymmetrical half grabens are characteristic of rift structures worldwide and are considered the basic building blocks of rifts (Rosendahl, 1987). The Sierra Cuchillo Range is an uplifted horst within the rift system, similar to a high-relief accommodation of Rosendahl (1987). The Black Range with its gently westward dipping strata is the uplifted shoulder of the rift.

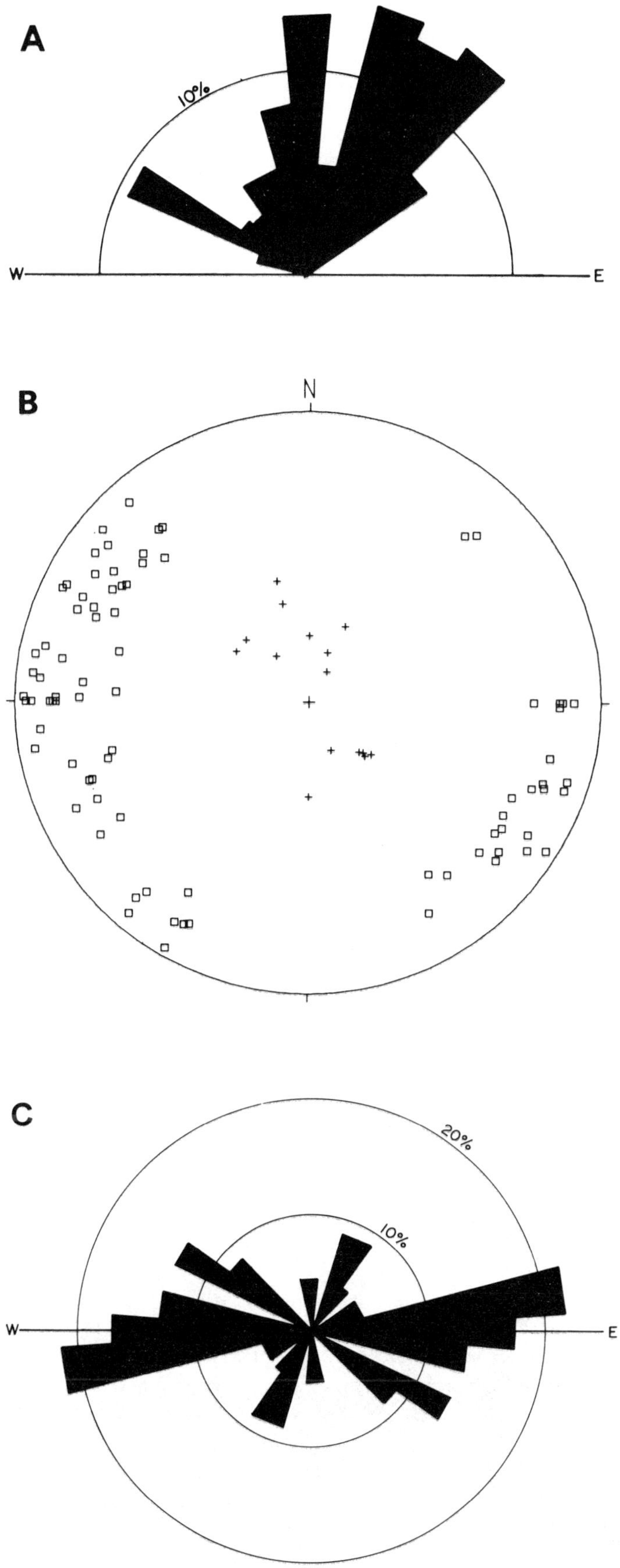

Figure 7. A, Compass-rose diagram in 10° groupings showing strikes for more than 400 km of Neogene, high-angle normal faults in the vicinity of the Winston graben; B, lower-hemisphere projection showing poles to observed Neogene fault surfaces (squares) and lines of slickenside striations and mullions (crosses); C, compass-rose diagram for horizontal, least principal stress directions determined from slickenside striations and mullions on Neogene faults in the vicinity of the Winston graben.

The odd structure in this tectonic framework is the very narrow, symmetrical Winston graben. Such structures are not part of Rosendahl's (1987) rift development scheme, where asymmetrical half grabens are the basic building blocks. Two possible solutions to this problem are suggested. The first possibility is derived from the work of Gibbs (1984). Based on seismic and well data, he developed an interpretive cross section across the central graben beneath the southern North Sea. It shows a symmetrical graben that develops when the dip of the detachment surface steepens along a basement ramp. In similar fashion to the Winston graben, this symmetrical graben formed down dip from an uplifted horst block (high-relief accommodation zone). The interpretive cross section of Figure 8 for the Rio Grande rift is modelled after Gibbs' (1988) cross section. The overall morphology of both cross sections is remarkably similar: shallow basin/listric fan (Engle basin); deep basin/tilted half graben (Palomas basin); central high/horst block (Sierra Cuchillo); symmetrical graben (Winston graben); minor to unfaulted rift shoulder (Black Range). The basic difference between the two cross sections is that in Gibbs' North Sea section the symmetrical graben lies directly above the basement ramp and the Rio Grande rift section the Winston graben occurs at a distance down dip from the known ramp.

The cross section in Figure 8 balances reasonably well, but only when the change in dip along the detachment zone is placed where a major listric fault separates the Palomas and Engle basins. Geologic mapping (R. W. Harrison and R. P. Lozinsky, unpublished data), well data (see Lozinsky, 1987), and unpublished seismic lines (R. Keller, personal communication, 1991) support the existence of this structure. The only minor problems encountered in balancing the cross section in Figure 8 are in the area of the eastern Winston graben and Sierra Cuchillo uplift. This is a complex area of both syn-rift magmatism and pre-rift (Laramide) deformation. It is important to note that much of the syn-rift volcanism along the cross section of Figure 8 occurred in or immediately adjacent to the Sierra Cuchillo uplift (high-relief accommodation zone). A similar location for volcanism was described for the East Africa rift by Rosendahl (1987) and elsewhere in the Rio Grande rift by Reynolds (1984).

A second possibility for the formation of a symmetrical graben is that of a rollover. Rollover structures are produced by the slumping or collapsing of hanging-wall fault blocks in re-

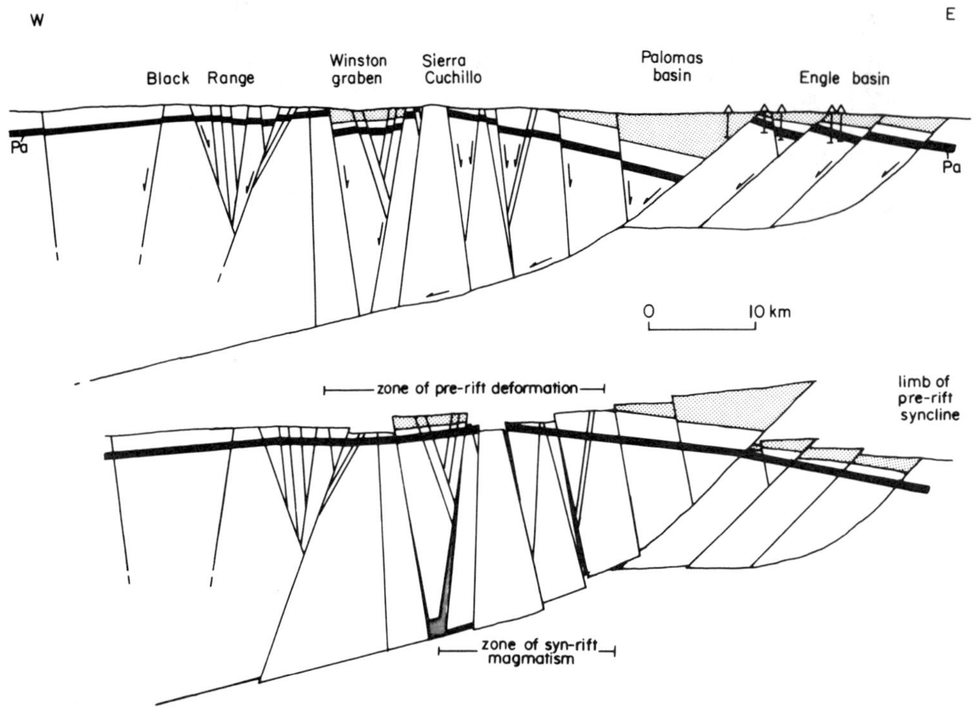

Figure 8. Interpretive cross section for the Rio Grande rift in south-central New Mexico and reconstructed cross section to pre-rift geometry. See Figure 1 for approximate location. Dark marker horizon is the Permian Abo Formation, dotted pattern is Santa Fe Group, the section between the Abo Formation and Santa Fe Group is comprised of Tertiary volcanic and volcaniclastic rocks. Eastern third of the section is adapted from Lozinsky (1987). It should be noted that excess space occurs in areas of syn-rift magmatism and space deficit occurs in areas of pre-rift (Laramide) deformation. Strata in the eastern part of the reconstructed cross section is on a limb of a pre-rift (Laramide) syncline.

sponse to motion on listric normal faults (Xiao and Suppe, 1992). If the Winston graben is a rollover structure, then its eastern boundary fault should dip at 67° or less and flatten with depth. However, this seems unlikely because: (1) the eastern boundary fault has a near-vertical, strike-slip ancestry; (2) where exposed at the surface, it has a dip of 75° or greater; (3) exploration drilling shows that it maintains a steep dip to depth of at least 540 m (1,500 ft) (P. Willard, personal communication, 1987); and (4) it has served as a conduit for magmatism, implying that its steep dip extends to considerable depth.

Whatever its true nature is at depth, it seems most likely that the location of the rather odd, symmetrical Winston graben has been controlled by preexisting structure, particularly Eocene strike-slip faults. Similar grabens occur west of Socorro in other areas where Eocene strike-slip faulting is suspected (Chapin, 1983).

SUMMARY: DEVELOPMENT OF THE WINSTON GRABEN

The earliest evidence of a depression in the area of the Winston graben is found in the occurrences of Vicks Peak Tuff (28.6 Ma, McIntosh, 1989) preserved in paleovalleys in the southern part of the graben. This tuff is not present in the adjacent Black Range. Whether or not these occurrences are related to faulting is unclear. The lower Santa Fe Group is interbedded with ash-flow tuffs and water-laid tuffs, one of which is believed to be distal South Canyon Tuff. If this correlation is correct, then the lower Santa Fe Group dates to approximately 27.3 Ma and indicates the graben's existence at that time. The zeolitic alteration of the lower Santa Fe Group and intercalated tuffs described by Bowie and Barker (1986) is most probably related to an ancient saline lake that ponded in the Winston graben.

Rotation of the lower Santa Fe Group is the first clear evidence of structural deformation in the graben. And while the exact age of this deformation is only loosely constrained, the rotation is most likely related to faulting associated with the vein deposits of the nearby Chloride mining district that are approximately 26.5 Ma (Harrison, 1986). Indeed, a northwest-striking segment of the western boundary fault (northwest of the town of Chloride) is along one of the Chloride district veins.

The next major event in development of the Winston graben was emplacement of the gravity slide block of Vicks Peak Tuff that became detached from the adjacent Sierra

Cuchillo uplift. The maximum age of this event is constrained by the 18.6-Ma andesite of the Winston graben that directly overlies the gravity slide block. Eruption of the andesite of the Winston graben represents the first and most significant volcanic event in the graben and is noteworthy because volcanism of this age is rare in the southern Rio Grande rift.

Subsequently, the graben received sediments of the Santa Fe Group, the uppermost portion of which records the initial exposure and erosion of Cretaceous and Paleozoic rocks from the adjacent Sierra Cuchillo. At approximately 4.8 Ma, the small basalt flow preserved on Tabletop Mountain was erupted onto an extensive pediment surface that covered the Winston graben and parts of the Black Range. The lack of offset on this surface and the basalt flow by boundary faults indicates that the Winston graben has been structurally inactive for at least the past 4.8 m.y. During the Pleistocene, parts of the Winston graben were incised and subsequently filled with Quaternary sediments.

ACKNOWLEDGMENTS

This research was part of a dissertation grant from the New Mexico Bureau of Mines and Mineral Resources. I am very grateful for this support and the extremely beneficial interaction with Bureau personnel, particularly C. E. Chapin. I also wish to thank Jud May and Betsy Julian for reviews and suggestions that greatly improved this paper.

REFERENCES CITED

Abitz, R. J., 1989, Geology and petrogenesis of the northern Emory caldera, Sierra County, New Mexico [Ph.D. thesis]: Albuquerque, University of New Mexico, 174 p.

Aldrich, M. J., Jr., 1972, Tracing a subsurface structure by joint analysis: Santa Rita–Hanover axis, southwestern New Mexico [Ph.D. thesis]: Albuquerque, University of New Mexico, 106 p.

Aldrich, M. J., Jr., 1974, Structural development of the Hanover-Fierro pluton, southwestern New Mexico: Geological Society of America Bulletin, v. 85, p. 963–968.

Aldrich, M. J., Jr., 1976, Geology and flow directions of volcanic rocks of the North Star Mesa Quadrangle, Grant County, New Mexico, *in* Elston, W. E., and Northrop, S. A., eds., Cenozoic volcanism in southwestern New Mexico: New Mexico Geological Society Special Publication 5, p. 79–81.

Bowie, M. R., and Barker, J. M., 1986, Clinoptilolite west of Cuchillo Negro Creek, New Mexico—Zeolite authigenesis of the tuff of Little Mineral Creek: New Mexico Geological Society Guidebook 37, p. 283–286.

Callendar, J. F., Seager, W. R., and Swanberg, C. A., 1983, Late Tertiary and Quaternary tectonics and volcanism: Geothermal resources of New Mexico: New Mexico State University Energy Institute, Scientific Map Series, scale 1:500,000, 1 sheet.

Cameron, K. L., Nimz, G. J., Kuentz, D., Niemeyer, S., and Gunn, S., 1989, Southern Cordilleran basaltic andesite suite, southern Chihuahua, Mexico: A link between Tertiary continental arc and flood basalt magmatism in North America: Journal of Geophysical Research, v. 94, no. B6, p. 7817–7840.

Chapin, C. E., 1983, An overview of Laramide wrench faulting in Southern Rocky Mountains with an emphasis on petroleum exploration: Rocky Mountain Association of Geologists, 1983 Field Conference, Guidebook, p. 169–179.

Chapin, C. E., 1989, Volcanism along the Socorro accommodation zone, Rio Grande rift, New Mexico, *in* Chapin, C. E., and Zirdek, J., eds., Field excursions to volcanic terranes in the western United States, Volume I: Southern Rocky Mountain region: New Mexico Bureau of Mines and Mineral Resources Memoir 46, p. 46–57.

Chapin, C. E., and Cather, S. M., 1981, Eocene tectonics and sedimentation in the Colorado Plateau–Rocky Mountain area: Arizona Geological Society Digest, v. 14, p. 173–198.

Chapin, C. E., and Seager, W. R., 1975, Evolution of the Rio Grande rift in the Socorro and Las Cruces areas: New Mexico Geological Society, Guidebook 26, p. 297–322.

Chapin, C. E., Chamberlain, R. M., Osburn, G. R., Sanford, A. R., and White, D. W., 1978a, Exploration framework of the Socorro geothermal area, New Mexico, *in* Chapin, C. E., Elston, W. E., and James, H. L., eds., Field guide to selected cauldrons and mining districts of the Datil-Mogollon volcanic field, New Mexico: New Mexico Geological Society Special Publication 7, p. 115–129.

Chapin, C. E., Jahns, R. H., Chamberlain, R. M., and Osburn, G. R., 1978b, First day road log: Truth or Consequences via Magdalena and Winston, *in* Chapin, C. E., Elston, W. E.,and James, H. L., eds., Field guide to selected cauldrons and mining district of the Datil-Mogollon volcanic field, New Mexico: New Mexico Geological Society Special Publication 7, p. 1–31.

Coney, P. J., 1976, Structure, volcanic stratigraphy, and gravity across the Mogollon Plateau, New Mexico, *in* Elston, W. E., and Northrop, S. A., eds., Cenozoic volcanism in southwestern New Mexico: New Mexico Geological Society Special Publication 5, p. 29–41.

Deal, E. G., and Rhodes, R. C., 1976, Volcano-tectonic structures in the San Mateo Mountains, Socorro County, New Mexico, *in* Elston, E. E., and Northrop, S. A., eds., Cenozoic volcanism in southwestern New Mexico: New Mexico Geological Society Special Publication 5, p. 51–56.

Eggleston, T. L., 1987, The Taylor Creek district, New Mexico: Geology, petrology, and tin deposits [Ph.D. thesis]: Socorro, New Mexico Institute of Mining and Technology, 473 p.

Elston, W. E., 1957, Geology and mineral resources of Dwyer Quadrangle, Grant, Luna, and Sierra Counties, New Mexico: New Mexico Bureau of Mines and Mineral Resources Bulletin 38, 86 p.

Elston, W. E., 1968, Terminology and distribution of ash flows of the Mogollon–Silver City–Lordsburg region, New Mexico: Arizona Geological Society , Guidebook III to Southern Arizona, p. 231–240.

Elston, W. E., 1976, Glossary of stratigraphic terms of the Mogollon-Datil volcanic province, New Mexico, *in* Elston,, W. E., and Northrop, S. A., eds., Cenozoic volcanism in southwestern New Mexico: New Mexico Geological Society Special Publication 5, p. 131–144.

Elston, W. E., and Bornhorst, T. J., 1979, The Rio Grande rift in context of regional post–40 m.y. volcanic and tectonic events, *in* Riecker, R. E., ed., Rio Grande rift: Tectonics and magmatism: Washington, D.C., American Geophysical Union, p. 416–438.

Elston, W. E., Coney, P. J., and Rhodes, R. C., 1968, A progress report on the Mogollon Plateau volcanic province, southwestern New Mexico, *in* Epis, R. C., ed., Cenozoic volcanism in the Southern Rocky Mountains: Colorado School of Mines Quarterly, v. 63, p. 261–287.

Elston, W. E., Coney, P. J., and Rhodes, R. C., 1970, Progress report on the Mogollon Plateau volcanic province, southwestern New Mexico, No. 2: New Mexico Geological Society, Guidebook 21, p. 75–86.

Elston, W. E., Damon, P. E., Coney, P. J., Rhodes, R. C., Smith, E. I., and Bikerman, M., 1973, Tertiary volcanic rocks, Mogollon-Datil province, New Mexico, and surrounding region: K-Ar dates, patterns of eruption, and periods of mineralization: Geological Society of America Bulletin, v. 84, p. 2259–2273.

Elston, W. E., Seager, W. R., and Clemons, R. E., 1975, Emory cauldron, Black Range, New Mexico: Source of Kneeling Nun Tuff: New Mexico

Geological Society, Guidebook 26, p. 283–292.
Ericksen, G. E., Wedow, H., Jr., Eaton, G. P., and Leland, G. R., 1970, Mineral resources of the Black Range Primitive Area, Grant, Sierra, and Catron Counties, New Mexico: U.S. Geological Survey Bulletin 1319-E, 162 p.
Fodor, R. V., 1975, Petrology of basalt and andesite of the Black Range, New Mexico: Geological Society of America Bulletin, v. 86, p. 295–304.
Fodor, R. V., 1978, Ultramafic and mafic inclusions and megacrysts in Pliocene basalt, Black Range, New Mexico: Geological Society of America Bulletin, v. 89, p. 451–459.
Gibbs, A. D., 1984, Structural evolution of extensional basin margins: Journal of the Geological Society of London, v. 141, p. 609–620.
Gilbert, G. K., 1875, Report of the geology of portions of New Mexico and Arizona examined in 1873: U.S. Geographic and Geological Surveys West of 100th Meridian (Wheeler), v. 3, pt. 5, p. 503–567.
Gordon, C. H., and Graton, L. C., 1907, Lower Paleozoic formations in New Mexico: Journal of Geology, v. 15, p. 91–92.
Harley, G. T., 1934, The geology and ore deposits of Sierra County, New Mexico: New Mexico Bureau of Mines and Mineral Resources Bulletin 10, 220 p.
Harrison, R. W., 1986, General geology of the Chloride mining district, Sierra and Catron Counties, New Mexico: New Mexico Geological Society, Guidebook 37, p. 265–272.
Harrison, R. W., 1988, Primary structural control of epithermal mineralized shoots in southeastern Chloride mining district, New Mexico: New Mexico Geology, v. 10, no. 4, p. 80–81.
Harrison, R. W., 1989a, Exotic blocks within the early Tertiary Rubio Peak Formation in the north-central Black Range, New Mexico: Occurrence, insights into post-emplacement tectonic activity, economic implications, and emplacement hypothesis: New Mexico Geological Society, Guidebook 40, p. 99–106.
Harrison, R. W., 1989b, Geology of Winston Quadrangle, Sierra County, New Mexico: New Mexico Bureau of Mines and Mineral Resources, Open-File Report 358, scale 1:24,000, 17 p., 2 sheets.
Harrison, R. W., 1990, Cenozoic stratigraphy, structure, and epithermal mineralization of the north-central Black Range, New Mexico, in the regional geologic framework of south-central New Mexico [Ph.D. thesis]: New Mexico Institute of Mining and Technology, Socorro, 402 p.
Harrison, R. W., and Chapin, C. E., 1990, A NNE-trending, dextral wrench-fault zone of Laramide age in southern New Mexico: Geological Society of America Abstracts with Programs, v. 22, no. 7, p. 327.
Jahns, R. H., 1955, Geology of Sierra Cuchillo, New Mexico: New Mexico Geological Society, Guidebook 6, p. 158–174.
Jicha, H. L., Jr., 1954, Geology and mineral deposits of the Lake Valley Quadrangle, Grant, Luna, and Sierra Counties, New Mexico: New Mexico Bureau of Mines and Mineral Resources Bulletin 37, 93 p.
Kottlowski, F. E., 1953, Tertiary-Quaternary sediments of the Rio Grande Valley in southern New Mexico: New Mexico Geological Society, Guidebook 4, p. 144–148.
Kottlowski, F. E., 1955, Cenozoic sedimentary rocks of south-central New Mexico: New Mexico Geological Society, Guidebook 6, p. 88–91.
Le Maitre, R. W., 1984, A proposal by the IUGS Subcommission classification of volcanic rocks based on the total alkali silica (TAS) system: Australian Journal of Earth Sciences, v. 31, p. 243–255.
Lozinsky, R. P., 1987, Cross section across the Jornada del Muerto, Engle, and northern Palomas basins, south-central New Mexico: New Mexico Geology, v. 9, no. 3, p. 55–57, 63.
Lozinsky, R. P., and Hawley, J. W., 1986, Upper Cenozoic Palomas Formation of south-central New Mexico: New Mexico Geological Society, Guidebook 37, p. 239–247.
McDowell, F. W., 1971, K-Ar ages of igneous rocks from the western United States: Isochron/West, no. 2, p. 1–16.
McIntosh, W. C., 1989, Ages and distribution of ignimbrites in the Mogollon-Datil volcanic field, southwestern New Mexico: A stratigraphic framework using $^{40}Ar/^{39}Ar$ dating and paleomagnetism [Ph.D. thesis]: Socorro, New Mexico Institute of Mining and Technology, 351 p.
Reynolds, D. J., 1984, Structural and dimensional repetition in continental rifting [M.S. thesis]: Durham, North Carolina, Duke University, 175 p.
Rosendahl, B. R., 1987, Architecture of continental rifts with special reference to East Africa: Annual Review of Earth and Planetary Science, v. 15, p. 445–503.
Scott, K. M., 1988, Lahars and Lahar-Runout flows in the Toutle-Cowlite river system, Mount St. Helens, Washington–Origins, behavior, and sedimentology: U.S. Geological Survey Professional Paper 1447-A, 76 p.
Seager, W. R., and Morgan, P., 1979, The Rio Grande rift in southern New Mexico, west Texas, and northern Chihuahua, *in* Riecker, R. E., ed., Rio Grande rift: Tectonics and magmatism: Washington, D.C., American Geophysical Union, p. 87–106.
Seager, W. R., Clemons, R. E., Hawley, J.W., and Kelley, R. E., 1982, Geology of northwest part of Las Cruces 1 × 2 degree sheet, New Mexico: New Mexico Bureau of Mines and Mineral Resources, Geologic Map 53, scale 1:125,000, 3 sheets.
Seager, W. R., Shafiqullah, M., Hawley, J. W., and Marvin, R. F., 1984, New K-Ar dates from basalts and the evolution of the southern Rio Grande rift: Geological Society of America Bulletin, v. 95, p. 87–99.
Smith, G. A., 1986, Coarse-grained nonmarine volcaniclastic sediment: Terminology and depositional process: Geological Society of America Bulletin, v. 97, p. 1–10.
Smith, G. A., 1987, The influence of explosive volcanism on fluvial sedimentation: The Duschutes Formation (Neogene) in central Oregon: Journal of Sedimentary Petrology, v. 57, p. 613–629.
Smith, G. A., 1988, Sedimentology of proximal to distal Ellensburg Formation (late Miocene), central Washington: Sedimentology, v. 35, p. 953–977.
Walker, G.P.L., 1985, Origin of coarse lithic breccias near ignimbrite source vents: Journal of Volcanology and Geothermal Research, v. 25, p. 157–171.
Xiao, H., and Suppe, J., 1992, Origin of rollover: American Association of Petroleum Geologists Bulletin, v. 76, no. 4, p. 509–529.

MANUSCRIPT ACCEPTED BY THE SOCIETY DECEMBER 22, 1993

Printed in U.S.A.

Geological Society of America
Special Paper 291
1994

Crustal structure and basin geometry in south-central New Mexico

Donald C. Adams and G. Randy Keller
Department of Geological Sciences, The University of Texas at El Paso, El Paso, Texas 79968

ABSTRACT

The region of central New Mexico where the Rio Grande rift begins to widen has been investigated via an integrated geological and geophysical analysis. Existing seismic refraction profiles and two seismic reflection profiles, when combined with gravity and drilling data, provide a good picture of basin geometry deep crustal structure in the region. The Palomas Basin is shown to be a half graben bounded by a listric-normal fault imaged by the seismic reflection data. The Engle Basin is also associated with a half graben. The structure of the Tularosa Basin is more symmetric with two grabens. An analysis of deeper structures indicates that the crust is 28 to 33 km thick in the rift with the greatest thickness being in the northern portion of the area.

INTRODUCTION

The Rio Grande rift extends southward from central Colorado to central New Mexico as a single physiographic rift valley. Near Socorro, New Mexico (Fig. 1), the rift widens eastward to include the Tularosa Basin, and widens farther as it extends southward into Mexico and west Texas. Cordell (1982) showed that this widening is part of a general southward increase in extension along the rift. However, south of Socorro, the physiographic widening is abrupt, the rift-basin geometry is known only in a few areas, and our knowledge of the rift margins at depth is very limited. Thus the region where the rift begins to widen is important in efforts to understand its overall structure and evolution.

In the Rio Grande rift, the basins have been the key features in defining its extent (e.g., Chapin, 1971; Seager and Morgan, 1979). However, the physiographic basins often contain complex subsurface structure (e.g., Ramberg et al., 1978; Cordell, 1978; Keller et al., 1984). In this study, we have investigated the basin geometry and crustal structure of the rift in south-central New Mexico where it widens significantly. We present a new seismic reflection profile and have employed gravity data to integrate the scattered control provided by seismic profiles, drill holes, and Paleozoic and older outcrops.

BACKGROUND

The geologic history of south-central New Mexico is dominated by several periods of major tectonic activity. The Precambrian assembly of the region imparted a northeast structural grain to the upper crust (Grambling et al., 1988); the building of the Ancestral Rocky Mountains (late Paleozoic) produced large basins and basement uplifts (Kluth and Coney, 1981); the Laramide orogeny (Cretaceous–early Tertiary) produced a variety of compressional structures and basins (Seager and Mack, 1986); and mid-Tertiary volcanism produced several large volcanic fields (Elston, 1976). In the study area, the Ancestral Rocky Mountains developed as a series of fault-block mountain ranges, including the Pedernal uplift on the eastern margin of the Rio Grande rift. Examples of Laramide structures are the Love Ranch basin and Rio Grande uplift in the southern portion of the study area (Seager and Mack, 1986). Following the Laramide orogeny, a period of volcanism in western New Mexico formed the Mogollon-Datil volcanic field west of the rift (Elston, 1976) and Tertiary intrusions formed in the Sacramento Mountains on the east.

Development of The Rio Grande rift started at about 30 Ma with rapid extension oriented northeast-southwest producing northwest-southeast–oriented basin structures characterized by low-angle normal faulting. Igneous activity during this

Adams, D. C., and Keller, G. R., 1994, Crustal structure and basin geometry in south-central New Mexico, *in* Keller, G. R., and Cather, S. M., eds., Basins of the Rio Grande Rift: Structure, Stratigraphy, and Tectonic Setting: Boulder, Colorado, Geological Society of America Special Paper 291.

period was characterized by alkali igneous rocks (Seager et al., 1984). This period of activity extended to about 18 Ma and was followed by a magmatically and tectonically quiet period that lasted to 10 Ma. At 10 Ma, a period of rapid east-west extension commenced, which produced north-south–oriented basins and high-angle normal faults overprinting the early rift structures. Igneous activity during this period was characterized by alkali-olivine basalts. This stage of rifting has continued to the present. The current physiographic expression of the rift is primarily a result of the second stage of rifting. Features created during this stage include the Jornada del Muerto, Palomas, and Tularosa Basins, which are targets of this investigation.

Seismic refraction data in the south-central New Mexico region provide a general picture of crustal structure but are scattered (Toppozada and Sanford, 1976; Cook et al., 1978; Jaksha, 1982; Sinno et al., 1986; Harder et al., 1986; Roberts et al., 1991; Schneider and Keller, this volume). The crust generally thins from north to south, and the region of crustal thinning widens also from north to south (Keller et al., 1990).

Gravity data in the Rio Grande rift region are generally well distributed (Keller and Cordell, 1984), but a few readings were added in this study in areas of sparse coverage. The primary sources of these data are the U.S. Geological Survey, University of Texas at Dallas, Defense Mapping Agency, and University of Texas at El Paso. A gravity anomaly map of the entire rift was compiled by Cordell et al. (1982) and a map of New Mexico was compiled by Keller and Cordell (1984). The gravity field in the Rio Grande rift region has been shown to contain information on three scales of structures (e.g., Cordell, 1982). Lithospheric thinning produces anomalies with wavelengths greater than 500 km (Ander, 1981). An axial gravity high with intermediate wavelengths (100 to 500 km) is associated with crustal thinning (Daggett et al., 1986). Short wavelength (<100 km) gravity anomalies are associated with rift-basin structures (Ramberg et al., 1978). Also present in the gravity field are features related to isostatic compensation and to the pre-rift tectonic history of the region such as mid-Tertiary batholiths (Schneider and Keller, this volume).

REGIONAL RIFT STRUCTURE

The gravity data for this study were obtained from a database maintained by the University of Texas at El Paso, Department of Geological Sciences. These data were gridded at a

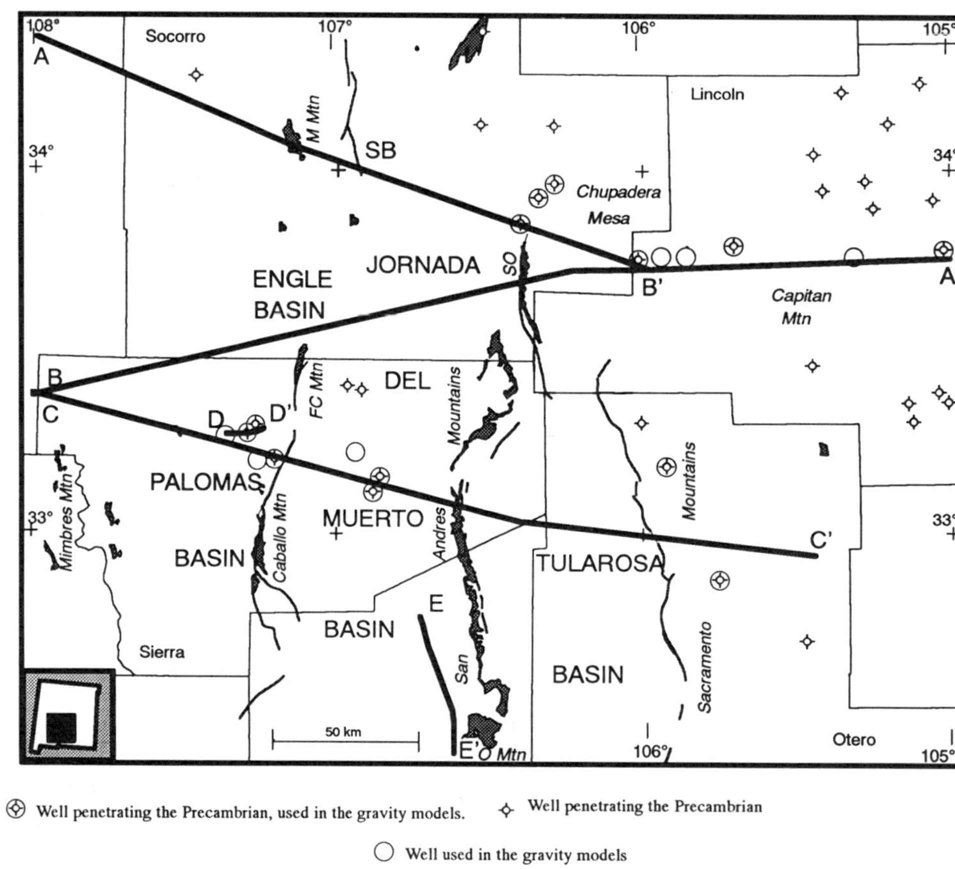

Figure 1. Index map. Heavy lines labeled A–A′ to C–C′ are the gravity profiles modeled. Heavy lines labeled D–D′ and E–E′ indicate seismic reflection profiles and corresponding gravity profiles. Stippled areas are outcrops of Precambrian rocks. FC Mtn, Fra Cristobal Mountains; M Mtn, Magdalena Mountains; O Mtn, Organ Mountains; SO, Sierra Oscura; SB, Socorro Basin.

4-km interval using a minimum curvature algorithm and contoured to produce the Bouguer gravity anomaly map shown in Figure 2. This map contains a northwest to southeast regional gradient that partially obscures the rift structures and was removed by subtracting the third-order polynomial surface shown in Figure 3 from the original grid. This surface approximates the mantle and isostatic effects of the transition from the Cordillera to the Great Plains. Removal of a polynomial surface is also necessary if wavelength filtering is to be effectively used. The resulting residual anomaly grid (Bouguer anomalies minus polynomial surface values) was wavelength filtered using a Hartley Transform filtering algorithm developed by Adams et al. (1991). The resulting grids were contoured at a 5-mgal interval. After producing a suite of maps with various filter parameters, the band-pass filters chosen for this study were 125 to 1,200 km (Fig. 4) and 20 to 125 km (Fig. 5). In the southern portion of the rift, the separation of wavelengths greater than 125 km from shorter wavelengths has been shown to approximately isolate contributions due to the deep crustal structure of the rift from the shallow structures related primarily to the basins (Daggett et al., 1986). We found this separation to also be satisfactory in this study.

The low-pass filtered map in Figure 4 shows a prominent north-northeast–trending gravity high indicating crustal thinning associated with the rift. This high is flanked by gravity lows on the east and west sides of the rift. The western gravity low corresponds to the Mogollon-Datil volcanic field (Schneider and Keller, this volume) and the combination of a batholith and crustal thickening. The gravity anomaly on the east is associated with the Sacramento Mountains and Paleozoic Pedernal uplift and is probably due to crustal thickening. Other studies have shown that the axial gravity high increases in width and amplitude to the south (Keller et al., 1990). Sinno et al. (1986) showed that increased gravity anomaly values correspond to increased crustal thinning in the southern rift.

Five gravity models were developed in this study, and represent an integration of all available data. Three intersecting models are regional in scope and extend east-west across the region (Fig. 1). Two additional models are coincident with seismic-reflection profiles. The models were produced using a 2.5-dimensional modeling algorithm (e.g., Cady, 1980). Zero depth in the gravity models corresponds to sea level.

Two types of gravity anomaly data were used in the gravity models. Free-air anomaly values were used in profiles A–A′, B–B′, and C–C′ so that topography could be incorporated in the models. Incorporation of topography allows the geologic section to be represented without reduction to a base elevation. These models incorporate 2.2 km of relief among the gravity profiles. Since profiles D–D′ and E–E′ have very little relief along the model, Bouguer gravity values were used

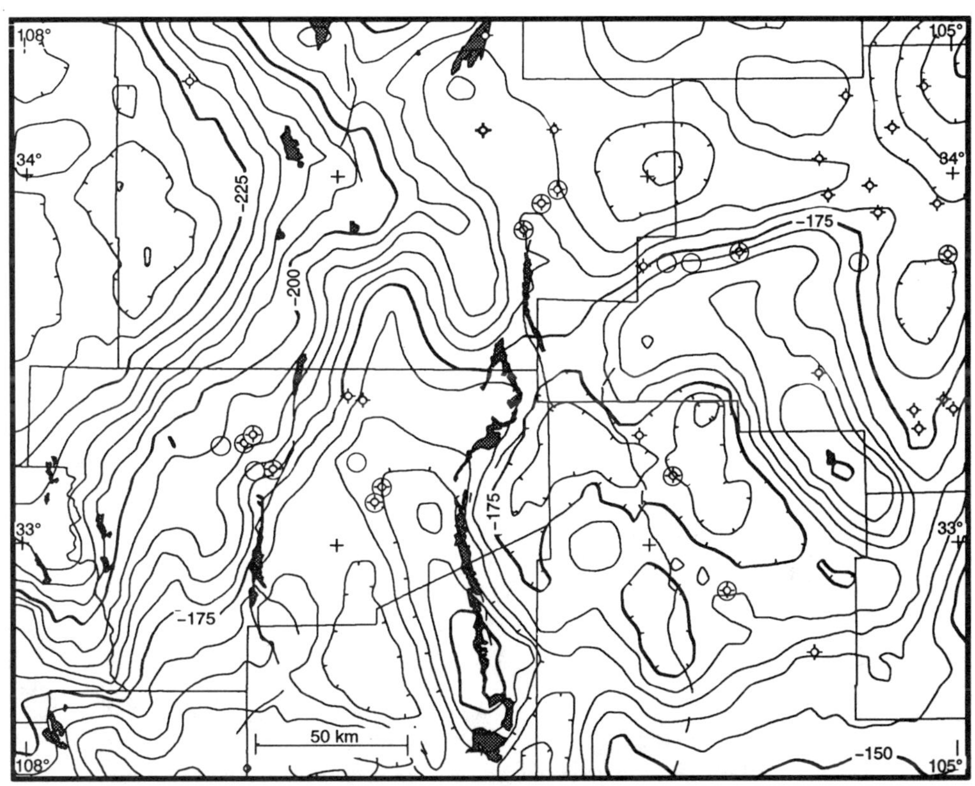

Figure 2. Bouguer gravity anomaly map. The gravity anomalies were produced with a reduction density of 2.67 g/cc and a sea-level datum. Also shown in this figure are Precambrian-penetrating well locations (Table 1), Precambrian outcrops (stippled areas), and county lines. Contour interval is 5 mgal. See Figure 1 for key to well data.

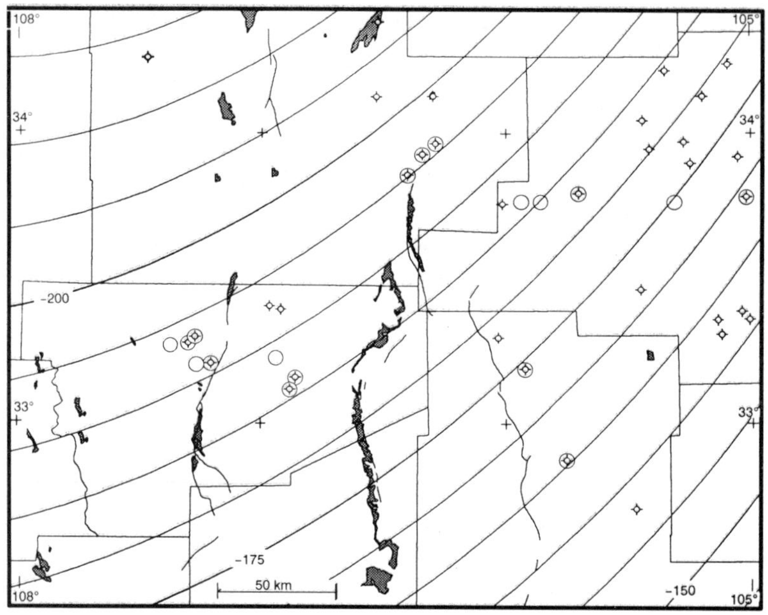

Figure 3. Third-order polynomial surface fitted to the grid of Bouguer gravity anomalies from Figure 2. The subtraction of this grid from the Bouguer gravity map produced the residual gravity anomaly grid used for the filtered maps and the Exxon and Texaco gravity models. Also shown are well locations, Precambrian outcrops (stippled areas), and county lines. Contour interval is 5 mgal. See Figure 1 for key to well data.

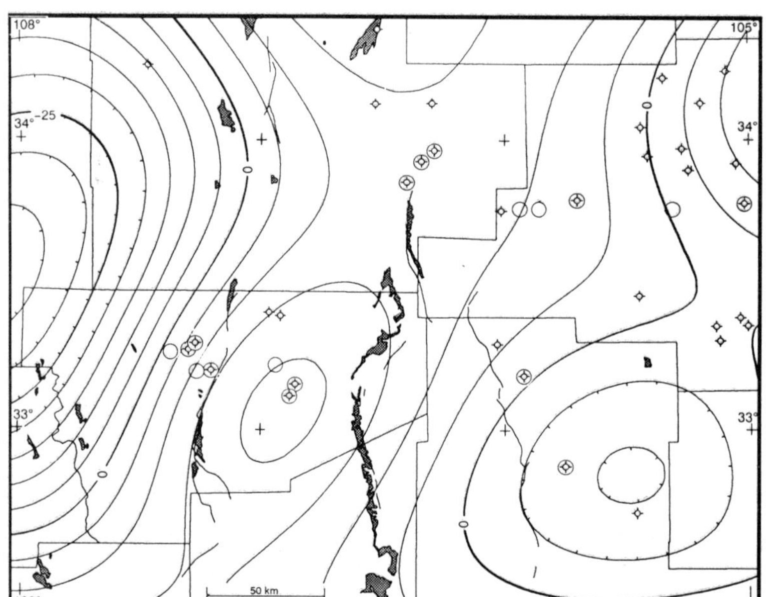

Figure 4. Low-pass filtered gravity map (125 to 1,200 km wavelengths passed). This map shows the north-northeast–trending gravity high associated with crustal thinning along the Rio Grande rift. Also shown are well locations, Precambrian outcrops (stippled areas), and county lines. Contour interval is 5 mgal. See Figure 1 for key to well data.

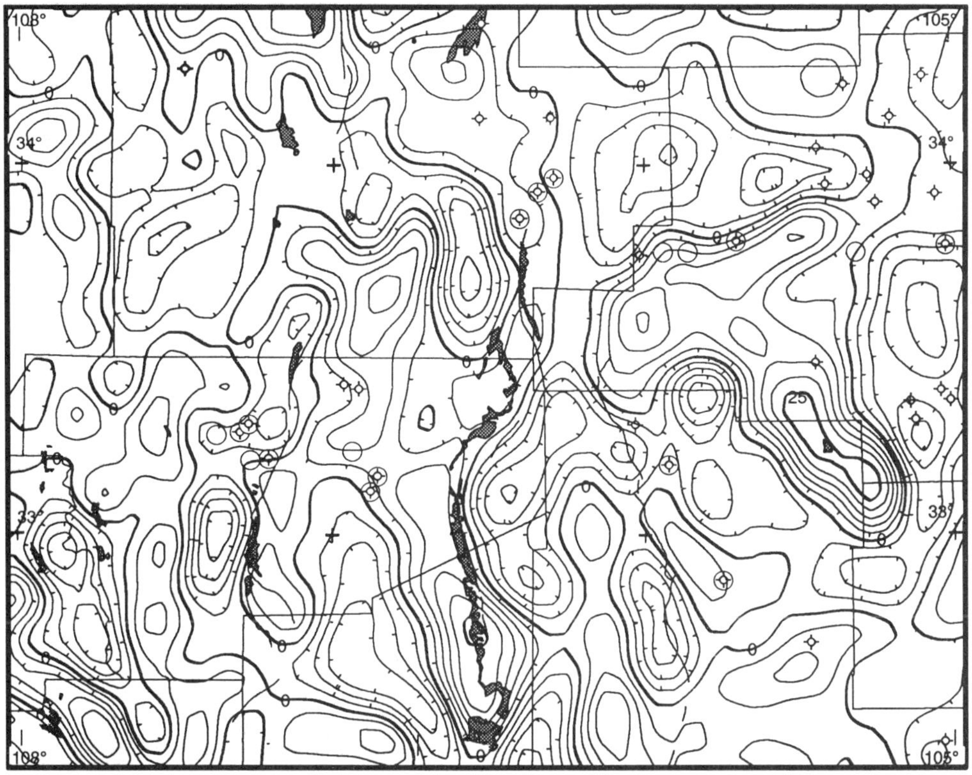

Figure 5. High-pass filtered gravity map (20 to 125 km wavelengths passed). This map accentuates gravity anomalies associated with horst and graben structures within the rift. Also shown are well locations, Precambrian outcrops (stippled areas), and county lines. Contour interval is 5 mgal. See Figure 1 for key to well data.

in these models. These models have a flat upper surface that represents the actual surface of the earth. The Bouguer gravity values used a reduction density of 2.67 g/cc.

The density values used in the gravity models were obtained from several sources. Density values for the mantle, lower crust, middle crust, and upper crust were obtained by the graphical relationship between density and P-wave seismic velocities illustrated in Dobrin and Savit (1988). The seismic velocities for these bodies are given in models by Cook et al. (1978), Jaksha (1982), Sinno et al. (1986), Harder et al. (1986), and Roberts et al. (1991). The density values for the Paleozoic rocks, volcanics, granite batholith, alluvium, and mafic intrusions were determined from density values given in the tables of Telford et al. (1984).

The regional gravity models (A–A′, B–B′, C–C′; Fig. 1) shown in Figures 6, 7, and 8 indicate a north-northeast–trending region of thinned crust and axial thinning of the crust toward the south. The deep crustal structure of the models is based on combination of the results from several sources. Part of the northern model (Fig. 6) is coincident with a long-range seismic refraction profile recorded by Roberts et al. (1991). Since this profile will ultimately extend from the Nevada Test Site (NTS) to Lubbock, Texas, as part of a continuing effort to study deep lithospheric structure, the distances on this model are relative to NTS. The middle profile (B–B′; Fig. 1) is coincident with a seismic profile by Jaksha (1982). This seismic profile was reinterpreted by Schneider and Keller (this volume), and the distances in Figure 7 are relative to the west end of the seismic profile at Morenci, Arizona. The southern profile (C–C′; Fig. 1) is coincident with a geologic cross section compiled by Lozinsky (1987). The regional models all show a long wavelength gravity high associated with the crustal thinning. The gravity high has a maximum amplitude of 25 mgal. All models were also constrained by the seismic results of Stewart and Pakiser (1962), Harder et al. (198), Keller et al. (1986), Schneider (1990), and Sinno et al. (1986). The pub-

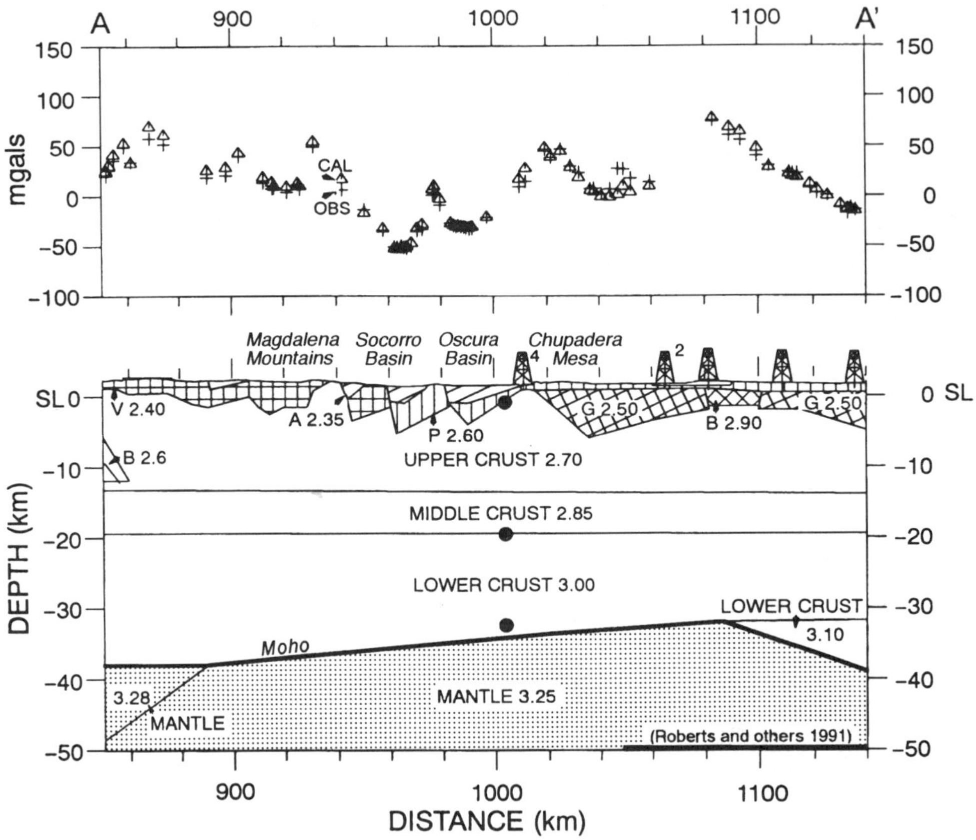

Figure 6. Gravity model for profile A–A′ (Fig. 1). The wells indicate the positions of the well control and the number of wells providing the control. This model is also constrained by the seismic models of Roberts et al. (1991) and Olson et al. (1979) represented by a heavy line and dots, respectively. Zero depth in the model represents sea level. Calculated anomalies are represented by triangles. The pluses are the observed free-air anomalies. V, volcanics; A, alluvium; P, Paleozoic strata; B, intrusive; G, graben. Numbers indicate density values (g/cc).

lished seismic refraction data provides constraints on the thickness of the crust at the intersection between the gravity and seismic profiles. By employing the gravity data, it is possible to model the geometry of the Moho between the seismic constraints and produce an overall picture of crustal structure. The gravity models collectively call for a crust that is over 35 km thick west of the rift, about 30 km thick in the rift, and about 40 km thick immediately east of the rift. The three-layer crust represented in the gravity models is due to the seismic results of Sinno et al. (1986) and is not required to explain the gravity anomalies. From north to south, the models have minimum crustal thicknesses beneath the rift of 33 km (A–A′), 30 km (B–B′), and 29 km (C–C′), respectively. Superimposed on these anomalies are shorter wavelength anomalies associated with the rift basins.

RIFT-BASIN STRUCTURE

The high-pass (Fig. 5) filtered map (wavelengths of 20 to 125 km passed) primarily shows the effects of the rift basins on the gravity field. The physiographic expression of the Tularosa, Jornada del Muerto, and Palomas Basins appears simple, but the gravity map in Figure 5 indicates that their subsurface expression is much more complex. An obvious interpretation is that relative gravity lows in the rift basins show the positions of graben and the relative gravity highs show the positions of buried horst blocks. This interpretation is generally consistent with known depths to the Precambrian basement from drill holes (Table 1) and seismic results, but the complex pre-rift history of the region is a complicating factor. Some of the short wavelength features on Figure 5 are undoubtedly due

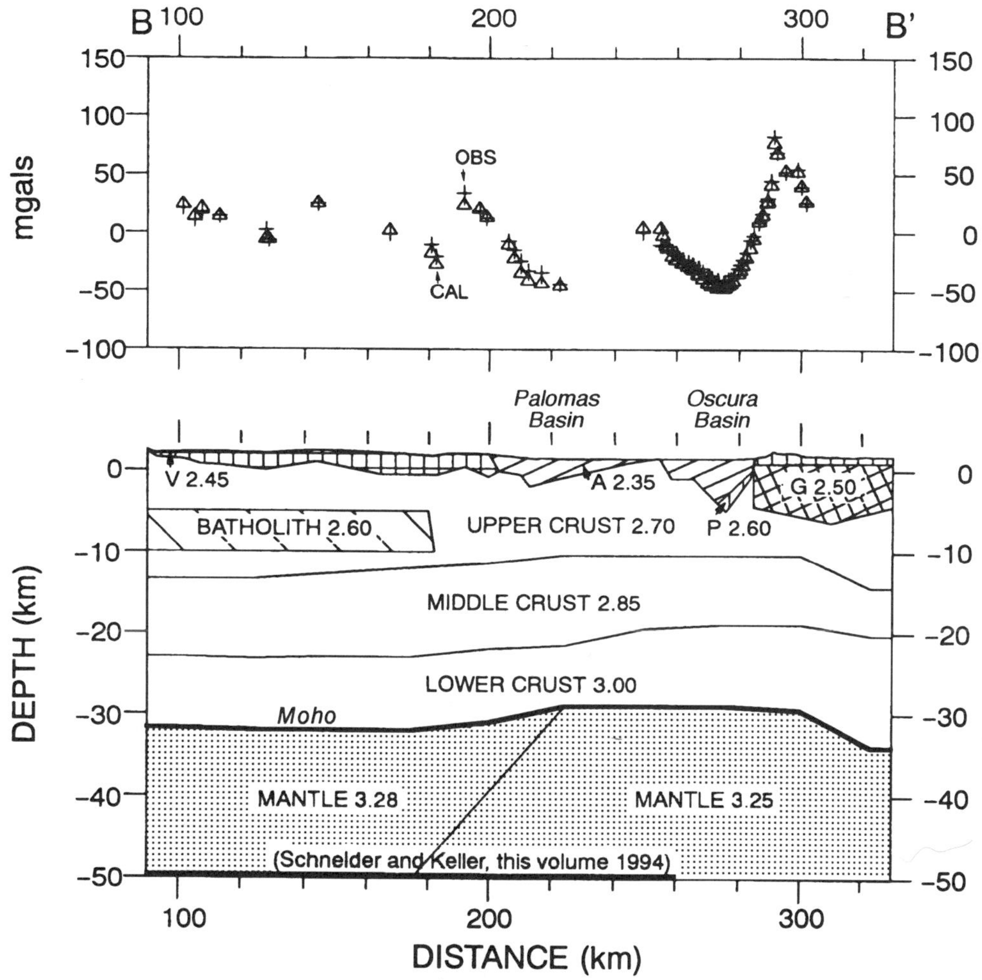

Figure 7. Gravity model for profile B–B′ (Fig. 1). This model is constrained by the seismic model of Schneider and Keller (this volume) for the crust below 7 km (heavy solid line). Zero depth in the model represents sea level. Calculated anomalies are represented by triangles. The pluses are the observed free-air anomalies. V, volcanics; A, alluvium; P, Paleozoic strata; G, graben. Numbers indicate density values (g/cc).

to older structures. However, the tendency for reactivation of older structures has resulted in many younger basins being at least partly coincident with older ones. Thus, although the basin fill may not all be rift related, the overall basin geometry is reflected in Figure 5. Also shown on this gravity map are the positions of Precambrian outcrops and bounding faults of the rift basins. The Precambrian outcrops are located on relative gravity; highs or in regions of high gravity gradients.

The correlation of the mountain ranges and Precambrian outcrops with steep gravity gradients indicates that they are associated with the locations of the master faults of the rift basins. An example of this arrangement is in the northern Tularosa Basin where a large graben depression is bordered on the west and north by the San Andres Mountains and on the east by the Sacramento Mountains. In the Palomas Basin, the eastern side is bounded by the Caballo Mountains and a master fault.

The dip on faults cannot be precisely determined from gravity data, but the shape of the gravity anomalies suggest that the basins are generally asymmetrical (Figs. 6, 7, 8, and 9). The surficial structure of the models is tied to the available wells based on published reports, drilling reports, and outcrops. The Phanerozoic rocks were divided into a Paleozoic section and combined Tertiary basin fill.

Model A–A′ in Figure 6 crosses the rift between Socorro

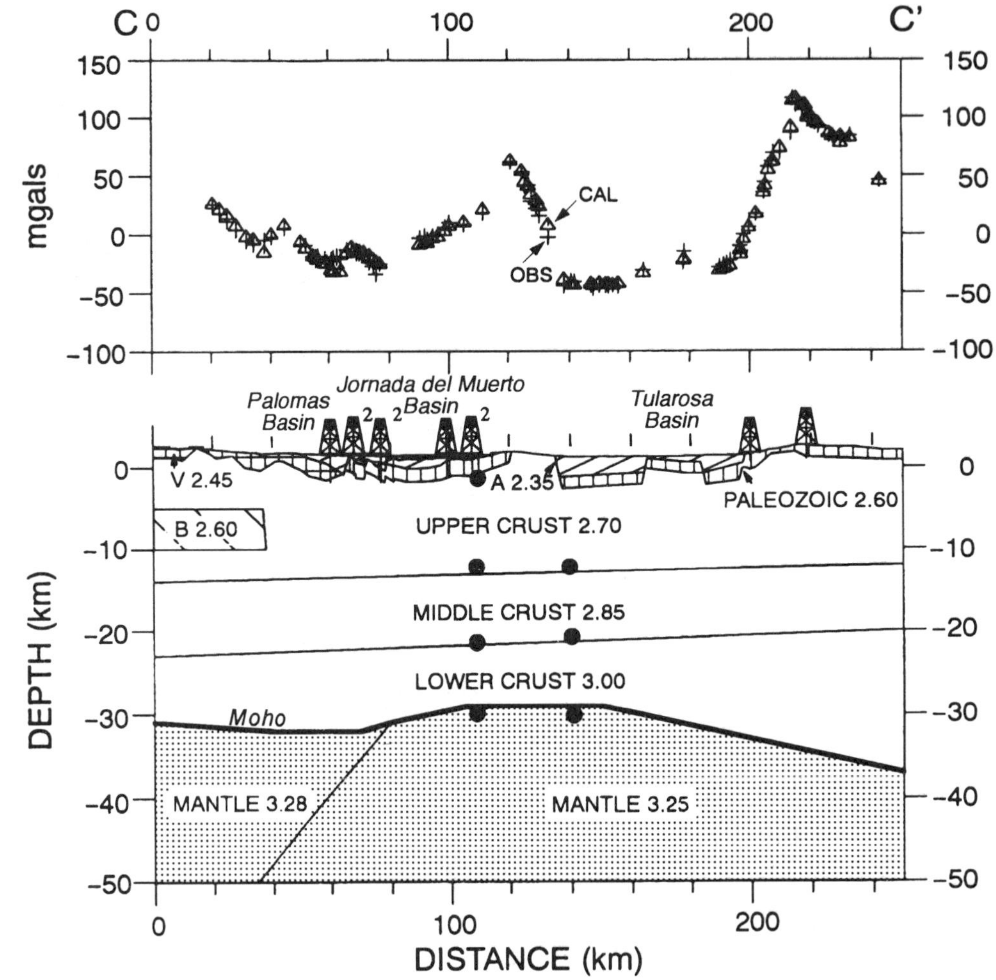

Figure 8. Gravity model for profile C–C′ (Fig. 1). The wells indicate the position of the well control and the number of wells providing the control. This model is also constrained by seismic models of Sinno et al. (1986) as represented by the heavy dots. Zero depth in the model represents sea level. Calculated anomalies are represented by triangles. The pluses are the observed free-air anomalies. V, volcanics; A, alluvium; B, intrusive. Numbers indicate density values (g/cc).

and Carrizozo. The rift basins in this model form asymmetric grabens with the west side down dropped mostly along high-angle normal faults. For the purposes of modeling, all of the faults were assumed to dip at 60° unless otherwise constrained. Volcanic rocks of the Mogollon-Datil volcanic field extend to the west side of the rift. On the east side of the rift, the gravity field and constraints on thickness of the Paleozoic section from well data require the presence of additional low-density material within the Precambrian basement flanking the Pajarito Mountain gravity high (Suleiman and Keller, 1985). The Pajarito Mountain gravity high is related to a 1.2-Ga mafic body that outcrops at Pajarito Mountain within the Sacramento Mountains (Bowsher, 1991). This body may be related to a Middle Proterozoic rifting event in southern New Mexico. The low-density material is possibly composed of sediments deposited in Proterozoic rift basins. The presence of sediments associated with this gravity low is supported by arkosic sediments drilled in the Stanolind #1 Picacho unit in Lincoln County reported by Flawn (1956). The sharpness of the gravity low and constraints on Moho structure from the seismic refraction data of Roberts et al. (1991) requires an upper-crustal explanation for this feature. This model ties with the end of B–B′ at 1,051 km on A–A′.

Model B–B′ in Figure 7 intersects A–A′ near Carrizozo,

TABLE 1. WELLS PENETRATING PRECAMBRIAN BASEMENT

Lat.	Long.	Name	Depth to Precambrian (m)	Lat.	Long.	Name	Depth to Precambrian (m)
32.066	105.334	Turner #1 State	1,520	33.921	105.019	Mesa Petroleum #1 Gallo	941
32.199	105.707	Transocean #1 G. J. Ablah	1,355	33.924	106.326	Lockhart Lockhart #2*	796
32.199	105.714	Turner #1 Evans	1,140	33.934	106.322	Lockhart Lockhart Federal #1*	905
32.302	104.993	Coral #1 Warren	709	33.936	106.314	Landreth #1 Panhandle A*	899
32.325	105.456	Hunt Mcmillian-Turner	624	33.944	105.408	Yates Pet. #1 Boggle ZH Federal	667
32.425	105.038	Standard of Texas #1 Scrap Unit	842	33.953	106.290	Landreth #1 Landreth*	1,044
32.704	105.464	Marathon #1 Mesa Verde Ranch	2,099	33.961	105.252	Yates Pet. #1 Asparas ABX Federal	721
32.856	105.739	Southern Prod #1 Cloudcroft Unit*	1,393	34.039	105.424	Yates Pet. #1 Bonita ZC Federal	515
33.052	105.928	Houston and Minerals #1 State 3724*	1,364	34.118	104.879	Mesa Pet. #1 Dunlap State	1,206
33.110	106.890	Exxon #1 Beard Federal*	2,648	34.118	106.511	Skelly Oil Goddard #1	1,023
33.150	106.857	Beard #3 Jornada Del Muerto*	703	34.119	104.917	Yates Pet. #1 Chisum AAW Federal	1,114
33.206	107.209	Getty Oil #2 W. Elephant Butte*	2,261	34.125	106.290	Yates Pet. #1 McCaw Federal	1,286
33.270	107.309	Gartland #1 Brister*	2,596	34.126	105.182	Yates Pet. 1-Y Cowboy Mesa AAY Fed.	741
33.287	107.263	Getty Oil #1 W. Elephant Butte*	2,142	34.127	105.182	Mesa Petroleum #1 Chadwick State	802
33.295	106.009	Houston #1 State L. G. 1453	2,980	34.131	104.886	Yates Pet. #1 Chisum YN State	1,016
33.302	105.134	Stanolind #1 Picacho	702	34.219	105.341	Rault Pot #1 Taylor State	659
33.352	105.141	National Expl #1 Picacho	664	34.238	105.094	Yates #1 Inexco YP State	1,016
33.373	105.015	Sanders Pet. #1 Koko State	902	34.261	107.461	Transocean #1 Henderson SFPRR	2,842
33.390	105.043	Sanders Pet. #1 Tecolote State	876	34.383	106.498	Laing Sanchez #2A	224
33.399	106.931	Sun Victoria L and C #1	1,815	34.513	107.693	L Mitchell Red Lake #1	1,216
33.403	106.956	Sun Victoria L and C #2	1,923	34.520	107.194	White and Mangels #1 State	61
33.465	105.432	Yates Pet. 1 Munoz Canyon AAN Fed.	781	34.528	107.517	Ohio McDonald State #1	605
33.560	104.894	Texam O and G #1 Boyle	753	34.531	107.483	Transocean #1 Major SFPRR	1,397
33.683	104.885	Yates Pet. #1 White Flat LO State*	1,608	34.576	107.364	Spanel and Heinze #1-9608 Santa Fe	1,438
33.754	105.992	Standard Oil #1 Heard Federal	2,348	33.951	106.296	Reeves Bros #1 Morrison Federal*	887
33.777	105.021	Yates Pet. #1 Macho Uno ABL Fed.*	997	33.769	105.927	Dogger #1*	379
33.788	105.67	Kincheloe Dalton #1 Arnold Federal*	1,079	33.765	105.324	Rault Pot #1 Cactus State*	719
33.856	106.393	Sun Oil CO Bingham State #1*	801	33.268	107.362	Gartland Gainer #1*	1,976
33.894	105.236	Elliott #1-10 Federal	688	33.197	107.266	Summit Mims #A1*	1,877
33.919	106.326	Lockhart Lockhart #3*	795	33.215	106.937	Gulf Oil Corp #1 K Sierra State*	2,382

*Wells used in gravity models.

New Mexico, and extends southwest across the rift. This model ties with C–C′ at 130 km. The upper-crustal structure is unconstrained by well data. Constraints are provided by its ties to the northern model, the seismic line of Jaksha (1982), and the information provided by geologic maps. Beginning on the west, the Mogollon-Datil volcanic field extends to 200 km and has been analyzed by Schneider and Keller (this volume). From 200 to 300 km, the profile extends across the Palomas and Jornada del Muerto Basins. The model terminates in the Sierra Oscura on the east. The Palomas Basin contains only a small amount of fill where it is crossed by this gravity profile. Just west of the Sierra Oscura, a prominent gravity low indicates the presence of a graben with a high-angle master fault on its eastern side. This deep feature was first recognized by Healey et al. (1978) and lies in the region of the northern Jornada del Muerto. It could represent a pre-rift feature, but from the isopach maps of Kottlowski (1963) this basin could contain only about 1,400 m of Paleozoic and 600 m of Cretaceous strata; thus it is probably filled mostly with Cenozoic and younger sediments. Here we will refer to this feature as the Oscura Basin because it is located at the foot of the Sierra Oscura. Further geological and geophysical studies are needed to confirm the interpretation of this feature as a rift basin.

Model C–C′ in Figure 8 has the best control of rift-basin

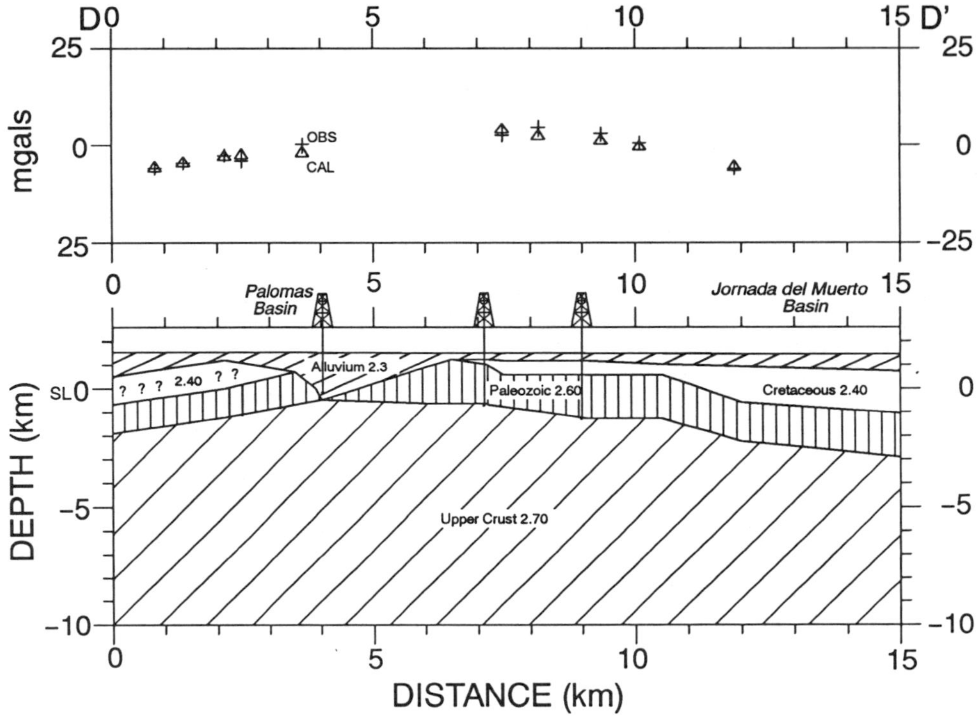

Figure 9. Gravity model for Texaco seismic profile (D–D′; Fig. 1). The wells indicate the position of the well control and the number of wells providing the control. Zero depth in the model represents sea level. Calculated anomalies are represented by triangles. The pluses are the observed Bouguer anomalies. Numbers indicate density values (g/cc).

structure provided by wells and a geologic cross section that extends between the distances of 50 and 110 km on the model (Lozinsky, 1987). In the western portion of this model, the volcanics of the Mogollon-Datil volcanic field dip under the Palomas Basin. The Palomas Basin has a half-graben geometry with a low-angle master fault on its east side. The next feature along the model is the Jornada del Muerto Basin. This basin has a synclinal shape overprinted by normal faults on the west side. Farther to the east, the model depicts the San Andres Mountains and the Tularosa Basin. The Tularosa Basin is associated with a nearly symmetric horst and graben-block structure. The graben blocks have slight dips to the west. On the east, constraints on the thickness of the Paleozoic section are provided by wells drilled to the Precambrian in the Sacramento Mountains.

REFLECTION SEISMIC AND GRAVITY PROFILES

Two seismic reflection profiles were used in this study. These profiles were provided by the Texaco and Exxon Corporations. Gravity models were also generated along both seismic profiles. The Texaco profile (D–D′; Fig. 1) is a 13-km-long profile oriented perpendicular to strike of the Palomas and Engle Basins. The seismic data are migrated and are of good quality. The Exxon profile (E–E′; Fig. 1) is 42 km long and oriented approximately parallel to the San Andres Mountains. This profile crosses the Jornada del Muerto Basin and Love Ranch basin. Vibroseis seismic sources were employed using sweeps from 9 to 36 Hz. The data were recorded with a 96-channel recording system using a 67-m group interval and a 4-ms sample rate (Keller et al., 1986). These parameters resulted in 48-fold coverage for the profile. The data in this profile are also of good quality.

The Texaco seismic profile and the corresponding gravity model in Figures 9 and 10 are constrained by three wells drilled within 1.6 km of the profile. Drilling reports were used to tie the wells to the seismic profile and gravity model. The tie was made to the seismic data by assuming the velocity of the Cretaceous rocks is 2.5 km/s and the velocity of the Paleozoic rocks is 4.5 km/s.

The Texaco seismic section has been interpreted (Fig. 10B) to show a low-angle, listric-normal fault complex. This fault dips westward forming the eastern margin of the Palomas

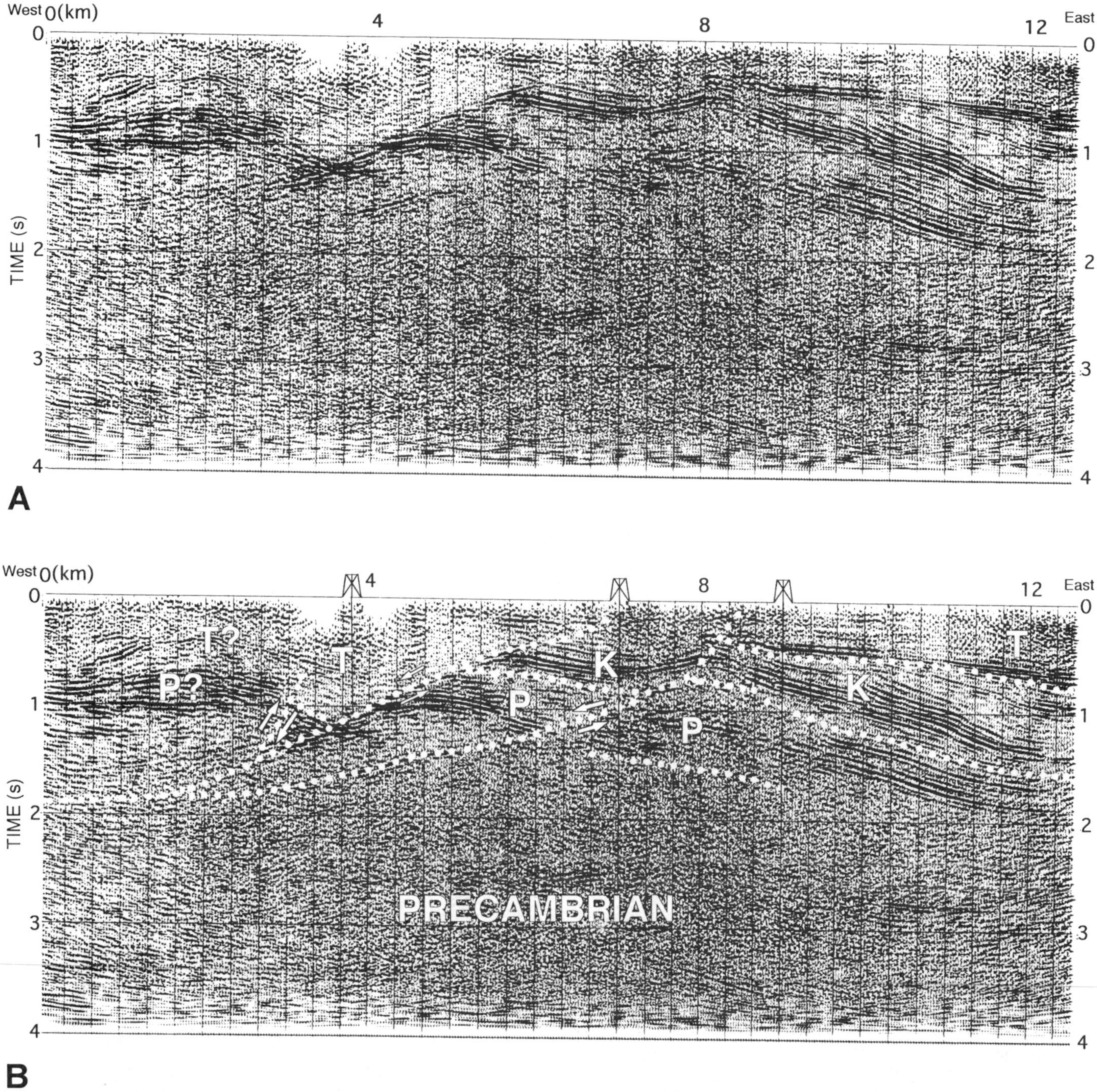

Figure 10. A, Uninterpreted Texaco seismic reflection profile (D–D′; Fig. 1). B, Interpreted Texaco seismic reflection profile (D–D′; Fig. 1). P, Paleozoic; K, Cretaceous; T, Tertiary.

Basin. The Cretaceous and Paleozoic reflectors offset by faulting also show evidence of drag folding along the faults. An unusual feature of the normal fault complex is a fault-bounded wedge of Paleozoic sediments at the base of the Palomas Basin. This wedge is not consistent with a simple pattern of normal faulting and could be due to oblique fault movements or pre-rift structures.

This section also shows evidence of erosion and possible tectonic inversion before rifting. Between 7 and 13 km, the top of the Cretaceous strata is marked by a well-defined unconformity. Between 0 and 7 km, the Cretaceous is interpreted to be absent with Tertiary rocks lying directly on Paleozoic strata. Between 7 km and 13 km the top of the Paleozoic is at 1.0 to 1.5 s travel time. The base of the Paleozoic where it is known from well data is at 1.5 to 1.75 s travel time. The interpreted structure is supported by gravity modeling (Fig. 9).

The Exxon profile stratigraphy (Fig. 11B) is based on picks provided by Exxon that are probably tied through other seismic profiles to the Exxon #1 Prisor well (Keller et al., 1986). The Cretaceous and Paleozoic sections in this well show the same high reflectivity patterns as the Texaco profile. The Exxon profile is situated along the high-gravity gradient along the eastern margin of the Love Ranch basin. The three-dimensional geometry of the gravity field along this profile makes 2.5-dimensional gravity modeling (Fig. 12) of questionable value. This profile crosses from the Love Ranch basin in the north onto the Rio Grande uplift at a distance of about 30 km. The seismic data reveal an overthrust between the Love Ranch basin and the Rio Grande uplift. The uplift and basin are Laramide in age (Seager and Mack, 1986). Aside from these Laramide structures, the Tertiary sedimentary section on this profile is nearly horizontal. The highly reflective packages in the section correspond to the Cretaceous and lower Paleozoic sections at 1.1 s and 1.75 s, respectively. This section is an example of Laramide tectonics within the physiographic boundaries of the rift.

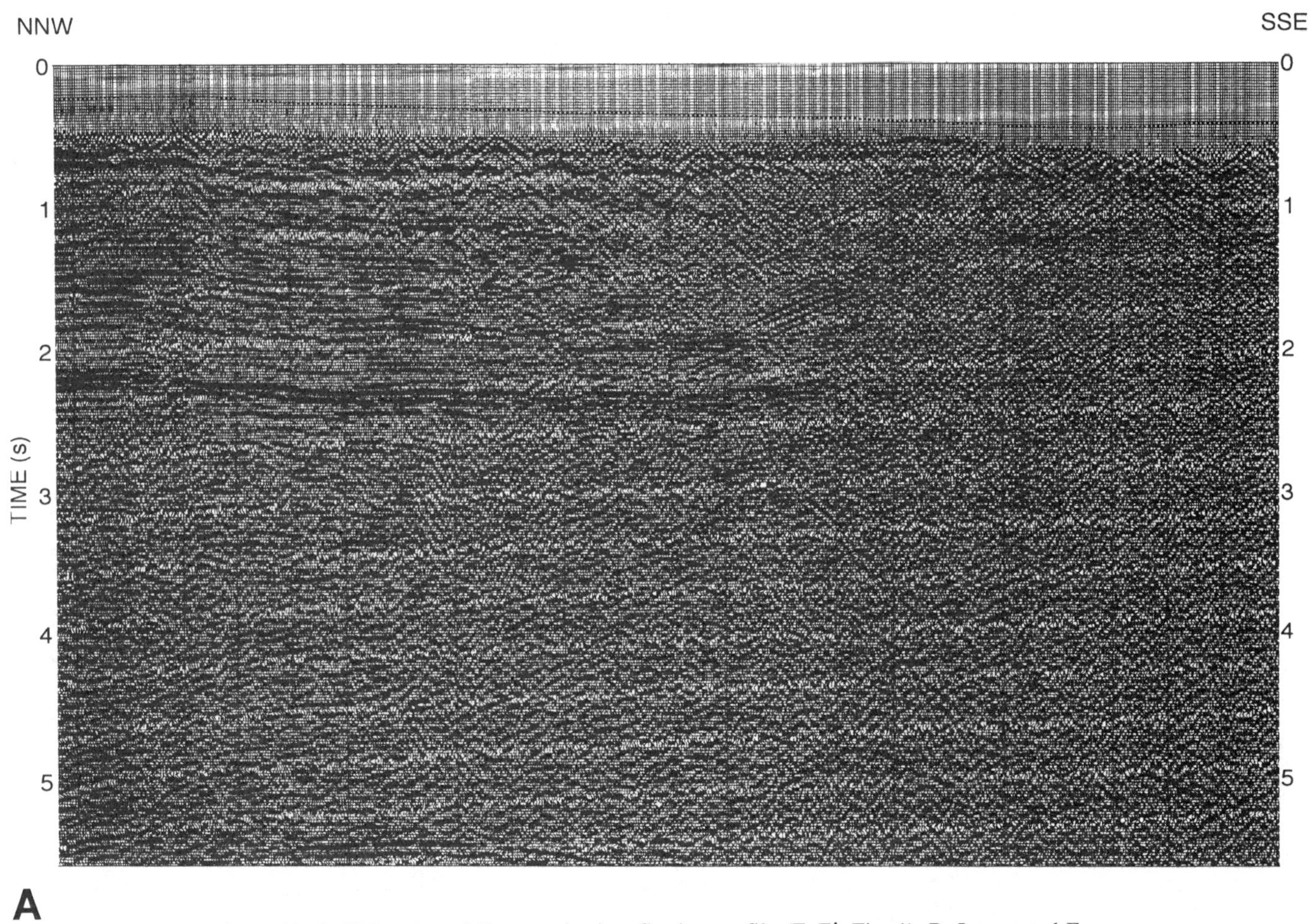

Figure 11. A, Uninterpreted Exxon seismic reflection profile (E–E′; Fig. 1). B, Interpreted Exxon seismic reflection profile (E–E′; Fig. 1). Tlr, Love Ranch Formation; Q, Quaternary; K, Cretaceous; Permian (unlabeled), Pennsylvanian (unlabeled); S, Silurian; P€, Precambrian.

CONCLUSIONS

Several conclusions may be drawn from this study. The crust along the axis of the rift is 28 to 33 km thick. Thickness increases to the north as indicated by the low-pass filtered map (Fig. 4), the regional gravity models, and the seismic data. The Jornada del Muerto, Tularosa, and Palomas Basins are approximately 80 to 150 km in length. These basins appear to be physiographically simple but the high-pass filtered gravity map reveals that the basins are complex, like the rift basins of Rosendahl (1987). Each basin is composed of a complicated set of half-graben and horst blocks as illustrated in Figure 13. The positions of high-angle faults, as deduced from the presence of sharp gravity gradients, usually bound the mountain ranges (Fig. 13) in this part of the rift.

The Palomas Basin has a half-graben structure controlled by a low-angle listric normal fault on the east side (Figs. 7, 8, and 9). The Tularosa Basin grabens have developed into slightly asymmetric full grabens (Fig. 8). The proposed Oscura Basin is also slightly asymmetric. The Exxon seismic reflection profile illustrates the influence of the Laramide orogeny on the structure of the rift.

ACKNOWLEDGMENTS

This study was partially supported by U.S. Air Force Office of Scientific Research Contract F49620-92-J-0438. Ron Broadhead and Richard Lozinsky were very helpful in providing well data and in sharing their geologic insight. Helpful reviews were provided by John Schlue and Allan Sanford.

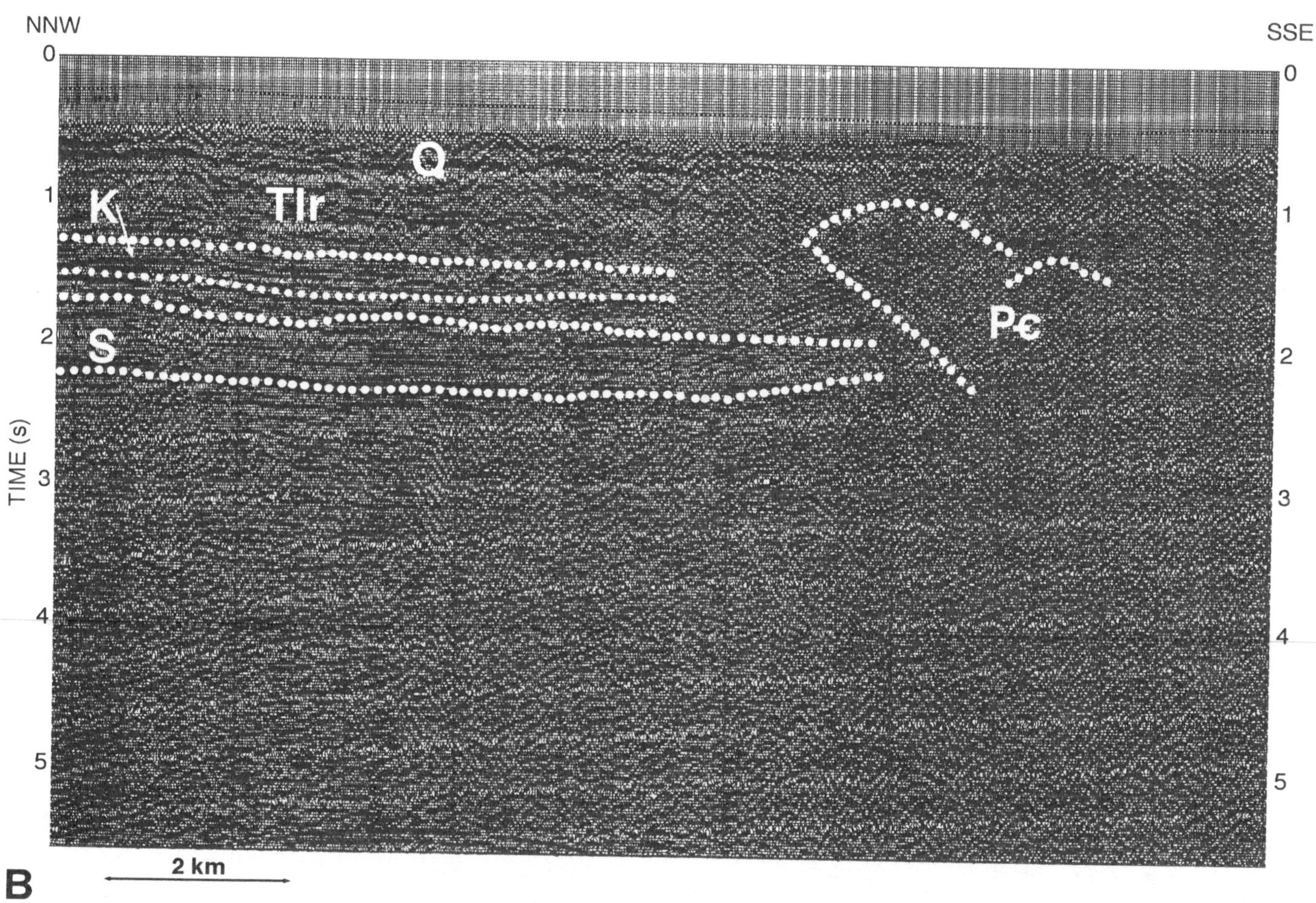

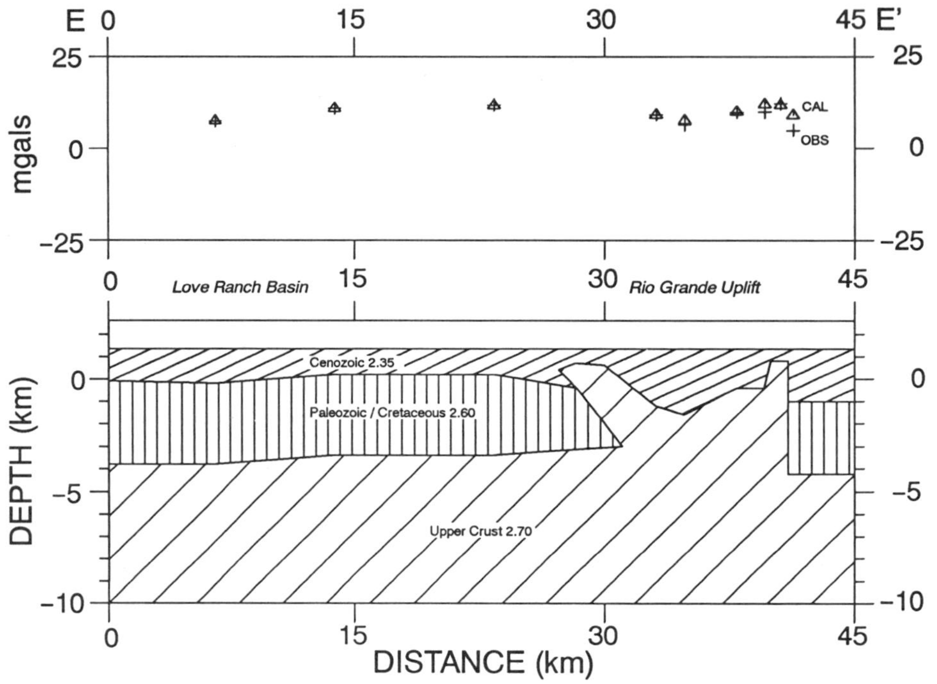

Figure 12. Gravity model for Exxon seismic profile (E–E′, Fig. 1). Zero depth in the model represents sea level. Calculated anomalies are represented by triangles. The pluses are the observed Bouguer anomalies.

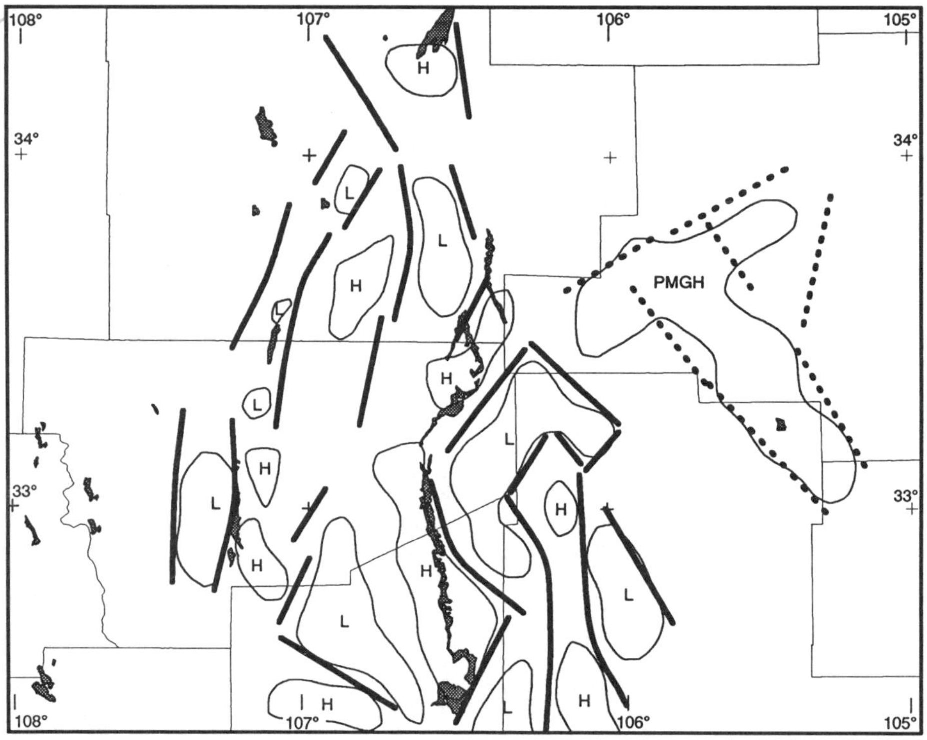

Figure 13. Map showing the positions of graben and horst blocks in the south-central Rio Grande rift. The closed outlines represent isolated gravity highs (H) and gravity lows (L) associated with horsts and grabens, respectively. The bold solid lines are faults interpreted from high gravity gradients and the gravity models. Also shown is the Pajarito Mountain gravity high (PMGH), and faults (dashed bold lines) associated with it. The shaded areas are Precambrian outcrops.

REFERENCES CITED

Adams, D. C., Gridley, J. M.,and Keller, G. R., 1991, Analysis of gravity data in the northern midcontinent region through filtering with Hartley Transforms: EOS Transactions of the American Geophysical Union, v. 72, p. 430.

Ander, M. F., 1981, Geophysical study of the crust and upper mantle beneath the central Rio Grande rift and adjacent Great Plains and Colorado Plateau: Los Alamos National Laboratory Report, LA-8676-T, 218 p.

Bowsher, A. L., 1991, Some effects of Precambrian basement on development of the Sacramento Mountains: New Mexico Geological Society, 42nd Annual Field Conference, Guidebook, p. 81–89.

Cady, J. W.,1980, Calculation of gravity and magnetic anomalies of finite-length right polygonal prisms: Geophysics, v. 45, p. 1507–1512.

Chapin, C. E., 1971, The Rio Grande rift; Part I, Modifications and additions: New Mexico Geological Society, 22nd Field Conference, Guidebook, p. 191–201.

Cook, F. A., Decker, E. R., and Smithson, S. B., 1978, Preliminary transient heat flow model of the Rio Grande rift in southern New Mexico: Earth and Planetary Science Letters, v. 40, p. 316–326.

Cordell, L., 1978, Regional geophysical setting of the Rio Grande rift: Geological Society of America Bulletin, v. 89, p. 1073–1090.

Cordell, L., 1982, Extension in the Rio Grande rift: Journal of Geophysical Research, v. 87, p. 8561–8569.

Cordell, L., Keller, G. R., and Hildenbrand, T. G., 1982, Bouguer gravity map of the Rio Grande rift: U.S. Geological Survey, Geophysical Investigations Map GP-949, scale 1:1,000,000.

Daggett, P. H., Keller, G. R., Morgan, P., and Wen, C. L., 1986, Structure of the southern Rio Grande rift from gravity interpretation: Journal of Geophysical Research, v. 91, p. 6157–6167.

Dobrin, M. B., and Savit, C. H., 1988, Introduction to geophysical prospecting (fourth edition): McGraw-Hill, 867 p.

Elston, W. E., 1976, Tectonic significance of mid-Tertiary volcanism in the Basin and Range province: A critical review with special reference to New Mexico: New Mexico Geological Society, Special Publication, no. 5, p. 93–102.

Grambling, J. A., Williams, M. L., and Mawer, C. K., 1988, Proterozoic tectonic assembly of New Mexico: Geology, v. 16, p. 724–727.

Harder, S. H., Keller, G. R., Daggett, P. H., and Sinno, Y. A., 1986, Cenozoic fill thickness estimates from P-wave delays in the Jornada del Muerto and Palomas Basins: New Mexico Geological Society, 37th Field Conference, Guidebook, p. 135–138.

Healey, D. L., Wahl, R. R., and Currey, F. E., 1978, Gravity survey of the Tularosa valley and adjacent areas, New Mexico: U.S. Geological Survey Open-File Report 78-309, 37 p.

Flawn, P. T., 1956, Basement rocks of Texas and southeast New Mexico: University of Texas, Bureau of Economic Geology, no. 5605, 259 p.

Jaksha, L. H., 1982, Reconnaissance seismic refraction-reflection surveys in southwestern New Mexico: Geological Society of America Bulletin, Part I, v. 90, p. 8–10.

Keller, G. R., and Cordell, L., 1984, Bouguer gravity anomaly map of New Mexico: National Oceanic and Atmospheric Administration, National Geophysical Data Center Map, scale 1:500,000.

Keller, G. R.,and Cordell, L., Davis, G. H., Peeples, W. J., and White, G., 1984, A geophysical study of the San Luis Basin: New Mexico Geological Society, 35th Field Conference, Guidebook, p. 51–47.

Keller, G. R., Seager, W. R., and Thompson, S., 1986, A seismic-reflection study of part of the southern Jornada del Muerto: New Mexico Geological Society, 37th Field Conference, Guidebook, v. 37, p. 139–142.

Keller, G. R., Morgan, P., and Seager, W. R., 1990, Crustal structure, gravity anomalies and heat flow in the southern Rio Grande rift and their relationship to extensional tectonics: Tectonophysics, v. 174, p. 21–37.

Kluth, C. F., and Coney, P. J., 1981, Plate tectonics of the Ancestral Rocky Mountains: Geology, v. 9, p. 10–15.

Kottlowski, F. E., 1963, Paleozoic and Mesozoic strata of southwestern and south-central New Mexico: New Mexico State Bureau of Mines and Mineral Resources Bulletin 79, 100 p.

Lozinsky, R. P., 1987, Cross section across the Jornada del Muerto, Engle, and northern Palomas Basins, south-central New Mexico: New Mexico Geology, p. 55–57.

Olsen, K. H., Keller, G. R., and Stewart, J. N., 1979, Crustal structure along the Rio Grande rift from seismic refraction profiles, *in* Riecker, R. E., ed., Rio Grande rift: Tectonics and magmatism, American Geophysical Union Special Publication, p. 127–143.

Ramberg, I. B., Cook, F. A., and Smithson, S. B., 1978, Structure of the Rio Grande rift in southern New Mexico and west Texas based on gravity interpretation: Geological Society of America Bulletin, v. 89, p. 107–123.

Roberts, D. G., Adams, D. C., and Keller, G. R., 1991, A geophysical analysis of crustal structure in the Ruidoso area: New Mexico Geological Society, 42nd Field Conference, Guidebook, p. 191–197.

Rosendahl, B. R., 1987, Architecture of continental rifts with special reference to east Africa: Annual Reviews of Earth and Planetary Sciences, v. 15, p. 445–503.

Schneider, R. V., 1990, An integrated geophysical and geological investigation of the transition zone between the Colorado Plateau, Rio Grande rift, and Basin and Range provinces: Arizona and New Mexico [Ph.D. thesis]: University of Texas at El Paso, 204 p.

Seager, W. R., and Mack, G. H., 1986, Laramide paleotectonics of southern New Mexico: American Association of Petroleum Geologists Memoir 41, p. 669–685.

Seager, W. R., and Morgan, P., 1979, Rio Grande rift in southern New Mexico, west Texas, and northern Chihuahua, *in* Reicker, R. E., ed., Rio Grande rift: Tectonics and magmatism: American Geophysical Union Special Publication, p. 87–106.

Seager, W. R., Shafiquallah, M., Hawley, J. W., and Marvin, R. F., 1984, New K-Ar dates from basalts and the evolution of the southern Rio Grande rift: Geological Society of America Bulletin, v. 95, p. 89–99.

Sinno, Y. A., Daggett, P. H., Keller, G. R., Morgan, P., and Harder, S. H., 1986, Crustal structure of the southern Rio Grande rift determined from seismic refraction profiling: Journal of Geophysical Research, v. 91, p. 6143–6156.

Stewart, S. W., and Pakiser, L. C., 1962, Crustal structure in eastern New Mexico interpreted from the GNOME explosion: Bulletin of the Seismological Society of America, v. 52, p. 1017–1030.

Suleiman, A. S., and Keller, G. R., 1985, A geophysical study of basement structure in northeastern New Mexico: New Mexico Geological Society, 36th Field Conference, Guidebook, p. 153–159.

Telford, W. W., Goldart, L. P., Sheriff, R. E., and Keys, D. A., 1984, Applied geophysics: Cambridge University Press, 860 p.

Toppozada, T. R., and Sanford, A. R., 1976, Crustal structure in central New Mexico interpreted from the GASBUGGY explosion: Bulletin of the Seismological Society of America, v. 66, p. 877–886.

Manuscript Accepted by the Society October 13, 1993

Printed in U.S.A.

Geological Society of America
Special Paper 291
1994

Late Pliocene and early Pleistocene sedimentation as influenced by intrabasinal faulting, southern Rio Grande rift

Greg H. Mack
Department of Geological Sciences, New Mexico State University, Las Cruces, New Mexico 88003
W. Calvin James
Department of Geological Sciences, Ohio State University, Columbus, Ohio 43210
Stephen L. Salyards
Department of Earth and Space Sciences, University of California at Los Angeles, Los Angeles, California 90024

ABSTRACT

In the Rio Grande rift of southern New Mexico, the intrabasinal East Robledo fault in the Mesilla basin and the Jornada fault in the Jornada del Muerto basin experienced hundreds of meters to kilometers of offset during late Miocene to early Pliocene time and tens of meters of offset since middle Pleistocene. Late Pliocene and early Pleistocene activity on the faults is assessed by comparing sedimentological characteristics of the Camp Rice Formation, which is correlated by reversal magnetostratigraphy, on either side of the faults.

The Jornada fault is interpreted to have been inactive from approximately 3.4 to 2.5 Ma, because Gauss-age fluvial strata at Rincon Arroyo, located on the hanging wall, have similar sediment accumulation rates, degree of development of calcic paleosols, and relative abundance of fluvial channel lithofacies as coeval footwall strata at Cedar Hill and Lucero Arroyo. In contrast, syndepositional movement on the Jornada fault along the northern flank of the Dona Ana Mountains from 2.5 to 0.7 Ma is suggested by a condensed Matuyama interval characterized by mature stage III and IV calcic paleosols at Lucero Arroyo.

The northern segment of the East Robledo fault was active during Gauss time, based on an abundance of relatively thick (59 m), fluvial-channel deposits in the hanging-wall section at Northeast Robledo. The southern segment of the East Robledo fault, however, was probably not active during most of Gauss time, because of the presence on the footwall of a thick (50 m) section of fluvial strata of Gauss age at Picacho Mountain. Major movement on the East Robledo fault near the end of Gauss time terminated sedimentation at Picacho Mountain and in the Corralitos basin, abandoning the upper La Mesa geomorphic surface.

INTRODUCTION

Crustal extension responsible for the modern distribution of uplifts and basins in the southern Rio Grande rift began in late Miocene or early Pliocene time and continues with decreased intensity to the present day (Seager et al., 1984). The most continuous sedimentary record of this phase of extension in the vicinity of Las Cruces, New Mexico, is the Plio-Pleistocene Camp Rice Formation, which is present in four contiguous basins all of which were traversed at one time or another by the Ancestral Rio Grande (Fig. 1). Two of the basins are bisected by intrabasinal faults, the East Robledo

Mack, G. H., James, W. C., and Salyards, S. L., 1994, Late Pliocene and early Pleistocene sedimentation as influenced by intrabasinal faulting, southern Rio Grande rift, *in* Keller, G. R., and Cather, S. M., eds., Basins of the Rio Grande Rift: Structure, Stratigraphy, and Tectonic Setting: Boulder, Colorado, Geological Society of America Special Paper 291.

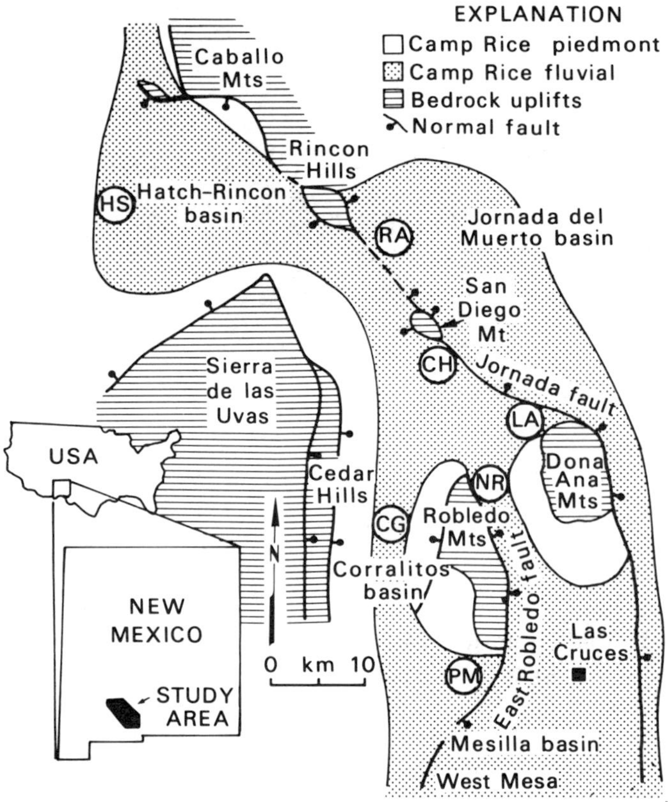

Figure 1. Generalized geologic map of the Rio Grande rift in south-central New Mexico, showing the location of measured sections of the Camp Rice Formation. RA, Rincon Arroyo; CH, Cedar Hill; LA, Lucero Arroyo; NR, Northeast Robledo; CG, Corralitos Graben; PM, Picacho Mountain; HS, Hatch Siphon.

fault in the Mesilla basin and the Jornada fault in the Jornada del Muerto basin (Fig. 1). Seager et al. (1984) documented late Miocene to early Pliocene and post-middle Pleistocene movement on these and associated faults. Less well understood is whether these faults were active in late Pliocene and early Pleistocene time during deposition of the Camp Rice Formation. This question can now be addressed because of the ability to correlate Camp Rice strata within and among sedimentary basins by means of high-resolution reversal magnetostratigraphy (Mack et al., 1993). The possibility of syndepositional movement on the intrabasinal faults can be evaluated by comparing sedimentological characteristics of coeval strata on the hanging wall and footwall of the faults. The criteria used in the comparison are (1) sediment accumulation rates, (2) cumulative thickness of fluvial channel sediment, (3) paleocurrent directions of fluvial-channel deposits, and (4) number and degree of development of calcic paleosols. The results of this study partially close a significant gap in our knowledge of the history of the Jornada and East Robledo faults, and aid in understanding the relationships between sedimentation and intrabasinal tectonic activity in rift basins.

STRATIGRAPHY

The Camp Rice Formation in the study area ranges in exposed thickness from 20 m at Corralitos Graben to 130 m at Northeast Robledo and can be separated into piedmont and axial-fluvial lithofacies (Fig. 1). With the exception of the basal 5 m of the Corralitos Graben section and the upper 71 m of the Northeast Robledo section, all of the strata considered in this study represent axial-fluvial lithofacies, which includes fluvial-channel and flood-plain deposits. Fluvial-channel sediment consists of trough cross-bedded, pebbly, medium-grained sand and sandstone, whereas beds of silt to fine sand and mudstone, many of which display calcic paleosols, constitute flood-plain deposits (Mack and Seager, 1990; Mack and James, 1992). Piedmont lithofacies at Corralitos Graben and Northeast Robledo are composed of a pebble, cobble, and boulder conglomerate that was deposited on alluvial fans (Mack and Seager, 1990).

Deposition of the Camp Rice Formation ended when the Ancestral Rio Grande and its tributaries began to entrench the basins (Hawley and Kottlowski, 1969; Hawley, 1981). The constructional top of the formation locally exists as a relatively flat geomorphic surface underlain by a thick calcic paleosol (Ruhe, 1964; Gile et al., 1966, 1981). Where this surface is developed upon axial-fluvial strata it is called the lower La Mesa surface, such as at the Rincon Arroyo and Lucero Arroyo sections (Figs. 1 and 2). Where the constructional surface is developed on piedmont lithofacies, as it is at the Northeast Robledo section, it is called the Jornada I surface (Figs. 1 and 3). Deposition ceased on the upper La Mesa surface, which is present at the Picacho Mountain and Corralitos Graben sections, long before cessation of deposition on the lower La Mesa surface, as a result of movement on the East Robledo fault (Seager et al., 1984). Abandonment of the upper La Mesa surface occurred no later than 1.2 Ma, based on a K-Ar date of a basalt that flowed onto the geomorphic surface south of the study area (Seager et al., 1984).

The age of the Camp Rice Formation within and south of the study area, as well as the age of the coeval Palomas Formation in the Palomas basin immediately north of the study area, are bracketed between 3.4 and 0.7 Ma by a combination of radiometric dates of basalts, tephrochronology of ash beds, vertebrate fauna, and reversal magnetostratigraphy (Strain, 1966; Hawley et al., 1969; Reynolds and Larsen, 1972; Bachman and Mehnert, 1978; Tedford, 1981; Kortemeier, 1982; Seager et al., 1984; Lozinsky and Hawley, 1986a, b; Lucas and Oakes, 1986; Repenning and May, 1986; Vanderhill, 1986; Mack et al., 1993). Polarity reversals and correlations shown in Figures 2 and 3 are based on Mack et al. (1993) and the reader is referred to this work for the details of the paleomagnetic record. The six stratigraphic sections in Figures 2 and 3 are separated into Gauss and Matuyama intervals based on the assumptions that (1) the lower La Mesa and Jornada I surfaces are approximately coeval and are near the

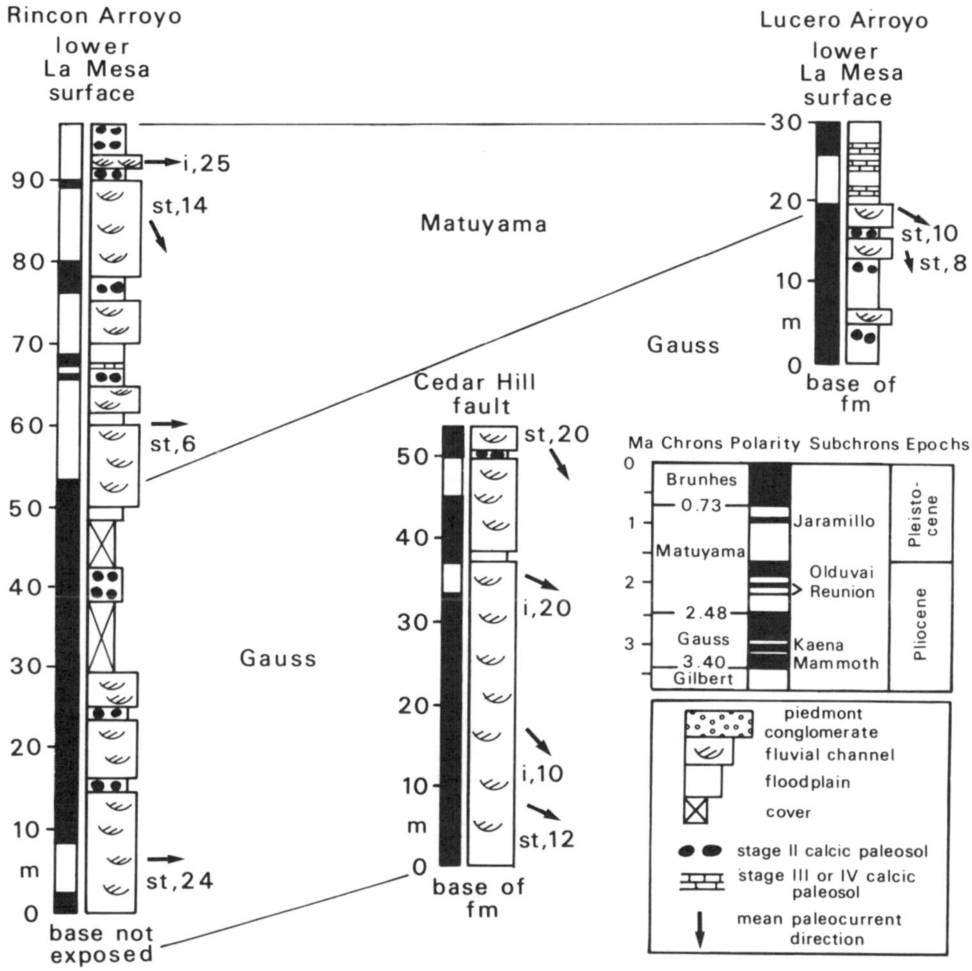

Figure 2. Lithostratigraphy and reversal magnetostratigraphy of the Camp Rice Formation adjacent to the Jornada fault. Reversal magnetostratigraphy and correlations are from Mack et al. (1993), and polarity reversal time scale is from Mankinen and Dalrymple (1979). Arrows represent mean paleocurrent direction of fluvial channel pebbly sand/sandstone; number by each arrow is the number of measurements of trough cross-beds (st), following the method of DeCelles et al. (1983), pebble-cobble imbrication (i), or parting lineation (pl). Thickness is in meters.

Matuyama-Brunhes boundary, (2) the upper La Mesa surface is near the Matuyama-Gauss boundary, and (3) the base of the Camp Rice Formation nearly coincides with the Gauss-Gilbert boundary (Mack et al., 1993). Gauss strata at Lucero Arroyo are interpreted to represent the interval from the top of the Kaena to the Gauss-Matuyama boundary, because very closely spaced sampling revealed no reversed intervals and there is no evidence that the section is condensed.

SEDIMENTOLOGICAL EVIDENCE OF SYNDEPOSITIONAL FAULTING

Syndepositional movement of the Jornada and East Robledo faults should be evident by comparing the sedimentological characteristics of fluvial strata that occupy opposite sides of the faults. Net displacement large enough to place the footwall above the level of flooding would, in effect, terminate net sedimentation on the footwall. Even lower-magnitude fault movements should result in significantly different depositional regimes on either side of the fault. Minor fault movement should be recognizable by differences across the fault in sediment accumulation rates, the relative abundance of fluvial channel deposits, degree of development of paleosols, and fluvial-channel paleocurrents.

Sediment accumulation rates are calculated for Gauss and Matuyama intervals of the Camp Rice Formation by dividing the thickness of strata within the respective chron by the duration of the chron (Figs. 2 and 3). Because the stratigraphic sections are not thicker than 130 m and the Camp Rice Formation was never buried by more than a few meters of younger sediment, the stratigraphic sections were not decompacted. Sediment accumulation rates are in most cases minimum val-

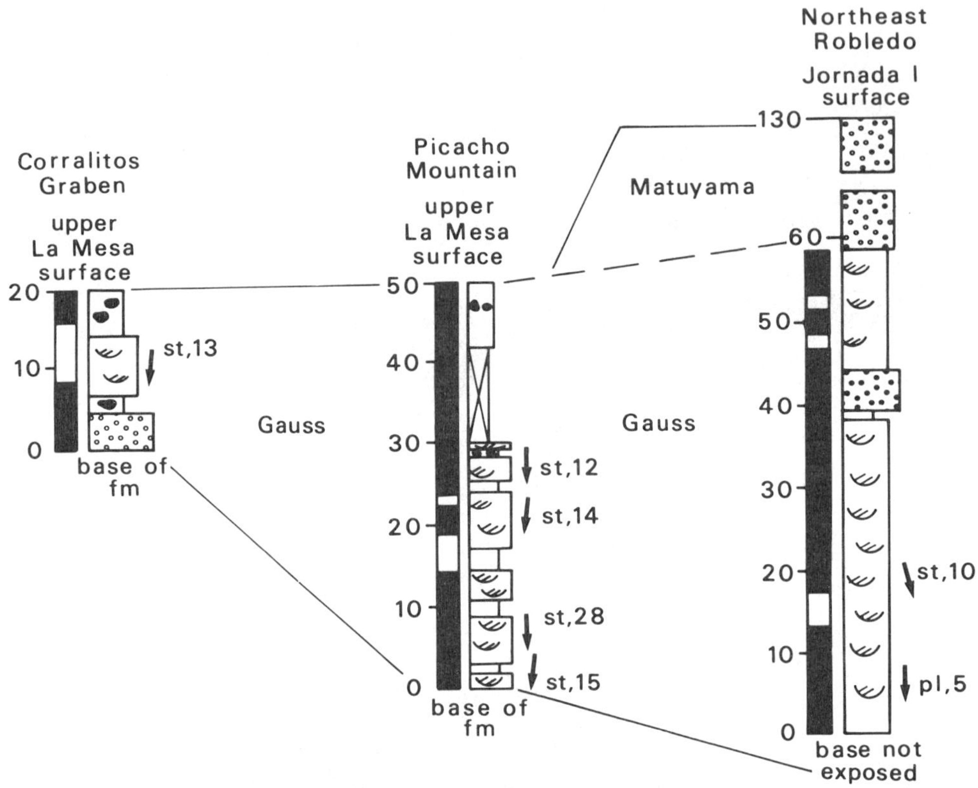

Figure 3. Lithostratigraphy and reversal magnetostratigraphy, based on data of Mack et al. (1993), of the Camp Rice Formation related to the East Robledo fault. See Figure 2 for polarity reversal time scale and description of symbols. Thickness is in meters.

ues, because chron boundaries are not present in each stratigraphic section. The values are, however, adequate for comparative purposes. Also shown in Figures 4 and 5 are sediment accumulation rates for well-dated fluvial strata that were unaffected by syndepositional movement of the Jornada and East Robledo faults. The Hatch Siphon section, which is located in the southwestern part of the Hatch-Rincon basin (Fig. 1), has a nearly complete Gauss interval (Mack et al., 1993). The West Mesa sediment accumulation rate on Figure 5 represents the average sediment accumulation rate of the Matuyama interval of three stratigraphic sections located 44 to 55 km south of the study area in the southern part of the Mesilla basin, based on the magnetostratigraphic data of Vanderhill (1986). Sediment accumulation rates for Hatch Siphon and West Mesa are presented for comparative purposes.

Sediment accumulation rates do not provide information on the style of fluvial deposition nor on the degree to which deposition was continuous. These sedimentologic parameters can be assessed by the relative abundance of channel deposits and the degree of development of paleosols. Areas of active basin subsidence should be frequently traversed by fluvial channels, resulting in a relatively high proportion of channel deposits compared to flood-plain deposits (Bridge and Leeder, 1979; Alexander and Leeder, 1987; Leeder and Gawthorpe, 1987). The percentage of the thickness of fluvial-channel strata to total fluvial sediment thickness for the six stratigraphic sections of interest in this study are shown in Figures 4 and 5, along with percentages for Gauss strata at Hatch Siphon and Matuyama strata along the West Mesa, which are presented for comparative purposes.

An indication of the continuity of deposition can be inferred from the number and degree of development of paleosols (Allen, 1974; Leeder, 1975; Bown and Kraus, 1987; Kraus, 1987; Kraus and Bown, 1988). Calcic paleosols are common in fluvial strata of the Camp Rice Formation and are described in detail by Mack and James (1992). Following the classification of Gile et al. (1966), calcic paleosols in flood-plain strata of the Camp Rice Formation have been separated into three morphological stages (Figs. 1, 2, 4, 5). Stage II calcic paleosols are characterized by scattered nodules and tubules of microcrystalline low-Mg calcite separated by noncalcareous parent sediment. Stage III morphology consists of massive beds of micrite with few or no patches of parent sediment, and stage IV morphology is represented by massive carbonate, similar to stage III, overlain by a horizon of laminar micrite. All three types of calcic horizons are commonly overlain by an argillic horizon, unless truncated by channels. Assuming other pedogenic variables are nearly constant, such

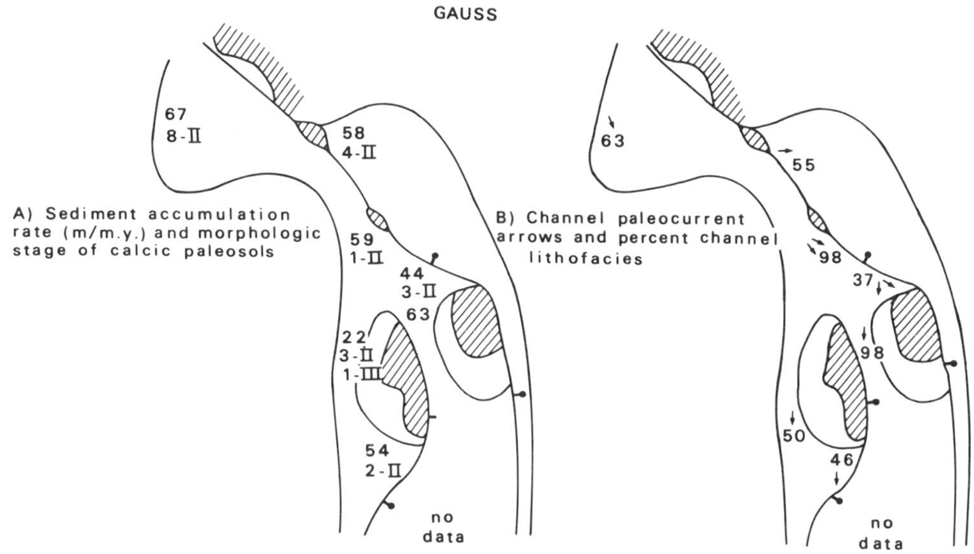

Figure 4. Sedimentologic data of Camp Rice strata deposited during the Gauss Chron (3.4 to 2.5 Ma). A) Estimated sediment accumulation rates in meters per million years shown by Arabic numerals, whereas associated Arabic and Roman numerals represent the number and morphologic stage of development of calcic paleosols. B) Numbers represent the percentage of fluvial-channel lithofacies, and arrows indicate selected channel paleocurrent azimuths. Diagonal pattern represents uplifts.

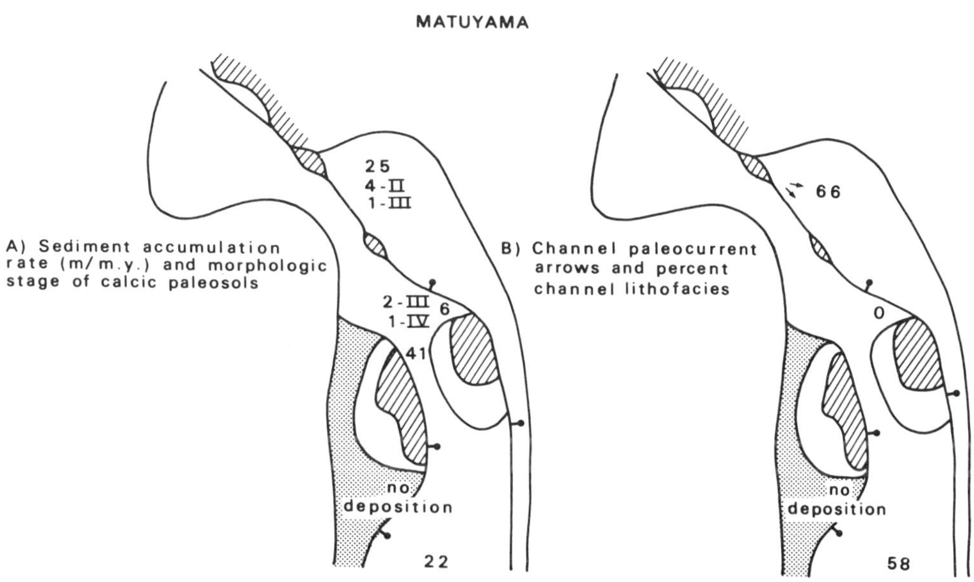

Figure 5. Sedimentologic data of Camp Rice strata deposited during the Matuyama Chron (2.5 to 0.7 Ma). A) Estimated sediment accumulation rates in meters per million years shown by Arabic numerals whereas associated Arabic and Roman numerals represent the number and morphologic stage of development of calcic paleosols. B) Numbers represent the percentage of fluvial-channel lithofacies, and arrows indicate selected channel paleocurrent azimuths. Diagonal pattern represents uplifts.

as composition and grain size of the parent sediment, slope, and vegetation, factors that vary little among Camp Rice paleosols, the morphological stage is directly related to the length of time in which the paleosol developed (Gile et al., 1966, 1981). Stage II calcic palesols require thousands of years to form, stage III calcic paleosols require tens of thousands to hundreds of thousands of years, and stage IV calcic paleosols require several hundred thousand to perhaps a half-million years (Gile et al., 1981). The number of paleosols and their morphological stage thus provide a measure of periods of nondeposition and landscape stability among the Camp Rice sections. These data for the six sections of interest in this study, along with data from Hatch Siphon, are shown in Figures 4 and 5. Paleosol data are not available for the West Mesa sections of Vanderhill (1986).

Finally, both small- and large-scale movement on intrabasinal faults may have influenced paleoflow directions of fluvial channels. Paleocurrents that cross the fault trace at high angles suggest there was no relief across the fault. In contrast, a sharp deviation in paleocurrents may reflect fault movement, although other explanations, such as random avulsions or natural bends in the channel, could also account for this relationship. Mean paleocurrent directions of fluvial channel sediment, based on trough cross-bed, imbrication, and parting lineation data, are shown as arrows on Figures 2, 3, 4, and 5.

HISTORY OF ACTIVITY OF THE INTRABASINAL JORNADA AND EAST ROBLEDO FAULTS

Jornada fault

The Jornada fault is the major boundary fault along the east and north flank of the Dona Ana Mountains and is one of a series of faults that bound San Diego Mountain and the Rincon Hills (Fig. 1; Seager and Hawley, 1973; Seager et al., 1971, 1976). Initial movement on the Jornada and associated faults at San Diego Mountain and the Rincon Hills postdates 9.6 Ma because the faults cut the Rincon Valley Formation, which contains in its upper part the 9.6-Ma Selden basalt (Seager et al., 1984). During late Miocene to early Pliocene time up to a kilometer or more of structural relief was created at San Diego Mountain and the Rincon Hills (Seager, 1975; Seager et al., 1984). By the time through-going drainage of the Ancestral Rio Grande was established in the region, at approximately 3.4 Ma, the relief created by thc preceding pulse of tectonism had been substantially reduced because fluvial deposits of the Camp Rice Formation locally onlap the Rincon Hills, San Diego Mountain, and the Dona Ana Mountains and there were no large alluvial fans along the flanks of the uplifts.

The Jornada fault has also experienced post-middle Pleistocene movement, because it locally offsets and/or deforms the Camp Rice Formation. North of the Dona Ana Mountains the constructional top of the Camp Rice Formation (lower La Mesa surface) is offset by about 45 m. The warped lower La Mesa surface can be traced northwestward from the Dona Ana Mountains to the Cedar Hill stratigraphic section, where the fault cuts Camp Rice strata. Along the north side of San Diego Mountain the lower La Mesa surface is offset approximately 18 m. Post-middle Pleistocene movement on the northern segment of the Jornada fault is also indicated by 10° of tilting of Camp Rice beds at the Rincon Arroyo section and adjacent to the margin of the Rincon Hills, although the absolute amount of displacement cannot be determined.

Activity of the Jornada fault during the period 3.4 to 0.7 Ma can be evaluated by comparing sedimentological characteristics of Camp Rice fluvial strata at Rincon Arroyo, located on the hanging wall, with footwall sections at Cedar Hill and Lucero Arroyo (Figs. 2, 4, 5). The Jornada fault is interpreted to have been inactive during Gauss time. The hanging-wall section of Gauss age at Rincon Arroyo and the roughly coeval footwall section at Cedar Hill have similar estimated sediment accumulation rates that are comparable to the Gauss reference section at Hatch Siphon (Figs. 2 and 4). An abundance of channel sandstones and the presence of relatively immature stage II calcic paleosols at Rincon Arroyo and Cedar Hill also indicate active sedimentation on both sides of the fault (Fig. 4). Furthermore, paleocurrents at Rincon Arroyo and Cedar Hill are at high angles to the trace of the fault, suggesting that there was no topographic impediment to current flow across the fault (Fig. 4). Although the footwall section of late Gauss age at Lucero Arroyo has a lower sediment accumulation rate than Cedar Hill and Rincon Arroyo, the presence at Lucero Arroyo of channel sandstones and stage II calcic paleosols, as well the fact that paleoflow in one of the channels was nearly perpendicular to the fault trace, suggest that the Jornada fault along the northern flank of the Dona Mountains was also inactive or experienced minimal activity.

During Matuyama time, however, there is evidence at the Lucero Arroyo section of syndepositional movement of the Jornada fault along the northern flank of the Dona Ana Mountains. The Matuyama interval at Lucero Arroyo has a sediment accumulation rate that is substantially less than that of the Rincon Arroyo section and of the Matuyama reference sections along the West Mesa (Figs. 2 and 5). That the Lucero Arroyo Matuyama interval is condensed is also indicated by the absence of channel sandstones and the presence of mature stage III and stage IV calcic paleosols (Figs. 2 and 5). Deposition during Matuyama time at Lucero Arroyo was characterized by short periods of flood-plain sedimentation followed by long periods of landscape stability and consequent pedogenesis. These sedimentologic characteristics are consistent with relative isolation of the Lucero Arroyo section on the footwall of the active Dona Ana Mountains segment of the Jornada fault. Note that no strata of Matuyama age are present at Cedar Hill.

The possibility of movement of the northern segment of the Jornada fault (between San Diego Mountain and the Rincon Hills) during Matuyama time is constrained only by the sedimentologic characteristics of the Matuyama interval at Rincon Arroyo (Figs. 2 and 5). Active sedimentation during Matuyama time at Rincon Arroyo, indicated by relatively high sediment accumulation rate, an abundance of channel sandstones, and the predominance of stage II calcic paleosols, could reflect "normal" subsidence and sedimentation rates within the Jornada del Muerto basin independent of movement on the Jornada fault or could reflect elevated subsidence and sedimentation rates on the hanging wall of an active Jornada fault. The absence of a nearby footwall section with which to compare makes it difficult to distinguish between these two alternative hypotheses. However, the fact that paleocurrent directions of two of the channels at Rincon Arroyo are at high angles to the trace of the fault and the sediment accumulation rate at Rincon Arroyo is comparable to those along the West Mesa suggests that the northern segment of the Jornada fault was inactive during Matuyama time.

East Robledo fault

Initial uplift of the northern end of the Robledo Mountains is bracketed between 9.6 and 7.1 Ma (Seager et al., 1984). As was the case at San Diego Mountain and the Rincon Hills, most of the structural relief of the Robledo Mountains, up to 700 m, was created in late Miocene or early Pliocene time (Seager et al., 1984). There has been approximately 100 m of uplift of the Robledo block since late Pliocene time, including a few tens of meters of post-middle Pleistocene offset (Seager et al., 1984).

The northern segment of the East Robledo fault was also active during Gauss time (3.4 to 2.5 Ma), based on active sedimentation at the Northeast Robledo section located on the hanging wall of the fault (Figs. 1, 3, and 4). The presence at Northeast Robledo of axial-fluvial lithofacies within a few hundred meters of the boundary fault was interpreted by Mack and Seager (1990) to indicate intermittent westward tilting of the northern Mesilla basin as a result of movement on the East Robledo fault. In contrast, the southern segment of the East Robledo fault was probably not active during Gauss time. This interpretation is based on the existence of a thick (50 m) section of fluvial strata of Gauss age at Picacho Mountain, which is located on the footwall of the East Robledo fault. Active subsidence and sedimentation at the Picacho Mountain section is indicated by a sediment accumulation rate comparable to that at Northeast Robledo and Hatch Siphon, a relatively high ratio of fluvial channel to total fluvial strata, and the presence of only two stage II calcic paleosols (Figs. 3 and 4). The southern segment of the East Robledo fault appears not to have been an impediment to fluvial sedimentation at Picacho Mountain. Sedimentation also occurred during Gauss time in the Corralitos basin, which at least once was occupied by the Ancestral Rio Grande (Fig. 3). Sedimentation in the Corralitos basin was only half as rapid as at Picacho Mountain, however (Fig. 4).

Deposition at the Picacho Mountain section ended as a result of major movement on the southern segment of the East Robledo fault. This period of movement occurred between 2.92 and 2.48 Ma, because the Picacho Mountain section contains the Kaena subchron but does not appear to cross the Gauss-Matuyama boundary (Fig. 3). This interpretation is consistent with the presence of a 1.2-Ma basalt flow that overlies a thick stage IV calcic paleosol capping the upper La Mesa surface south of the study area (Seager et al., 1984). This relationship suggests that the upper La Mesa surface had been abandoned for hundreds of thousands of years, the time necessary to develop the stage IV paleosol, before eruption of the 1.2-Ma basalt.

The upper La Mesa surface is also present at the Corralitos Graben section, suggesting that termination of deposition in the Corralitos basin was contemporaneous with that at Picacho Mountain. It is possible that at this time the Robledo Mountains and Corralitos basin acted as a single block during movement on the northern segment of the East Robledo fault. The northern segment of the fault continued to be active throughout the Matuyama, as indicated by a thick (71 m) accumulation of piedmont conglomerate beneath the Jornada I surface at the Northeast Robledo section (Fig. 3).

CONCLUSIONS

High-resolution reversal magnetostratigraphy allows a comparison of sedimentological characteristics of the Plio-Pleistocene Camp Rice Formation in the southern Rio Grande rift in order to evaluate syndepositional activity on the intrabasinal Jornada and East Robledo faults. Gauss-age (3.4 to 2.5 Ma) fluvial strata on the hanging wall of the Jornada fault have similar sediment accumulation rates, degree of development of calcic paleosols, and relative abundance of channel lithofacies to coeval footwall strata, suggesting the Jornada fault was inactive. In contrast, activity on the Dona Ana Mountains segment of the Jornada fault during Matuyama time (2.5 to 0.7 Ma) is indicated by a condensed footwall section characterized by mature calcic paleosols. The northern segment of the East Robledo fault was active during Gauss time; however, the southern segment was inactive because a hanging-wall section adjacent to the northern segment of the fault and a footwall section adjacent to the southern segment of the fault have similar high sediment accumulation rates, abundant fluvial channel deposits, and few immature calcic paleosols. Fluvial sedimentation on the footwall block of the southern segment of the East Robledo fault ceased near the end of the Gauss time, whereas fluvial sedimentation continued throughout Matuyama time on the hanging wall.

ACKNOWLEDGMENTS

W. R. Seager, Kathleen Marsaglia, and J. A. Raney read an earlier version of this manuscript and made helpful suggestions for its improvement. This study was funded in part by the New Mexico Bureau of Mines and Mineral Resources and by the Petroleum Research Fund of the American Chemical Society (ACS-PRF 24378-B8-C).

REFERENCES CITED

Alexander, J. A., and Leeder, M. R., 1987, Active tectonic control on alluvial architecture: Society of Economic Paleontologists and Mineralogists, Special Publication 39, p. 243–252.

Allen, J.R.L., 1974, Studies in fluviatile sedimentation: Implications of pedogenic carbonate units, Lower Old Red Sandstone, Anglo-Welsh outcrop: Geological Journal,v. 9, p. 181–208.

Bachman, G. O., and Mehnert, H. H., 1978, New K-Ar dates and late Pliocene to Holocene geomorphic history of the Rio Grande region, New Mexico:

Geological Society of America Bulletin, v. 89, p. 283–292.

Bown, T. M., and Kraus, M. J., 1987, Integration of channel and floodplain suites in aggrading alluvial systems. I. Developmental sequence and lateral relations of lower Eocene alluvial paleosols, Willwood Formation, Bighorn basin, Wyoming: Journal of Sedimentary Petrology, v. 57, p. 587–601.

Bridge, J. S., and Leeder, M. R., 1979, A simulation model of alluvial stratigraphy: Sedimentology, v. 26, p. 617–644.

DeCelles, P. G., Langford, R. P., and Schwartz, R. K., 1983, Two new methods of paleocurrent determination from trough cross-stratification: Journal of Sedimentary Petrology, v. 53, p. 629–642.

Gile, L. H., Peterson, F. F., and Grossman, R. B., 1966, Morphological and genetic sequences of carbonate accumulation in desert soils: Soil Science, v. 101, p. 347–360.

Gile, L. H., Hawley, J. W., and Grossman, R. B., 1981, Soils and geomorphology in the Basin and Range area of southern New Mexico—Guidebook to the Desert Project: New Mexico Bureau of Mines and Mineral Resources Memoir 39, 222 p.

Hawley, J. W., 1981, Pleistocene and Pliocene history of the international boundary area, southern New Mexico: El Paso Geological Society, Geology of the Border–Southern New Mexico–Northern Chihuahua, p. 25–32.

Hawley, J. W., and Kottlowski, F. E., 1969, Quaternary geology of the south-central New Mexico border region: New Mexico Bureau of Mines and Mineral Resources Circular 104, p. 89–104.

Hawley, J. W., Kottlowski, F. E., Seager, W. R., King, W. E., Strain, W. S., and LeMone, D. V., 1969, The Santa Fe Group in the south-central New Mexico border: New Mexico Bureau of Mines and Mineral Resources Circular 104, p. 52–76.

Kortemeier, C. P., 1982, Occurrence of Bishop Ash near Grama, New Mexico: New Mexico Geology, v. 4, p. 22–24.

Kraus, M. J., 1987, Integration of channel and floodplain suites in aggrading alluvial systems, II. Vertical relations of lower Eocene paleosols, Willwood Formation, Bighorn basin, Wyoming: Journal of Sedimentary Petrology, v. 57, p. 602–612.

Kraus, M. J., and Bown, T., 1988, Pedofacies analysis: A new approach to reconstructing ancient fluvial sequences: Geological Society of America Special Paper 216, p. 143–152.

Leeder, M. R., 1975, Pedogenic carbonates and floodplain accretion rates: A quantitative model for alluvial arid-zone lithofacies: Geological Magazine, v. 112, p. 257–270.

Leeder, M. R., and Gawthorpe, R. L., 1987, Sedimentary models for extensional tilt block/half-graben basins: Geological Society of London Special Publication 28, p. 139–152.

Lozinsky, R. P., and Hawley, J. W., 1986a, The Palomas Formation of south-central New Mexico—A formal definition: New Mexico Geology, v. 8, p. 73–78, 82.

Lozinsky, R. P., and Hawley, J. W., 1986b, Upper Cenozoic Palomas Formation of south-central New Mexico: New Mexico Geological Society Guidebook 37, p. 239–247.

Lucas, S. G., and Oakes, W., 1986, Pliocene (Blancan) vertebrates from the Palomas Formation, south-central New Mexico: New Mexico Geological Society Guidebook 37, p. 249–255.

Mack, G. H., and James, W. C., 1992, Calcic paleosols of the Plio-Pleistocene Camp Rice and Palomas Formations, southern Rio Grande rift, U.S.A.: Sedimentary Geology, v. 77, p. 89–109.

Mack, G. H., and Seager, W. R., 1990, Tectonic control on facies distribution of the Camp Rice and Palomas Formations (Pliocene-Pleistocene) in the southern Rio Grande rift: Geological Society of America Bulletin, v. 102, p. 45–53.

Mack, G. H., Salyards, S. L., and James, W. C., 1993, Magnetostratigraphy of the Plio-Pleistocene Camp Rice and Palomas Formations in the Rio Grande rift of southern New Mexico: American Journal of Science, v. 293, p. 49–77.

Mankinen, E. A., and Dalrymple, G. B., 1979, Revised geomagnetic polarity time scale for the interval 0–5 m.y. B.P.: Journal of Geophysical Research, v. 84, p. 615–626.

Repenning, C. A., and May, S. R., 1986, New evidence for the age of lower part of the Palomas Formation, Truth or Consequences, New Mexico: New Mexico Geological Society Guidebook 37, p. 257–259.

Reynolds, R. L., and Larsen, E. E., 1972, Paleomagnetism of Pearlette-like air-fall ash in the midwestern and western United States—A means of correlating Pleistocene deposits: Geological Society of America Abstracts with Programs, v. 4, p. 405.

Ruhe, R. V., 1964, Landscape morphology and alluvial deposits in southern New Mexico: Association of American Geographic Annals, v. 54, p. 147–159.

Seager, W. R., 1975, Cenozoic tectonic evolution of the Las Cruces area, New Mexico: New Mexico Geological Society Guidebook 26, p. 241–250.

Seager, W. R., and Hawley, J. W., 1973, Geology of Rincon Quadrangle, New Mexico: New Mexico Bureau of Mines and Mineral Resources Bulletin 101, 42 p.

Seager, W. R., Hawley, J. W., and Clemons, R. E., 1971, Geology of San Diego Mountain area, Dona Ana County, New Mexico: New Mexico Bureau of Mines and Mineral Resources Bulletin 97, 38 p.

Seager, W. R., Kottlowski, F. E., and Hawley, J. W., 1976, Geology of Dona Ana Mountains, New Mexico: New Mexico Bureau of Mines and Mineral Resources Circular 147, 36 p.

Seager, W. R., Shafiqullah, M., Hawley, J. W., and Marvin, R., 1984, New K-Ar dates from basalts and the evolution of the southern Rio Grande rift: Geological Society of America Bulletin, v. 95, p. 87–99.

Strain, W. S., 1966, Blancan mammalian fauna and Pleistocene formations, Hudspeth County, Texas: Texas Memorial Museum Bulletin 10, 55 p.

Tedford, R. H., 1981, Mammalian biochronology of late Cenozoic basins of New Mexico: Geological Society of America Bulletin, v. 92, p. 1008–1022.

Vanderhill, J. B., 1986, Lithostratigraphy, vertebrate paleontology, and magnetostratigraphy of Plio-Pleistocene sediments in the Mesilla basin, New Mexico [Ph.D. thesis]: Austin, University of Texas, 305 p.

Manuscript Accepted by the Society October 13, 1993

Printed in U.S.A.

Geological Society of America
Special Paper 291
1994

Tertiary and Quaternary tectonics of the Hueco bolson, Trans-Pecos Texas and Chihuahua, Mexico

Edward W. Collins and Jay A. Raney
Bureau of Economic Geology, The University of Texas at Austin, Austin, Texas 78713

ABSTRACT

The Hueco bolson of Trans-Pecos Texas formed in response to Cenozoic extensional tectonism and lies within the southern Rio Grande rift near the poorly defined boundary between the rift and the southern Basin and Range province. The bolson is composed of a northwest subbasin that contains north-striking normal faults and a southeast subbasin that contains northwest-striking normal faults. Cenozoic basin fill is thin (less than 150 to 200 m) on the east and northeast bolson margins and is thick (as much as 2,850 m) in the central bolson and on the west and southwest bolson margins where major normal faults bound a graben that is 15 to 25 km wide. Major faults bounding the graben on the west and southwest have been more active and exhibit greater offset than do boundary faults on the east. This disparity in displacement between the graben margins has resulted in an asymmetric graben. Isopach maps of lower and upper basin fill sequences, differentiated from seismic data, indicate that much of the southeast Hueco bolson subsided more than the northwest Hueco bolson during deposition of the lower basin fill. Thicker upper basin fill within the northwest basin indicates that the basin subsided more in the northwest than in the southeast during deposition of the upper basin fill.

The two major faults that bound the Hueco graben on the west and southwest, the East Franklin Mountains fault and Amargosa fault, respectively, have had the most recent (late Pleistocene–Holocene) surface ruptures. Scarp-slope angles of these faults are commonly steeper than 20°, and middle Pleistocene surficial deposits that contain indurated calcic soils having stage IV to V morphology are offset between 24 and 32 m. Maximum throw on these faults during single surface-rupture events has been between 1.6 and 3 m. Major faults bounding the southeast Hueco graben on the northeast (Campo Grande, Caballo, and an unnamed fault) had their most recent surface ruptures during the late Pleistocene. Scarp-slope angles of these faults are rarely as much as 15° and more commonly between 4 and 7°. Middle Pleistocene surficial deposits that contain indurated calcic soils having a stage IV to V morphology are offset between 1.6 and 24 m. Maximum throw on these faults during single surface-rupture events has been between 0.6 and 2 m.

INTRODUCTION

The Hueco bolson lies within the northern Chihuahuan Desert in El Paso and Hudspeth Counties, Trans-Pecos Texas, and Chihuahua, Mexico (Fig. 1). The bolson, similar to other intermontane basins of central New Mexico, Trans-Pecos Texas, and Chihuahua, Mexico, formed in response to Cenozoic extensional tectonism. Published information on the Tertiary and Quaternary fault tectonics of the Hueco bolson is limited. However, subsurface data, local outcrops, and fault scarps in Quaternary sediments provide a good data base to evaluate the bolson's overall structural framework including

Collins, E. W., and Raney, J. A., 1994, Tertiary and Quaternary tectonics of the Hueco bolson, Trans-Pecos Texas and Chihuahua, Mexico, *in* Keller, G. R., and Cather, S. M., eds., Basins of the Rio Grande Rift: Structure, Stratigraphy, and Tectonic Setting: Boulder, Colorado, Geological Society of America Special Paper 291.

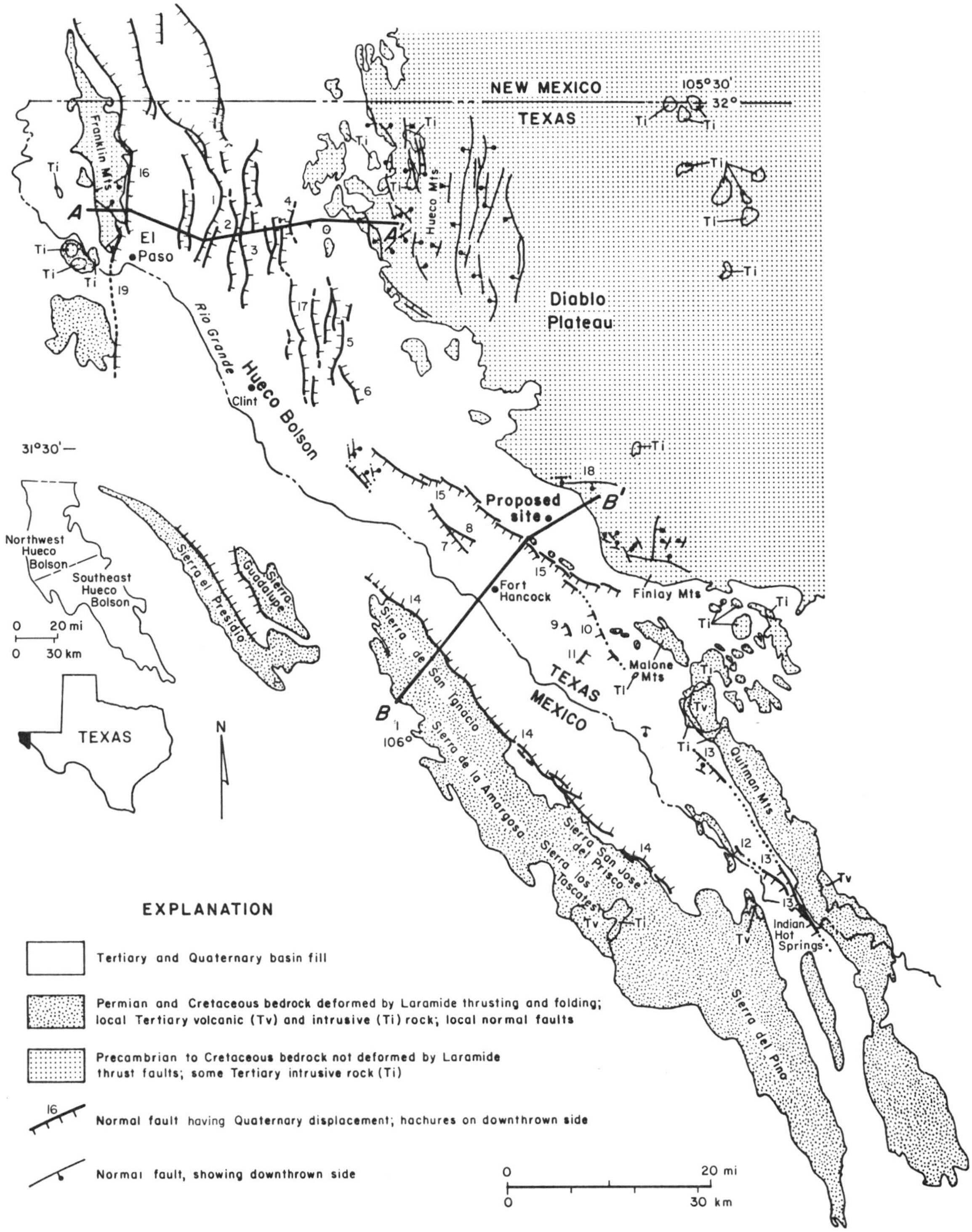

Figure 1. Regional map of surface faults, Hueco bolson, Trans-Pecos Texas and Chihuahua, Mexico. Numbers identify faults discussed in this paper. Numbered faults that have informal names are Fault 13 (Caballo fault), Fault 14 (Amargosa fault), Fault 15 (Campo Grande fault), and Fault 16 (East Franklin Mountains fault). Hachures identify faults that offset Quaternary sediments. It is unknown whether some of the other normal faults (bars) in the region have moved during the Quaternary. Map was compiled from Albritton and Smith (1965), Jones and Reaser (1970), Woodward and others (1978), Seager (1980), Dietrich and others (1983), Henry and others (1985), and field and aerial-photograph mapping. See Figure 7 for cross sections A–A′ and B–B′. Slightly modified from Collins and Raney (1991c).

the timing and rates of late Cenozoic fault movements. Even though historical seismicity within the bolson is low, there is a need to evaluate the Quaternary paleoseismicity of the area because of the presence of fault scarps in Quaternary deposits of the bolson and the proximity of these faults to El Paso, Texas; Juarez, Mexico; and to the proposed Eagle Flat low-level radioactive waste repository (about 10 km east of the Hueco bolson). In this paper we describe the bolson's structure and late Cenozoic fault history. For young faults that have surface expression in the basin, we describe scarp morphologies, fault geometries (strikes, dips, and lengths), histories of fault movements, amounts of offset on strata of different ages, and amounts of offset during single surface-rupture events.

METHODS

Our results were derived from interpretations of subsurface data (proprietary seismic lines and available boreholes), aerial-photo and surface geologic mapping, field observations, measurements of fault scarps, and descriptions of outcrops and excavation (locations of the seismic lines and boreholes are shown in the inset map of Fig. 8). Structure-contour and isopach maps and selected cross sections (scaled using no vertical exaggeration) were constructed from the subsurface data. A time-versus-depth curve for the Chevron U.S.A. State No. 359 exploration borehole (borehole 7 on Fig. 8 inset map) aided our conversion of depths in time on the seismic lines to depths in feet/meters on the maps and cross sections. Aerial photographs used in this study were (1) black-and-white, low-sun-angle (morning and afternoon) photographs at approximate scales of 1:12,000 and 1:6,000, taken during December 1985; (2) black-and-white, low-sun-angle (afternoon) photographs at an approximate scale of 1:10,000, taken during March 1989; (3) black-and-white photographs at an approximate scale of 1:25,000 taken during November 1979; and (4) black-and-white Army Mapping aerial photographs at an approximate scale of 1:62,000, taken in 1953. We also flew over parts of the area at low altitudes in fixed-wing aircraft and a helicopter to view surface expressions of the faults.

PREVIOUS WORK

Much of the previous work on Cenozoic faulting of the Heuco bolson involved regional mapping of faults but contained few detailed fault descriptions. Albritton and Smith (1965) and Dietrich and others (1983) produced regional geologic maps of the Texas part of the Hueco bolson. Regional geologic maps (scale 1:250,000) of Chihuahua, Mexico, were compiled by Coordinacin General de los Servicios Nacionales de Estadstica, Geografa e Informatica. Published studies of regional and local areas that illustrate mapped Quaternary faults within the Hueco bolson include the investigations of Muehlberger and others (1978), Woodward and others (1978), Seager (1980), Henry and Gluck (1981), and Henry and others (1985). Some specific Quaternary faults of the Hueco bolson were discussed by Lovejoy (1971, 1972), Machette (1987), Barnes and others (1989a, b), Collins and Raney (1990, 1991a, b, c), and Keaton and others (1989). Other researchers who have studied the slip rates of Quaternary faults in other parts of Trans-Pecos Texas and the Rio Grande rift of New Mexico include Goetz (1980), Machette (1982, 1987, 1988), Muehlberger and others (1985), Personius and Machette (1984), Gile (1987), and Menges (1990). Studies that describe the general structure of the northwest Hueco bolson and the bolson's asymmetric geometry were done by Mattick (1967), Ramberg and others (1978), and Seager (1980). Regional gravity and magnetic investigations by Wen (1983), Keller and Peeples (1985), and Daggett and others (1986) identify areas of thick basin-fill deposits in the Hueco bolson. Seismicity of the region, including the 1931 Valentine earthquake (magnitude 6.4), was discussed by Sanford and Toppozada (1974), Dumas (1980), Reagor and others (1982), Doser (1987, 1990), and Davis and others (1989).

REGIONAL TECTONIC SETTING

The Hueco bolson lies within the southern extension of the Rio Grande rift where the rift merges with the southern Basin and Range province (Chapin, 1971, 1979; Seager and Morgan, 1979). The rift may include related parts of the southern province in Trans-Pecos Texas and northern Mexico, although Cenozoic faulting in Chihuahua, Mexico, has only been documented locally. Gries (1979, 1980) suggested that Cenozoic extension may have reactivated some Laramide and older structures and caused flowage of thick evaporite sequences.

The bolson also lies on the projection of a broad, northwest-trending regional structure zone called the "Texas Lineament" (Muehlberger, 1980). Intermittent deformation from the Precambrian to the present has been inferred to be associated with sections of the Texas Lineament (Horak, 1985), with major phases of activity probably occurring during the Precambrian and the late Paleozoic. In Trans-Pecos Texas the Texas Lineament may be indicated by such features as (1) a major Precambrian discontinuity, (2) the northwest arm of the Paleozoic Marfa basin, (3) the axis or east flank of the Mesozoic Chihuahua trough, (4) the west margin of the Diablo platform, and (5) the approximate, leading edge of Laramide (80 to 50 Ma) thrusting (the margin of the Chihuahua tectonic belt).

The Hueco bolson is within the northeast part of an older, deep sedimentary basin, the Chihuahua trough, which developed during the Jurassic in westernmost Trans-Pecos Texas and in Chihuahua, Mexico (Henry and Price, 1985). The northeast margin of the trough approximately parallels the present Rio Grande on the Texas side of the river and probably consisted of Mesozoic down-to-the-southwest normal faults. Jurassic evaporites are the oldest deposits in the Chihuahua trough. Cretaceous marine sediments filled this basin and buried the trough-bounding normal faults (Henry and Price,

1985). The subsurface Clint fault, interpreted by Uphoff (1978) from borehole data collected 30 km southeast of El Paso, may be one of these faults that bound the Chihuahua trough. Subsequent Laramide deformation thrust Cretaceous rocks northeastward along a décollement zone of Jurassic evaporites and produced northwest-trending thrust faults, folds, and monoclines along the northeast margin of the Chihuahua trough (Gries and Haenggi, 1971; Gries, 1980; Henry and Price, 1985). The timing of Laramide deformation is not closely constrained, but it began no earlier than the Late Cretaceous, perhaps 80 Ma. Laramide thrust faulting and folding appears to have ceased by about 50 Ma (Price and Henry, 1985), and Laramide deformation appears to have waned by about 30 Ma (Henry and Price, 1985).

Volcanic activity in the Trans-Pecos region occurred from 48 to 17 Ma, although most of the activity was between 38 and 28 Ma (Henry and Price, 1984, 1985; Henry and others, 1986). Most of the volcanism in Trans-Pecos Texas and the Hueco bolson study area occurred while the area was under regional east-northeast compression (Price and Henry, 1984).

A transition to regional extension occurred about 30 Ma, and subsequent normal faulting associated with Basin and Range extension was well developed by about 24 Ma (Henry and Price, 1985, 1986; Stevens and Stevens, 1985). Basin and Range faulting and magmatism in Trans-Pecos Texas and southern New Mexico have been episodic (Seager and others, 1984; Henry and Price, 1985; Stevens and Stevens, 1985; Mack and Seager, 1990). Precise times of accelerated fault movement and sediment deposition in the Hueco bolson are unknown. In Trans-Pecos Texas, periods of accelerated fault movement and sediment deposition in structural troughs may have occurred between 24 and 17 Ma, at about 10 Ma, and after 7 Ma (Stevens and Stevens, 1985). Similar periods of deformation within the southern Rio Grande rift in southern New Mexico were also reported by Seager and others (1984), Morgan and others (1986), and Mack and Seager (1990). In Trans-Pecos Texas, field data indicate that the early regional least principal horizontal stress was oriented east-northeast–west-southwest and later was oriented northwest-southeast (Henry and Price, 1985), but the time of this shift has not been well established. A similar change in stress-field orientation occurred in other parts of the Rio Grande rift and Basin and Range province about 10 Ma (Henry and Price, 1985; Aldrich and others, 1986; Morgan and others, 1986).

Currently the regional least principal horizontal stress orientation of the Basin and Range/Rio Grande rift stress province is west-northwest–east-southeast although the stress field may vary locally (Zoback and Zoback, 1980, 1989). In situ stress has not been measured in the Hueco bolson. Furthermore, the influence of preexisting zones of crustal weakness on the occurrence and geometry of Quaternary faults in the Hueco bolson is unknown.

Most of the historical seismicity of Trans-Pecos Texas has occurred near the north-trending Salt basin region, located east of the Hueco bolson, and near El Paso at the west edge of the Hueco bolson (Sanford and Toppozada, 1974; Dumas, 1980; Reagor and others, 1982; Davis and others, 1989). The largest event near El Paso was a Modified Mercalli (MM) intensity VI. The 1931 Valentine earthquake (Doser, 1987), a MM intensity VII (M = 6.4), is the largest recorded earthquake in Trans-Pecos Texas. The epicenter was located about 130 km southeast of the study area in the southern Lobo Valley. Until recently, seismic recording stations had not been located near the central Hueco bolson, and the frequency of microseismic events in the area is unknown.

CENOZOIC STRATIGRAPHY

The Cenozoic stratigraphy of the Hueco bolson (Fig. 2) comprises mostly basin-fill siliciclastic deposits, although local intrusive and extrusive igneous rocks are present. Little is known about the character of most of the Tertiary basin-fill sequence in the Hueco bolson because outcrops are limited to the upper few tens of meters of the basin fill and because few stratigraphic data have been reported from drill holes. Studies of seismic data indicate a distinct reflector within the Tertiary basin fill that can be traced throughout the deeper, central basin. Locally this reflector is discordant with overlying reflectors; we speculate that the reflector may record a regional unconformity, although no physical description of this horizon exists. We have used this horizon to map two subsurface basin-fill units, informally called the lower basin fill and upper basin fill units (Figs. 2 and 3) on the seismic section, but the age of this horizon is unknown. The lower basin fill, locally thicker than 1,830 m in the deepest parts of the basin, overlies pre-Tertiary bedrock and fills only the axial part of the central basin (Figs. 3 and 4). The upper basin fill is locally thicker than 1,220 m, although at the northeast and southeast basin margins the upper basin fill is less than 300 m thick and overlies pre-Tertiary bedrock rather than older basin fill (Figs. 3 and 5).

Only about 30 m of the youngest sediments of the upper basin fill unit is exposed in outcrops. Strain (1966) studied outcrops within the Hueco bolson and defined the Pliocene to lower Pleistocene Fort Hancock and Camp Rice Formations. The Albritton and Smith (1965) study of these deposits coincided with that of Strain (1966), although Albritton and Smith informally referred to these formations as the older and younger basin fill. More recent investigations that more fully describe the sedimentation and facies of these units in Trans-Pecos Texas and south-central New Mexico include descriptions by Hawley and others (1969), Strain (1971), Hawley (1975), Stuart and Willingham (1984), Vanderhill (1986), and Gustavson (1991). The Fort Hancock and Camp Rice Formations compose only the upper part of the upper basin fill unit that we identified from seismic data.

The Fort Hancock Formation sediments were deposited in a bolson setting. These sediments are mostly lacustrine clay, bedded gypsum, and silt and alluvial-fan gravel, sand, silt, and clay, although fluvial deposits appear locally. The Camp Rice

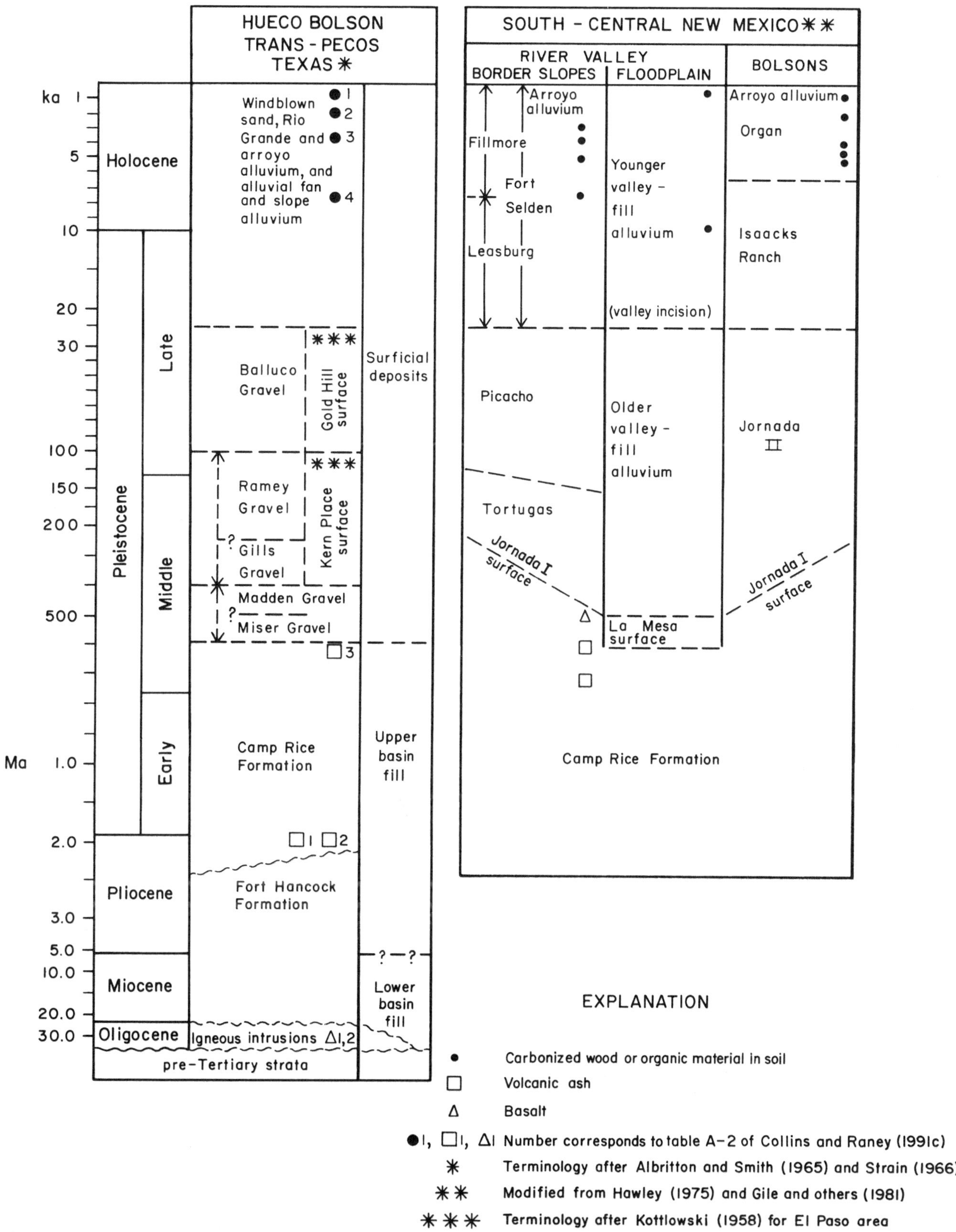

Figure 2. Stratigraphy of the Hueco bolson study area and south-central New Mexico.

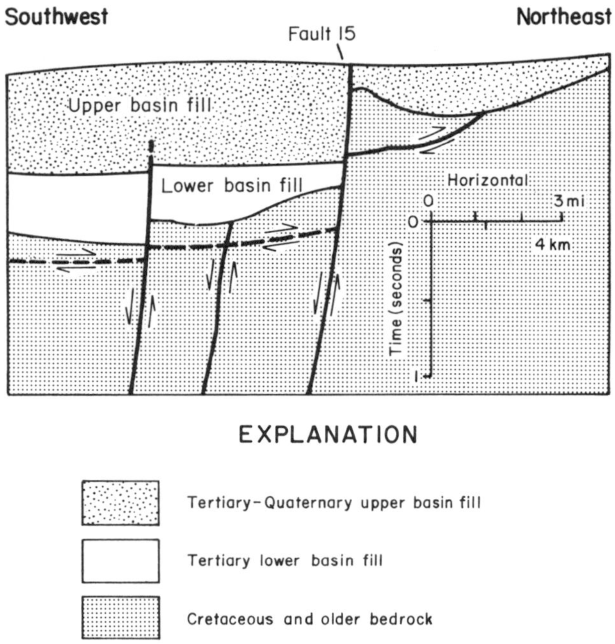

Figure 3. Cross section of pre-Cenozoic bedrock, upper-basin fill, lower-basin fill, and normal faults determined on seismic line E. Location of seismic line E shown on index map of Figure 8. Thrust fault is of Laramide age. Time is two-way travel time.

Formation, composed of sand and gravel and lesser amounts of silt and clay, represents alluvial-fan, fluvial, minor lacustrine, and flood-plain deposition. Camp Rice braided-stream deposits near the basin axis were deposited by the ancestral Rio Grande after it developed as a through-flowing stream (Albritton and Smith, 1965).

Camp Rice sediments unconformably overlie the Fort Hancock Formation. Near the Camp Rice Formation type locality, Vanderhill (1986) determined the contact to be about 2.5 Ma (late Pliocene) on the basis of paleomagnetic studies. This age may be locally correct, but the Fort Hancock/Camp Rice contact is a time-transgressive unconformity, and the hiatus represented by the contact is an unknown time span that may vary regionally. Blancan vertebrate fossils were identified in the Fort Hancock Formation by Strain (1966) and Vanderhill (1986), indicating a late Pliocene to early Pleistocene age for the upper Fort Hancock Formation. Volcanic ash lenses within the Camp Rice have aided in determining its age (Fig. 2). Ash present in arroyo outcrops east of Fort Hancock is within the lower part of the Camp Rice Formation and was reported as the 2.1-Ma Huckleberry Ridge ash by Izett (1981) and Izett and Wilcox (1982). The 0.6-Ma Lava Creek B ash was reported by these researchers to crop out at the top of the Camp Rice Formation near El Paso, Texas. These ashes were assigned as Huckleberry Ridge and Lava Creek B by Izett (1981) and Izett and Wilcox (1982) on the basis of the similarity of the petrologic properties of the Trans-Pecos Texas ashes with isotopically dated Huckleberry Ridge and Lava Creek B ashes from the western United States (Izett, 1981). Isotopic ages of the Trans-Pecos Texas ashes have not been determined.

Middle and upper Pleistocene deposits overlying the Fort Hancock and Camp Rice Formations in the southeast Hueco bolson were described by Albritton and Smith (1965; Fig. 2). They defined and mapped five Pleistocene gravel units and named them (oldest to youngest) Miser, Madden, Gills, Ramey, and Balluco gravels. These Pleistocene gravel units were deposited on piedmont slopes and on terraces of arroyos that were incised into older sediments. The gravels are locally derived and generally consist of limestone, sandstone, andesite, and chert clasts. These units are typically 1.5 to 2.5 m thick, and the calcic soils and petrocalcic soils that are developed within and that cap these units are 0.5 to 1.5 m thick. Geologic mapping by Albritton and Smith (1965) indicates that Madden, Ramey, and Balluco gravels are regionally extensive deposits, whereas the Miser and Gills gravels are present only locally. Miser gravel occurs only as well-dissected remnant deposits adjacent to the Quitman Mountains, and we interpret this unit as a fan facies of the Madden piedmont gravel.

Because they have few materials suitable for accurate dating, these Pleistocene gravel units (Miser, Madden, Gills, Ramey, and Balluco) have had relative age estimates made on the basis of field stratigraphic relationships, the degree of calcic soil development (Gile and others, 1966, 1981; Machette, 1985), and correlation with similar units in New Mexico. The regionally extensive Madden gravel probably correlates with the La Mesa and Jornada I surfaces of south-central New Mexico (Fig. 2), and the 0.6-Ma Lava Creek B ash in the upper Camp Rice Formation provides a maximum age. In south-central New Mexico the La Mesa surface is overlain by basalt dated at about 0.5 Ma (Gile and others, 1981; Seager and others, 1984). Calcic soils of the Madden, Gills, Ramey, and Balluco gravels exhibit progressively decreasing development of $CaCO_3$ morphologic stages. Following Machette's (1985) six-stage morphologic classification for calcic soils, the Madden gravel caliche is mostly stage IV. The Madden caliche has thin to thick laminae in the upper part of the horizon and laminae that drape over fractured surfaces. Stage V, which is characterized by thick laminae and carbonate-coated fractures, is developed locally. Ramey gravel caliche is mostly stage IV to III, which is determined by the massive $CaCO_3$ accumulations existing between clasts where gravel content is high and by a matrix that is firmly to moderately cemented where gravel content is low. Sparse laminae also occur in the upper part of the Ramey gravel caliche. Locally, Ramey gravel caliche is less developed and is stage III. It is characterized by a firmly cemented matrix with slight to massive $CaCO_3$ accumulations between clasts and by coatings on tops and undersides of pebbles or, where gravel content is low, by a matrix that is weakly cemented. Balluco gravel calcic soils are usually stage II and are characterized by continuous, thin to thick $CaCO_3$ coatings on the tops and undersides of pebbles. Field relationships indicate that the Balluco gravel, the youngest gravel unit, is older

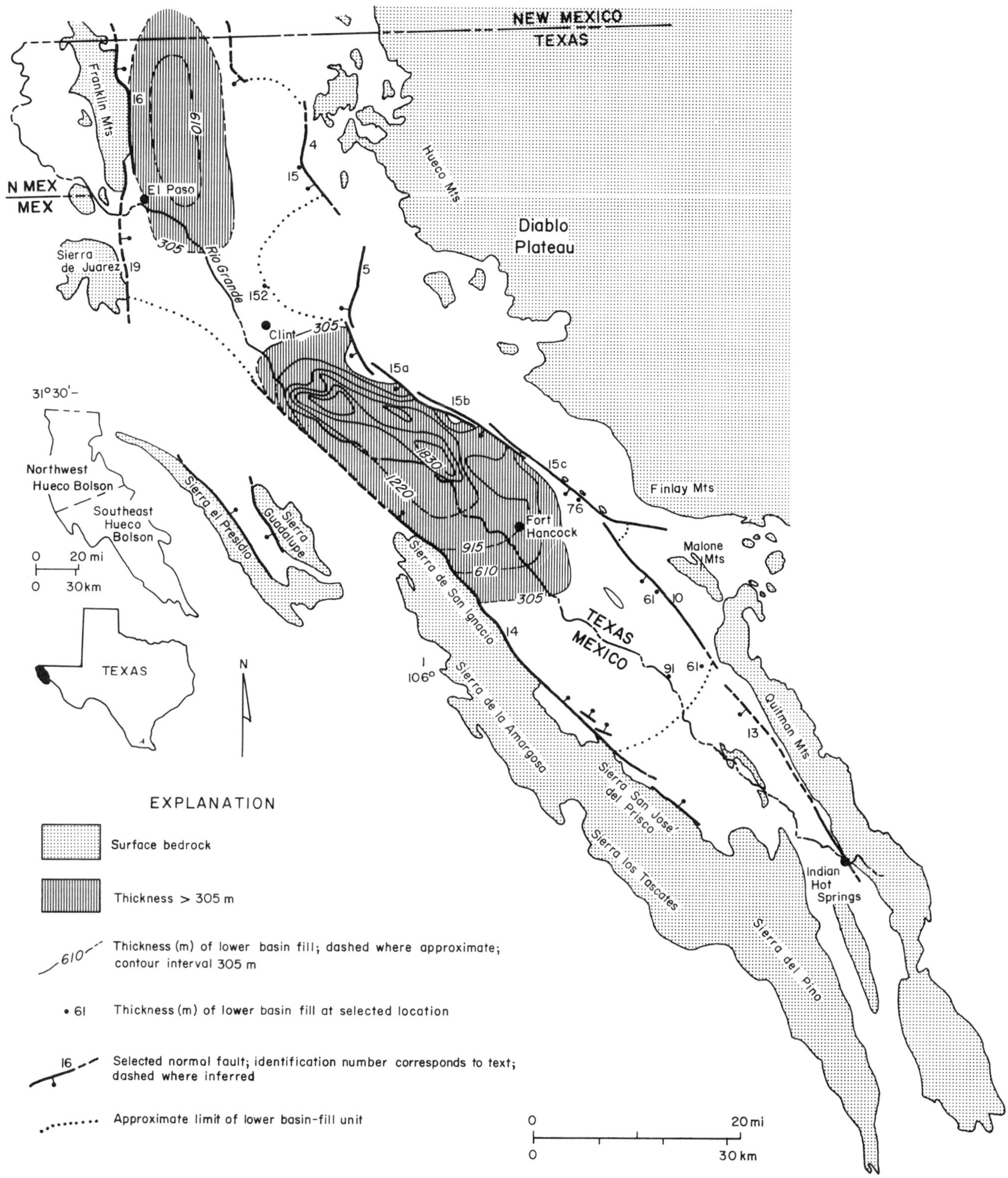

Figure 4. Isopach map of Cenozoic lower basin fill. Interpretations of seismic data were used to construct map.

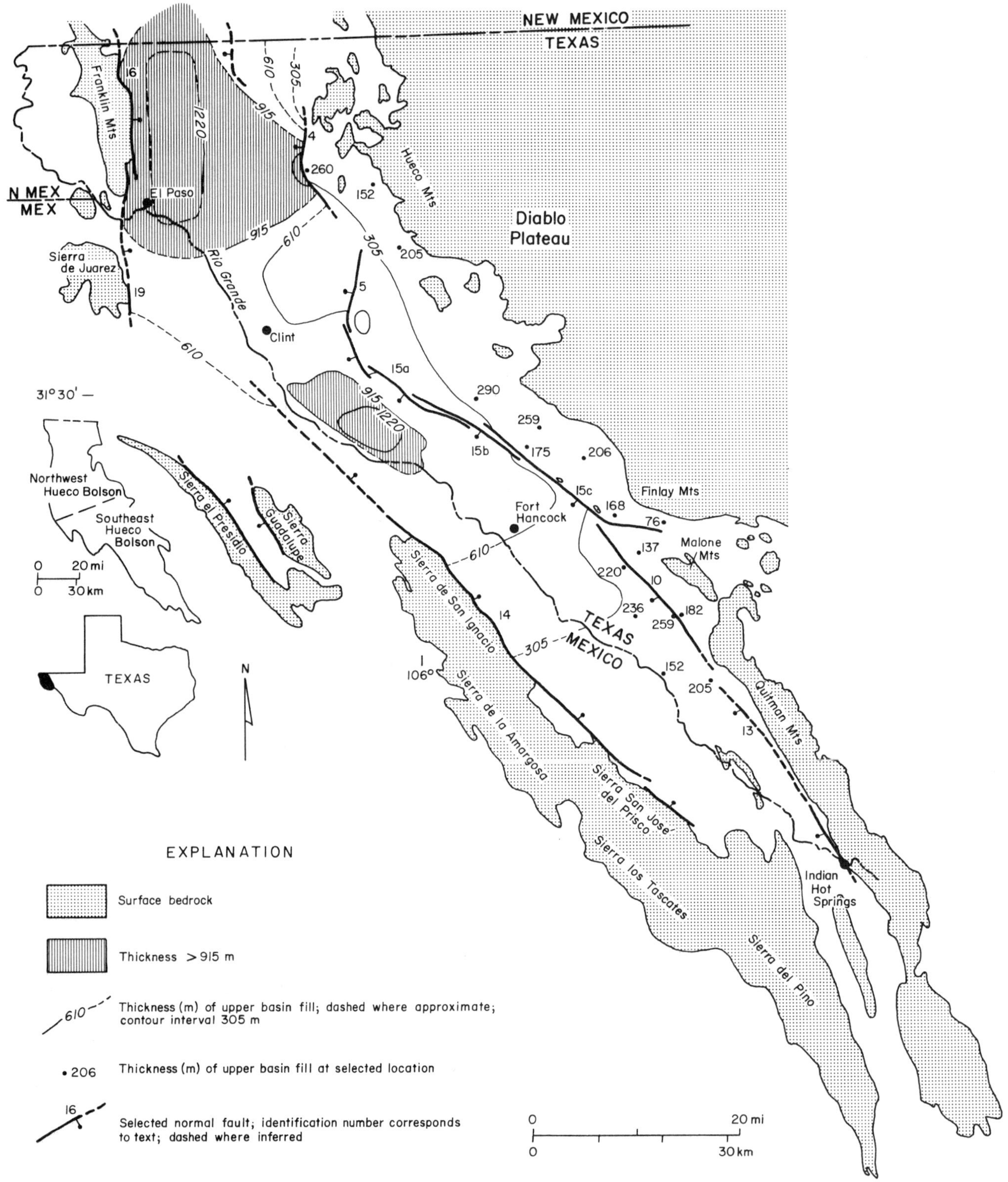

Figure 5. Isopach map of Cenozoic upper basin fill. Interpretations of seismic data were used to construct map.

than terraces of the present Rio Grande. The last major episode of Rio Grande entrenchment occurred approximately 25,000 to 10,000 yr ago (Gile and others, 1981).

Pleistocene geomorphic surfaces adjacent to the Franklin Mountains were discussed by Kottlowski (1958), who named the Kern Place and Gold Hill surfaces (Fig. 2). The Pleistocene and Holocene stratigraphy of south-central New Mexico and adjacent areas of Trans-Pecos Texas, including the Hueco bolson, were described by Hawley (1975) and Gile and others (1981), and much of the stratigraphy shown in Figure 2 is based on the work of these researchers.

STRUCTURE AND FAULT HISTORY OF THE HUECO BOLSON

The Hueco bolson is composed of two subbasins (Fig. 6). The northwest part of the bolson contains faults that generally strike northward, whereas faults of the southeast Hueco bolson strike northwestward. Seager (1980) described the northwest Hueco bolson as an asymmetric, west-tilted graben (Fig. 7A). Mattick (1967) and Ramberg and others (1978) also recognized the structural asymmetry of the northwest bolson, and calculated as much as 2,740 m and 2,000 to 3,000 m, respectively, of basin fill within the bolson's deepest part. The southeast Hueco bolson may have a similar, asymmetric geometry, although no subsurface data from the southwest margin of the southeast bolson in Mexico are available for interpretation (Fig. 7B). We conclude from seismic data that the deepest part of the southeast Hueco bolson has as much as 2,850 m of Cenozoic basin fill. Furthermore, the Chevron U.S.A. No. 359 State and Humble Oil Co. No. 1 University "DW" exploration boreholes (Fig. 8, index map, boreholes 7 and 6) penetrated about 2,280 m and 2,716 m of basin fill, respectively.

The deep part of both asymmetric subbasins are bound by major faults that define large graben 15 to 25 km wide (Figs. 6 and 7). Older, lower basin fill deposits are preserved only within the main graben (Fig. 4). The major faults that bound the Hueco graben on the west and southwest have more Cenozoic throw than the major faults that bound the graben on the east and northeast (Fig. 7). Intragraben faults also exist (Fig. 7), and normal faults are common outside the main graben, particularly in the mountains that bound the Hueco bolson. Fault geometries within the bolson have only been interpreted for depths as much as 3 to 3.5 km, and the cross sections we constructed from seismic data (Fig. 8) illustrate a predominant horst-and-graben style of normal faulting containing some tilted fault blocks. The faults generally dip between 50 and 80°. We do not know if the faults become listric at depths greater than 3.5 km because of the depth limit of seismic resolution. Previous studies of Cenozoic faulting of the Rio Grande rift indicate that some normal faults become listric at depths of 5 km and deeper (Cape and others, 1983; de Voogd and others, 1988). We identified some shallow (<2 km deep) listric faults that lie within the Cenozoic basin fill, but these faults do not intersect pre-Cenozoic bedrock, and they overlie older, graben-forming block faults (Fig. 8C). Early Tertiary paleotopography of the region is unknown, although Cenozoic faulting may have been a significant tectonic process that has uplifted the fault-bounded mountain ranges relative to the downdropped basin. Maximum topographic relief of the mountains bounding the bolson varies from 150 to 820 m.

Faulting and concurrent subsidence and infilling of the Hueco bolson probably occurred episodically, similar to the development of other basins in the region (Seager and others, 198; Henry and Price, 1985; Stevens and Stevens, 1985; Morgan and others, 1986; Mack and Seager, 1990). The lower basin fill unit fills parts of both the northwest and southeast Hueco bolson, although the southeast basin was the larger depocenter (Fig. 4). Subsurface data of the west and southwest sides of the Hueco bolson are unavailable, but cross sections constructed from seismic data of the central basin and the basin's east and northeast sides show the lower basin fill unit abutting major faults that bond the Hueco graben on the east and northeast (Fig. 3 and 8). In the north-trending northwest Hueco bolson, the thickest lower basin fill sequence is on the west-central side of the basin (Fig. 4). The north-striking East Franklin Mountains fault (Fig. 4, Fault 16), a west-boundary fault of the basin, probably was active during lower basin fill deposition in the northwest basin. The north-striking, down-to-the-west faults that bound the graben on the east have little to no lower basin fill preserved on the downthrown-fault block, indicating that these faults were mostly inactive during deposition of the lower-basin fill (Figs. 4 and 8A).

In the northwest-trending southeast Hueco bolson, the thickest lower basin fill sequence is between the Campo Grande fault sections (Fig. 4, Faults 15a, b, c) and the northwest extension of the Amargosa fault (Fig. 4, Fault 14). Of the faults that bound the southeast Hueco bolson on the northeast, the northwest and middle Campo Grande fault sections (Fig. 4, Faults 15a and 15b, respectively) have as much as 1,190 and 930 m, respectively, of lower basin fill on their downthrown sides (Fig. 4). Thickness of the lower basin fill decreases southeastward (Fig. 4), and the southeast Campo Grande fault section and Fault 10 have 145 and 25 m, respectively, of lower basin fill in the hanging-wall fault block. The lower basin fill unit is not present adjacent to the Caballo fault (Fig. 4, Fault 13), indicating that this part of the basin was not actively subsiding during the lower basin fill phase of deposition and associated deformation. Nevertheless, the relief (maximum of 640 m) of the Quitman Mountains on the footwall block suggests that some faulting, along with relative uplift of the footwall, could have occurred coincident with subsidence in other parts of the basin. Intragraben faults of both the northwest and southeast Hueco bolson have thicker sections of lower-basin fill on the hanging-wall fault blocks than the footwall blocks, indicating that these faults were active during deposition of the lower-basin fill (Fig. 8).

Near the boundary between the northwest and southeast

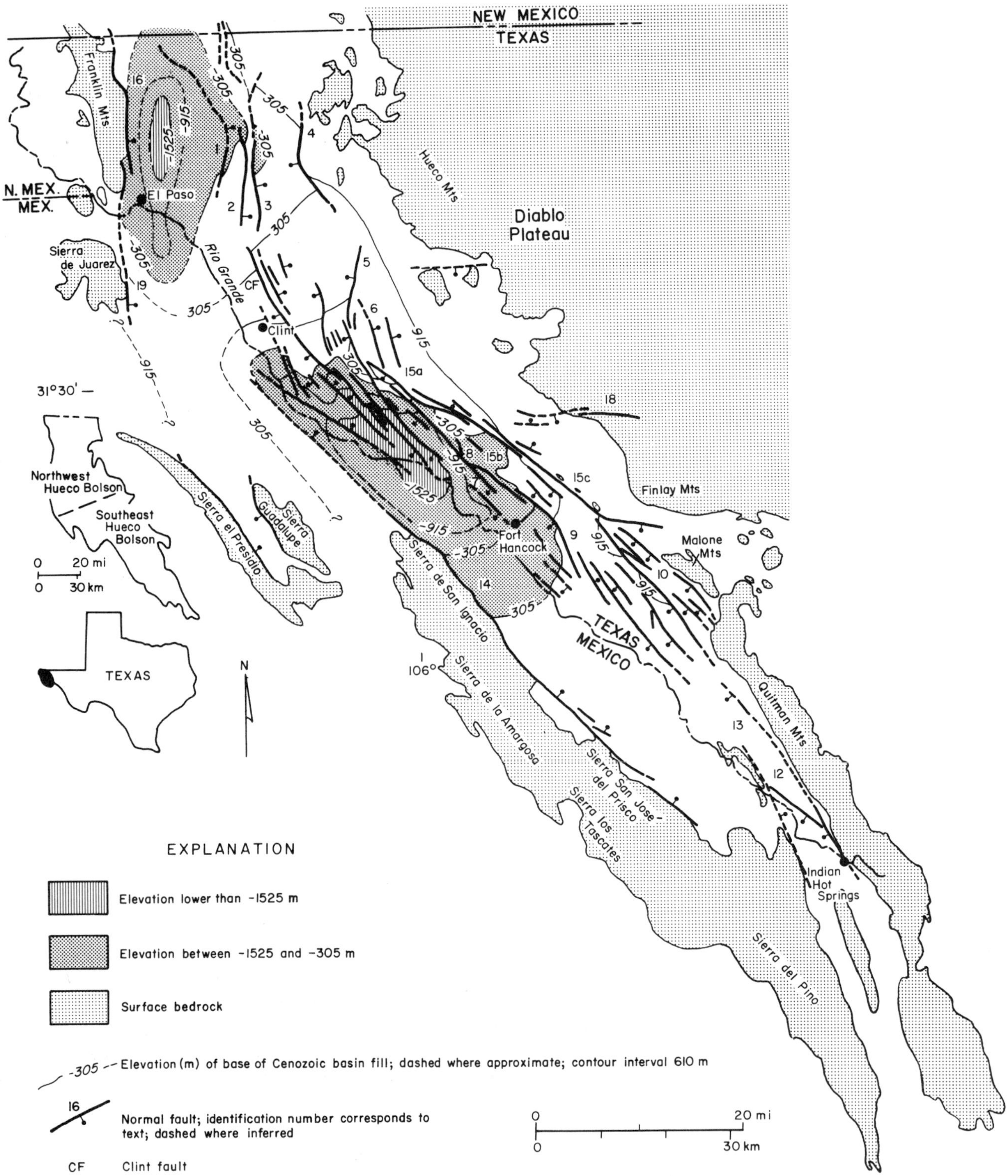

Figure 6. Structure-contour map of the base of Cenozoic basin fill (top of pre-Cenozoic bedrock). Interpretations of seismic and borehole data were used to construct map.

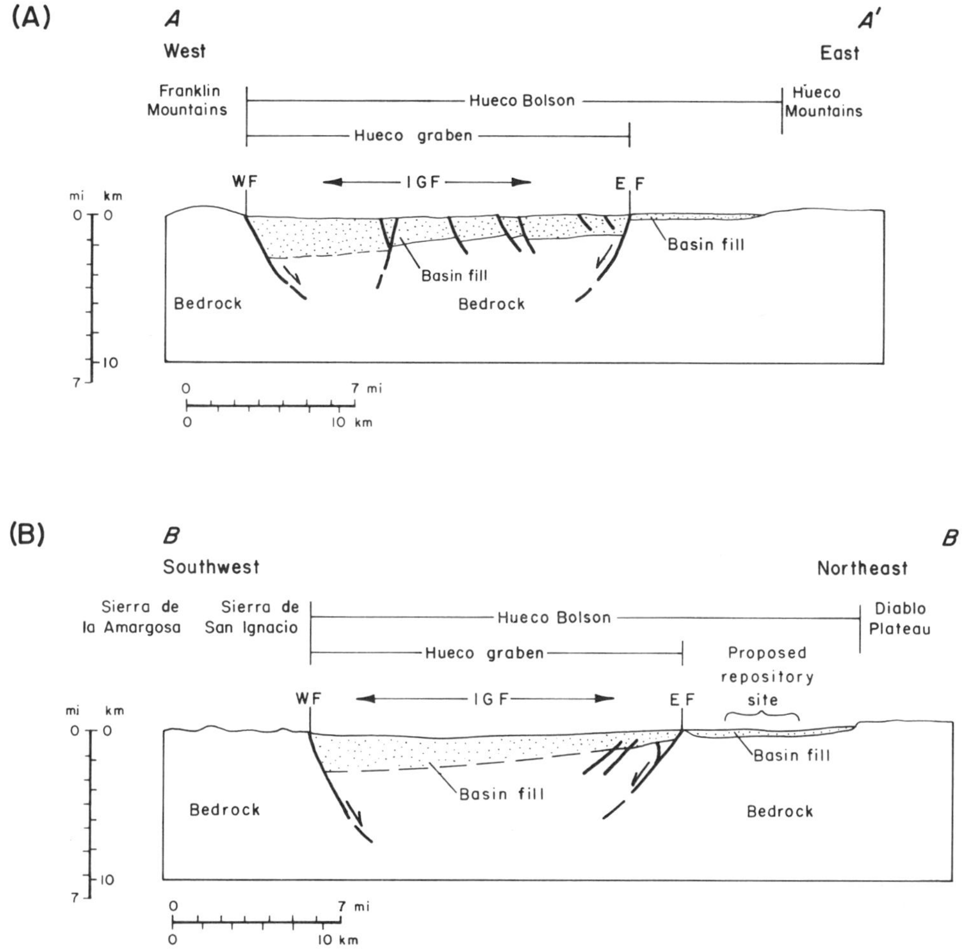

Figure 7. Cross sections A–A′ and B–B′ of (A) northwest Hueco bolson (constructed from seismic data, see Fig. 8, index map, seismic line A, interpreted for this study and from previous work of Seager, 1980 and (B) southeast Hueco bolson (Fig. 8 index map, seismic line G). WF, major fault that bounds graben on the west or southwest; IGF, intragraben fault; EF, major fault that bounds graben on the east or northeast. Laramide folds and thrust faults in bedrock not shown. Section lines shown in Figure 1.

basins, the lower-basin fill is between 150 and 300 m thick. A cross section across this area (Fig. 8B) shows the lower-basin fill abutting a fault-bounded bedrock high. The main down-to-the southwest fault illustrated in Figure 8B is probably the Clint fault (Uphoff, 1978). Our data indicate that the Clint fault was a main fault that bounded the Hueco graben on the east during deposition of the lower-basin fill. But in contrast to other faults that bound the main graben on the east, such as the Campo Grande fault, the Clint fault has been inactive since about the middle period of lower basin-fill deposition. Other faults marking the east edge of lower basin fill deposition continued to be active during the later deposition of the upper basin fill unit.

The upper basin fill unit is present throughout the Hueco topographic basin, although this unit is thicker within the area defined by the bounding faults of the main Hueco graben (Fig. 5). The north-striking, down-to-the west faults that bound the northwest Hueco graben on the east (Fig. 5, Faults 4 and 5) were much more active during deposition of the upper-basin fill than they were during deposition of the lower basin fill unit, as evidenced by the thick upper basin fill sequence preserved on the downthrown-fault block (Figs. 5 and 8A). Also, these major bounding faults on the east side of the northwest Hueco graben, Faults 4 and 5, were more active during deposition of the upper-basin fill than some of the northwest-striking faults that bound the northeast side of the southeast Hueco graben. Faults 4 and 5 (northwest Hueco Bolson) have about 1,230 and 350 m, respectively, of upper-basin fill on their downthrown fault blocks, whereas the Campo Grande fault sections (Faults 15a, b, and c) and Fault 10 (southeast Hueco bolson) have between 720 and 50 m, respectively, of upper-basin fill on their downthrown-fault blocks. Intragraben faults

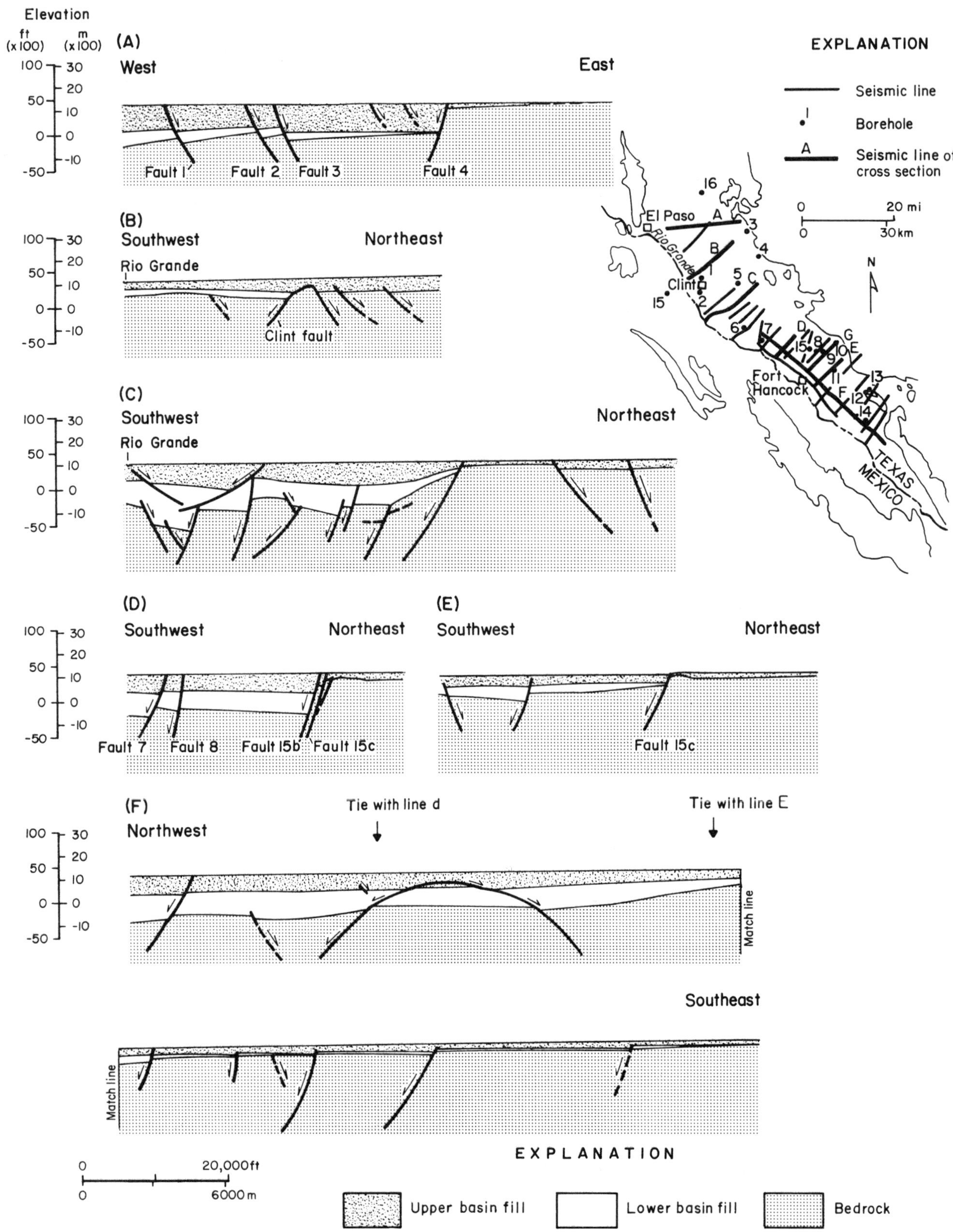

Figure 8. Cross sections interpreted from seismic data. Sections A through F correspond to seismic-line locations A through F on index map. Data from lines A, G, and E, were used to construct Figures 7A, 7B, and 3, respectively. Laramide folds and thrust faults in bedrock not shown.

of both basin areas have similar amounts of displacement but exhibit less offset of upper basin fill deposits than do the main graben-bounding faults.

QUATERNARY FAULT MOVEMENT

Of the faults that offset Pleistocene sediments in the Hueco bolson, the East Franklin Mountains and Amagosa faults (Fig. 1; Faults 16 and 14, respectively) have been the most active since the middle Pleistocene and probably since Cenozoic faulting began. These down-to-the-east and down-to-the-northeast faults bound the Hueco graben on the west and southwest, respectively, and may have had surface ruptures as recently as the Holocene. Down-to-the-west and down-to-the-southwest faults that bound the Hueco graben on the east and northeast, respectively, have not been as active as the east- and northeast-dipping faults. The most recent surface ruptures on the down-to-the southwest faults that bound the Hueco graben on the northeast occurred during the late Pleistocene. Some of the intragraben faults expressed at the surface also most recently ruptured the surface during the middle to late Pleistocene.

Northwest Hueco bolson

The East Franklin Mountains fault (Fig. 1, Fault 16) bounds the west edge of the northwest Hueco bolson. This fault was called the East Boundary fault of the Franklin Mountains by Lovejoy (1971, 1972) and more recently Machette (1987) referred to this fault as the East Franklin Mountains fault. Characteristics of this fault and the local geology were described by many other workers, including Sayre and Livingston (1945), Harbour (1972), Lovejoy (1975), LeMone (1984) and Dyer (1989). En echelon to the East Franklin Mountains fault is another boundary fault (Fault 19; Fig. 1), which trends along the southeastern Franklin Mountains and probably strikes southward toward the east flank of Sierra de Juarez in northern Chihuahua. Fault 19 was not studied in detail because its surface expression has been disturbed by construction in El Paso,and it is covered by apparently unfaulted Rio Grande alluvium south of the Franklin Mountains.

The East Franklin Mountains fault, composed of several fault strands and having a surface trace of about 23 km, strikes N10°W to N10°E and dips eastward. Its surface expression is the most distinct of the faults within the northwest Hueco bolson. Scarps commonly have compound-slope angles, and the steepest slope angles are between 13 and 23°. Multiple rupture events have caused older Quaternary surfaces to be offset more than younger surfaces, and scarp heights generally range between 5.5 and 39 m, reflecting the different ages of the faulted Quaternary surfaces along the fault trace. The scarp heights generally do not indicate the true offset of the Quaternary surfaces because of the steep depositional slopes (commonly 5 to 6.5°) of the alluvial surfaces at the margin of the mountains and because post-faulting deposition of variable amounts typically covers the faulted surface on the downthrown block. Lovejoy and Hawley (1978) reported 100 to 120 m of offset on the fluvial facies of the basin fill (Camp Rice Formation). Middle Pleistocene sediments (Jornada I surface) are offset at least 32 m (Fig. 9). Sediments of late Pleistocene age compose two surfaces that are locally called the Kern Place and Gold Hill surfaces (Fig. 2); the older Kern Place surface is offset at least 8.5 m, and the younger Gold Hill surface is faulted at least 4.5 to 6 m in some places. Machette (1987) reported that sediments of late Pleistocene to Holocene age (Isaacks Ranch or younger alluvium; Fig. 2) are offset by the East Franklin Mountains fault across a 3.7-m-high scarp. The approximate offset during the last rupture event ranged between 1.7 and 3 m (assuming that the steep

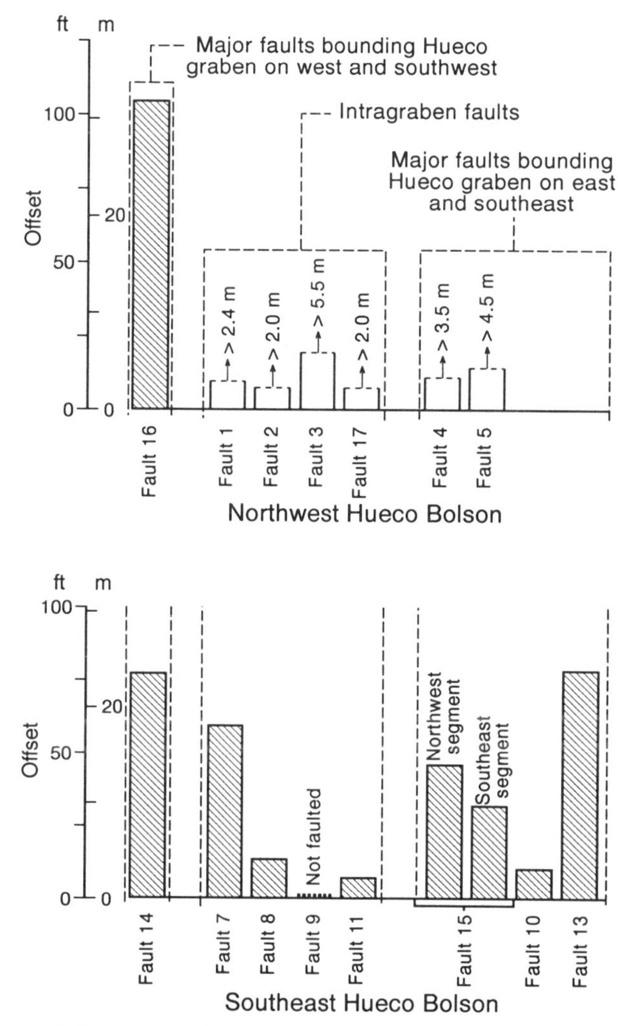

Figure 9. Bar graphs illustrating fault offset on middle Pleistocene surficial sediments for selected Quaternary faults of the Hueco bolson. Faults that have been informally named are Fault 13 (Caballo fault), Fault 14 (Amargosa fault), Fault 15 (Campo Grande fault), and Fault 16 (East Franklin Mountains fault). Middle Pleistocene surficial sediments studied at the different faults most likely are not precisely the same age, although all have stage IV to V calcic horizons. Faults shown in Figure 1.

parts of compound scarps reflect the latest single rupture event). Machette (1987) suggested that the youngest scarps along the East Franklin Mountains fault are early Holocene or latest Pleistocene in age (about 10,000 to 5,000 yr old) on the basis of scarp morphology.

Faults that are east of the East Franklin Mountains fault are mostly covered by windblown sand and have subtle surface traces (Seager, 1980; Henry and Gluck, 1981). Two unnamed faults, Faults 4 and 5 of Figure 1, bound the east side of the Hueco graben. These west-dipping faults are 10 and 11 km long, respectively, and strike northward at N40°W to N20°E and N20°W to N10°E, respectively. The fault scarps have single scarp-slope angles of about 3° and scarp heights between 2.5 and 5 m. Seismic data indicate that fault offsets of the base of the basin-fill deposits (top of bedrock) are about 1,070 m across Fault 4 and 870 m across Fault 5. Fault offsets of middle Pleistocene deposits (Jornada I surface; Fig. 2) that are capped by stage IV to V calcrete are unknown because the eolian deposits generally cover the scarps. Scarp heights of Faults 4 and 5 and the projection of the faulted surface indicate that minimum throw of middle Pleistocene sediments are greater than 3 and 4 m, respectively.

Intragraben faults that we studied (Faults 1, 2, 3, and 17; Fig. 1) strike northward between N35°W and N30°E, dip eastward, and have subtle surface traces between 12.5 and 21 km long. Scarps are subtle because they are covered by windblown sand, like the scarps of Faults 4 and 5. Scarp-slope angles of intragraben faults are between 2 and 4.5° and scarp heights are between 2.2 and 7.2 m. Throw of the base of the basin-fill deposits (top of bedrock) are about 150 m across Fault 1, 140 m across Fault 2, and 330 m across Fault 3. These offsets, calculated from seismic data, are considerably less than the offsets of the base of the basin fill across the faults that bound the northwest Hueco graben on the east and west. Similar to those of Faults 4 and 5, the offsets of the middle Pleistocene deposits (Jornada I surface; Fig. 2) that have stage IV to V caliche are unknown because windblown sand mostly covers the scarps, although the scarp heights of Faults 1, 2, 3, and 17, and the projection of the faulted surface indicate that minimum throws of middle Pleistocene sediments are greater than 2, 2, 5, and 2 m, respectively (Fig. 9).

Southeast Hueco bolson

The Amargosa fault (Fault 14, Fig. 1), recently named by Barnes and others (1989b) and Keaton and others (1989), bounds the southwest edge of the Hueco bolson at the base of Sierra de San Ignacio, Sierra de la Amagosa, and Sierra San Jose del Prisco. The Amargosa fault is composed of en echelon fault strands, and it has a surface trace of about 70 km and a regional strike of N40 to 50°W. It dips northeastward between 75 and 80° at the surface (Collins and Raney, 1991c). The fault appears to exhibit mostly vertical offset where we studied it, although Barnes and others (1989b) and Keaton and others (1989) reported grabenlike extensional features along the fault as evidence of lateral components of fault slip. Scarp-slope angles of the Amargosa fault are between 19 and 27°. Scarp heights range between 2.8 and 32 m, and the range is due to multiple rupture events during the Quaternary that caused older Quaternary surfaces to be offset more than younger surfaces. At one locality, piedmont deposits with stage IV to V indurated calcic soil, estimated to be middle Pleistocene age, are offset about 24 m (Fig. 9). Three younger, possibly upper Pleistocene, arroyo terraces in this area are also faulted. The two highest arroyo terraces have stage III to IV calcic soils and are vertically offset 4 and 6.5 m. The youngest arroyo terrace has a stage II calcic soil and is offset 2.5 m. The varying amounts of offset of these three arroyo terraces indicate at least three fault events since the highest arroyo terrace was formed. Surface ruptures at this locality have had a maximum offset per event of between 1.5 and 2.5 m.

It is likely that different parts of the Amagosa fault zone have had different rupture histories. At another locality, possibly upper Pleistocene deposits with stage II calcic soil are vertically offset as much as 4.5 m, and younger, possibly Holocene, sediments are not faulted. However, aerial photograph mapping suggests that young (possible Holocene) deposits may be offset elsewhere along the fault zone.

Scarps of intragraben faults and major faults that bound the northeastern side of the southeast Hueco graben are not as distinct as the Amargosa fault scarp. The Campo Grande fault, an unnamed fault, and the Caballo fault (Faults 15, 10, and 13, respectively; Fig. 1) are the main faults that bound the northeastern side of the southeast Hueco graben. The Campo Grande fault has a surface trace of about 45 km, and at the surface it is composed of en echelon fault strands that are 1.5 to 10 km long and that have strikes of N25 to 75°W (Collins and Raney, 1990). Dips are between 60 and 89° southwest, and grooves on fault planes indicate mostly dip-slip movement. Interpretations of seismic data (Collins and Raney, 1991a, c) indicate that the surface trace of the Campo Grande fault represents three en echelon subsurface fault sections, called the northwest, middle, and southeast sections (Faults 15a, b, and c of Fig. 4), that are approximately 21, 21, and 34 km long, respectively. The greatest amount of offset is at the central part of each fault section, and the amount of offset decreases toward the terminations of the faults. The northwest section of the Campo Grande fault has a maximum offset of about 1,220 m on the base of the basin-fill deposits (top of bedrock). Windblown sand mostly covers the surface scarps of this northwest fault section, although locally the scarp is distinct and the offset of middle Pleistocene deposits (Madden Gravel or Jornada I surface; Fig. 2) is about 14 m. The middle section of the Campo Grande fault has a maximum offset of about 1,290 m on the base of the basin-fill deposits (top of bedrock). Windblown sand covers the entire surface expression of this middle fault segment, and the offset on middle Pleistocene deposits (Madden Gravel or Jornada I surface) is unknown.

The surface expression of the southeast section of the Campo Grande fault is more distinct than that of the northwest and middle sections because less windblown sand covers this fault section and because this section crops out in arroyos and gullies. Fault scarps have been modified by erosion of the footwall and deposition on the hanging wall, and erosion-resistant caliche (stage IV to V) at the surface aids in preserving scarp heights of between 1.5 and 11.5 m and scarp slopes of 4 to 17° (Collins and Raney, 1990). Successively younger units cut by the southeast section of the Campo Grande fault have less displacement. This southeast section has a maximum offset of about 460 m on the base of the basin-fill deposits (top of bedrock). Offset of the upper Pliocene contact (approximately 2.5 Ma; Vanderhill, 1986) of the Fort Hancock and Camp Rice Formations is about 44 m. The middle Pleistocene Madden gravel is displaced by all of the fault strands that compose the southeast section of the Campo Grande fault, and the Madden gravel has a maximum vertical offset of about 10 m (Fig. 9). The middle to upper Pleistocene Ramey gravel is offset as much as 3 m by several surface fault strands of the Campo Grande fault's southeast section, although some Ramey deposits overlie fault strands and are not faulted. The younger upper Pleistocene Balluco gravel and Holocene sediments are not offset by the Campo Grande fault. On the downthrown block of one surface fault strand of the southeast segment, faulted calcic soil horizons (0.5 to 1.0 m thick; stage III morphology) that have vertical separations of 1 to 2 m indicate at least five episodes of movement, deposition, and surface stabilization since the middle Pleistocene (Collins and Raney, 1990). Maximum vertical offsets during single faulting events have been as much as 2 m, judging from the vertical separations of these faulted, calcic soil horizons. Maximum vertical offset during the last faulting event was about 1 to 1.5 m (Collins and Raney, 1990). Doser (1990) determined that earthquakes having magnitudes as great as 6.2 to 7.2 may have occurred during surface ruptures of the Campo Grande fault during the Quaternary.

Fault 10 (Fig. 1), another major fault that bounds the northeastern side of the Hueco graben, is en echelon to the southeast section of the Campo Grande fault (Collins and Raney, 1991b, c). Fault 10 has a locally covered surface trace of 15 km and strikes N30 to 60°W. In outcrop it dips 60 to 80° southwestward. Much of the trace of Fault 10 is covered by unfaulted sediments of late Pleistocene and Holocene age. Fault 10's scarp is preserved in the middle Pleistocene Madden gravel and the scarp has compound slopes that are as steep as 15°. Throw on the middle Pleistocene Madden gravel is as much as 3 m. The approximate vertical offset on Fault 10 during the last surface rupture was about 0.6 m, assuming that the steep parts of compound scarps reflect the latest single rupture event. Maximum throw of the base of the basin-fill deposits is 240 m, much less than the offsets of the Campo Grande fault sections.

The Caballo fault (Fig. 13; Fig. 1), another main fault that bounds the northeastern side of the southeast Hueco graben, has a surface trace of about 48 km, strikes N20 to 40°W and dips southwestward. The scarp of the Caballo fault is well dissected and difficult to distinguish along most of the fault trace, although along the northwestern part of the fault a probable compound fault scarp in the middle to upper Pleistocene Ramey gravel was measured. Excavations have not been made to verify the location and continuity of the fault or to describe it in detail. As much as 24 m of possible throw was estimated for the middle Pleistocene Madden gravel, although erosion of sediments along the fault trace makes precise measurements difficult. Offset of the middle to upper Pleistocene Ramey gravel is 7 m, and the younger upper Pleistocene Balluco gravel is not faulted. The approximate amount of vertical offset during the last surface rupture was about 1.7 m, assuming that the steep part of the compound scarp reflects the latest single rupture event. The southeastern part of the Caballo fault is not expressed by a scarp on middle Pleistocene fan surfaces and does not displace colluvium that covers the base of the mountains and edge of the basin.

Intragraben faults (Faults 7, 8, 9, 11, and 12; Fig. 1) of the southeast Hueco bolson do not have scarps that are as distinct as scarps of the main faults bounding the graben. Fault 7 strikes N40 to 50°W, dips southwestward, and the surface trace of its scarp is about 10 km long. Wind-blown sand obscures the surface expression of this fault in some areas. The scarp has a single-slope angle of about 9° (maximum), and the scarp is about 2.2 m high. Erosion-resistant Madden gravel caliche (stage IV to V) helps preserve the scarp on the upthrown block, and young alluvium and wind-blown sand covers the Madden gravel on the downthrown block adjacent to the fault. The middle Pleistocene Madden gravel is estimated to be displaced about 18 m by Fault 7 and offset of the base of the Cenozoic basin-fill deposits (top of bedrock) is 300 m. Fault 8 strikes N60 to 70°W, dips southwestward, and has a surface trace of about 3 km. The middle Pleistocene Madden gravel is displaced 4 m by Fault 8 and offset of the base of the Cenozoic basin-fill deposits (top of bedrock) is approximately 90 m.

CONCLUSIONS

1. The northwest-trending Hueco bolson is composed of two distinct subbasins. The northwest basin was developed along north-striking normal faults, and the southeast basin was developed along northwest-striking normal faults. A major 15- to 20-km-wide graben, the Hueco graben, lies on the east-central to west side of the bolson. Cenozoic basin fill is generally thin (not more than 150 to 200 m) at the east and northeast basin margins outside the main graben, and is thick (locally as much as 2,850 m) in the central basin area, toward the west and southwest parts of the basin, within the graben. Large down-to-the-east and down-to-the-northeast faults bound the graben on the west and southwest and have been more active than the down-to-the-west and down-to-the-southwest faults

bounding the graben on the east since the middle to late Tertiary when normal faulting began.

2. The Cenozoic basin-fill sequence has been informally divided into lower and upper units on the basis of a distinct seismic reflector within the sedimentary fill. The areal distribution of the lower-basin fill outlines a paleo-Hueco basin bounded by the major faults of the Hueco graben. No lower basin fill deposits are preserved outside the limits of the paleo-Hueco basin. Thick (greater than 305 m) lower basin fill deposits lie in parts of both the northwest and southeast basins, indicating that both north- and northwest-striking faults were active during lower basin fill deposition. The down-to-the-west faults that bound the northwest part of the Hueco graben on the east were not as active during deposition of the lower-basin fill as were the down-to-the-southwest faults that bounded the southeast part of the graben. The thickest lower basin fill of the northwest basin is bounded by the north-striking, down-to-the-east East Franklin Mountains fault (Fault 16; Fig. 1), which was probably as active as any of the northwest-striking faults. Upper basin fill deposits pervade the Hueco topographic basin. Thick (>915 m) upper basin fill deposits are more extensive in the northwest Hueco Bolson than in the southeast Hueco bolson because the bounding faults of the northwest Hueco graben generally were more active during deposition of the upper-basin fill.

3. Faults that offset Quaternary sediments are common within the Hueco bolson (Fig. 1). Faults that have had the most recent surface ruptures are the down-to-the-east East Franklin Mountains fault and the down-to-the-northeast Amargosa fault (Faults 16 and 14, respectively; Fig. 1). These faults bound the west and southwest edges of the Hueco bolson, and the footwall fault blocks are mountain ranges having maximum relief between 650 and 820 m. The most recent surface ruptures on both of these faults occurred during the late Pleistocene-Holocene, and the maximum amount of vertical offset per rupture was between 1.6 and 3 m. The most recent surface ruptures on the down-to-the-southwest surface faults that bound the southeast Hueco graben on the northeast (Campo Grande fault, Fault 10, and Caballo fault) occurred during the late Pleistocene. The maximum amount of throw during surface ruptures on these faults was between 0.6 and 2 m.

4. The East Franklin Mountains fault, the Amagosa fault, and possibly the northwestern part of the Caballo fault (Faults 16, 14, and 13; Fig. 1) have been the most active faults since middle Pleistocene. Middle Pleistocene deposits are offset between 23 and 32 m across these faults. During pre-Pleistocene bolson development, the Caballo fault was not as active as these other two faults as evidenced by a shallower basin and the absence of lower basin fill deposits on the hanging-wall block of the Caballo fault (Figs. 4 and 6). The Campo Grande fault (Fault 15; Fig. 1), which was active enough during pre-Pleistocene to help form the deepest part of the southeast Hueco graben (Figs. 4 and 6), displaces middle Pleistocene deposits 10 to 14 m. Upper Pleistocene deposits having stage II calcic soils are displaced 2.5 to 6 m across the East Franklin Mountains and Amagosa faults.

ACKNOWLEDGMENTS

Fieldwork for this research was funded by the Texas Low-Level Radioactive Waste Disposal Authority under Interagency Contract Number IAC(90-91)-0268. During this study, we benefited from helpful comments from R. W. Baumgardner, Jr., T. C. Gustavson, C. D. Henry, M. P. A. Jackson, T. F. Hentz, S. E. Laubach, D. D. Schultz-Ela, and A. R. Scott. We also thank M. E. Milling and S. J. Seni for their help in collecting and analyzing the proprietary seismic data and J. W. Hawley for his useful discussions on the stratigraphy of the region. J. H. Coggin, W. Kolb, J. L. Lardon, and P. A. Porter drafted the figures under the direction of R. L. Dillon. This paper also benefited from editing comments by Lana Dieterich on early drafts of this work. Helpful comments were made by G. H. Mack and D. Moss.

REFERENCES CITED

Albritton, C. C., Jr., and Smith, J. F., Jr., 1965, Geology of the Sierra Blanca area, Hudspeth County, Texas: U.S. Geological Survey Professional Paper 479, 131 p.

Aldrich, M. J., Jr., Chapin, C. E., and Laughlin, A. W., 1986, Stress history and tectonic development of the Rio Grande rift, New Mexico: Journal of Geophysical Research, v. 91, no. B6, p. 6199–6211.

Barnes, J. R., Peizhen, Z., and Slemmons, D. B., 1989a, The Campo Grande fault zone: A source for focused ground motion at the proposed low-level radioactive waste disposal site, Hudspeth County, Texas: Association of Engineering Geologists Abstracts and Program, 32nd Annual Meeting, p. 50.

Barnes, J. R., Shlemon, R. J., and Slemmons, D. B., 1989b, The Amargosa fault: A previously unstudied major active fault in northern Chihuahua, Mexico: Association of Engineering Geologists Abstracts and Program, 32nd Annual Meeting, p. 50–51.

Cape, C. D., McGeary, S., and Thompson, G. A., 1983, Cenozoic normal faulting and the shallow structure of the Rio Grande rift near Socorro, New Mexico: Geological Society of America Bulletin, v. 94, no. 1, p. 3–14.

Chapin, C. E., 1971, The Rio Grande rift, part 1: Modifications and additions: New Mexico Geological Society, Annual Field Conference, Guidebook No. 22, p. 191–201.

Chapin, C. E., 1979, Evolution of the Rio Grande rift, *in* Reicker, R. E., ed., Rio Grande rift: Tectonics and magmatism: Washington, D.C., EOS Transactions of the American Geophysical Union, p. 1–5.

Collins, E. W., and Raney, J. A., 1990, Neotectonic history and structural style of the Campo Grande fault, Hueco Basin, Trans-Pecos Texas: The University of Texas at Austin, Bureau of Economic Geology, Report of Investigations No. 196, 39 p., 3 Pl.

Collins, E. W., and Raney, J. A., 1991a, Neotectonic history and geometric segmentation of the Campo Grande fault: A major structure bounding the Hueco basin, Trans-Pecos Texas: Geology, v. 19, no. , p. 493–496.

Collins, E. W., and Raney, J. A., 1991b, Quaternary faults near a proposed low-level radioactive waste repository in the Hueco bolson of Trans-Pecos Texas: West Texas Geological Society Bulletin, v. 30, no. 6, p. 5–11.

Collins, E. W., and Raney, J. A., 1991c, Tertiary and Quaternary structure and paleotectonics of the Hueco basin, Trans-Pecos Texas and Chihuahua, Mexico: The University of Texas at Austin, Bureau of Economic Geology, Geological Circular 91-2, 44 p.

Daggett, P. H., Keller, G. R., Morgan, P., and Wen, C.-L., 1986, Structure of the southern Rio Grande rift from gravity interpretation: Journal of Geophysical Research, v. 91, no. B6, p. 6157–6167.

Davis, S. D., Pennington, W. D., and Carlson, S. M., 1989, A compendium of earthquake activity in Texas: The University of Texas at Austin, Bureau of Economic Geology Geological Circular 89-3, 27 p., 4 microfiche.

de Voogd, B., Serpa, L., and Brown, L., 1988, Crustal extension and magmatic processes: COCORP profiles from Death Valley and the Rio Grande rift: Geological Society of America Bulletin, v. 100, no. 10, p. 1550–1567.

Dietrich, J. W., Owen, D. E., and Shelby, C. A., 1983, Van Horn–El Paso Sheet, Barnes, V. E., project director: The University of Texas at Austin, Bureau of Economic Geology, Geologic Atlas of Texas, scale 1:250,000.

Doser, D. I., 1987, The 16 August 1931 Valentine, Texas, earthquake: Evidence for normal faulting in west Texas: Bulletin of the Seismological Society of America, v. 77, p. 2005–2017.

Doser, D. I., 1990, Seismological studies for the proposed low-level radioactive waste repository, Hudspeth County, Texas, *in* Keller, G. R., Doser, D. I., and Baker, M., eds., Geophysical studies related to the proposed low-level radioactive waste repository, Hudspeth County, Texas: The University of Texas at El Paso, contract report prepared for the Texas Low-Level Radioactive Waste Disposal Authority under interagency contract no. IAC (90-91)-0268, 27 p.

Dumas, D. B., 1980, Seismicity in the Basin and Range province of Texas and northeastern Chihuahua, Mexico, *in* Dickerson, P. W., Hoffer, J. M., and Callender, J. F., eds., Trans-Pecos region, southeastern New Mexico and west Texas: New Mexico Geological Society, Annual Field Conference, Guidebook No. 31, p. 77–81.

Dyer, J. R., 1989, Structural geology of the Franklin Mountains, West Texas, *in* Muehlberger, W. R., and Dickerson, P. W., eds., Structure and stratigraphy of Trans-Pecos Texas: 28th International Geological Congress Field Trip, Guidebook T317, p. 65–70.

Gile, L. H., 1987, Late Holocene displacement along th Organ Mountains fault in southern New Mexico: New Mexico Bureau of Mines and Mineral Resources Circular 196, 43 p.

Gile, L. H., Peterson, F. F., and Grossman, R. B., 1966, Morphological and genetic sequences of carbonate accumulation in desert soils: Soil Science, v. 101, n. 5, p. 347–360.

Gile, L. H., Hawley, J. W., and Grossman, R. B., 1981, Soils and geomorphology in the Basin and Range area of southern New Mexico—Guidebook to the Desert Project: New Mexico Bureau of Mines & Mineral Resources memoir 39, 222 p.

Goetz, L. K., 1980, Quaternary faulting in Salt Basin graben, west Texas, *in* Dickerson, P. W., Hoffer, J. M., and Callender, J. F., eds., Trans-Pecos region, southeastern New Mexico and west Texas: New Mexico Geological Society, Annual Field Conference, Guidebook No. 31, p. 83–92.

Gries, J. G., 1979, Problems of delineation of the Rio Grande rift into the Chihuahua tectonic belt of northern Mexico, *in* Ricker, R. E., ed., Rio Grande rift: Tectonics and magmatism: Washington D.C., EOS Transactions of the American Geophysical Union, p. 107–113.

Gries, J. G., 1980, Laramide evaporite tectonics along the Texas–northern Chihuahua border, *in* Dickerson, P. W., Hoffer, J. M., and Callender, J. F., eds., Trans-Pecos region, southeastern New Mexico and west Texas: New Mexico Geological Society, Annual Field Conference, Guidebook No. 31, p. 93–101.

Gries, J. G., and Haenggi, W. T., 1971, Structural evolution of the eastern Chihuahua tectonic belt, in Seewald, K. O., and Sundeen, D., eds., The geologic framework of the Chihuahua tectonic belt: West Texas Geological Society, p. 119–138.

Gustavson, T. C., 1991, Arid basin depositional systems and paleosols: Fort Hancock and Camp Rice Formations (Pliocene-Pleistocene), Hueco Bolson, west Texas and adjacent Mexico: The University of Texas at Austin, Bureau of Economic Geology, Report of Investigations No. 98, 49 p.

Harbour, R. L., 1972, Geology of the northern Franklin Mountains, Texas and New Mexico: U.S. Geological Survey Bulletin 1298, 129 p.

Hawley, J. W., 1975, Quaternary history of Dona Ana County region, south-central New Mexico, *in* Seager, W. R., Clemons, R. E., and Callender, J. F., eds., Las Cruces country: New Mexico Geological Society, Annual Field Conference, Guidebook No. 26, p. 139–140.

Hawley, J. W., Kottlowski, F. E., Seager, W. R., King, W. E., Strain, W. S., and LeMone, D. V., 1969, The Santa Fe Group in south-central New Mexico border region: New Mexico Bureau of Mines and Mineral Resources Circular 104, p. 52–76.

Henry, C. D., and Gluck, J. K., 1981, A preliminary assessment of the geologic setting, hydrology, and geochemistry of the Hueco Tanks geothermal area, Texas and New Mexico: The University of Texas at Austin, Bureau of Economic Geology Geological Circular 81-1, 48 p.

Henry, C. D., and Price, J. G., 1984, Variations in caldera development in the Tertiary volcanic field of Trans-Pecos Texas: Journal of Geophysical Research, v. 89, no. B10, p. 8765–8786.

Henry, C. D., and Price, J. G., 1985, Summary of the tectonic development of Trans-Pecos Texas: The University of Texas at Austin, Bureau of Economic Geology Miscellaneous Map No. 36, 8 p.

Henry, C. D., and Price, J. G., 1986, Early Basin and Range development in Trans-Pecos Texas and adjacent Chihuahua: Magmatism and orientation, timing, and style of extension: Journal of Geophysical Research, v. 91, no. B6, p. 6213–6224.

Henry, C. D., Gluck, J. K., and Bockoven, N. T., 1985, Tectonic map of the Basin and Range province of Texas and adjacent Mexico: The University of Texas at Austin, Bureau of Economic Geology, Miscellaneous Map No. 36, scale 1:500,000.

Henry, C. D., McDowell, F. W., Price, J. G., and Smyth, R.E.C., 1986, Compilation of potassium-argon ages of Tertiary igneous rocks, Trans-Pecos Texas: The University of Texas at Austin, Bureau of Economic Geology Geological Circular 86-2, 34 p.

Horak, R. L., 1985, Trans-Pecos tectonism and its effect on the Permian basin, *in* Dickerson, P. W., and Muehlberger, W. R., eds., Structure and tectonics of Trans-Pecos Texas: West Texas Geological Society Field Conference Publication No. 85-81, p. 81–87.

Izett, G. A., 1981, Volcanic ash beds: Recorders of upper Cenozoic silicic pyroclastic volcanism in the western United States: Journal of Geophysical Research, v. 86, no. B11, p. 10200–10222.

Izett, G. A., and Wilcox, R. E., 1982, Map showing localities and inferred distributions of the Huckleberry Ridge, Mesa Falls, and Lava Creek ash beds (Pearlette Family ash beds) of Pliocene and Pleistocene age in the western United States and southern Canada: U.S. Geological Survey Miscellaneous Investigations Map I-1325, scale 1:4,000,000.

Jones, B. R., and Reasor, D. F., 1970, Geology of southern Quitman Mountains, Hudspeth County, Texas: The University of Texas at Austin, Bureau of Economic Geology Geologic Quadrangle Map No. 39, scale 1:48,000, text 24 p.

Keaton, J. R., Shlemon, R. J., Slemmons, D. B., Barnes, J. R., and Clark, D. G., 1989, The Amargosa fault: A major late Quaternary intra-plate structure in northern Chihuahua, Mexico: Geological Society of America Abstracts with Programs, v. 21, no. 6, p. A148.

Keller, G. R., and Peeples, W. J., 1985, Regional gravity and aeromagnetic anomalies in west Texas, *in* Dickerson, P. W., and Muehlberger, W. R., eds., Structure and tectonics of Trans-Pecos Texas: West Texas Geological Society Publication 85-81, p. 101–105.

Kottlowski, F. E., 1958, Geologic history of the Rio Grande near El Paso: West Texas Geological Society Field Trip Guidebook, p. 4654.

LeMone, D. V., 1984, Aspects of border region geology from the Tom Lea Park overlook, El Paso, Texas, *in* Clark, K. F., Gerald, R. E., Bridges,

L. W., Dyer, J. R., and LeMone, D. V., leaders, Geology and petroleum potential of Chihuahua, Mexico: West Texas Geological Society, Field Trip Guidebook, Publication No. 84-80, p. 104–111.

Lovejoy, E.M.P., 1971, Tectonic implications of high level surfaces bordering Franklin Mountains, Texas: Geological Society of America Bulletin, v. 82, no. 2, p. 433–445.

Lovejoy, E.M.P., 1972, Basin and range faulting rates, Franklin Mountains, Texas: Geological Society of America Abstracts with Programs, v. 4, no. 6, p. 387–388.

Lovejoy, E.M.P., 1975, An interpretation of the structural geology of the Franklin Mountains, Texas, *in* Seager, W. R., Clemons, R. E., and Callender, J. F., eds., Las Cruces country: New Mexico Geological Society, Annual Field Conference, Guidebook No. 26, p. 261–268.

Lovejoy, E.M.P., and Hawley, J. W., 1978, El Paso to New Mexico Texas state line, *in* Hawley, J. W., ed., Guidebook to the Rio Grande rift in New Mexico and Colorado: New Mexico Bureau of Mines and Mineral Resources Circular 163, p. 57–68.

Machette, M. N., 1982, Quaternary and Pliocene faults in the La Jencia and southern part of the Albuquerque-Belen basins, New Mexico: Evidence of fault history from fault-scarp morphology and Quaternary geology, *in* Wells, S. G., Grambling, J. A., and Callender, J. F., eds., Albuquerque country II: New Mexico Geological Society, Annual Field Conference, Guidebook No. 33, p. 161–169.

Machette, M. N., 1985, Calcic soils of the southwestern United States, *in* Weide, D. L., ed., Soils and Quaternary geology of the southwestern United States: Geological Society of America Special Paper 203, p. 121.

Machette, M. N., 1987, Preliminary assessment of paleoseismicity at White Sands Missile Range, southern New Mexico: Evidence for recency of faulting, fault segmentation, and repeat intervals for major earthquakes in the region: U.S. Geological Survey Open-File Report 87-444, 46 p.

Machette, M. N., 1988, Quaternary movement along the La Jencia fault, central New Mexico: U.S. Geological Survey Professional Paper 1440, 82 p.

Mack, G. H., and Seager, W. R., 1990, Tectonic control on facies distribution of the Camp Rice and Palomas Formations (Pliocene-Pleistocene) in the southern Rio Grande rift: Geological Society of America Bulletin, v. 102, no. 1, p. 45–53.

Mattick, R. E., 1967, A seismic and gravity profile across the Hueco bolson, Texas: U.S. Geological Survey Professional Paper 575-D, p. 85–91.

Menges, C. M., 1990, Late Quaternary fault scarps, mountain-front landforms, and Pliocene-Quaternary segmentation on the range-bounding fault zone, Sangre de Cristo Mountains, New Mexico, *in* Krinitzsky, E. L., and Slemmons, D. B., eds., Neotectonics in earthquake evaluation: Geological Society of America Reviews in Engineering Geology, v. 8, p. 131–156.

Morgan, P., Seager, W. R., and Golombek, M. P., 1986, Cenozoic thermal, mechanical, and tectonic evolution of the Rio Grande rift: Journal of Geophysical Research, v. 91, no. B6, p. 6263–6276.

Muehlberger, W. R., 1980, Texas lineament revisited, *in* Dickerson, P. W., Hoffer, J. M., and Callender, J. F., eds., Trans-Pecos region, southeastern New Mexico and west Texas: New Mexico Geological Society, Annual Field Conference, Guidebook No. 31, p. 113–121.

Muehlberger, W. R., Belcher, R. C., and Goetz, L. K., 1978, Quaternary faulting in Trans-Pecos Texas: Geology, v. 6, no. 6, p. 337–340.

Muehlberger, W. R., Belcher, R. C., and Goetz, L. K., 1985, Quaternary faulting in Trans-Pecos Texas, *in* Dickerson, P. W., and Muehlberger, W. R., eds., Structure and tectonic of Trans-Pecos Texas: West Texas Geological Society Publication 85-81, p. 21.

Personius, S. F., and Machette, M. N., 1984, Quaternary and Pliocene faulting in the Taos Plateau region, northern New Mexico, *in* Baldridge, W. S., Dickerson, P. W., Riecker, R. E., and Zidek, J., eds., Rio Grande rift: Northern New Mexico: New Mexico Geological Society, Annual Field Conference, Guidebook No. 35, p. 83–90.

Price, J. G., and Henry, C. D., 1984, Stress orientations during Oligocene volcanism in Trans-Pecos Texas: Timing the transition from Laramide compression of Basin and Range tension: Geology, v. 12, no. 4, p. 238–241.

Price, J. G., and Henry, C. D., 1985, Summary of Tertiary stress orientations and tectonic history of Trans-Pecos Texas, *in* Dickerson, P. W., and Muehlberger, W. R., eds., Structure and tectonics of Trans-Pecos Texas: West Texas Geological Society Publication No. 85-81, p. 149–151.

Ramberg, I. B., Cook, F. A., and Smithson, S. B., 1978, Structure of the Rio Grande rift in southern New Mexico and West Texas based on gravity interpretation: Geological Society of America Bulletin, v. 89, no. 1, p. 107–123.

Reagor, B. G., Stover, C. W., and Algermissen, S. T., 1982, Seismicity map of the State of Texas: Denver, Colorado, U.S. Geological Survey Miscellaneous Field Studies Map No. MF-1388, scale 1:1,000,000.

Sanford, A. R., and Toppozada, T. R., 1974, Seismicity of proposed radioactive waste disposal site in southeastern New Mexico: New Mexico Bureau of Mines and Mineral Resources Circular 143, 15 p.

Sayre, A. N., and Livingston, P., 1945, Ground-water resources of the El Paso area, Texas: U.S. Geological Survey Water-Supply Paper 919, 190 p.

Seager, W. R., 1980, Quaternary fault system in the Tularosa and Hueco basins, southern New Mexico and west Texas, *in* Dickerson, P. W., Hoffer, J. M., and Callender, J. F., eds., Trans-Pecos region, southeastern New Mexico and west Texas: New Mexico Geological Society, Annual Field Conference, Guidebook No. 31, p. 131–135.

Seager, W. R., and Morgan, P., 1979, Rio Grande rift in southern New Mexico, west Texas, and northern Chihuahua, *in* Riecker, R. E., ed., Rio Grande rift: Tectonics and magmatism: Washington, D.C., EOS Transactions of the American Geophysical Union, p. 87–106.

Seager, W. R., Shafiqullah, M., Hawley, J. W., and Marvin, R. F., 1984, New K-Ar dates from basalts and the evolution of the southern Rio Grande rift: Geological Society of America Bulletin, v. 95, no. 1, p. 87–99.

Stevens, J. B., and Stevens, M. S., 1985, Basin and Range deformation and depositional timing, Trans-Pecos Texas, *in* Dickerson, P. W., and Muehlberger, W. R., eds., Structure and tectonics of Trans-Pecos Texas: West Texas Geological Society Field Conference Publication 85-81, p. 157–163.

Strain, W. S., 1966, Blancan mammalian fauna and Pleistocene formations, Hudspeth County, Texas: Texas Memorial Museum Bulletin No. 10, 55 p.

Strain, W. S., 1971, Late Cenozoic bolson integration in the Chihuahua tectonic belt, *in* Hoffer, J. M., ed., Geologic framework of the Chihuahua tectonic belt: West Texas Geological Society Publication 71-59, p. 167–173.

Stuart, C. J., and Willingham, D. L., 1984, Late Tertiary and Quaternary fluvial deposits in the Mesilla and Hueco bolsons, El Paso area, Texas: Sedimentary Geology, v. 38, no. 1/4, p. 120.

Uphoff, T. L., 1978, Subsurface stratigraphy and structure of the Mesilla and Hueco bolsons, El Paso region, Texas and New Mexico [M.A. thesis]: The University of Texas at El Paso, 65 p.

Vanderhill, J. B., 1986, Lithostratigraphy, vertebrate paleontology, and magnetostratigraphy of Plio-Pleistocene sediments in the Mesilla Basin, New Mexico [Ph.D. thesis]: The University of Texas at Austin, 305 p.

Wen, C.-L., 1983, A study of bolson fill thickness in the southern Rio Grande rift, southern New Mexico, west Texas and northern Chihuahua [M.A. thesis]: The University of Texas at El Paso, 74 p.

Woodward, L. A., and 6 others, 1978, Tectonic map of Rio Grande rift region in New Mexico, Chihuahua, and Texas, *in* Hawley, J. W., ed., Guidebook to Rio Grande rift in New Mexico and Colorado: New Mexico Bureau of Mines and Mineral Resources Circular 163, sheet 2, scale 1:1,000,000.

Zoback, M. L., and Zoback, M. D., 1980, State of stress in the conterminous United States: Journal of Geophysical Research, v. 85, n. B11, p. 6113–6156.

Zoback, M. L., and Zoback, M. D., 1989, Tectonic stress field of the continental United States, in Pakiser, L. C., and Mooney, W. D., eds., Geophysical framework of the continental United States: Geological Society of America Memoir 172, p. 523–539.

Manuscript Accepted by the Society October 13, 1993

Printed in U.S.A.

Geological Society of America
Special Paper 291
1994

Basins in the Big Bend segment of the Rio Grande rift, Trans-Pecos Texas

Patricia Wood Dickerson and William R. Muehlberger
Department of Geological Sciences, 61120, The University of Texas at Austin, Austin, Texas 78712

ABSTRACT

In the Big Bend segment of the Rio Grande rift (Trans-Pecos Texas), the formation of deep grabens began in early Miocene time (about 23 Ma) and was accompanied by basaltic volcanism of small volume. Graben faults that cut Quaternary deposits attest that extensional deformation continues.

Volcanism accompanied only the early phase of rifting, between 23.3 and 21 Ma. Alkaline basalt flows, dikes, and sills predominate, and the felsic suites of the northern Rio Grande rift are not present. The only voluminous rift volcanism was in Black Gap graben in easternmost Big Bend and in the Bofecillos Mountains on the west. The young eruptive centers of the northern rift are absent, but heat flow is nonetheless elevated in both rift basins and transverse structural zones in the Big Bend. Although geophysical data are far sparser here, there is no persuasive evidence for a massive mafic intrusive complex beneath the southern rift. Rupture of the entire lithosphere is suggested by the relatively primitive compositions of the basalts, several of which bear lower crustal or mantle xenoliths.

Rift basins include Presidio (northern and southern segments), Redford, Santana, and Black Gap grabens, as well as the Sunken Block (centered on Big Bend National Park), which encompasses Castolon and Tornillo grabens and their continuations into Mexico. Basal sedimentary deposits in earlier formed grabens are dated at 23 to 22 Ma and rest unconformably across older rocks. Younger grabens of more northerly orientation began to develop at about 10 Ma.

Bounding faults strike northwest to north-northwest, and basins of various sizes terminate or change size and/or geometry at west-trending discontinuities. The Shafter, Presidio, and Terlingua transverse zones are long-lived fault complexes that link and/or segment Big Bend rift grabens. None of these westerly structural corridors has yet functioned as a transform. All predate late Cenozoic extensional deformation, all extend well beyond their intersections with rift-basin-bounding faults, and none shows significant lateral offset, although right-lateral movement is documented. These west-trending zones subdivide the region into smaller structural units with steeper bounding faults than is typical of the northern rift.

Deep early Miocene grabens, synextensional volcanism, elevated heat flow, Quaternary normal faulting, and evidence for full lithospheric rupture indicate that the Rio Grande rift extends at least as far south as the continuations of the Castolon and Tornillo grabens into Chihuahua and Coahuila, Mexico. The southern terminus of the rift, however, remains undefined.

Dickerson, P. W., and Muehlberger, W. R., 1994, Basins in the Big Bend segment of the Rio Grande rift, Trans-Pecos Texas, *in* Keller, G. R., and Cather, S. M., eds., Basins of the Rio Grande Rift: Structure, Stratigraphy, and Tectonic Setting: Boulder, Colorado, Geological Society of America Special Paper 291.

INTRODUCTION

"The Trans-Pecos rift valley province of the United States has a length—from the northern end of the San Luis Valley of southern Colorado to the southern Big Bend of the Rio Grande—of 700 miles. It has a farther, but still unknown, extent into Mexico" (Baker, 1934, p. 170).

Subsequent to Baker's early structural summary, the defining geological studies were in New Mexico from Socorro northward into southern Colorado by Kirk Bryan (1938) and his students at Harvard. Kelley (1952) introduced the term *Rio Grande rift,* which has taken precedence as the name for the >1,126-km-long (700 mi) rift system.

Although Baker's closing statement is as true now as it was 60 years ago, data from a variety of disciplines—sedimentology, vertebrate paleontology, geophysics, K-Ar geochronology, and structural and igneous geology—have since been brought to bear upon the question of southward extent of the rift. Here we summarize data on the grabens and transverse fault zones of the Big Bend segment. Much of this information is in guidebooks of limited distribution or in unpublished theses and dissertations; the rest is drawn from our ongoing field work.

In the Big Bend of Trans-Pecos Texas, as in the northern rift of Colorado and New Mexico (Fig. 1), the formation of deep grabens began in early Miocene time (about 23 Ma) and was accompanied by basaltic volcanism of comparatively small volume; timing of basin foundering has become increasingly well established on the basis of both vertebrate paleontology and K-Ar geochronology (Stevens and Stevens, 1985, 1990a, b; Henry and McDowell, 1986; Henry and Aranda-Gomez, 1990). Extensional faulting continues today, as indicated by normal faults that cut Quaternary deposits (Dickerson, 1966; Dietrich, 1966; Groat, 1972; Muehlberger et al., 1978).

The principal differences between the southern and northern segments of the rift system appear to be in volume, composition, and age of rift volcanism. When the critical geochronological and geochemical data were lacking in the southern rift, the paucity of volcanism was thought to indicate either that the rift did not extend into the Big Bend region or that rifting was in a nascent state (Chapin and Seager, 1975; Seager and Morgan, 1979; Olsen et al., 1987).

Through the work of Henry, McDowell, and others, the ages of rift-related igneous bodies are now known to group around 23.3 to 22 Ma, with one as young as 17 Ma (Henry and McDowell, 1986; James and Henry, 1991; C. D. Henry, oral communication, 1992). However, younger volcanic centers so common in the northern rift are absent from this region. With the change to northwesterly extension—probably about 10 Ma, as documented elsewhere in the rift (e.g., Baldridge and Olsen, 1989)—the already well established northwesterly to north-northwesterly faults of Big Bend were oriented to respond in right transtension. The extensional component may not have been sufficient to allow ascent of magma to shallow crustal levels.

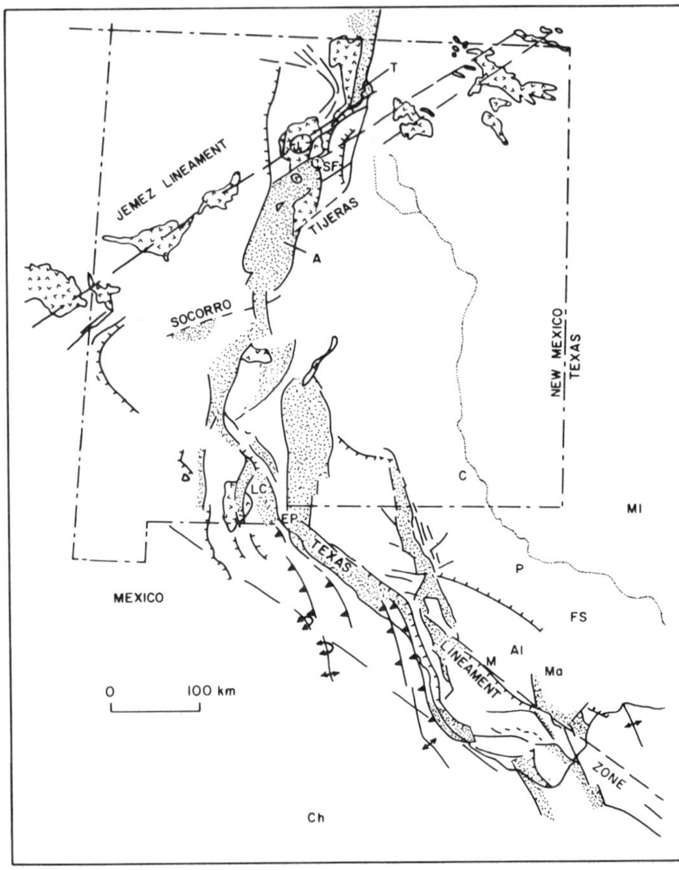

Figure 1. Rio Grande rift and other late Cenozoic tectonic elements of New Mexico, Trans-Pecos Texas and adjacent Mexico. Major grabens are stippled. Areas with random v's are Pliocene to recent volcanic centers. Normal faults are lines with ticks. Dashed lines mark principal lineaments. Laramide thrusts (teeth on hanging wall) and folds are also shown. South boundary of 80-km-wide Texas lineament zone (Muehlberger, 1980) is the dashed line that coincides with the Rio Grande along the southwest margin of the Big Bend region. It continues southeastward into Mexico as the La Babia lineament (Charleston, 1981). Dotted line is Pecos River. Population centers: A, Albuquerque; Al, Alpine; C, Carlsbad; Ch, Chihuahua; EP, El Paso; FS, Fort Stockton; J, Jemez Mountains; LC, Las Cruces; M, Marfa; Ma, Marathon; Mi, Midland; P, Pecos; SF, Santa Fe; T, Taos.

The small flows, dikes, stocks, and, rarely, sills are dominantly alkaline basalts that are notably more primitive than those from older vents nearby (James and Henry, 1991). Several 22-Ma sills and dikes bear lower crustal and/or mantle xenoliths; such bodies were intruded along deeply rooted long-lived fault zones and are evidence for deep crustal rupture in this part of the rift. No felsic lavas or intrusions are known from the Miocene Big Bend rift suite, but age determinations for newly recognized rhyolitic ash-flow tuffs associated with 22-Ma basalts (C. D. Henry, written communication, 1992; Dickerson, analyses in progress) may require revision of that statement.

Volcanism is now recognized as the most variable element of rift tectonics with respect to volume, composition, and age relative to structural development (Rosendahl, 1987). As Mohr (1982) has illustrated, volcanism can occur at virtually any stage of rift evolution and can differ markedly from one segment of a rift to another—well exemplified by the Eastern and Western branches of the East African rift system (Mohr and Wood,1976; Dunkelman, 1986; Rosendahl, 1987).

In the Colorado and New Mexico portions of the Rio Grande rift, Baldridge et al. (1984) and Olsen et al. (1987) comment that no single igneous composition is dominant; in addition, there appears to have been no temporal or spatial evolution in rift magma composition. They account for the compositional variability by means of unintegrated heat sources beneath the rift that produced localized melting. A "magma gap" between 26 and 13 Ma is documented in that region (Seager and Morgan, 1979).

In contrast, in Trans-Pecos Texas rift-related igneous bodies are alkalic basalts (James and Henry, 1991; Henry and McDowell, 1986). Volumetrically significant volcanism in Big Bend is restricted to 23 to 21 Ma and to only two eruptive centers: the Bofecillos Mountains and in Black Gap graben. Here, as farther north, there is no persuasive evidence for a massive mafic intrusive complex in the lower crust beneath the Rio Grande rift (Baldridge et al., 1984). Seismic refraction and reflection data are sparse to nonexistent in the Big Bend, but there is no indication of such a feature in gravity or magnetic data (Mraz, 1977; Mraz and Keller, 1980; La Freniere, 1977; Keller and Peeples, 1985).

Although young volcanic centers are absent from this region, heat-flow values are nonetheless elevated in both rift basins and transverse structural zones of the Big Bend, extending into eastern Chihuahua (Kopp, 1977; Hoffer, 1979; Smith and Jones, 1979; Taylor and Roy, 1980; La Freniere, 1977; Henry et al., 1983). Warm to hot springs are found along boundary faults of deep rift basins or at intersections of graben-bounding faults and transverse zones; examples of both modes of occurrence are well documented in Presidio graben (Kopp, 1977; Swanberg, 1979; La Freniere,1977; Henry and Price, 1985). The hottest temperature recorded (90°C) for springs in Chihuahua and west Texas is at Ojos Calientes, Chihuahua, on one of the western bounding faults of the northern Presidio graben (Henry et al., 1983).

Henry et al. (1983, p. 116) comment that, except for the Valles caldera system in New Mexico, temperatures at Ojos Calientes are higher than those for any other purely convective geothermal system in the Rio Grande rift of Colorado or New Mexico. The highest apparent heat flow is along the Rio Grande in Chihuahua, Texas, and westernmost Coahuila, consistent with the continuation of the rift through the Big Bend.

Deep early Miocene grabens, synextensional volcanism, elevated heat flow, and evidence for rupture of the entire lithosphere indicate that the Rio Grande rift extends at least as far south as the continuations of Castolon and Tornillo grabens into Chihuahua and Coahuila, Mexico (Figs. 2, 3).

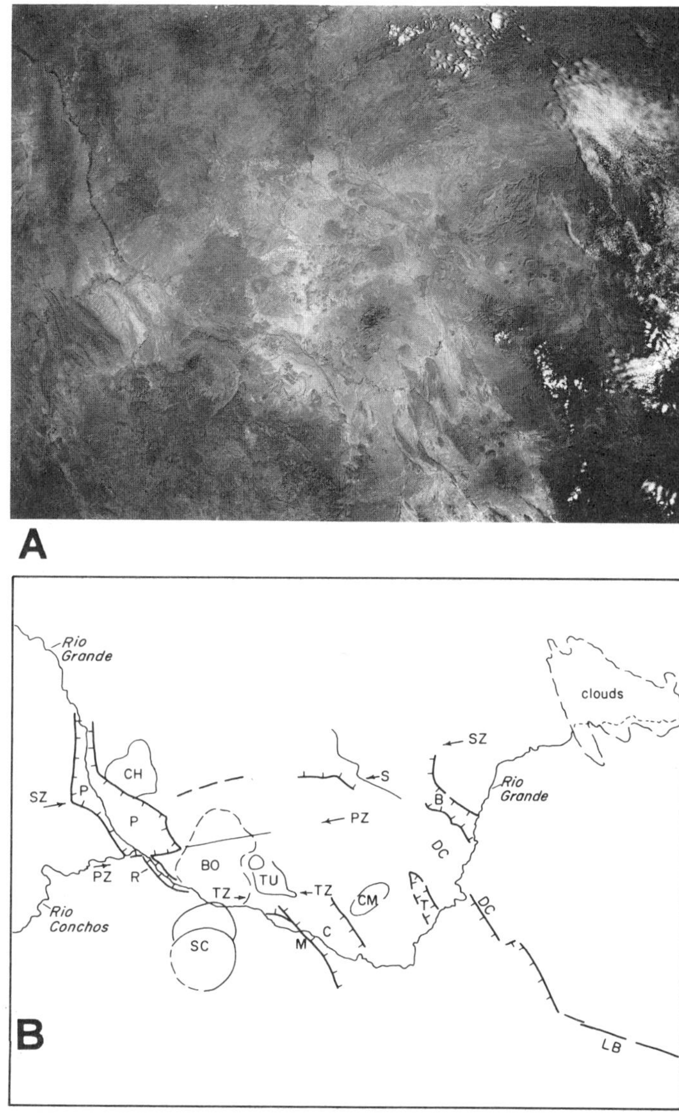

Figure 2. A, Part of Space Shuttle hand-held photograph of Big Bend; view is north (Frame STS52-153-054). B. Sketch map of A; abbreviations: B, Black Gap graben; BO, Bofecillos Mountains; C, Castolon graben; CH, Chinati Mountains; CM, Chisos Mountains; DC, Sierra del Carmen; LB, La Babia lineament, M, Mesa de Anguila; P, Presidio graben; PZ, Presidio zone; R, Redford graben; S, Santiago Mountains; SC, Santana/San Carlos calderas; SZ, Shafter zone; T, Tornillo graben; TU, Terlingua uplift with the Solitario laccolith (circled) in northern half of uplift; TZ, Terlingua zone.

RIFT BASINS

Introduction

Basins of the Big Bend segment of the rift include the Presidio (northern and southern segments), Redford, Santana, Black Gap, Castolon, and Tornillo grabens (Figs. 2, 3), as well as extensions of the latter two into Mexico. The largest, deep-

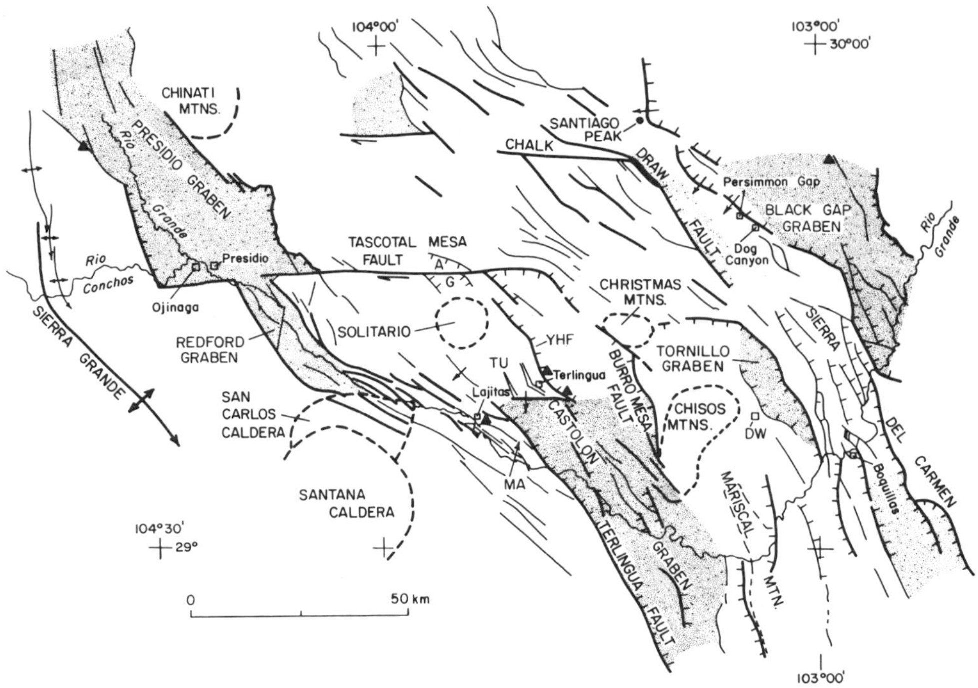

Figure 3. Big Bend segment of Rio Grande rift. Late Cenozoic grabens are stippled and major mid-Cenozoic igneous centers are shown. The Sunken Block (Udden, 1907) extends southward from the Chalk Draw fault zone and encompasses the structurally low block between the Terlingua fault on the west and the Santiago Peak–Sierra del Carmen on the east. AG, Alazan graben; DW, Dugout Wells; MA, Mesa de Anguila; TU, Terlingua uplift; YHF, Yellow Hill fault. Laramide monoclines (arrow points to down side) have been reactivated as late Cenozoic normal (oblique slip) faults (tick marks on downthrown side). Triangles indicate Miocene basalts that locally contain lower crust or mantle xenoliths; age at the Presidio graben locale is not known. Tornillo Creek flows east, then southeast along bounding faults of Tornillo graben; Tornillo Flats and Grapevine Hills are within the bend of the creek and faults. Dugout Wells fault defines east side of Tornillo graben.

est basin—Presidio—is on the west and is superposed upon the easternmost margin of the Mesozoic Chihuahua trough. South of Presidio graben the Redford, Santana, and Castolon grabens appear to have developed along what was the platform-basin hinge of the older trough.

Bounding faults strike north-northwest, and basins of varying sizes terminate or change geometry at westerly transverse fault zones, many of which have undergone right-lateral movement (Dickerson, 1980, 1983; Ewing, 1985).

Foundering of the basins, based upon vertebrate remains recovered from the sediments and K-Ar dating of igneous units interlayered with the fill, began in early Miocene time (Stevens and Stevens, 1989, 1990a; Henry and McDowell, 1986; Henry and Price, 1985). Basal deposits in the earlier formed grabens are dated at 23 to 22 Ma and rest with slight to moderate unconformity upon Eocene-Oligocene volcanic and volcaniclastic material (Stevens and Stevens, 1990a).

A later episode of graben formation occurred at about 10 Ma; younger basins (e.g., Tornillo; Figs. 3, 4) trend north-northwest, whereas those of the earlier generation (e.g., southern Presidio; Figs. 3, 5) trend northwest. In Tornillo basin the oldest graben fill contains Clarendonian (late Miocene) vertebrates, which restricts the age of the sediments to between 11 and 9 Ma (Stevens and Stevens, 1990a). No synextensional igneous units that might be dated isotopically have been found in the 10-Ma basins. In the Castolon graben there is an angular unconformity between lower Miocene and upper Miocene graben fill, attesting to structural activity there prior to deposition of the upper Miocene sequence; the exact timing of pre-late Miocene tilting has not yet been established (Stevens and Stevens, 1990a).

Despite the varied sediment source areas and lithologies, bolson fill from basin to basin consists of similar-appearing suites of pink-tan silt, clay, and sandstone, and gray conglomerate. Presidio graben is the only one in this part of the rift in which highly saline lakes are known to have persisted (Dickerson, 1966; Groat, 1972). The term *bolson* refers to the internally drained basins that, in this context, formed during rifting. The rift basins ceased to be bolsons when through drainage became established by the Rio Grande, probably in latest Pliocene to earliest Pleistocene time.

In the rift basins of the Big Bend, both margins are faulted; half grabens with ramp margins, such as described by Rosendahl (1987) in the East African rift, are not observed.

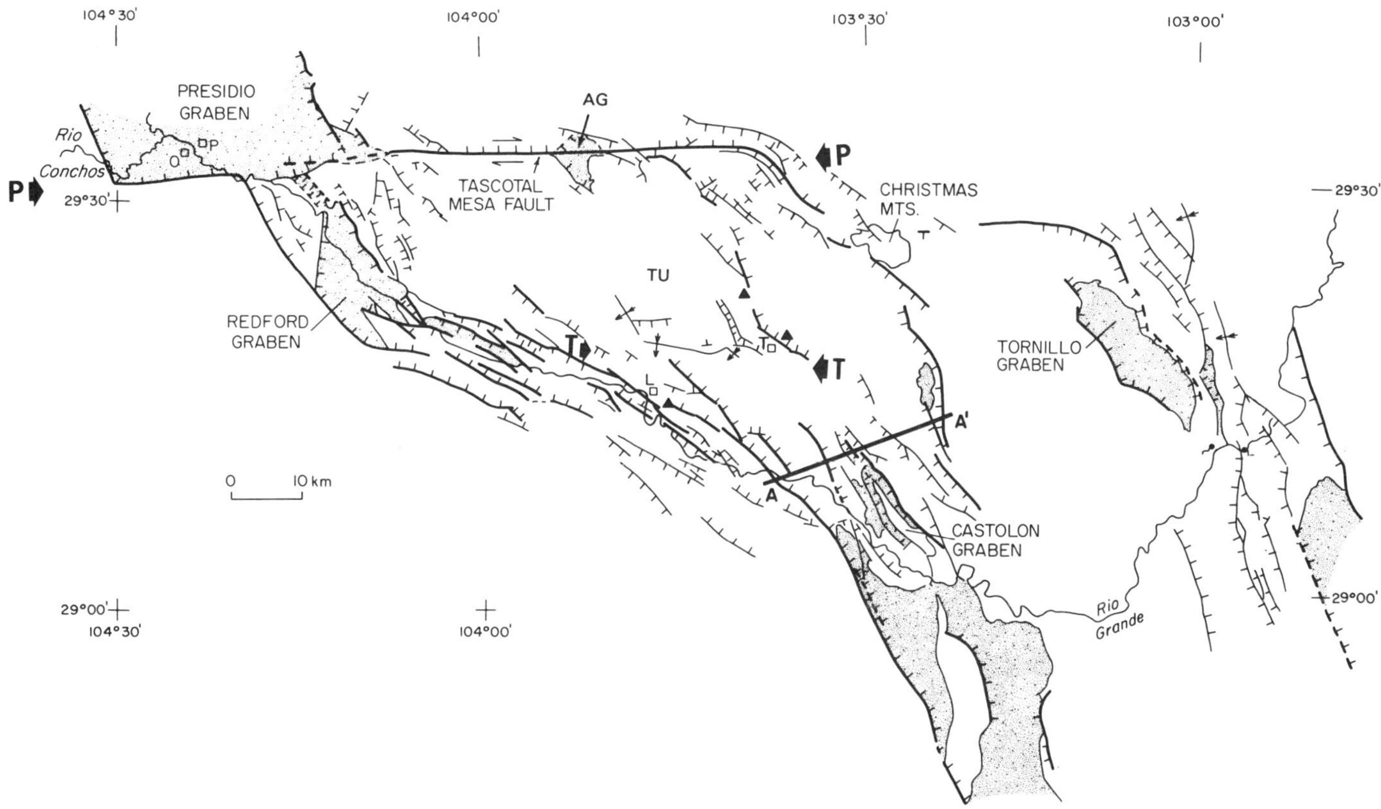

Figure 4. Presidio and Terlingua transverse structural zones, rift grabens, and late Cenozoic faults. Large arrows labeled **P** and **T** indicate extent of the zones. A–A′ is line of cross section, Figure 6. AG, Alazan graben, astride Tascotal Mesa fault, contains conglomerates mapped as Quaternary (Erickson, 1953), as well as an older suite along the southwest graben margin (Dickerson, field work in progress). TU, Terlingua uplift. Triangles, early Miocene basalts that locally contain lower crust or mantle xenoliths. Tadpole symbols, hot springs. Late Cretaceous–Eocene monoclinal flexures (Terlingua block and Sierra del Carmen) discussed in text. Population centers (squares): O, Ojinaga; P, Presidio; L, Lajitas; T, Terlingua. Age of graben fill in the Mexican extension of Castolon graben and in that farther east in Coahuila has not been established.

Bounding faults are steep, with dips greater than 60° and usually greater than 70°. No listric faulting is known and hanging-wall strata have undergone relatively little tilting or rotation.

The degree of asymmetry of many of the basins is not known, as most are not eroded sufficiently to expose deep basin architecture; borehole and geophysical data are sparse to nonexistent. Only Castolon graben has been eroded deeply enough to reveal that it is not markedly asymmetrical (Fig. 6). However, gravity data in Presidio bolson (Fig. 7) are indicative of asymmetry.

Stevens and Stevens (1990) comment that basinal depositional geometries in this region agree with patterns suggested for rift and pull-apart basins by Blair and Bilodeau (1988). Further, they observe that the grabens in Big Bend, within the Texas lineament zone (Muehlberger, 1980; Muehlberger and Dickerson, 1989), are generally narrower and more steeply faulted than those outside. They attribute the differences in basin size and geometry to transtensional deformation in the Texas lineament corridor, contrasted with more distinctly extensional deformation outside the zone—the northern Hueco and Salt basins, for example (Seager, 1980; Goetz, 1980).

Rio Grande rift basins that are within the Texas lineament zone step to the left; in contrast, they step to the right upon crossing the Jemez lineament in the New Mexico portion of the rift (Fig. 1). Such changes are common as, for example, in the East African rift (Rosendahl, 1987).

Presidio graben

Presidio graben is by far the largest and deepest in the Big Bend region. It comprises two major subbasins. The northern subbasin is 20 by 40 km, and is north trending. The southern subbasin is 30 by 40 km, and is northwest trending (Fig. 5). Presidio graben lies astride an older tectonic discontinuity that separates the thick Mesozoic sedimentary sequence of the Chihuahua trough from a thin Mesozoic platform section on the Texas side.

The northerly subbasin is bounded on the east by Sierra

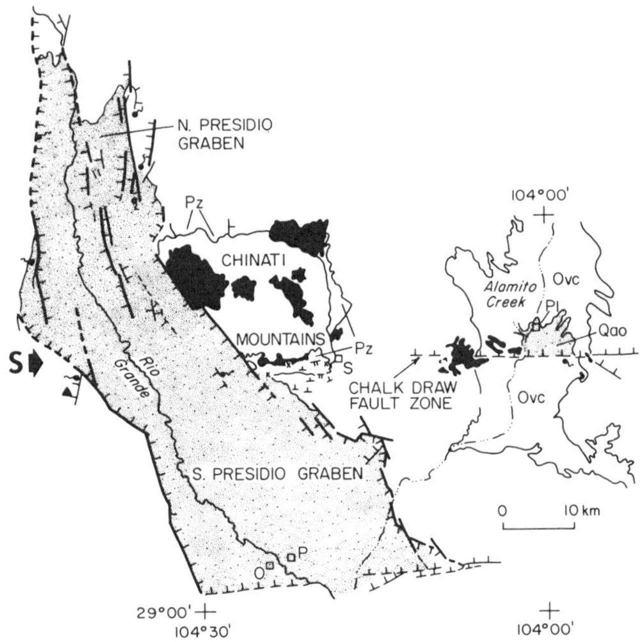

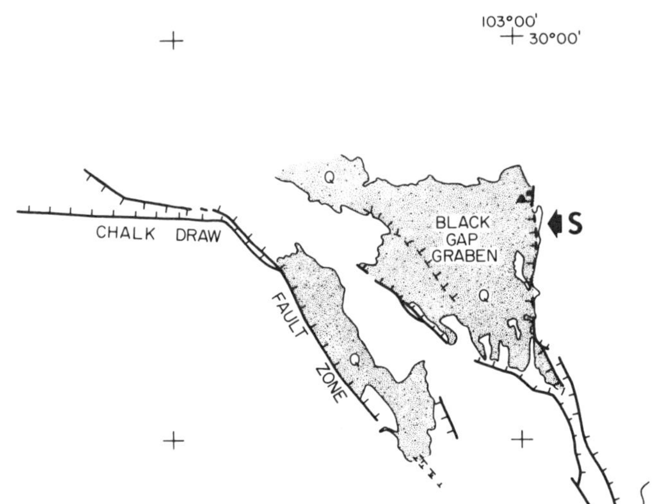

Figure 5. Shafter transverse structural zone, rift grabens, and late Cenozoic faults. Large arrows labeled **S** indicate extent of the zone. Population centers (squares): O, Ojinaga; P, Presidio; Pl, Plata; S, Shafter. Stratigraphy: Ovc, Oligocene volcaniclastics; Pz, Paleozoic strata; stipple, graben fill; black, Eocene-Oligocene intrusions. Basin sediments in Black Gap graben, the unnamed graben west of Black Gap, and the small Plata subbasin are undated but in all cases include post-Oligocene sediments of at least two generations; these deposits are shown as Quaternary on earlier maps (Henry and Price, 1985; Barnes, 1979). Age relations are discussed in detail in text. Triangles, intrusions bearing lower crust or mantle xenoliths; age of basalt at Black Gap is early Miocene, diorite at Presidio graben is undated. Tadpole symbols, hot springs or wells yielding hot waters.

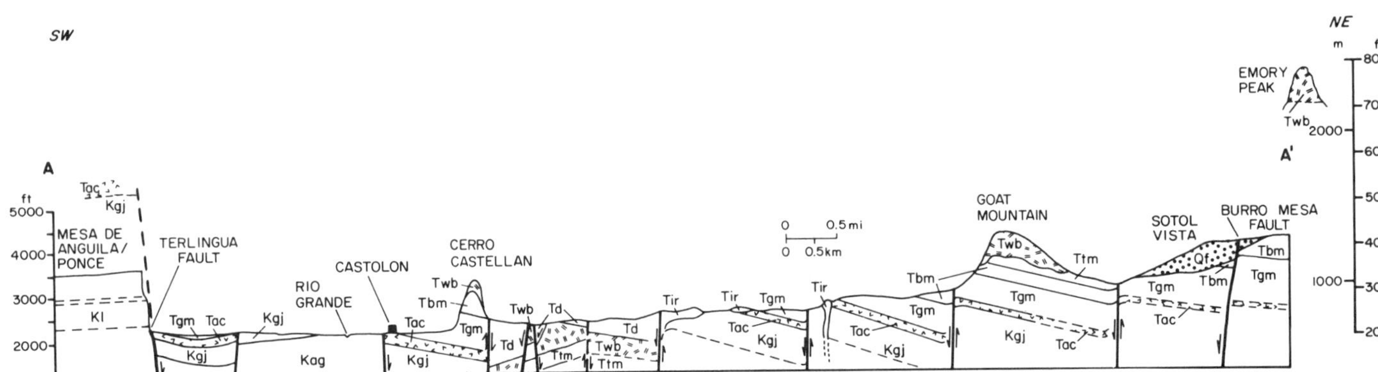

Figure 6. Cross section A–A′ across Castolon graben (location on Fig. 4; vertical exaggeration 5.1×). Arrow indicates fault discussed in text as having undergone several episodes of displacement. Symbols: Kl, undifferentiated Lower Cretaceous rocks (mainly Albian); Kag, Aguja Formation (Campanian); Kgj, Javelina Formation (Maastrichtian). Members of Chisos Formation (Eocene-Oligocene) include: Tac, Alamo Creek basalt (43 to 42 Ma); Tgm, Goat Mountain Member of Ash Spring basalt (34.5 Ma); Tbm, Bee Mountain basalt and Mule Ear Spring Tuff (34.5 to 34 Ma); Ttm, Tule Mountain trachyandesite. Members of South Rim Formation (Oligocene) include Twb, Wasp Spring Member and Burro Mesa Rhyolite (32 to 31 Ma). Tir, intrusive/extrusive rhyolite; Td, Delaho Formation, lower member (lower Miocene, 23 Ma); Qf, Fingers Formation (Pleistocene). (Modified from Stevens, 1969.)

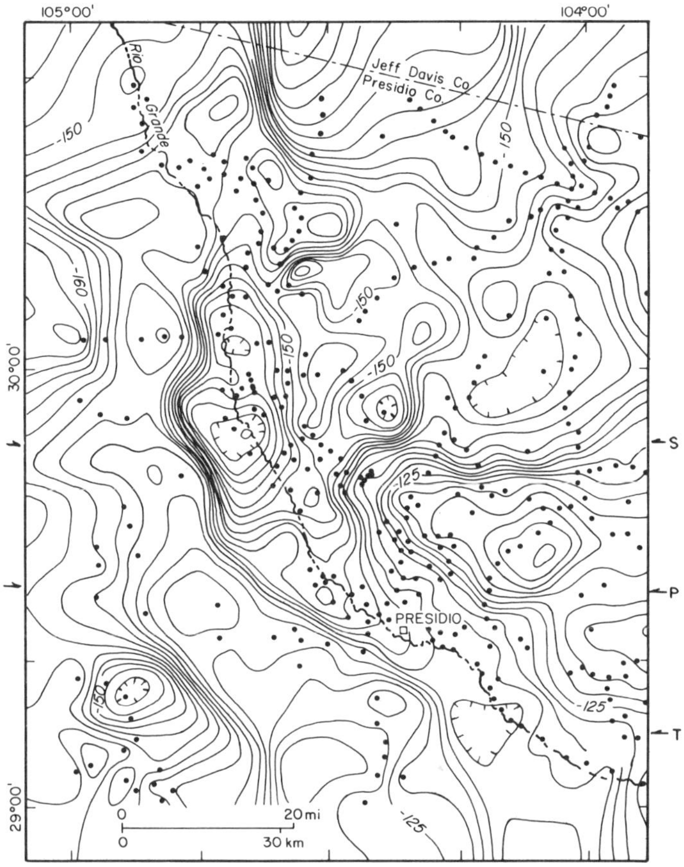

Figure 7. Bouguer gravity map of Presidio and Redford grabens, adjacent western Big Bend and Mexico. Contour interval is 2.5 mgal; dots, gravity stations. Shafter, Presidio, and Terlingua transverse zones are expressed as pronounced gradients labeled S, P, and T; only westernmost Terlingua zone is shown. (Modified from La Freniere, 1977).

Vieja and the Chinati Mountains and on the west by Sierra de Pilares/northern Sierra Grande in Chihuahua. Sierra Vieja exposes Eocene-Oligocene extrusive rocks and a thin Cretaceous platform section. The Chinati Eocene-Oligocene caldera complex is flanked by Permian and Pennsylvanian cratonal sedimentary rocks and by Cretaceous platform carbonates.

Across the graben, thick Cretaceous basinal shales and limestones and Jurassic evaporites are exposed in the great folded and thrust-faulted ranges of the early Cenozoic Chihuahua tectonic belt (Gries and Haenggi, 1970). High-angle normal faults on both sides of the graben juxtapose the older sedimentary and igneous strata with Pliocene and younger basin deposits. The presence of Miocene rocks is likely but has not yet been established. At least one graben-margin fault cuts young alluvial-fan deposits shed from the Chinati Mountains (Groat, 1972). The western margin of the northern subbasin is marked by a steep 25-mgal gravity gradient (Fig. 7). Assuming a 0.2-mgal density contrast between Mesozoic sedimentary rocks and generally unconsolidated fill, then structural relief across the western boundary fault is about 3.5 km. The eastern margin has a relatively gentle 10 to 15 mgal gravity gradient into the basin. These gravity data suggest that Presidio graben is asymmetrical.

The greatest structural relief in the southern subbasin is also along the western margin where Cretaceous basinal sedimentary strata are in fault contact with the Pliocene and younger fill. The northwesterly graben-bounding faults on the east side put Eocene-Oligocene volcanic and volcaniclastic strata erupted from the Chinati center against the younger fill.

Subsurface data indicate that the southern basin is not as deep as the northern portion. Bolson fill in the south exceeds 415 m in thickness (Groat, 1972). Gravity differential across the western boundary fault suggests a thickness of fill of about 1.75 km. The ARCO #1 Presidio State well, about 11 km northwest of Presidio, penetrated 1.04 km of Cenozoic deposits (including an unknown thickness, but probably small, of early Cenozoic prebasinal volcanic and clastic rocks) before reaching Cretaceous rocks (DeJong and Addy, 1992). The fill comprises fine-grained, basin-center sediments and coarser, more angular basin-margin debris; mud, silt, and gypsum accumulated in axial high-salinity lakes.

Elevated heat flow has been documented in Presidio graben and an alignment of higher temperatures is coincident with the trough. Several hot springs rise along border faults in the north; Ojos Calientes, on one of the western bounding faults in Chihuahua, yields the hottest waters of any springs or wells in the region (Henry et al., 1983). Warm to hot springs are also found at the intersection of bounding faults with Shafter zone (Fig. 5; Kopp, 1977; Swanberg, 1979; La Freniere, 1977; Henry et al., 1983).

Redford graben

Redford graben is a maximum of 30 km long and 1 km wide (Figs. 3, 4). On the northeast latest Oligocene to earliest Miocene volcanic flows of the Bofecillos center are in fault contact with basin fill (McKnight, 1970). Both Cretaceous rocks and flows that erupted from San Carlos caldera (late Oligocene) are faulted against fill along the southwestern boundary (Groat, 1972).

The southern Presidio and Redford grabens are distinct on the gravity map (Fig. 7) and are separated by a saddle of relatively higher values at the position of the Presidio zone (Mraz, 1977; La Freniere, 1977). The Redford graben is seen as a shallow depression on the gravity contours.

Groat (1972) describes most of the fill of Redford bolson as interbedded, poorly sorted pebbly sandstone and sandy conglomerate with siltstone and mudstone near the basin center. Thickness of the fill remains to be determined but it is probably thinner than that of the southern Presidio graben. Southern Presidio and Redford basins were not connected during bolson filling; conglomerate and pebbly sandstone were deposited along an alluvial divide that separated the two depressions where finer-grained sediments accumulated. The divide coincided with what is expressed at present as the saddle in the gravity data.

Santana graben

Santana graben is superposed on the platform hinge of the Mesozoic Chihuahua trough; between the Solitario on the western Terlingua uplift and the northeastern margin of Santana caldera (Fig. 3) there is a five-fold thickening of the Lower Cretaceous sedimentary section (Muehlberger, 1989). During latest Oligocene (28 to 27.4 Ma; Henry and McDowell, 1986) volcanism in the Santana and Bofecillos eruptive centers this area was an intervening depression into which the Santana tuff thickened markedly.

Two bolsons—the Santana and another unnamed one—formed after the tuff was deposited, in response to right transtension along west-northwesterly faults. Timing of faulting and onset of bolson filling in both is well established based upon early Miocene fossils within the basal fill and, in the case of Santana bolson, on isotopic age determinations on a 22-Ma basalt intercalated with the sediments (Stevens and Stevens, 1985, 1990a). Right-lateral movement may have been of greater magnitude here than in basins to the north and to the south, owing to its more west-northwesterly orientation during early Miocene east-northeasterly extension (Henry and others, 1983, 1991).

Sunken Block

The Sunken Block of Udden (1907) is a broad structural and topographic low area, as wide as the Albuquerque basin (Fig. 1), that straddles most of the southern Big Bend. The block consists of the Castolon and Tornillo grabens and the intervening horst occupied by the Eocene-Oligocene Chisos Mountains eruptive centers (Figs. 1 through 3). The architecture of this composite graben is similar to that of San Luis Basin in the northern rift, although the central San Luis Basin is somewhat wider. The structural continuity of this block with the rest of the Rio Grande rift (Trans-Pecos rift system of Baker, 1934; Pearson, 1981) can be seen in Figure 1. In contrast, Olsen et al. (1987), in their review of the Rio Grande rift, terminate the rift with the Presidio graben.

Structural relief of Sunken Block grabens is not as great as for the Presidio and others farther upriver, where the Rio Grande has yet to erode to the bottom of the rift fill. These smaller basins have been excavated deep into the fill and locally to the substrate. Because only remnants of the bolson sediments are preserved, the principal faults active during different stages of graben foundering are harder to define. However, deep erosion, as in Castolon graben has exposed the steeply dipping, basin-bounding faults.

The eastern boundary of the Sunken Block is defined by the Texas segment of the Sierra del Carmen (also called Sierra del Caballo Muerto), which consists of north-trending Laramide monoclines that are offset or terminate at northwesterly left shears. These, in turn, are cut by superposed north-trending Rio Grande rift-related normal faults; the normal faults terminate at the northwesterly faults that acted as right shears during late Cenozoic extension (Fig. 3; Moustafa, 1988).

The eastern boundary is not a single fault but is a series of major faults that step southeastward as one goes south into Mexico (Fig. 3). Along the west base of the Mexican segment of Sierra del Carmen the highest fault scarp attains more than 1.8 km of relief; it persists southward into Mexico many tens of kilometers to its intersection with the La Babia lineament (Fig. 2; Charleston, 1981), where it steps westward and continues farther south. The southern limit of the Sunken Block has not been documented.

Chalk Draw fault forms the northern margin of the block (Udden, 1907). Structural relief from north to south in this graben complex is substantial: in the 65 km from Chalk Draw fault south to Lajitas, the Upper Cretaceous Boquillas Formation is dropped progressively down to the south about 536 m (Stevens and Stevens, 1990b). No single, areally extensive mid-Cenozoic unit covers this expanse, so it is not possible to state how much of this structural relief is attributable to Laramide and how much to late Cenozoic tectonism; however, available data permit the interpretation that most is a result of the later episode.

The Terlingua fault zone, along the east face of Mesa de Anguila, defines the western limit of the Sunken Block; it continues more than 30 km into Mexico as the east-facing scarp of Sierra Ponce/Sierra de Santa Elena. At the entrance to Santa Elena Canyon on the Rio Grande, stratigraphic throw is about 855 m on the main fault. Stratigraphic throw increases to at least 1,350 m about 15 km south of the canyon. Throw then decreases so that in another 35 km it appears from photo interpretation to have decreased to zero.

Castolon graben

The west-trending monocline that separates Terlingua uplift from Mesa de Anguila (Fig. 3) marks the northern end of Castolon graben. Near the intersection of graben faults with the westerly zone are syn-rift basalt dikes that bear mantle xenoliths (Figs. 3, 4; Nelson and Schieffer, 1990).

The Terlingua fault, described above, is the western bounding fault of Castolon graben, which extends well into Chihuahua and Coahuila. The Burro Mesa fault zone defines the eastern limit of the 1,000-m-deep basin (Figs. 3, 4, 6; Stevens, 1969, 1988). A series of west-dipping normal faults drops successive east-dipping blocks into the graben; in contrast, throw on the Terlingua fault alone equals the sum of the distributed slip across the remainder of the graben.

Until recently we believed that structural relief across the graben could be approximated by using the surfaces developed on the 31.7-Ma Burro Mesa rhyolite on Emory Peak, Burro

Mesa, Goat Mountain, and the lowest preserved remnant on Cerro Castellan (Twb on Fig. 6). The rhyolite had been interpreted as having been erupted from the Pine Canyon caldera in the Chisos Mountains. Tsai (1992), however, has demonstrated that compositionally similar rhyolites mapped as Burro Mesa were erupted not just from Pine Canyon caldera but from other vents as well. Although extrusion of the several flows occurred over a brief time interval, continuity of the unit cannot be assumed and thus the unit is no longer a reliable stratigraphic datum for late Tertiary structural reconstructions in the Castolon basin.

Stevens and Stevens (1990b, p. 44) describe a sequence of events that suggests several episodes of normal faulting, the earliest being earlier than 28 Ma. The principal fault on which episodic movement is interpreted marks the east side of the deepest downthrown fault block lying east of Cerro Castellan (arrow, Fig. 6). There, 28-Ma rhyolite (Tir) intrudes the fault and lies with angular unconformity across Cretaceous rocks and the lowest units of the Eocene-Oligocene volcanic sequence. On the downthrown block the lower Miocene fill (age well established based on vertebrate fauna), with a 23-Ma basalt flow near the base of the unit, rests on strata nearly 305 m higher in the mid-Cenozoic volcanic sequence. An additional 244 m of down-to-the-southwest offset on that fault occurred after eruption of the 23-Ma basalt and deposition of the lower graben fill (Stevens and Stevens, 1990b).

Thus, Stevens and Stevens suggest that this fault and others in the vicinity underwent significant normal displacement earlier than the 24-Ma time assigned to the beginning of Basin and Range events in Trans-Pecos Texas by Price and Henry (1984). They further suggest that this faulting may be an early expression of the approximately 31-Ma reorientation of regional stresses from east-northeast maximum compression to east-northeast minimum compression (Price and Henry, 1984; Muehlberger and Dickerson, 1989).

The coarser upper Miocene fill lies unconformably across the finer earlier Miocene fill as well as on units lower in the underlying volcanic sequence, indicating continuing differential subsidence and erosion of high-standing blocks. It also covers a wider area and extends south across the Rio Grande. At the foot of the Terlingua fault scarp, the upper interval contains the only limestone pebbles known from any of the graben fill, presumably derived from the upthrown side of the fault (Stevens, 1969).

In this deeply eroded graben can be seen the fault pattern that may also lie at depth within the bigger and less deeply eroded basins, such as Presidio. A maximum of 340 m of graben-filling sediments is measurable within a fault block (Stevens, 1969). For the most part, the lower Miocene fill crops out north of the exposures of the upper Miocene unit, so that areas of maximum thickness are not superposed. Faults die out along strike but are spaced about 1 km apart across the 23-km-wide graben (Figs. 4, 6). In the eastern half of the graben, the faults are down to the west with east-tilted blocks, whereas the western half is deeply downthrown on a master border fault with a central deep (area east of Cerro Castellan; Fig. 6). The only faults with stratigraphic separations of more than 200 m are the faults of the western boundary, on which throws are 1,000 m.

Consistently tilted blocks and faults of the opposite dip, such as those of the eastern half, have been interpreted in other areas as evidence that the master fault is listric. There is no master fault along the eastern boundary of the Castolon graben. We can only show that the faults of the entire graben are all steeply dipping (greater than 60°, usually greater than 70°), and that the en echelon and curving pattern of faults suggests a strike-slip (simple shear) component to extensional deformation in the region. The extensive low-angle normal faulting that is common elsewhere in the rift during this early phase of deformation is not seen here (cf. Olsen and others, 1987).

Tornillo graben

The name *Tornillo graben* as used in this paper refers to the normal-fault-bounded basin in which Tornillo Creek flows and which includes Tornillo Flat farther north (Figs. 3, 4). This usage contrasts with that of Lehman (1986, 1988, 1991) who applies the name to the Late Cretaceous to Eocene foredeep that extended southeastward across much of the southwestern part of Trans-Pecos Texas. The late Cenozoic graben overlies what had been the deepest part of the older basin (Lehman, 1991).

Tornillo Creek marks the eastern boundary of the structurally lowest segment of the Tornillo graben (= Estufa graben of Stevens and Stevens, 1990a, b), as well as the eastern margin of the Sunken Block. Quaternary movement on this fault is indicated by offset caliche horizons (Stevens and Stevens, 1990a). The north-northwest–trending escarpment just east of Dugout Wells (Fig. 3) is along an active fault that forms the western graben edge. Stevens (oral communication, 1986) has traced this fault south to the Rio Grande and at least as far north as the Grapevine Hills. Hot springs in the Rio Grande rise where Tornillo graben-forming faults either end or step eastward.

The southern half of the 20-km-long by 8-km-wide graben is covered by fan and pediment deposits that obscure its termination at or south of the Rio Grande. The northern half is divided into two north-trending segments. The eastern portion of the northern segment exposes the upper 500 m of upper Miocene alluvial fan deposits, the oldest of which is dated at about 10 Ma (Stevens and Stevens, 1990b) and which has a southern source, presumably from Mexico (Thurwachter, 1984). The western portion exposes the upper few meters of a Pleistocene alluvial fan sequence derived from the Chisos Mountains on the west.

Two kilometers north of the preserved graben fill Tornillo Creek turns abruptly westward at Tornillo Flat toward its headwaters in the Christmas Mountains. West-striking dikes and normal faults, along with the straight westerly course of the creek, are interpreted as marking the northern margin of the Tornillo graben. The southern edge is undefined at present.

The lower Tornillo Creek fault zone is roughly aligned with the north-northwest-oriented segment of Chalk Draw fault (farther north; Fig. 3); however, throw on the two faults is opposite, changing where they encounter the easterly discontinuity. Quaternary movement is not yet documented on that part of Chalk Draw fault.

No detailed gravity or heat-flow investigations have been conducted in this area as far as we are aware. Geophysical and heat-flow studies would bear significantly on the structural setting of the hot springs along the Rio Grande near Boquillas (Fig. 4), and would help define the southeastward extent of Tornillo graben into Coahuila.

Unnamed basin north of Tornillo graben

The northern basin margin is complex, marked by several small igneous bodies and by an abrupt jog (northwest to north-northwest) in the Santiago Mountains (Figs. 3, 5). The north-northwesterly section of Chalk Draw fault defines the west side of the graben, where basin fill is juxtaposed against generally flat-lying Cretaceous strata. This basin has not been studied, but the presence of scattered outcrops of Cretaceous rocks suggests that the depth of fill is small.

The east side is flanked by the Santiago Mountains, where deformed Paleozoic and Cretaceous sedimentary rocks are exposed, and the northernmost Sierra del Carmen, consisting mostly of Cretaceous rocks (Fig. 3). Normal faults are not mapped along the eastern boundary north of Dog Canyon (Cobb and Poth, 1980); to the south, however, north-northwesterly faults of the northern Sierra del Carmen drop Cretaceous bedrock down to the west.

Although the contact of Cretaceous rock with both older and younger fill is quite sharp and straight, it is not mapped as a fault (Barnes, 1979; Henry and Price, 1985). Areas of older and younger alluvium have been delineated but age distinctions among the sedimentary suites have not been drawn, nor are estimates of thickness of the fill available.

Black Gap graben

Black Gap graben is east of the Santiago Mountains (Figs. 3, 5). In contrast with the northwesterly trend of the western graben boundary and its Laramide flower structures (Cobb and Poth, 1980), faults of the eastern margin are northerly and juxtapose relatively undeformed Cretaceous strata and graben fill. Along one of the eastern border faults mantle xenoliths have been brought up in a young basalt plug dated at 23 Ma (Dahl and Lambert, 1986).

Early synextensional basaltic volcanism appears to have been more voluminous in this graben than elsewhere in Trans-Pecos Texas. St. John (1965) showed that Laramide monoclines had been eroded to a level surface prior to extrusion of 22-Ma basalts. As much as 140 m of basalt is preserved on various fault blocks throughout the 22- by 50-km graben and in a tight V-shaped syncline within a deep, narrow, extensional basin. The basalt flows serve as markers for 1,000 m of structural relief across the late Cenozoic graben complex. A conglomerate containing clasts of Paleozoic rocks derived from the Marathon Basin to the north underlies the basalt near Black Gap.

In the southernmost part of the graben, which extends no more than 3 to 4 km into Coahuila, St. John (1965) mapped more than 50 m of Pliocene-Pleistocene(?) bolson deposits consisting predominantly of pinkish orange silty clay similar to the lower fine-grained fill in other grabens. Interbedded with the silty clay are pebbly and cobbly gravels that contain no Marathon Basin facies clasts. The sediments are interpreted to have been deposited in a small basin imponded by a fault block of Cretaceous limestone, through which the Rio Grande has since cut.

In northern Black Gap graben, as in the basin on the opposite side of the Santiago Range, areas of older and younger alluvial fill have been mapped (Barnes, 1979), but neither litho- nor biostratigraphic studies of the sediments have been undertaken. Contacts between different generations of fill are straight, sharp, and in places coincident with stream courses (e.g., Maravillas Creek), and there is the possibility of structural control (inferred faults, Fig. 5)

Small basins within Sierra del Carmen

A number of narrow, north-trending basins lie in downthrown blocks within the Texas portion of the Sierra del Carmen (Moustafa, 1988, Fig. 7). None has been studied to ascertain age, thickness, or geometry of the fill. It is likely, however, that all contain only minor thicknesses of locally derived sediment—probably younger than Miocene. In one, traversed by the national park road to Boquillas Canyon, playa-lake deposits (Pliocene-Pleistocene?) are well exposed. The Rio Grande flows through the basin, most of which is in Mexico, and imposing fault scarps bound the graben farther downstream. Study of both faults and fill could bear significantly upon the continuation of the rift into Mexico.

TRANSVERSE STRUCTURAL ZONES

North-northwest–trending rift basins of the Big Bend are linked or segmented by westerly structural discontinuities of varying antiquity, relief, and displacement. All the transverse zones demonstrably predate Miocene and younger extensional/transtensional deformation and foundering of the basins.

Graben-bounding faults terminate, change strike, or step laterally at the westerly structural zones (Figs. 4, 5—particularly the change from southern Presidio graben structures to those of Redford graben). Hot springs and/or syn-rift basalt bodies are commonly found at boundary-fault/transverse-zone intersections.

Displacement on Big Bend westerly transverse fault zones is dominantly vertical with a minor component of right slip, which is consistent with determinations of Neogene stress orientations for the region (Henry et al., 1983, 1991; Henry and Price, 1985).

In this part of the southern Rio Grande rift, early Miocene syn-rift basalts have intruded two of the transverse zones, as well as one of the graben-bounding faults; several contain lower crustal and/or mantle xenoliths (C. D. Henry, written communication, 1992; Dahl and Lambert, 1986; Nelson and Schieffer, 1990).

Rift-basin terminations or offsets are commonly attributed to transform-fault activity (see numerous sources cited in Rosendahl, 1987). Preliminary observations on the westerly structural zones of this region led Dickerson (1980) to speculate that at least some might be transforms within the Rio Grande rift system.

Several lines of evidence, however, indicate that the structural discontinuities discussed here, although active during Rio Grande rifting, have not operated as transforms. First, the transverse structures long predate late Cenozoic rifting. Second, the westerly zones extend well beyond their intersections with the several rift basins. Finally, they have produced negligible lateral offset. Locations and orientations of the transverse zones are inherited from prior deformational episodes and are not products of the rifting process. If extension continues, however, those parts of the zones that are within the rift could eventually serve as transforms.

Analogous transverse fault zones have been documented by Bartley (1990, 1992) and Taylor et al. (1989) within the broader extensional context of the Basin and Range province of east-central Nevada. There late Cenozoic basin-bounding normal faults terminate, splay, or step laterally at the mid-Cenozoic Blue Ribbon and Tempiute transverse zones. The westerly transverse zones predate Miocene extensional basin development and, as appears to be the case in the Big Bend region, are interpreted to have served as propagation barriers to the north-striking, basin-bounding faults.

Shafter zone

Shafter zone (Fig. 5) segments Presidio graben into two major subbasins of differing trend and width, as discussed previously. Hot springs mark intersections of the transverse zone with graben-bounding faults. La Freniere (1977) concluded that hot springs and zones of elevated thermal gradients in Presidio graben are primarily associated with west-trending structural zones and secondarily with graben-bounding faults (as at Ojos Calientes).

The Shafter zone–Sierra Grande junction in Chihuahua (Fig. 5) is of additional interest, as gneiss and phyllite xenoliths have been brought up in a diorite plug at that locale (Cerro Mal Nombre; Heiken, 1966). Predominantly west-striking as well as west-southwest–striking basalt dikes (undated) crop out in the same area.

About 38 km east of Presidio graben, the zone forms the southern boundary of a small (6 by 7 km) post-Oligocene graben, herein named the Plata basin, which is superposed upon a Late Pennsylvanian to Early Permian basin that received thousands of meters of sediment (Ammon, 1981; Pearson, 1978, 1981; Dickerson, 1980, 1983). In the subbasin, south of Plata (Fig. 5), alluvium mapped as Quaternary is downwarped and may be downfaulted against Oligocene volcaniclastic deposits (W. B. Bourbon, unpublished field data, 1983).

The late Paleozoic basin and the Plata subbasin share a common southern boundary, the long-lived Chalk Draw fault zone. Although subtly expressed in places or discontinuous at the surface, it is manifest across the region as a continuous and steep down-to-the-north, east-trending gradient in both gravity (approximately 10 mgal) and magnetic (approximately 150 gamma) data (Keller and Peeples, 1985; Dickerson, 1980, 1983). As mentioned previously, the fault zone defines the northern limit of the Sunken Block graben complex that encompasses most of the southern Big Bend.

At present, vertical relief along the Plata subbasin margin at the Chalk Draw fault is 25 to 30 m; topographic and drainage offsets suggest but do not prove about 2 km of post-Oligocene right slip. A well just south of the fault yielded hot water (Fig. 5; Kopp, 1977; Hoffer, 1979) with a subsurface temperature of 85°C determined by means of silica geothermometry. That value may be too high, however, as the water-bearing volcaniclastic sediments are quite siliceous. Surface-water temperature at that well was 32°C, and temperatures in springs along the fault are somewhat elevated (Kopp, 1977).

From 50 to 90 km still farther east the Shafter zone defines the northern margins of two additional late Cenozoic basins, the unnamed one lying east of the Chalk Draw fault and Black Gap graben, which lie on either side of the Santiago Mountains (Fig. 3). Structural relief in each is greatest where lower Paleozoic Marathon Basin strata are exposed in the Santiago Mountains that separate the two later basins. The only known Miocene or younger basalts in the Shafter zone or its conjoined basins are in Black Gap graben.

Presidio zone

The Presidio zone (Fig. 4) intervenes between the 30-km-wide southern Presidio graben on the north and the 10-km-wide Redford graben on the south. In addition to its present surface structural expression, the zone coincides with an alluvial divide preserved in bolson-stage sediments of the two basins, and it is expressed in gravity data as a saddle of relatively higher values separating the negative anomalies of the basins (Fig. 7; Mraz, 1977; Mraz and Keller, 1980; La Freniere, 1977). The zone continues eastward as a distinct gradient in the regional gravity data suggesting that it, like the Shafter zone, is a structural corridor with deep roots.

Lead isotopic data indicate that the zone transects an area underpinned by three crustal types—North American, transitional, and southeastern—which likely became juxtaposed during late Paleozoic Ouachita-Marathon tectonism (James and Henry, 1990). These data bear significantly on the configuration of the pre–late Paleozoic craton margin, and the relation of the Presidio zone to that ancient boundary is under investigation.

Mid-Cenozoic volcanism along the zone is represented by some of the oldest (44 to 40 Ma) and some of the youngest (27.4 Ma) Eocene-Oligocene igneous rocks in the region (Henry and McDowell, 1986; Henry, oral communication, 1992). A small 35-Ma rhyolite intrusion just north of the Solitario has been offset right-laterally about 1 km by Tascotal Mesa fault (C. D. Henry, written communication, 1992).

Onset of late Cenozoic extension is recorded in the zone by a marked change to notably more primitive basalt compositions at 24 Ma (James and Henry, 1991). Probable syntectonic conglomerates overlie and, in places, are intercalated with 23.3- to 17-Ma basalts. At least one 22-Ma basalt has ascended along a southeasterly splay of Tascotal Mesa fault east of the Terlingua uplift (McCulloh, 1977).

Comparatively recent movement within the Presidio zone is seen along the Tascotal Mesa fault north of the Solitario where the small Alazan graben contains Quaternary gravels that are interpreted to be faulted (Fig. 4; Erickson, 1953). In addition, Quaternary deposits are displaced where Presidio zone intersects northwesterly structures of the southern Presidio graben (Dietrich, 1966).

Terlingua zone

The Terlingua zone links the Santana graben and at least one other unnamed graben with Castolon graben (Fig. 4). The Castolon basin complex, with its southeastward continuation into Mexico, is about 20 km wide and at least 50 km long.

During Late Cretaceous to Eocene, compressional deformation throughout the region the Terlingua block began to be uplifted (Erdlac, 1990) and, locally, coarse detritus eroded from the block was deposited southwest of the uplift as conglomerates laid down prior to Eocene-Oligocene volcanism (C. D. Henry, oral communication, 1992). Along the northern margin of Mesa de Anguila, northwesterly faults that formed during the early Cenozoic deformational episode change strike to west-northwest, then die out in low-amplitude folds within Terlingua zone (DeCamp, 1985; Muehlberger and Dickerson, 1989).

Emplacement of the mercury deposits that are arrayed along the Terlingua zone has been attributed to intrusion of Eocene-Oligocene igneous bodies, although geochemical and temporal details of emplacement have yet to be established. J. C. Price comments (written communication, 1992) that mid-Cenozoic igneous activity provided the heat to mobilize and concentrate the mercury.

Commercial accumulations were generally in solution-collapse zones along faults or in solution-collapse breccia pipes (Yates and Thompson, 1959). Muehlberger (1989) noted that the collapse zones are concentrated in small pull-apart grabens that lie along crests of monoclines; graben orientations and shapes are appropriate for having formed in response to late Cenozoic transtension. If the pull-apart grabens are Miocene or younger, mercury accumulation may be more recent than previously interpreted. Price differs (written communication, 1992), stating that vein orientations in the Terlingua district are more consistent with an origin related to mid-Tertiary volcanism.

Between Lajitas and Terlingua the west-trending Terlingua zone is coincident with a linear steep gradient in aeromagnetic data that is unrelated to surface and shallow near-surface strata, based upon magnetic depth estimates (Keller and Peeples, 1985; Dickerson, 1980, 1983). Whatever the origins of this long-lived discontinuity, the magnetic data indicate that it is a deeply rooted crustal flaw.

The structural belt was a locus for faulting and igneous activity once again during early Miocene time. Alkaline basalts emplaced along the zone brought up lower crust/upper mantle xenoliths (lherzolites, amphibolites, spinel peridotites, pyroxenites). Where dated, the host basalts yield ages from 24 to 20 Ma, similar to those of other extension-related basalts of the southern rift (Nelson and Schieffer, 1990; Dahl and Lambert, 1986).

The Shafter, Presidio, and Terlingua zones are old, repeatedly reactivated structural corridors that have strongly influenced—and, in places, perhaps controlled—graben development in the Big Bend segment of the Rio Grande rift.

NORTHERN AND SOUTHERN RIO GRANDE RIFT—SIMILARITIES AND DIFFERENCES

Manifestations of Rio Grande rifting in the Big Bend of west Texas are similar in many regards to those farther north along the rift. Graben formation began in early Miocene and has occurred in recent times as indicated by graben faults that cut Quaternary deposits. Initial development of deep grabens was accompanied by volcanism; timing of basin foundering (23 to 21 Ma) has become increasingly well established on the basis of both vertebrate paleontology and K-Ar geochronology. A later episode of graben formation, sans volcanism, began at about 10 Ma.

As farther north, there is no evidence in Big Bend for a massive mafic intrusive complex beneath the rift. Rupture of the entire lithosphere is suggested by the relatively primitive rift basalt compositions (James and Henry, 1991); several of the basalts have carried lower crustal or mantle xenoliths to the surface.

Linking rift grabens across the Big Bend region are west-trending transverse structural discontinuities that long predate late Cenozoic rifting; these zones have participated in and strongly influenced development of the rift structures. Across them graben-bounding faults terminate, change strike, or step laterally. Sizes and orientations of basins also change. The generally smaller size of Big Bend rift basins, relative to those farther north in the rift, probably reflects the influence of the transverse faults.

Heat-flow values are elevated both in rift basins and along transverse fault zones of the Big Bend. Warm to hot springs are found at intersections of graben-bounding faults and transverse zones, as well as along some boundary faults of deep rift basins.

There are also distinct differences, arising in large part from the crustal and structural antecedence of the region: Rio Grande rift basins are superposed upon preexisting northwesterly structures. Basin-bounding faults strike northwest to north-northwest, and grabens of varying sizes terminate or change geometry at westerly transverse faults, which have commonly undergone right-lateral movement.

In contrast with the rest of the Rio Grande rift system, the rift in Trans-Pecos Texas lies mostly within the 80-km-wide, N60°W belt of reactivated structures of the Texas lineament zone (Muehlberger, 1980; Fig. 1). Grabens of the Big Bend, within the Texas lineament zone, are narrower and more steeply faulted than those outside that belt. Long-established west-trending transverse faults subdivide this region into smaller fault blocks than are seen in the northern rift.

Differences in basin size and geometry are attributed by Stevens and Stevens (1990a) to transtensional deformation within that corridor, as contrasted with more distinctly extensional deformation outside (Fig. 1). The size of the largest and deepest rift basin of the Big Bend region—Presidio—is probably due in part to its superposition upon the easternmost margin of the Jurassic to Early Cretaceous Chihuahua trough.

Westerly discontinuities segment or link rift grabens in the Big Bend. None has yet functioned as a transform. All predate late Cenozoic extensional/transtensional deformation, all extend well beyond their intersections with rift-basin-bounding faults, and none shows significant lateral offset.

Volcanism accompanied only the early phase of rifting, between 23 and 21 Ma; in contrast with the northern rift, there are no known young eruptive centers in th Big Bend. Alkaline basalts—principally widely scattered plugs, dikes, and sills—predominate, and the felsic eruptive suites of the northern rift are not yet documented here.

Voluminous basaltic rift volcanism has occurred in only two parts of Big Bend—in the northern Bofecillos Mountains on the west and at Black Gap on the east. No vents are yet known for the 23- to 21-Ma basalts. The one 17-Ma basalt was erupted from a vent on the eastern margin of Presidio graben near the intersection with the Presidio transverse fault zone.

Acknowledging the generally noncoaxial correlation between stress and strain, and the uncertainties that this introduces in fault-slip analysis (e.g., Edelman,1989; Ring et al., 1992), some statements can nonetheless be made regarding stress orientations during rifting in the Big Bend.

Henry et al. (1983, 1991) have determined Late Cenozoic extension directions for the region from dike, vein, fault, and slickenline orientations. These data reflect an early Miocene west-southwesterly phase of extension, followed by a change to northwesterly extension. The more northerly orientations of smaller grabens containing fill dated as 10 Ma and younger probably reflect the shift to more northwesterly extension, a shift that is widely recognized in the southwestern United States (Zoback and Thompson, 1978; Eaton, 1979; Baldridge and Olsen, 1989).

The lack of volcanism in Big Bend during the 10-Ma rifting episode, contrasted with that from 23 to 21 Ma, could be attributable to the shift in extensional stress orientations. When the direction of maximum extension was northwest (Henry et al., 1983, 1991), the long-established northwesterly faults of the region would have been parallel to the greatest extensional stress—that is, under some degree of compression—and not as likely to permit magma ascent to shallow crustal levels.

In summary, deep early Miocene grabens, synextensional volcanism, elevated heat flow, Quaternary normal faulting, and evidence for full lithospheric rupture distinguish the Rio Grande rift from the Basin and Range province and indicate that the rift extends at least as far south as the continuations of the Castolon and Tornillo grabens into Chihuahua and Coahuila, Mexico. Although late Cenozoic Basin and Range faults have been documented as far south as the Trans-Mexican volcanic belt (Henry and Aranda-Gomez, 1990), the other characteristics of rifting appear restricted to the Rio Grande corridor between Texas, Chihuahua, and westernmost Coahuila (Henry et al., 1983). Interpretation of satellite imagery suggests that structures specific to the Rio Grande rift persist 150 km farther south-southeast of the Big Bend. The southern terminus of the rift, however, remains undefined.

ACKNOWLEDGMENTS

This paper has benefitted from thoughtful reviews by T. H. Anderson, F. E. Julian, F. W. McDowell, and G. E. Murray. The thorough evaluation by J. G. Price resulted in significant revisions and, we hope, improvements; we particularly appreciate his efforts.

REFERENCES CITED

Ammon, W. L., 1981, Geology and plate tectonic history of the Marfa basin, Presidio County, Texas, *in* Marathon-Marfa region of west Texas: Society of Economic Paleontologists and Mineralogists Permian Basin Section, Publication 81-20, p. 75–102.

Baker, C. L., 1934, Major structural features of Trans-Pecos Texas, Part 2—Structural and economic geology: Austin, University of Texas Bulletin 3401, p. 137–214.

Baldridge, W. S., and Olsen, K. H., 1989, The Rio Grande rift: American Scientist, v. 77, p. 240–247.

Baldridge, W. S., Olsen, K. H., and Callender, J. F., 1984, Rio Grande rift: Problems and perspectives, *in* Baldridge, W. S. Dickerson, P. W., Riecker, R. E., and Zidek, J., eds., Rio Grande rift: Northern New Mexico: New Mexico Geological Society, 35th Field Conference Guidebook, p. 1–12.

Barnes, V. E., 1979, Emory Peak—Presidio sheet: Bureau of Economic Geology, Geologic Atlas of Texas, 1 map, scale 1:250,000.

Bartley, J. M., 1990, Hypothetical mid-Tertiary synvolcanic stress rotation in the southeastern Great Basin: Geological Society of America Abstracts with Program, v. 22, no. 3, p. 6.

Bartley, J., 1992, Transverse structures in east-central Nevada: American Chemical Society, Petroleum Research Fund, Reports of Research, 1991, p. 291–292.

Blair, T. C., and Bilodeau, W. L., 1988, Development of tectonic cyclothems in rift, pull-apart and foreland basins; Sedimentary response to episodic tectonism: Geology, v. 16, p. 517–520.

Bryan, K., 1938, Geology and ground-water conditions of the Rio Grande depression in Colorado and New Mexico, *in* [U.S.] Natural Resources Planning Board, The Rio Grande Joint Investigations in the Upper Rio Grande Basin: Washington, D.C., U.S. Government Printing Office, v. 1, pt. 2, p. 197–225.

Chapin, C. E., and Seager, W. R., 1975, Evolution of the Rio Grande rift in the Socorro and Las Cruces areas, *in* Seager, W. R., Clemons, R. E., and Callender, J. F., eds., Las Cruces country: New Mexico Geological Society, 26th Field Conference Guidebook, p. 297–322.

Charleston, S., 1981, A summary of the structural geology and tectonics of the state of Coahuila, Mexico, *in* Lower Cretaceous stratigraphy and structure, northern Mexico: West Texas Geological Society Publication 81-74, p. 28–36.

Cobb, R. C., and Poth, S., 1980, Superposed deformation in the Santiago and northern Del Carmen Mountains, Trans-Pecos Texas, *in* Dickerson, P. W., and Hoffer, J. M., eds., The Trans-Pecos region: New Mexico Geological Society, 31st Field Conference Guidebook, p. 71–76.

Dahl, D. A.,and Lambert, D. D., 1986, Petrology of the Black Hills, Brewster County, Texas, *in* Price, J. G., Henry, C. D., Parker, D. F., Jr., and Barker, D. S., eds., Igneous geology of Trans-Pecos Texas: Texas Bureau of Economic Geology, Guidebook 23, p. 320–332.

DeCamp, D. W., 1985, Structural geology of Mesa de Anguila, Big Bend National Park, Texas: West Texas Geological Society Publication 85-81, p. 127–136.

DeJong, H. W., and Addy, S. K., 1992, Only relatively small production seen in basins of far West Texas: Oil and Gas Journal, v. 90, no. 3, p. 59–65 and no. 4, p. 97–102.

Dickerson, E. J., 1966, Bolson fill, pediment and terrace deposits of Hot Springs area, Presidio County, Trans-Pecos Texas [M.A. thesis]: University of Texas at Austin, 100 p.

Dickerson, P. W., 1980, Structural zones transecting the southern Rio Grande rift, *in* Dickerson, P. W., and Hoffer, J. M., eds., The Trans-Pecos region: New Mexico Geological Society, 31st Field Conference Guidebook, p. 63–70.

Dickerson, P. W., 1983, Evidence for basement control of structural zones transected by the southern Rio Grande rift: Proceedings, 4th International Basement Tectonics Conference, p. 103–114.

Dietrich, J. W., 1966, Geology of Presidio area, Presidio County, Texas: Texas Bureau of Economic Geology, Geologic Quadrangle Map No. 28, scale 1:48,000.

Dunkelman, T. J., 1986, The structural and stratigraphic evolution of Lake Turkana, Kenya, as deduced from a multichannel seismic survey [M.S. thesis]: Durham, North Carolina, Duke University, 64 p.

Eaton, G. P., 1970, A plate-tectonic model for late Cenozoic crustal spreading in the western United States, *in* Riecker, R. E., ed., Rio Grande rift: Tectonics and magmatism: American Geophysical Union, p. 7–32.

Edelman, S. H., 1989, Limitations of the concept of stress in structural analysis: Journal of Geological Education, v. 37, p. 102–106.

Erdlac, R. J., Jr., 1990, A Laramide-age push-up block: The structures and formation of the Terlingua-Solitario structural block, Big Bend region, Texas: Geological Society of America Bulletin, v. 102, p. 1065–1076.

Erickson, R. L., 1953, Stratigraphy and petrology of the Tascotal Mesa Quadrangle, Texas: Geological Society of America Bulletin, v. 64, p. 1353–1386, map.

Ewing, T. E., 1985, Westward extension of the Devil's River uplift—Implications for the Paleozoic evolution of the southern margin of North America: Geology, v. 13, p. 433–436.

Goetz, L. K., 1980, Quaternary faulting in Salt Basin graben, west Texas, *in* Dickerson, P. W. and Hoffer, J. M., eds., The Trans-Pecos region: New Mexico Geological Society, 31st Field Conference Guidebook, p. 83–92.

Gries, J. C., and Haenggi, W. T., 1970, Structural evolution of the Chihuahua tectonic belt, *in* The geological framework of the Chihuahua tectonic belt: West Texas Geological Society Publication 70-12, p. 119–137.

Groat, C. G., 1972, Presidio bolson, Trans-Pecos Texas and adjacent Mexico—Geology of a desert aquifer system: Texas Bureau of Economic Geology, Report of Investigations 76, 46 p.

Heiken, G. H., 1966, Geology of Cerros Prietos, Municipio de Ojinaga, Chihuahua [M.A. thesis]: University of Texas at Austin, 101 p.

Henry, C. D., and Aranda-Gomez, J. J., 1990, The *real* southern Basin and Range: Mid-late Cenozoic extension in Mexico: Geological Society of America Abstracts with Program, v. 22, no. 7, p. 228.

Henry, C. D., and McDowell, F. W., 1986, Geochronology and magmatism in the Tertiary volcanic field, Trans-Pecos Texas, *in* Price, J. G., Henry, C. D., Parker, D. F., Jr., and Barker, D. S., eds., Igneous geology of Trans-Pecos Texas: Texas Bureau of Economic Geology, Guidebook 23, p. 99–122.

Henry, C. D., and Price, J. G., 1985, Summary of the tectonic development of Trans-Pecos Texas: Texas Bureau of Economic Geology, Miscellaneous Map 36, 8 p. text, scale 1:500,000.

Henry, C. D., Price, J. G., and McDowell, F. W., 1983, Presence of the Rio Grande rift in west Texas and Chihuahua, *in* Geology and mineral resources of north-central Chihuahua: El Paso Geological Society, Field Conference Guidebook, p. 108–119.

Henry, C. D., Price, J. G., and James, E. W., 1991, Mid-Cenozoic stress evolution and magmatism in the southern Cordillera, Texas and Mexico: Transition from continental arc to intraplate extension: Journal of Geophysical Research, v. 96, p. 13545–13560.

Hoffer, J. M., 1979, Geothermal exploration of western Trans-Pecos Texas: Texas Western Press, 50 p.

James, E. W., and Henry, C. D., 1990, Accreted late Paleozoic terrane in Trans-Pecos Texas: Isotopic mapping of the buried Ouachita front: Geological Society of America Abstracts with Programs, v. 22, p. 329.

James, E. W., and Henry, C. D., 1991, Compositional changes in Trans-Pecos Texas magmatism coincident with Cenozoic stress realignment: Journal of Geophysical Research, v. 96, p. 13561–13575.

Keller, G. R., and Peeples, W. J., 1985, Regional gravity and aeromagnetic anomalies in west Texas, *in* Dickerson, P. W., and Muehlberger, W. R., eds., Structure and tectonics of Trans-Pecos Texas: West Texas Geological Society Publication 85-81, p. 101–105.

Kelley, V. C., 1952, Tectonics of the Rio Grande depression of central New Mexico, *in* Southeastern Sangre de Cristo Mountains, New Mexico: New Mexico Geological Society, 7th Field Conference Guidebook, p. 109–114.

Kopp, R. A., 1977, Geothermal exploration of Presidio County, Texas [M.S. thesis]: University of Texas at El Paso, 75 p.

La Freniere, J. E., 1977, A gravity and geothermal study of the Presidio Bolson, Presidio County, Texas [M.S. thesis]: University of Texas at El Paso, 63 p.

Lehman, T. M., 1986, Late Cretaceous sedimentation in Trans-Pecos Texas: West Texas Geological Society Publication 86-82, p. 35–54.

Lehman, T. M., 1988, Stratigraphy of the Cretaceous-Tertiary and Paleocene-Eocene transition rocks of Big Bend National Park, Texas, a discussion: Journal of Geology, v. 96, p. 627–631.

Lehman, T. M., 1991, Sedimentation and tectonism in the Laramide Tornillo basin of west Texas: Sedimentary Geology, v. 75, p. 9–28.

McCulloh, R. P., 1977, Geology of Yellow Hill Quadrangle, Brewster County, Texas [M.A. thesis]: University of Texas at Austin, 126 p., 2 pls.

McKnight, J. F., 1970, Geology of Bofecillos Mountains area, Trans-Pecos

Texas: Texas Bureau of Economic Geology, Geologic Quadrangle Map 37, 36 p. text.

Mohr, P. A., 1982, Musing on continental rifts, *in* Palmason, G., ed., Continental and oceanic rifts: Washington, D.C., American Geophysical Union, Geodynamics Series, v. 8, p. 293–309.

Mohr, P., and Wood, C. A., 1976, Volcano spacings and lithospheric attenuation in the Eastern rift of Africa: Earth and Planetary Science Letters, v. 33, p. 126–144.

Moustafa, A. R., 1988, Structural geology of Sierra del Carmen, Trans-Pecos Texas: Texas Bureau of Economic Geology, Geological Quadrangle Map 54, scale 1:48,000.

Mraz, J. R., 1977, A gravity and subsurface investigation of the Presidio Bolson area, Texas [M.S. thesis]: University of Texas at El Paso, 49 p.

Mraz, J. R., and Keller, G. R., 1980, Structure of the Presidio bolson area, Texas, interpreted from gravity data: Texas Bureau of Economic Geology, Geological Circular 79-13, 20 p.

Muehlberger, W. R., 1980, The Texas lineament revisited, in Dickerson, P. W., and Hoffer, J. M., eds., The Trans-Pecos region: New Mexico Geological Society, 31st Field Conference Guidebook, p. 113–121.

Muehlberger, W. R., 1989, Summary of structural geology of Big Bend National Park and vicinity, *in* Muehlberger, W. R., and Dickerson, P. W., eds., Structure and stratigraphy of Trans-Pecos Texas: American Geophysical Union, 28th International Geological Congress, Field Trip Guidebook T-317, p. 179–197.

Muehlberger, W. R., and Dickerson, P. W., 1989, A tectonic history of Trans-Pecos Texas, *in* Muehlberger, W. R., and Dickerson, P. W., eds., Structure and stratigraphy of Trans-Pecos Texas: American Geophysical Union, 28th International Geological Congress, Field Trip Guidebook T-317, p. 35–54.

Muehlberger, W. R., Belcher, R. C., and Goetz, L. K., 1978, Quaternary faulting in Trans-Pecos Texas: Geology, v. 6, p. 337–340.

Nelson, D. O., and Schieffer, J. H., 1990, Ultramafic inclusions and megacrysts in mafic rocks of the Terlingua district, west Texas, *in* Dickerson, P. W., Stevens, M. S., and Stevens, J. B., eds., Geology of the Big Bend and Trans-Pecos region: South Texas Geological Society, 1989 Annual Meeting of American Association of Petroleum Geologists, Field Trip Guidebook, p. 211–249.

Olsen, K. H., Baldridge, W. S., and Callender, J. F., 1987, Rio Grande rift: An overview, *in* Ramberg, I. B., Milanovsky, E. E., and Qvale, G., eds., Continental rifts—Principal and regional characteristics: Tectonophysics, v. 143, p. 119–139.

Pearson, B. T., 1978, Ouachita-Marathon geosynclinal facies in the Green Valley area, Brewster County, Texas, *in* Mazzullo, S. J., ed., Tectonics and Paleozoic facies of the Marathon geosyncline, west Texas: Society of Economic Paleontologists and Mineralogists, Permian Basin Section, Field Conference Guidebook, p. 95–100.

Pearson, B. T., 1981, Some structural problems of the Marfa basin area, *in* Johns, R. D., and French, L. R., Jr., eds., Marathon-Marfa region of west Texas: Society of Economic Paleontologists and Mineralogists, Permian Basin Section, Publication 81-20, p. 59–73.

Price, J. G., and Henry, C. D., 1984, Stress orientations during Oligocene volcanism in Trans-Pecos Texas: Timing the transition from Laramide compression to Basin and Range tension: Geology, v. 12, p. 238–241.

Ring, U., Betzler, C., and Delvaux, D., 1992, Normal vs. strike-slip faulting during rift development in East Africa: The Malawi rift: Geology, v. 20, p. 1015–1018.

Rosendahl, B. R., 1987, Architecture of continental rifts with special reference to East Africa: Annual Reviews in Earth and Planetary Science, v. 15, p. 445–503.

Seager, W. R., 1980, Quaternary fault system in the Tularosa and Hueco basins, southern New Mexico and west Texas, *in* Dickerson, P. W., and Hoffer, J. M., eds., The Trans-Pecos region: New Mexico Geological Society, 31st Field Conference Guidebook, p. 131–136.

Seager, W. R., and Morgan, P., 1979, Rio Grande rift in southern New Mexico, west Texas, and northern Chihuahua, *in* Riecker, R. E., ed., Rio Grande rift—Tectonics and magmatism: American Geophysical Union, p. 87–106.

Smith, D. L., and Jones, R. L., 1979, Thermal anomaly in northern Mexico: An extension of the Rio Grande rift?, *in* Riecker, R. E., ed., Rio Grande rift—Tectonics and magmatism: American Geophysical Union, p. 269–278.

St. John, B. E., 1965, Structural geology of Black Gap area, Brewster County, Texas [Ph.D. thesis]: University of Texas at Austin, 132 p., 2 maps.

Stevens, J. B., 1969, Geology of the Castolon area, Big Bend National Park, Brewster County, Texas [Ph.D. thesis]: The University of Texas at Austin, 129 p.

Stevens, J. B., 1988, Mid-Tertiary and Pleistocene sections, Sotol Vista to Cerro Castellan, Big Bend National Park, Trans-Pecos Texas, *in* Hayward, O. T., ed.: Geological Society of America, Centennial Field Guide, v. 4, South-Central Section, p. 429–434.

Stevens, J. B., and Stevens, M. S., 1985, Basin and Range deformation and depositional timing, Trans-Pecos Texas, *in* Dickerson, P. W., and Muehlberger, W. R., eds., Structure and tectonics of Trans-Pecos Texas: West Texas Geological Society, Publication 85-81, p. 157–164.

Stevens, J. B., and Stevens, M. S., 1990a, Stratigraphy and major structural-tectonic events along and near the Rio Grande, Trans-Pecos Texas and adjacent Chihuahua and Coahuila, Mexico, *in* Dickerson, P. W., Stevens, M. S., and Stevens, J. B., eds., Geology of the Big Bend and Trans-Pecos region: South Texas Geological Society, 1989 Annual Meeting of American Association of Petroleum Geologists, Field Trip Guidebook, p. 73–116.

Stevens, J. B., and Stevens, M. S., 1990b, Road logs, field trip to Big Bend region, Trans-Pecos Texas, *in* Dickerson, P. W., Stevens, M. S., and Stevens, J. B., eds., Geology of the Big Bend and Trans-Pecos region: South Texas Geological Society, 1989 Annual Meeting of American Association of Petroleum Geologists, Field Trip Guidebook, p. 1–72.

Stevens, M. S., and Stevens, J. B., 1989, Neogene-Quaternary deposits and vertebrate faunas, Trans-Pecos Texas, *in* Wilson, J. A., Busbey, A. B., III, and Lehman, T. M., eds., Guidebook for the Big Bend field trip: Society of Vertebrate Paleontology, 1989 Annual Meeting, Austin, Texas, p. 67–90.

Swanberg, C. A., 1979, Chemistry of thermal and nonthermal ground waters in the Rio Grande rift and adjacent tectonic provinces, *in* Riecker, R. E., ed., Rio Grande rift—Tectonics and magmatism: American Geophysical Union, p. 279–288.

Taylor, B., and Roy, R. F., 1980, A preliminary heat-flow map of west Texas, *in* Dickerson, P. W., and Hoffer, J. M., eds., The Trans-Pecos region: New Mexico Geological Society, 31st Field Conference Guidebook, p. 137–140.

Taylor, W. J., Bartley, J. M., Lux, D. R., and Axen, G. J., 1989, Timing of Tertiary extension in the Railroad Valley—Pioche transect, Nevada: Constraints from $^{40}Ar/^{39}Ar$ ages of volcanic rocks: Journal of Geophysical Research, v. 94, B-6, p. 7757–7769.

Tsai, H., 1992, Rb-Sr isotope systematics of the Burro Mesa rhyolite, Big Bend National Park, Texas [M.A. thesis]: University of Texas at Austin, 107 p.

Thurwachter, J. E., 1984, Sedimentology of Neogene basin-fill deposits, Lower Tornillo Creek area, Big Bend National Park, Texas [M.A. thesis]: University of Texas at Austin, 87 p.

Udden, J. A., 1907, A sketch of the geology of the Chisos country: Austin, University of Texas Bulletin 93, 101 p.

Yates, R. G., and Thompson, G. A., 1959, Geology and quicksilver deposits of the Terlingua district, Texas: U.S. Geological Survey, Professional Paper 312, 114 p.

Zoback, M. L., and Thompson, G. A., 1978, Basin and Range rifting in northern Nevada; Clues from a mid-Miocene rift and its subsequent offsets: Geology, v. 6, p. 111–116.

Manuscript Accepted by the Society October 13, 1993

Printed in U.S.A.

Index

[Italic page numbers indicate major references]

T

U

V

W

Y

Z

Typeset by WESType Publishing Services, Inc., Boulder, Colorado
Printed in U.S.A. by Johnson Printing, Boulder, Colorado